d Titles

, E. G., Lapiga, E. J.

ration of Multiphase, Multicomponent Systems

978-3-527-40612-8

ouet, J.-M.

drodynamics of Free Surface Flows

delling with the Finite Element Method

7
BN: 978-0-470-03558-0

lah, G. A., Prakash, G. K. S. (eds.)

Carbocation Chemistry

2004
ISBN: 978-0-471-28490-1

Olah, G. A., Molnár, Á.

Hydrocarbon Chemistry

2003
ISBN: 978-0-471-41782-8

Relate

Sinaisk

Sepa

2007

ISBN:

Herv

Hy

Mo

200

IS

C

Emmanuil G. Sinaiski and Leonid I. Zaichik

Statistical Microhydrodynamics

WILEY-VCH

WILEY-VCH Verlag GmbH & Co. KGaA

The Authors

Prof. Dr. Emmanuil Sinaiski
An der Kotsche 12
04207 Leipzig
Germany

Prof. Dr. Leonid I. Zaichik
Nuclear Safety Institute
Russian Academy of Science
B. Tulskaya 52
115191 Moscow
Russia

Library of Congress Card No.:
applied for

British Library Cataloguing-in-Publication Data
A catalogue record for this book is available from the British Library.

Bibliographic information published by the Deutsche Nationalbibliothek
Die Deutsche Nationalbibliothek lists this publication in the Deutsche Nationalbibliografie; detailed bibliographic data are available in the Internet at <http://dnb.d-nb.de>.

Typesetting Thomson Digital, India
Printing betz-druck GmbH, Darmstadt
Book Binding Litges & Dopf GmbH, Heppenheim

Printed in the Federal Republic of Germany
Printed on acid-free paper

ISBN: 978-3-527-40656-2

Contents

Statistical Microhydrodynamics. Emmanuil G. Sinaiski and Leonid Zaichik

ISBN: 978-3-527-40656-2

Preface

Practically all fluids in our everyday surroundings are to some degree disperse systems one of the following types: liquid-solid particles, liquid-gas particles, gas-liquid particles, liquid-liquid particles. They are called accordingly suspensions, liquid-gas and gas-liquid mixtures, emulsions and consist of continuous or carrying phase, which could be liquid or gaseous, and dispersed phase, which contains solid particles of different size and form, bubbles and droplets. The latter will be henceforward called simply particles, adding when needed words solid, liquid or gaseous.

Particle size changes in broad range from microscopic (submicron, micron) to macroscopic (millimeter, centimeter). Particles, which sizes do not exceed several microns, are called Brownian, since they are subjected to Brownian (thermal) motion. Small particles, which density slightly differs from that of surrounding medium, are called inertialess and passive when in addition their volume concentration is negligible small, so that they do not affect the motion of carrying flow. Increasing of particle size and (or) particle density relative the outer medium leads to rise of particle inertia, while buildup of volume concentration enhances particle-fluid and particle-particle interactions, formation of particle aggregates and in the presence of external oriented force field production of oriented particle structures. The latter exerts influence on rheological properties and sometimes causes the non-Newtonian behavior of the medium.

Particle shape may also be multiform. Sometimes it could be approached by form close to canonical: spherical, ellipsoidal with different ratio of semi axes (in limiting case oblate disk-shaped, prolate cigar-shaped) and cylindrical.

Prime interest in engineering applications attracts determination of properties of dispersive media, separation of multiphase multicomponent mixtures subjected to different force fields (gravitational, centrifugal, electrical, magnetic) and motion of dispersive media in tubes, channels and through porous medium. To solve these problems it is necessary to know behavior of separate particle as well as ensemble of particles.

One of the basic parameter affecting both properties and dynamic characteristics of medium is volume content (volume concentration) φ of disperse phase. Disperse medium with $\varphi \ll 1$ are called dilute. For such a medium average spacing between particles is much more than the particle mean size and each particle in limiting case

Statistical Microhydrodynamics. Emmanuil G. Sinaiski and Leonid Zaichik

ISBN: 978-3-527-40656-2

of infinite diluted solution, that is at $\varphi \to 0$, behaves as a single one, its motion is completely determined by external forces, including forces acting on particles from surrounding medium. To the latter belong regular (mean) viscous drag force and random force due to collisions of molecules of surrounding medium with the particle, which causes Brownian motion of the particle. Brownian motion is noticeable only for Brownian particles and does not substantially affect movement of larger particles. Random force arises besides by chaotic fluctuations of carrying fluid, when particle moves in turbulent flow.

Enhancement of φ reduces the mean distance between particles and requires to take into account interactions between particles. The motion of particle subject to interaction forces is called hindered motion. Among interaction forces are distinguished hydrodynamic, molecular and electrostatic forces. The first one is characterized by long-range interaction and depends on hydrodynamic parameters, geometrical properties of particles (size, shape, orientation in space) and mutual arrangement of particles in space (configuration). Hydrodynamic forces are most pronounced when the distance between particle surfaces (clearance) is equal or less than particle linear size. The molecular interparticle force (Van der Waals attractive force) manifests itself distinctly only when the clearance between approaching particles becomes much less than the particle size. This force keeps particles together and promotes coagulation of rigid or coalescence of liquid and gaseous particles. Electrostatic force is repulsion force due to thin charged double layers on particle surfaces. This force prevents particle collisions and stabilizes dispersive medium. The range of action of electrostatic force is small compared with hydrodynamic force, and so it is short-range as well as molecular force. When particle volume concentration exceeds 10% the average interparticle spacing is not great and combined action of all mentioned forces can lead to formation of ordered structures of particles causing anisotropy of transport coefficients and non-Newtonian properties of dispersive system.

Great influence on particle behavior exert flow conditions of carrying phase. Particle in quiescent fluid settles under gravity with constant velocity called sedimentation velocity. At $\varphi \to 0$ this velocity is determined by Stokes formula for rigid spherical particle and Hadamar-Rybczynski formula for liquid one. Increasing of φ leads to noticeable influence of surrounding particles on sedimentation velocity. If besides the particle size is enough small, the Brownian motion also affects the sedimentation velocity. Inhomogeneity of velocity distribution in laminar flow can also influence particle motion. Since the particle size is small compared to characteristic linear scale of the flow region, the flow in the vicinity of the particle could be considered as sheared. Such a flow induces translation and rotation particle motions, which with regard to interaction forces brings to mutual approach and further collision and coagulation of particles.

Particle motion in turbulent flow is more complicated problem, while particle random motion due to effects of chaotic turbulent fluctuations, which enhances collision frequencies, is superimposed on regular transport with mean velocity under the action of carrying medium and external forces. Hence the rate of particle coagulation increases and processes of heat and mass exchange become more intensive.

The size of particles and ratio of densities of particle and carrying medium determine the particle inertia. Dimensionless parameter that allows to distinguish inertial particle from inertialess one, is Stokes number St equal to the ratio of particle dynamic relaxation time to characteristic time of exposure to environmental factors on the particle. Inertialess particles ($St \ll 1$, small particles which density differs slightly from that of carrying medium) are fully involved in the movement of the carrying flow and its motion is on the whole determined by the characteristics of continuous phase. Inertial particles ($St \gg 1$, relative big particles which density considerably deviates from the density of carrying medium) are only partially involved in the motion of continuous phase. All this makes difficult to investigate dynamics of such particles, since it requires to take into account interparticle collisions (for high concentrated disperse medium this concerns inertialess particles too) and to recruit kinetic theory of gases.

Presence of enormous amount of particles in a unit volume of disperse medium, action of random fluctuations of environmental factors and inverse influence of random motion of particles on the surrounding medium makes impossible the description of dispersive medium behavior through deterministic method. The most fruitful and productive method is statistical method. This method examines not the behavior of each particle but the behavior of particle ensemble by means of probability distribution function (PDF), which, is able to describe the change of particle ensemble configuration in space-time with regard for particle relative motion under action of different forces. Statistical characteristics of PDF permit to determine the macroscopic properties of dispersive medium.

The content of the book stems from lecture courses given to students of Moscow State University of Oil and Gas. The aim of the book is to give foundation of statistical methods used in hydrodynamics of micro particles that is hydrodynamics of suspension, which contains suspended in fluid micro particles. The first two chapters provide an introduction to probability theory and microhydrodynamics. The theory of Brownian motion of micro particles taking into account particle-particle and particle-fluid interactions are described in chapter 3. The fourth chapter contains necessary information about turbulent flow and its statistical description. The motion of micro particles in turbulent flow forms the subject of chapter 5. Chapters 6 and 7 deals with interactions of inertialess and inertial particles.

It should be made a remark about the title of the book. The notion of microhydrodynamics was first introduced by G.K. Batchelor (see Batchelor G.K. Developments in Microhydrodynamics/In theoretical and applied Mechanics. Ed. W.T. Koiter. - Amsterdam: North Holland, 1976. P. 33-55) and was defined as a part of hydrodynamics, which studies the motion of particles in fluid under low Reynolds numbers. In the book this notion is extended not only to small but also to finite Reynolds numbers, in order to cover micro particle motion in turbulent flow. It seems that this naturally reflects the increasing interest to the topic of the book.

December 2007

Emmanuil G. Sinaiski and
Leonid I. Zaichik

Nomenclature

a	particle radius
a_0	coefficient in the expression for correlation between acceleration fluctuations
a_1	empirical constant
a_i	semi-axes of an ellipsoid
a_F	empirical constant in the e-F model
a_ε	constant in the k-ε model
A	set of events
A	parameter in the equation of motion of a particle
$\boldsymbol{A}$	reaction constant
$\boldsymbol{A}$	drift vector
$\overline{A}$	(absolute) complement of a set A
$A, B, \ldots$	components (reactants) of a chemical reaction
$A + B \rightarrow$products	second order reaction
$A + B \rightarrow R$, $R + B \rightarrow P$	two-step reaction
A_u	particle acceleration
b	particle radius
$\boldsymbol{b}_{ij}$	mobility tensor
$b_{ijk}(\boldsymbol{r}, t)$	third order structure function
$b_{k_1 k_2 \ldots k_N}$	central moment
$b_{Lij}(\boldsymbol{r}, \tau)$	Lagrangian two-point structure function
$b_{LL}(r, t)$	longitudinal second-order structure function
$b_{LLL}(r)$	longitudinal third-order scalar function
$b_{NN}(r, t)$	transverse second-order structure function
$b_{uu}(M_1, M_1)$	two-point central moment of the second order; correlation function of fluctuations
B	parameter in the equation of motion of a particle
$\lVert \boldsymbol{B} \rVert$	mobility matrix
$\boldsymbol{B}$	mobility tensor
$B(0)$	total energy of the field
$\boldsymbol{B}(M_1, M_2)$	correlation tensor

Statistical Microhydrodynamics. Emmanuil G. Sinaiski and Leonid Zaichik

ISBN: 978-3-527-40656-2

$B(r)$	correlation function of an isotropic random field
$B_1,\ B_2,\ B_3$	empirical constants; dimensionless parameters
B_{ab}	correlation of concentration fluctuations at points $\boldsymbol{X_a}$ and $\boldsymbol{X_b}$
$B_{ij}(M_1,\ M_2)$	correlation function; two-point second-order moment
$B_{ijk}(\boldsymbol{r},\ t)$	third-rank correlation tensor of isotropic turbulence
$B_{k_1k_2\ldots k_N}$	moment of the order $k_1 + k_2 + \ldots + k_N$
$B_{La,b}$	triple correlation of longitudinal velocity and concentration fluctuations at points $\boldsymbol{X_a}$ and $\boldsymbol{X_b}$
$B^0_{Li\,j}(\tau)$	Lagrangian single-point correlation function
$B_{Lij}(\boldsymbol{r},\ \tau)$	Lagrangian two-point correlation function
$B_{LL}(r,\ t)$	longitudinal correlation function of isotropic turbulence
$B_{NN}(r,\ t)$	transverse correlation function of isotropic turbulence
$B_{LL,L}(r,\ t)$	scalar function of isotropic turbulence
$B_{LN,N}(r,\ t)$	scalar function of isotropic turbulence
$B_{NN,L}(r,\ t)$	scalar function of isotropic turbulence
$B_{uu\ldots u}(M_1,\ M_2,\ \ldots\ M_N)$	k-point moment
$B_{uv}(M_1,\ M_2)$	mutual correlation function; two-point mixed moment
c	parameter of the distribution $p(\overline{\varepsilon}_{cr}, \overline{\varepsilon})$
c_p	specific heat capacity at constant pressure
c_v	specific heat capacity at constant volume
C	concentration of passive impurity
$C'(\boldsymbol{X},\ t)$	concentration fluctuation
$\langle (C')^2 \rangle$	intensity of concentration fluctuations
C_0	characteristic value of passive impurity concentration
$C_1,\ C_2$	empirical constants; constants in the k-e model
C'_a	concentration fluctuation at point X_a
C_a	resistance factor in the Archimedes force
C_d	viscous resistance factor of a particle
C_d	empirical constant
$C_E,\ C'_E$	empirical constants
$\langle C'_i\,\Delta C'_i \rangle$	moment that characterizes micromixing
$\langle (C'_i)^2 \rangle$	variance of concentration distribution of i-th component
C_L	resistance factor in the lifting force arising due to the velocity shear
C_m	average mass concentration
C^N_m	binomial coefficient
C_p	resistance coefficient of a particle
C_w	surface concentration
C_z	molar concentration
$\pm dC_z/dt$	specific rate of increase/decrease of matter concentration in a chemical reaction
C_ε	concentration measurement error
C_μ	parameter in the Reynolds stress equation
Ca	modified capillary number

d	particle diameter
$\tilde{d}$	ratio of particle diameter to pipe diameter
d_0^2	dispersion of the average concentration distribution relative to the initial position of the source
$\langle d_c^2 \rangle$	dispersion of the concentration distribution relative to the center of gravity
$d_{cc}(r)$	structural concentration function of isotropic turbulence
$d_{ij}(\tau)$	components of the fluid particle's displacement dispersion tensor; components of the correlation tensor of displacement fluctuations
$d_{ij}^{(r)}$	components of the tensor of relative dispersion of two fluid particles
d_p	particle diameter
D	drag force on a particle
D	diffusion coefficient
$\boldsymbol{D}$	dispersion tensor
D	pipe diameter
$\boldsymbol{D}_{br}$	tensor of Brownian diffusion
D_{br}	coefficient of Brownian diffusion
D_{br}^0	coefficient of unhindered Brownian diffusion
D_{br}^r	coefficient of rotational Brownian diffusion
D_F	effective diffusion coefficient in equations for the *e*-*F* model
D_E	effective diffusion coefficient in equations for the *k*-*e* model
D_{ij}	components of the dispersion tensor
$\boldsymbol{D}_{ij}$	tensor of hindered relative (mutual) diffusion of two particles
$\boldsymbol{D}_{ij}^{(0)}$	tensor of unhindered relative (mutual) diffusion of two particles
$\boldsymbol{D}_{ij}^r$	tensor of relative diffusion of two fluid particles
D_{LL}^r	longitudinal component of the relative diffusion coefficient
D_m	coefficient of molecular diffusion
D_{NN}^r	transverse component of the relative diffusion coefficient
D_{pij}	components of the tensor of turbulent diffusion of particles
D_t	coefficient of turbulent diffusion
$D_t^{(0)}$	coefficient of unhindered mutual turbulent diffusion
D_ε	function in the *k*-*e* model
D_Φ	parameter in the three-parametric model of turbulence
D_Φ^*	parameter in the three-parametric model of turbulence
Da	Damköhler number
DNS	direct numerical simulation
e	coefficient of restitution (COR)

$\boldsymbol{e}$	***unit vector***
$\boldsymbol{e}$	unit vector along the line of centers of two particles
$(\boldsymbol{e}_1, \boldsymbol{e}_2, \boldsymbol{e}_3)$	basis of a Cartesian coordinate system
$e_i = \langle u'_k v'_k \rangle / 2$	fluctuation energy of interfacial interaction
$e_k = \rho u_k u_k / 2$	kinetic energy density in the continuous phase
$e_k = \langle u'_k u'_k \rangle / 2$	intensity of turbulence; mean kinetic energy of fluctuational motion per unit mass in the continuous phase
$e'_k = \rho \langle u'_k u'_k \rangle / 2$	kinetic energy density of fluctuational motion
$e_{pk} = \langle v'_k v'_k \rangle / 2$	turbulent energy of the disperse phase
$e_p = \langle v'_k v'_k \rangle / 2$	fluctuational energy of a particle
$e_s = \rho \langle u_i \rangle \langle u_i \rangle / 2$	kinetic energy density of averaged turbulent flow
E	internal energy
$\boldsymbol{E}$	vector of electric field strength
$\boldsymbol{E}$	rate-of-strain tensor
$E(k, t)$	spectrum of average energy
$E(\omega)$	energy distribution over the frequency spectrum
E_1	maximum principal value of rate-or-strain tensor
E_A	activation energy
$\boldsymbol{E}^{\infty}$	rate-of-strain tensor at the infinity
$E_c(k)$	spectral density of concentration fluctuations
$E_c(k, t)$	intensity spectrum of concentration fluctuations
E_{cr}	critical value of external electric field strength
E_{ij}	rate-of-strain tensor components
E_{ij}^{∞}	rate-of-strain tensor components at the infinity
$\langle E_{ik}(\boldsymbol{x}, t) E_{jn}(\boldsymbol{x} + \boldsymbol{r}, t) \rangle$	correlation function of rate-of-strain tensor
$E, F, \ldots$	components of the product of chemical reaction
f	coefficient of particle's involvement into fluctuational motion of the carrier flow (involvement coefficient)
$\boldsymbol{f}$	stochastic force
$\boldsymbol{f}$	external force density
$\boldsymbol{f}$	random force exerted by the flow on a fluid particle
$\boldsymbol{f}(\boldsymbol{X}_p(t), t)$	random Brownian force acting on a particle
$\langle f \rangle$	average (mean) value
$f(V)$	particle breakup frequency
$\langle f(\boldsymbol{X}, t) \rangle$	expectation value; ensemble average
$\langle f(X, t) \rangle_T$	time average
$\langle f(X, t) \rangle_V$	spatial average
$\langle f(X, t) \rangle_{TV}$	spacetime average
f_1, f_2	functions in the k-e model
$\widehat{f}_1, \widehat{f}_2$	dimensionless coefficients
$\boldsymbol{f}_A$	force acting on a particle due to the particle acceleration
f_A	dimensionless force of molecular attraction between spherical particles
$\boldsymbol{f}_B$	Basset force
$f_{e\Theta_1}$	resistance coefficient for particle's translational motion

f_{er}	resistance coefficient for particle's translational motion
f_{sr}	resistance coefficient for particle's translational motion
$f_{s\Theta}$	resistance coefficient for particle's translational motion
$\langle f'_{Ai}\, p\rangle$	correlation between Archimedes force fluctuations and probability density function (PDF) of particles
$\langle f'_{Bi}\, p\rangle$	correlation between Basset force fluctuations and PDF of particles
f_i	dimensionless coefficients in the expression for the force of electrical interaction between particles
f_j	j-th component of the force acting on the unit surface area of a particle
$\boldsymbol{f}_L$	lifting force acting on a particle due to the velocity shear of the carrier flow
$\langle f'_{Li}\, p\rangle$	correlation between lifting force fluctuations and PDF of particles
f_r	coefficient of involvement of a pair of particles separated by the distance r into fluctuational motion of the continuous phase
f_u	coefficient of involvement of a particle into fluctuational motion of the turbulent carrier flow
$f_\mu(\mathrm{Re})$	parameter in the equation for Reynolds stresses in two-parametric turbulence models
F	function in the e-F model
$\boldsymbol{F}$	systematic force
$\boldsymbol{F}$	particle interaction force
$\mathbf{F}$	generalized force vector
$F(\boldsymbol{k})$	spectrum of a homogeneous random field
$F[\varphi]$	functional
$\boldsymbol{F}(\boldsymbol{X}_p(t), t)$	external force acting on a particle
F_{2x}	component of interaction force between two conducting charged particles perpendicular to their line of centers
F_{2z}	component of interaction force between two conducting charged particles along their line of centers
$\lVert \mathrm{F}_0 \rVert$	generalized force matrix
$\boldsymbol{F}^E$	external force
$\boldsymbol{F}^B$	stochastic Brownian force
$\boldsymbol{F}_e$	external force
$\boldsymbol{F}_e$	component of the force exerted on a particle by the surrounding fluid; hydrodynamic force resulting from the particle's proper motion
$\boldsymbol{F}^{fl}$	fluctuating force
$\boldsymbol{F}^g$	gravitational force
$\boldsymbol{F}^m$	molecular interaction force
F_A^s	force of molecular attraction between two spherical particles

$F_{ab}(k,\ t)$	spectrum of concentration fluctuations
$\boldsymbol{F}_{cap}$	surface tension force (capillary force)
$\boldsymbol{F}_g$	gravitational force
$\boldsymbol{F}_h$	force exerted on a particle by the surrounding fluid
$\boldsymbol{F}_h$	hydrodynamic resistance force acting on a particle
$\boldsymbol{F}_{ij}(\boldsymbol{k})$	spectral tensor
$F_{ij,k}(k)$	spectrum of the third-rank correlation tensor of isotropic turbulence
$F_{ik}(\boldsymbol{r},\ t)$	integral of two-point structure function of velocity fluctuations in a continuous medium, taken along the trajectory that describes the relative motion of a pair of particles
$\boldsymbol{F}_k^{(i)}$	force acting on *i*-th particle
$F_{LL}(k,\ t)$	longitudinal spectrum of isotropic turbulence
$\boldsymbol{F}_{Mi}$	migration force caused by the interaction of particles with turbulent eddies of the carrier flow
$F_{NN}(k,\ t)$	transverse spectrum of isotropic turbulence
$\boldsymbol{F}_{\boldsymbol{R}}^s$	electrostatic repulsion force between two spherical particles
F_s	Stokesian component of the force exerted on a particle by the carrier flow
$\boldsymbol{F}_{ih}$	thermodynamic force
$\boldsymbol{F}_\nu$	viscous friction force
$\boldsymbol{g}$	acceleration of gravity
g	coefficient of involvement of particles into fluctuational motion of the carrier flow
$g_i(V/\omega)$	single-modal density distribution of particle volume ratios in the course of particle breakup
$\boldsymbol{g}^\infty$	velocity gradient tensor at the infinity
g_{ij}^∞	components of velocity gradient tensor at the infinity
$\lvert g_{jk}\rvert$	determinant of the matrix $\lVert g_{jk}\rVert$
$\lVert g_{jk}\rVert$	matrix
$\lVert g_{jk}\rVert^T$	transposed matrix
g^{fl}	fluctuation of the rate of matter production or consumption in the course of a chemical reaction
g_u	coefficient of involvement of a particle into fluctuational motion of the carrier turbulent flow
G	dimensionless parameter
G	Gibbs free energy
$\boldsymbol{G}$	spatial gradient of a physical parameter
G	empirical function
$G(C,\ C',\ C'')$	kernel of the micromixing operator in coalescence-dispersion models; distribution density of intermediate concentrations that are formed during an elementary mixing event
$S(v,\ \omega)$	collision cross section of two particles of volumes V and ω; normal cross-sectional area of the limiting flow tube

$\langle G \rangle_1$	conditional average
$G^{fl}(\boldsymbol{X}, t)$	fluctuational term in the diffusion equation
$G_{ik}(\boldsymbol{r}, t)$	integral of the two-point structure function of velocity fluctuations in the continuous phase, taken along the trajectory of the relative motion of two particles
G_j	parameter in the expression for coagulation kernel
Ga	Galilei number
h	coefficient of hydrodynamic resistance
h	coefficient of hydrodynamic resistance for particles approaching each other along their line of centers
h	distance between two planes
h_0	minimum gap between surfaces of spherical particles
h^0	hydrodynamic resistance coefficient for unhindered particle motion
h_i	hydrodynamic resistance coefficient for i-th particle
h_u	coefficient of particle's involvement into fluctuational motion of a turbulent carrier flow
h_δ	asymptotic expression for the hydrodynamic resistance coefficient of a particle at small clearances between particles
H	enthalpy
$H(X)$	Saffman function
$\mathrm{H}(X)$	Heaviside function
i	parameter of the gamma distribution
$(\boldsymbol{i}_1, \boldsymbol{i}_2, \boldsymbol{i}_3)$	basis of a local Cartesian coordinate system
$\boldsymbol{i}, \boldsymbol{j}$	mutually orthogonal unit vectors in a plane perpendicular to the line of centers of two particles
$\boldsymbol{I}[p]$	Boltzmann collision operator in the Enskog form
$\boldsymbol{I}$	unit tensor
I	number of particle collisions in the absence of the force of electrostatic repulsion
I_k	collisional term in the kinetic coagulation equation
I_R	number of particle collisions, with electrostatic repulsion taken into consideration
$\boldsymbol{j}$	diffusion flux
$j(R_1, R_2)$	diffusion flux of particles of radius R_2 toward a particle of radius R_1
J_0	diffusion flux of unhindered particles
$J_0(0)$	diffusion flux per unit solid angle
$\boldsymbol{j}^{fl}$	vector of diffusion flux fluctuation
$\boldsymbol{J}$	displacement vector
J	dimensionless diffusion flux of particles of radius R_2 toward a particle of radius R_1
J	particle flux toward a test particle
J_c	collision operator in the kinetic equation for PDF of velocity of a single particle

J_f	incident particle flux
J_i	flux density of impurity along the X_i-axis
J_i	moment of inertia of i-th particle
J_{ij}	term in the balance equation for turbulent stresses in the disperse phase, which are caused by particle collisions
J_{ij}^a	components of the Oseen tensor (stokeslet) for a-th particle
J_i^k	i-th component of the diffusion flux of k-th component of the impurity
J_r	particle flux reflected from the wall
J_w	flux of matter toward the wall
J_w	flux of particles depositing on the wall
k	ratio of particle radii
k	wave number
k	permeability of a porous medium
k	reaction rate constant
$\mathbf{k}$	unit vector along the line of centers of two colliding particles
k_B	parameter in the equation of motion of a particle in a turbulent flow
k_c	wave number of a scalar field
k_f	coefficient of aerodynamic resistance
k	the Boltzmann constant
$\mathbf{k}_{ab}^c$	conjugate mobility tensor
$\mathbf{k}_{ab}^r$	rotational mobility tensor
$\mathbf{k}_{ab}^t$	translational mobility tensor
k_b	wave number corresponding to the Batchelor scale
k_K	wave number corresponding to the Kolmogorov scale
k_m	wave number corresponding to the maximum size of inhomogeneity (the right end of the inertial-convective range, starting from which micromixing commences)
K	empirical coefficient
$\mathbf{K}$	transport coefficient
K	moment's order
$K(\mathrm{V},\omega)$	kernel of the kinetic coagulation equation; collision frequency of two particles of volumes ν and ω in a unit disperse phase volume at unit particle concentrations; coagulation constant
$\boldsymbol{K}$	micromixing operator
$\boldsymbol{K}_0^c$	conjugate translational resistance tensor
$\boldsymbol{K}_0^r$	rotational resistance tensor
$\boldsymbol{K}^t$	translational resistance tensor
$\boldsymbol{K}_0^{c*}$	conjugate rotational resistance tensor
$\boldsymbol{K}^t$	third-rank tensor of translational shear resistance
$\boldsymbol{K}_{ijk}^t$	components of translational shear resistance tensor
$\boldsymbol{K}^r$	third-rank tensor of rotational shear resistance

$\|\|\mathbf{K}_0\|\|$	generalized resistance matrix
$\|\|\boldsymbol{K}_0^{t,r,c}\|\|$	resistance matrices
$\boldsymbol{K}_{eff}$	macroscopic (effective) transport coefficient
K_g	collision kernel in the gravitational field
K_{ijk}^r	components of third-rank rotational shear resistance tensor
$\boldsymbol{K}_{pr}$	permeability matrix
K_s	collision kernel in a shear flow
K_t	collision kernel in a turbulent flow
K_{tg}	collision kernel in a turbulent flow with gravity taken into account
K_{ts}	collision kernel in a turbulent shear flow
l	step length for a particle's random walk
l	Prandtl mixing length
l	Prandtl–Nikuradze mixing length
l_1	Taylor mixing length
l_{ij}	components of symmetric scale tensor
l_m	mean free path of molecules
l_u	coefficient of particle's involvement into fluctuational motion of turbulent carrier flow
$\boldsymbol{l}$	unit vector characterizing particle's orientation
L	average distance between particles; characteristic linear size; spatial macroscale of turbulence; Eulerian spatial macroscale; integral spatial scale of turbulence
$\boldsymbol{L}$	linear operator
$\boldsymbol{L}$	torque
L	linear macroscale of concentration field
L	empirical function
L	parameter in the equation of motion of a particle in a turbulent flow
L_c	integral scale (macroscale) of concentration field
$\boldsymbol{L}$	hydrodynamic torque exerted on a particle by the surrounding fluid
$\boldsymbol{L}_k^{(i)}$	torque acting on i-th particle
L_{ki}	associated Laguerre polynomial
L_L	longitudinal integral scale
L_N	transverse integral scale
$L_q^{(r)}(j+\xi)$	Lagrange interpolation polynomial
$\boldsymbol{L}_s$	Stokesian component of torque exerted on a fixed particle by the surrounding fluid
LES	large eddy simulation
m	mass
m	empirical constant in the e-F model
m	particle surface mobility parameter
m	dimensionless parameter

m	virtual mass of a particle
$\widehat{m}$	dimensionless moment of distribution
m_0	zero order moment; number concentration of particles
m_1	first order moment; volume concentration of particles
m_i	mass of i-th particle
m_i	mass of i-th component produced or consumed in the course of chemical reaction
m_k	k-th order moment of the volume distribution of particles
m_u	coefficient of particle's involvement into fluctuational motion of turbulent carrier flow
M	migration coefficient
M_i	molecular mass of i-th component
N_i	number of moles of i-th component
$n(\boldsymbol{X}, a)$	distribution of particles over their radii
$n(\mathrm{V}, \boldsymbol{X}, t)$	distribution of particles over their volumes
n	empirical constant in the e-F model
n	total reaction order; kinetic order of reaction
$\boldsymbol{n}$	normal vector
n	number concentration of particles
n_0	concentration of particles far from the test particle
$n_0(\mathrm{V})$	initial distribution of particles over volumes
n_i	order of i-th reaction
N	spacetime point
N	number of particles in a given volume
$N(t)$	number of particles in a unit volume (number concentration)
N	number of steps (displacements) in particle's random walk
N	dimensionless parameter
$\langle N \rangle$	dissipation of concentration inhomogeneity
$\overline{N}$	scalar dissipation of concentration inhomogeneity
N_a	number of particles of type a in a unit volume
$\langle \overline{N}_c \rangle$	conditional average rate of scalar field dissipation at fixed concentration C
N_D	ratio between the characteristic time of a process and the characteristic time of impurity transport via molecular diffusion
N_t	ratio between the characteristic time of a process and the characteristic time of convective transport of impurity via turbulent motion
N_W	ratio between the characteristic time of a process and the characteristic time of matter production or consumption in a chemical reaction
Oh	Ohnesorge number
p	pressure
p	dimensionless parameter

$\boldsymbol{p}$	unit vector
p	dynamic probability density for particle ensemble
p'	perturbation (fluctuation) of pressure
$\overline{p}$	pressure averaged over the cross section
$p(\boldsymbol{r}\vert\xi_1, \boldsymbol{r}_0, \tau, t_0)$	PDF of distance between two particles; function of distance between neighbors
$p(\boldsymbol{u}, C_1, C_2, \ldots, C_N; \boldsymbol{X}, t)$	joint single-point PDF of velocity and scalar quantities
$C_1, C_2, \ldots, C_N$	(concentrations of different passive impurities) at time t
$p(\boldsymbol{X})$	probability density function, PDF
$p(\boldsymbol{X}_1, \boldsymbol{X}_2, \ldots, \boldsymbol{X}_N)$	joint multi-particle PDF
$p(\boldsymbol{X}, V\vert\boldsymbol{\xi}_1, \boldsymbol{\xi}_2, \ldots, \boldsymbol{\xi}_n; t_1, t_2, \ldots, t_m)$ $p(X, V\vert\boldsymbol{\xi}_1, \boldsymbol{\xi}_2, \ldots, \boldsymbol{\xi}_n; t_1, t_2, \ldots, t_m)$	joint multi-dimensional (multi-particle) PDF of coordinates $\boldsymbol{X}$ and Lagrangian velocities $\boldsymbol{V}$ for a given set of time values $t_1, t_2, \ldots, t_m$ and initial particle positions $\boldsymbol{\xi}_1, \boldsymbol{\xi}_2, \ldots, \boldsymbol{\xi}_m$; the main Lagrangian statistical characteristic of turbulence
$p(\boldsymbol{X}, \boldsymbol{V}, \boldsymbol{u}\vert\boldsymbol{\xi}_1, \boldsymbol{\xi}_2, \ldots, \boldsymbol{\xi}_n; \boldsymbol{X}_1, \boldsymbol{X}_2, \ldots, \boldsymbol{X}_n; t_1, t_2, \ldots, t_m)$	joint PDF of Lagrangian (V) and Eulerian (u) quantities
$p(\boldsymbol{u}, t\vert\boldsymbol{u}_0, t_0)$	conditional probability density
$p(V_1, V_2, \ldots, V_n\vert \boldsymbol{\xi}_1, \boldsymbol{\xi}_2, \ldots, \boldsymbol{\xi}_n; t_1, t_2, \ldots, t_n)$	joint PDF of Lagrangian velocities $\boldsymbol{V}(\boldsymbol{\xi}_1, \boldsymbol{t}_1)$, $\boldsymbol{V}(\boldsymbol{\xi}_2, t_2), \ldots$ $\boldsymbol{V}(\boldsymbol{\xi}_n, t_n)$ of n fluid particles at different moments of time
$p(\boldsymbol{Y}\vert\tau; \boldsymbol{\xi}, t_0)$	probability density distribution of particle displacements from the initial point $\boldsymbol{\xi}$ during the time τ
$p(\overline{\varepsilon})$	one-dimensional distribution of random specific energy dissipation
$p\left(\overline{\varepsilon} + \frac{\Delta t}{2}\dot{\overline{\varepsilon}}, \overline{\varepsilon} - \frac{\Delta t}{2}\dot{\overline{\varepsilon}}\right)$	joint distribution density of random values $\overline{\varepsilon}$ and $\dot{\overline{\varepsilon}}$ at different moments of time
$p(\overline{\varepsilon}, \dot{\overline{\varepsilon}})$	joint distribution density of random values $\overline{\varepsilon}$ and $\dot{\overline{\varepsilon}}$ at one and the same moment of time
$p(\overline{\varepsilon}_{cr}, \dot{\varepsilon})$	joint distribution density of $\overline{\varepsilon}_{cr}$ and $\dot{\overline{\varepsilon}}$
$\langle p'\boldsymbol{u}'\rangle$	mixed correlation (correlator) of pressure and velocity fluctuations
p^{∞}	static pressure at the infinity
$p_c(\boldsymbol{u}, C, \boldsymbol{X})$	joint PDF in a turbulent mixing layer
p_{eq}	equilibrium PDF
$p_{ij}(\boldsymbol{r})$	pair PDF
$p_p(\boldsymbol{X}, \boldsymbol{v}, t)$	dynamic probability density in the phase space of particle coordinates and velocities
P	probability
$P = \langle p\rangle$	velocity PDF of a system of particles
$P(A\vert B)$	conditional probability
$P(A \cap B)$	probability of a joint event

$P(\boldsymbol{r}, \boldsymbol{w}, t)$	PDF of a particle pair (probability of the relative velocity w separated by the distance r at the time t)
$P(\mathrm{V}, \omega)$	probability of formation of a droplet with volume in the range (v, v + dv) in the process of breakup of a droplet with volume in the range (ω, $\omega + d\omega$)
$P(w_r)$	probability distribution of fluctuational component of the relative radial velocity
$P(\Theta_N)$	probability distribution of a certain N-particle configuration
$P_0(\boldsymbol{v})$	Maxwellian velocity distribution
$\boldsymbol{P}_0^r$	associated vector of dynamic pressure in rotational motion
$P_2(\boldsymbol{v}, \boldsymbol{v}_1)$	PDF of velocities of two particles
P_n^m	associated Legendre polynomial
$\boldsymbol{P}^t$	associated vector of dynamic pressure in translational motion
Pe	Peclet number
Pe_D	diffusional Peclet number
Pe_T	thermal Peclet number
Pr	Prandtl number
Pr_p	Prandtl number of the disperse phase
Pr_t	turbulent Prandtl number of the carrying phase
$q(\boldsymbol{r}, \tau)$	Richardson function
q	particle migration coefficient
$\boldsymbol{q}$	heat flux vector
q	particle charge
q_ε	density of internal heat sources
Q	dynamic pressure
Q	directed flux of matter
Q_c	intensity of fluctuational energy dissipation via inelastic collisions of particles
$Q_{ij,n}^a$	*n-th* order moment (multipole) of surface forces acting on the particle *a*
r	distance between the centers of two particles
r_w	pipe radius
$\boldsymbol{r}$	unit vector along the line of centers of two particles
(r, Θ, z)	cylindrical coordinates
(r, Θ, Φ)	spherical coordinates
$\boldsymbol{r}_0$	initial radius vector of the particle center
$\boldsymbol{r}_0$	radius vector of a point relative to the center O.
(r_0, Θ_0, Φ_0)	initial coordinates of the particle center
$\boldsymbol{r}_i$	radius vector of the center of *i-th* particle
R	gas constant
$\boldsymbol{R}$	resistance tensor
$\boldsymbol{R}$	tensor of generalized resistance due to the translational and rotational motion of the particle
$\boldsymbol{R}$	grand resistance matrix

R	particle's center of reaction
R	particle radius
R	pipe radius
R	correlation coefficient of colliding particles
$R/(R+P)$	selectivity parameter of a two-stage chemical reaction
R_{12}	velocity correlation coefficient of two particles at the point of impact
R_a	radius of a particle of type a
R_c	coagulation radius
R_{ij}	components of the resistance tensor
$\boldsymbol{R}_{FE}$	component of the resistance tensor of a particle
$\boldsymbol{R}_{FU}$	component of the resistance tensor of a particle
$\boldsymbol{R}_{F\Omega}$	component of the resistance tensor of a particle
$\boldsymbol{R}_{LE}$	component of the resistance tensor of a particle
$\boldsymbol{R}_{LU}$	component of the resistance tensor of a particle
$\boldsymbol{R}_{L\Omega}$	component of the resistance tensor of a particle
$\boldsymbol{R}_{SE}$	component of the resistance tensor of a particle
$\boldsymbol{R}_{SU}$	component of the resistance tensor of a particle
$\boldsymbol{R}_{S\Omega}$	component of the resistance tensor of a particle
R_{ij}^{a}	rotlet of particle a
R_m	minimum radius of a breaking droplet
$\boldsymbol{R}_p$	coordinate vector of the particle's center
Re	Reynolds number
Re_{cr}	critical value of the Reynolds number
Re_p	Reynolds number of a moving particle
Re_t	Reynolds number corresponding to the spatial Taylor microscale
Re_λ	local Reynolds number for fluctuations having scale λ.
s	dimensionless distance between particle centers
s	minimum separation from a plane
s	distance traveled by a particle in time t
s	variable in the image domain of the Laplace transformation
$\mathbf{s}$	unit vector along the particle's displacement vector r
$\langle s^2 \rangle$	mean-square particle displacement
S	surface
S	entropy
S	sedimentation coefficient
$\mathbf{S}$	symmetric part of the velocity gradient tensor
$S(\boldsymbol{k})$	element of sphere surface area $\lvert\boldsymbol{k}\rvert = \boldsymbol{k}$
$d\boldsymbol{S}$	oriented surface element
S_A	parameter of molecular interaction
S_{ij}	sedimentation coefficients
S_{ij}	intensity of force dipoles distributed over the particle surface
S_{ij}^{a}	surface force dipole (stresslet) of particle a

S_{ij}^{B}	contribution of Brownian diffusion to the sedimentation coefficient
S_{ij}^{G}	contribution of gravity to the sedimentation coefficient
S_{ij}^{I}	contribution of molecular interaction to the sedimentation coefficient
S_k	specific source term of k-th component produced/consumed in a chemical reaction in the mass conservation equation for passive impurity components
$S_{k_1,k_2,\ldots,k_N}$	cumulant (semi-invariant)
$S_{\mathbf{R}}$	electrostatic interaction parameter
$\mathbf{S}_{T}^{B}$	contribution of particles' Brownian motion to the stress tensor
$\mathbf{S}_{T}^{E}$	contribution of particles' motion driven by non-hydrodynamic forces to the stress tensor
$\mathbf{S}_{T}^{H}$	contribution of particles' motion driven by interparticle hydrodynamic interactions to the stress tensor
Sc	Schmidt number
Sc_{br}	Schmidt number for Brownian diffusion
Sc_t	turbulent Schmidt number
St	Stokes number calculated with respect to the integral macroscale
St_0	Stokes number calculated with respect to the Kolmogorov microscale
t_0	initial moment of time
t_{12}	time required for a perturbation to go through the wave number range (k_1, k_2)
t_{12}	times of collisions of particle 1 with particles 2
t_{21}	times of collisions of particle 2 with particles 1
t_A	characteristic time of particle enlargement that takes into account molecular interactions between particles
t_c	effective time between particle collisions
t_{com}	characteristic time of change of system configuration
t_D	characteristic time of a diffusion process
t_E	characteristic enlargement time for conducting droplets in an electric field
$t_{e\varphi}$	rotational resistance coefficient of a particle
$t_{e\varphi_1}$	rotational resistance coefficient of a particle
$t_{s\varphi}$	rotational resistance coefficient of a particle
t_E^t	characteristic time of enlargement of conductive droplets in a turbulent flow in an external electric field
t_E^g	characteristic time of enlargement of conductive droplets undergoing gravitational sedimentation in an electric field
t_{eff}	effective relaxation time of a particle
t_{ik}	components of the Maxwellian stress tensor
t_L	characteristic non-stationarity time for an averaged flow

t_T	characteristic thermal relaxation time for a particle
t_u	effective relaxation time that takes into account the effect of virtual mass
t_ν	characteristic time of dynamic relaxation for a particle (depends on Re_p)
t_V	characteristic time of viscous relaxation for a particle
t_λ	characteristic period of fluctuations having the scale λ
T	characteristic time scale
$\boldsymbol{T}$	stress tensor
T	turbulent fluctuation period
T_0	parameter of the distribution $p(\bar{\varepsilon}_{cr}, \dot{\bar{\varepsilon}})$
$T^{(E)}$	Eulerian time scale
$\boldsymbol{T}^{fl}$	fluctuational component of the stress tensor
$T^{(\boldsymbol{L})}$	Lagrangian correlation time; Lagrangian integral time scale of turbulence
$T_\infty^{(L)}$	asymptotic expression for the Lagrangian integral scale of turbulence at high Re
$\boldsymbol{T}^s$	systematic component of the stress tensor
$\boldsymbol{T}_e$	component of hydrodynamic torque exerted on a particle by the surrounding fluid and caused by the particle's proper motion
$\boldsymbol{T}_{ij}^p$	contribution of particles suspended in the fluid to the stress tensor
T_{L0}	parameter in the parabolic exponential approximation of the autocorrelation function
$T_{Lp}^{(L)}(t)$	Lagrangian longitudinal (parallel to the average relative velocity of particles) integral time scale of fluid velocity fluctuations along the particle trajectory
$T_{Np}^{(L)}(t)$	Lagrangian transverse (perpendicular to the average relative velocity of particles) integral time scale of fluid velocity fluctuations along the particle trajectory
$T_p^{(L)}$	Lagrangian integral time scale
$T_r^{(L)}$	two-point integral time scale
$\boldsymbol{u}$	velocity vector
$\boldsymbol{u}$	relative velocity of two particles approaching each other along their line of centers
$\boldsymbol{u}$	carrier flow velocity
$\boldsymbol{u}(\boldsymbol{X}, t)$	Eulerian velocity field at the point $(\boldsymbol{X}, t)$
(u, v, w)	velocity components
$\boldsymbol{u}'$	velocity perturbation
$\boldsymbol{u}'$	fluctuational component of the carrier flow velocity
$\boldsymbol{u}'(\boldsymbol{X}, t)$	Eulerian field of velocity fluctuations
$\bar{\boldsymbol{u}}$	mean flow rate velocity
$\boldsymbol{u}_0$	initial velocity
u_0	characteristic turbulent velocity

$\boldsymbol{u}_0$	velocity of Stokesian flow around a particle
$u_0\psi(t)$	effective path length of a particle in fluctuational motion
$u_2^0 = \langle u'_k u'_k \rangle / 3$	turbulence intensity; intensity of velocity fluctuations of the continuous phase
u^*	dynamic velocity
$\boldsymbol{u}^\infty$	velocity at the infinity
$\Delta u_i(\boldsymbol{r}, t)$	difference of velocity fluctuations at two points
$u_i(S)$	Laplace transform image of carrying phase velocity
u_L	projection of velocity onto $\boldsymbol{r}$
u_m	average molecular velocity
u_N	projection of velocity onto a direction perpendicular to $\boldsymbol{r}$
u_λ	rate of fluctuations of scale λ
u_{λ_0}	rate of fluctuations on the Kolmogorov microscale λ_0
$\langle u'_i f'_{Bk} \rangle$	correlation between fluid velocity fluctuation and Basset force fluctuation
$\langle u'_i f'_{Lk} \rangle$	correlation between fluid velocity fluctuation and lifting force fluctuation
$\langle u'_i u'_k \rangle$	components of the correlation tensor of velocity fluctuations
$\langle u'_i u'_j \rangle$	specific Reynolds stresses of the carrying phase
$\rho\langle u'_i u'_j \rangle$	Reynolds stresses in the carrying phase
$\langle u'_i \vartheta' \rangle$	turbulent heat flux in the carrying phase
U	velocity; averaged velocity; average velocity of turbulent flow
$\mathbf{u}$	generalized velocity vector, which includes both translational and rotational components of velocity
(U, V, W)	velocity components
U_i	degree of immiscibility of i-th reagent
U	degree of immiscibility of the total reaction
$\overline{U}$	mean-flow-rate velocity
$\Vert\mathbf{u}_0\Vert$	matrix of generalized velocity
$\boldsymbol{U}_a^r$	angular velocity of particle a
$\boldsymbol{U}_a^t$	translational velocity of particle a
U_m	average mass velocity
U_v	sedimentation velocity of a particle of volume v
$\boldsymbol{v}$	velocity
$\boldsymbol{v}$	translational velocity of the particle center
$\boldsymbol{v}^\infty$	velocity at the infinity
$\boldsymbol{v}_p$	particle velocity
$\boldsymbol{v}_p, \boldsymbol{v}_{p_1}$	velocities of two particles before the collision
$\boldsymbol{v}'_p, \boldsymbol{v}'_{p_1}$	velocities of two particles after the collision
$\tilde{v}_{pi}(s)$	Laplace transform image of particle's velocity
V	particle volume
V_0	parameter of the volume distribution of particles
V_{av}	average particle volume

V_k	propagation speed of perturbations along the wave number axis
V_m	minimum volume of breaking droplets
(v_r, v_θ, v_Φ)	components of the velocity vector in a spherical coordinate system
V	spatial volume
V	total interaction potential of two spherical particles
ΔV	volume element
V	Velocity
$\boldsymbol{V}(\xi, t)$	Lagrangian velocity
$\boldsymbol{V}'$	particle velocity fluctuation
V_1, V_2	potentials of charged particles
$\boldsymbol{V}_0^r$	rotational velocity tensor
$\boldsymbol{V}^t$	translational velocity tensor
V_A^p	molecular interaction potential of two planes
V_A^s	molecular interaction potential of two spherical particles
$\boldsymbol{V}_{ij}$	relative velocity of two particles
$\boldsymbol{V}_{ij}^{(0)}$	relative velocity of unhindered motion of two particles
V_{ij}^m	Lenard–Jones potential
$\langle V_i V_j \rangle$	Lagrangian correlation function of components of the vector $\boldsymbol{V}$
$\langle V_i' V_j' \rangle$	turbulent stresses in the disperse phase
$\langle V_i' \vartheta_j' \rangle$	turbulent heat flux in the disperse phase
$\boldsymbol{V}_p$	particle velocity
V_R^s	electrostatic interaction potential of two particles
$\boldsymbol{w} = \boldsymbol{v}_p - \boldsymbol{v}_{p_1}$	relative velocity of colliding particles
$\boldsymbol{w}$	relative velocity of two particles
w	term in the particle's equation of motion that helps account for the hydrodynamic (collisionless) interactions of particles; represents a continuous random process
w_k'	random field of hydrodynamic interactions
w_r	radial component of relative velocity of a particle pair
w_r'	fluctuational part of the radial component of relative velocity of a particle pair
$\langle w_r' 2 \rangle$	intensity of fluctuations of the radial relative velocity of a particle pair
W	stability factor of a dispersed system
W	collisional term in the Langevin equation of motion of a particle; discontinuous random process
W	distribution of discontinuous jumps
$\lvert W_0 \rvert$	characteristic rate of chemical reaction
W_g	difference of two particle's sedimentation velocities
$\langle W_i \rangle$	average chemical reaction rate of i-th component

W_i	source term in the diffusion equation, which results from a chemical reaction
W_r	radial component of the average relative velocity of a particle pair
We	Weber number
x_1	dimensionless distance from the particle surface
(X, Y, Z)	Cartesian coordinates
$\boldsymbol{X}$	generalized vector that includes spatial coordinates and orientation angles relative to a given coordinate system
$\boldsymbol{X}(X_1, X_2, \ldots, X_n)$	point in space
$\boldsymbol{X}(\boldsymbol{\xi}, t)$	random function – Eulerian coordinates of a fluid particle
$\boldsymbol{X}_a^0$	vector of spatial coordinates (position vector) of the center of particle a in a given coordinate system
$\boldsymbol{X}_a^s$	vector of orientation angles of particle a in a given coordinate system
$\tilde{X}_{ab}^r, \tilde{x}_{ab}^r$	dimensionless resistance coefficients for rotational motion of particles about the line of centers
$\tilde{X}_{ab}^t, \tilde{x}_{ab}^t$	dimensionless resistance coefficients for translational motion of particles along the line of centers
$X_0(t)$	coordinate of the center of mass
$\langle X_c^2 \rangle$	variance of concentration distribution relative to a source
$\boldsymbol{X}_p(t)$	particle coordinates at the moment t
y_+	dimensionless distance to a wall
$\tilde{Y}_{ab}^c, \tilde{y}_{ab}^c$	dimensionless resistance coefficients for rotational motion of particles about the j-axis
$\tilde{Y}_{ab}^r, \tilde{y}_{ab}^r$	dimensionless resistance coefficients for rotational motion of particles about the i-axis
$\tilde{Y}_{ab}^t, \tilde{y}_{ab}^t$	dimensionless resistance coefficients for translational motion of particles perpendicular to the line of centers
$\mathbf{Y}(\tau)$	vector of particle's displacement from the initial position during the time interval τ
$\mathbf{Y}'$	fluctuation of particle displacement
Z_1, Z_2, Z_3	conservative variables; Schwab–Zeldovich variables
α	empirical constant
α	weight function
α	parameter of the distribution $p(\bar{\varepsilon}_{cr}, \dot{\bar{\varepsilon}})$
α_E	constant in the three-parametric turbulence model
α	empirical constant in the e-F model
α_j	exponent in the coagulation kernel expressed as a power function of particle volumes
α_L	constant dissipation rate of hydrodynamic field; inverse relaxation time
α_τ	constant in the three-parametric turbulence model
α_ε	constant in the k-ε model
β	correction factor

β	dimensionless parameter
β	empirical constant
β	parameter defining the condition for heterogeneous reaction
β_1	square of skewness
β_2	excess (kurtosis) of the distribution
β_E	constant in the three-parametric turbulence model
β_j	exponent in the coagulation kernel expressed as a power function of particle volumes
β_L	constant dissipation rate of a scalar field; inverse relaxation time
β_τ	constant in the three-parametric turbulence model
γ	empirical constant
γ	dimensionless parameter
γ	density ratio between a particle and the surrounding fluid
$\gamma(t)$	Green's function of the particle's equation of motion
γ	function in the *e*-*F* model
$\dot{\gamma}$	shear rate
$\dot{\gamma}^*$	dimensionless hydrodynamic interaction parameter
γ_f	particle shape parameter
γ'	constant in the three-parametric turbulence model
$\gamma_{ij}(t)$	components of velocity fluctuation gradient tensor
$\dot{\gamma}_t$	average shear rate in small-scale fluctuations
γ_τ	constant in the three-parametric turbulence model
γ_Φ	parameter in the three-parametric turbulence model
Γ	Hamaker constant
$\Gamma(x)$	complete gamma function
Γ_Φ	constant in the three-parametric turbulence model
δ	gap (clearance) between particle surfaces
δ	dimensionless parameter
δ	thickness of the viscous boundary layer
$\delta(X)$	delta function of a scalar quantity
$\delta(\boldsymbol{X})$	delta function of a vector quantity
δ_D	thickness of the diffusion boundary layer
Δ	dimensionless gap (clearance) between particle surfaces
$\Delta(\tau)$	effective diameter of a particle cloud
Δ_1, Δ_2	dimensionless parameters
ε	dielectric permittivity
ε	relative fluctuation of distributed particle density
ε	rate of turbulent energy dissipation
ε	dissipation function
ε	small parameter
$\boldsymbol{\varepsilon}$	permutation symbol
ε	parameter in the Lennard–Jones potential
$\bar{\varepsilon}$	specific dissipation of turbulence energy

$\bar{\varepsilon}_0$	specific dissipation rate of a scalar field
$\bar{\varepsilon}_{cr}(V)$	critical value of specific dissipation of energy
$\langle \varepsilon_k \rangle$	mean specific dissipation of energy of fluctuational motion
ε_{ijk}	components of the permutation (Levi–Civita) symbol
ε_s	specific dissipation of energy of averaged motion
ζ	drift parameter characterizing the effect of intersection of particle trajectories
η	porosity of a medium
ϑ	absolute temperature
ϑ'	temperature perturbation (fluctuation)
$\langle \vartheta' \boldsymbol{u}' \rangle$	mixed correlation of temperature fluctuations and velocity fluctuations
ϑ	temperature of the carrying medium
$\vartheta_p(t)$	particle temperature
θ	orientation angle of the particle pair relative to the external electric field
$\theta^*(\Phi)$	angle between the radius vector $\boldsymbol{r}_0$ of the particle center and a point on the limiting trajectory
κ	coefficient of heat conductivity
κ	empirical constant
κ	dimensionless parameter in the expression for the coefficient of mutual turbulent diffusion
κ	dimensionless parameter characterizing droplet stability in an external electric field
κ_p	coefficient of turbulent thermal conductivity
λ	dimensionless resistance coefficient
λ	scale of velocity fluctuations
λ_0	inner scale of turbulence; Kolmogorov spatial microscale
λ_c	differential scale of fluctuations
$\lambda_c^{(0)}$	inner scale of concentration; Batchelor microscale
$(\lambda_c)_i$	microscale of *i*-th component concentration
λ_D	Debye radius
λ_L	London wavelength
λ_L	longitudinal differential scale
λ_N	transverse differential scale
λ_t	Taylor spatial microscale
$\boldsymbol{\Lambda}$	vorticity tensor
Λ	Loitsyansky invariant
$\boldsymbol{\Lambda}^{\infty}$	vorticity tensor at the infinity
$\boldsymbol{\Lambda}^{\infty}$	antisymmetric part of the undisturbed velocity gradient tensor
Λ_{ij}	components of the tensor of relative dispersion of particle cloud
Λ_{ij}	components of the vorticity tensor
Λ_{ij}^{∞}	components of the vorticity tensor at the infinity

$\langle \Lambda_{ik}(\boldsymbol{x}, t)\Lambda_{jn}(\boldsymbol{x} + \boldsymbol{r}, t)\rangle$	correlation function of the vorticity tensor
μ	chemical potential
$\overline{\mu}$	ratio of viscosity coefficients of the internal and external fluids
μ_e	dynamic viscosity coefficient of the external fluid
μ_{eff}	effective dynamic viscosity coefficient
μ_i	dynamic viscosity coefficient of the internal fluid
μ_t	turbulent dynamic viscosity
ν	number of steps in particle's random walk
$\nu_A, \ldots, \nu_F$	stoichiometric coefficients
$\nu_A A + \nu_B B + \ldots \rightarrow \nu_E E + \nu_F F \ldots$	kinetic scheme of chemical reaction
ν_e	kinematic viscosity coefficient of the external fluid
ν_i	kinematic viscosity coefficient of the internal fluid
ν_p	turbulent viscosity coefficient of the disperse phase
ν_t	turbulent viscosity coefficient of the continuous phase
ξ	dimensionless gap (clearance) between particles
$\xi(t)$	degree of completeness of a chemical reaction
ξ	ratio of particle diffusion fluxes with and without electric field
ξ_j	Lagrangian coordinates; initial coordinates of a particle
Ξ_i	coefficient of virtual turbulent diffusion of a particle cloud in the X_i direction
$\boldsymbol{\Pi}$	triadic (third-rank) tensor
$\Pi(\boldsymbol{X}, t)$	intermittency function
ρ^0	density ratio of the carrying medium and the particle
ρ_e	external fluid density
ρ_i	internal fluid density
ρ_p	particle density
σ	surface charge density
σ	parameter in the Lennard–Jones potential
σ	root-mean-square deviation
σ^2	distribution variance
σ_{ij}	components of the viscous stress tensor
Θ_N	element of the set of configurations of an N-particle ensemble
Σ	specific surface area of the boundary between regions of different concentrations
Σ	surface tension coefficient
Σ	parameter characterizing the relative influence of gravity and turbulence on the collision kernel
τ	characteristic time of a process
τ	stress
τ	dimensionless time
τ	ratio of the particle radius to the double layer thickness

τ_E	characteristic time of decrease of the rate-of-strain tensor
τ'	turbulent component of stress
τ_c	relaxation time of a concentration field; characteristic time of micromixing
$(\tau_c)_1$	time of micromixing in the inertial-convective region
$(\tau_c)_2$	time of micromixing in the viscous-convective region
τ_D	characteristic time of impurity transport via molecular diffusion
τ_f	friction force per unit surface area
$\tau_i(\boldsymbol{r}_i)$	probability density distribution of i-th displacement
τ_{ij}	components of the stress tensor
$\tau_{ij}^{(1)}$	components of the Reynolds stress tensor
τ_m	empirical constant
τ_t	characteristic time of turbulent convective impurity transport
τ_t	Taylor time microscale; characteristic time of hydrodynamic relaxation
τ_w	characteristic time of matter production/consumption in a chemical reaction
τ_{λ_0}	characteristic period of fluctuations on scale λ_0; the Kolmogorov time microscale
τ_Λ	characteristic time of decrease of the vorticity tensor
υ	particle mobility
φ	volume concentration (content) of particles
$\varphi(\rho)$	characteristic function; moment generating function
φ_1	empirical constant
φ_m	limiting volume concentration of closely packed particles
ϕ	force field/electric field potential
$\phi(s)$	function in the Laplace transform image of particle's velocity
ϕ_1, ϕ_2	particle surface potentials
ϕ_{ij}	molecular interaction potential between i-th and j-th particles
$\lVert\boldsymbol{\Phi}\rVert$	shear resistance matrix
$\Phi(p)$	correction factor introduced into the expression for the molecular attraction force in order to account for electromagnetic retardation
$\Phi[(\rho, x)]$	characteristic functional
$\Phi_p(\boldsymbol{r}, t)$	probability density of particle's displacement by the distance $\boldsymbol{r}$ during the time t
χ	reflection coefficient of a particle
χ	thermal diffusivity
χ	inverse Debye radius
χ_t	turbulent thermal diffusivity
ψ	cumulant-generating function
ψ	stream function

$\psi(t)$	function characterizing the particle's effective free path resulting from its involvement into fluctuational motion of the carrier flow
$\Psi(\tau)$	autocorrelation function
$\Psi^{(E)}(\boldsymbol{r})$	Eulerian spatial autocorrelation function
$\Psi^{(E)}(t)$	Eulerian time autocorrelation function
$\Psi^{(L)}(t)$	Lagrangian autocorrelation function
Ψ_{cc}	correlation coefficient of concentration fluctuations
$\Psi_L^{(E)}(\boldsymbol{r},t)$	longitudinal (parallel to the average relative velocity vector) Eulerian spacetime autocorrelation function of fluid velocity fluctuations along the particle trajectory
$\Psi_L(\tau)$	dimensionless autocorrelation function characterized by Lagrangian integral time scale
$\Psi_{Lp}(t - t_1)$	two-time correlation function of carrier flow velocity fluctuations along the particle trajectory
$\Psi_{Lp}^{(L)}(t)$	longitudinal (parallel to the average relative velocity vector) Lagrangian time autocorrelation function of fluid velocity fluctuations along the particle trajectory
$\Psi_{Lr}(\tau\|r)$	Lagrangian autocorrelation function characterizing the relative motion of two particles initially separated by the distance $r = \|\boldsymbol{r}\|$
$\Psi_N^{(E)}(\boldsymbol{r},t)$	transverse (perpendicular to the average relative velocity vector) Eulerian spacetime autocorrelation function of fluid velocity fluctuations along the particle trajectory
$\Psi_{Np}^{(L)}(t)$	transverse (perpendicular to the average relative velocity vector) Lagrangian time autocorrelation function of fluid velocity fluctuations along the particle trajectory
$\Psi_u(\xi)$	autocorrelation function of fluid velocity fluctuations along the particle trajectory
Ψ_{uu}	correlation coefficient
Ψ_{uv}	correlation coefficient
Ψ_v	two-time autocorrelation function of fluid velocity fluctuations along the particle trajectory
Ψ_{Tv}	two-time autocorrelation function of fluid temperature fluctuations
Ψ_{vt}	two-time autocorrelation function of fluid velocity fluctuations
Ψ	function appearing in the e-F and k-ε models
$\|\|\boldsymbol{\Psi}\|\|$	shear resistance matrix
ω	element of the event set
$\boldsymbol{\omega}$	vorticity vector
$\langle\boldsymbol{\omega}\rangle$	average vorticity vector
$\omega(X)$	weight function
ω_{12}	frequency of collisions between particles of types 1 and 2
ω_λ	frequency of fluctuations on scale λ

ω'_λ	frequency; time period of fluid particle velocity recurrence
Ω	set of all events
$\mathbf{\Omega}$	angular velocity
$\mathbf{\Omega}$	vorticity vector
Ω_i	vorticity vector component
$\mathbf{\Omega}^\infty$	vorticity vector at the infinity
Ω_i^∞	vorticity vector component at the infinity
Ω_u	inertness parameter for a particle

Some mathematical notations

$\cup$	union of sets
$\cap$	intersection of sets
$\varnothing$	empty set
∇	gradient operator
$\nabla\cdot$	divergence operator
$\nabla\times$	curl operator
$\nabla\delta$	gradient of the delta function
$\Delta = \lvert\partial X_i/\partial\xi_j\rvert$	determinant of a Jacobian
$\boldsymbol{rr} = r_i r_j$	dyadic
$\boldsymbol{ab}\ldots\boldsymbol{c} = a_i b_j \ldots c_k$	polyadic
$\boldsymbol{Aa} = A_{ij} a_k$	product of tensor and vector
$\boldsymbol{A}\cdot\boldsymbol{a} = A_{ij} a_j$	scalar product of tensor and vector
$\boldsymbol{A}:\boldsymbol{B} = A_{ij} B_{ij}$	double-dot scalar product of tensors
$\dot{\boldsymbol{X}} = \frac{d\boldsymbol{X}}{dt}, \ddot{\boldsymbol{X}} = \frac{d^2\boldsymbol{X}}{dt^2}$	time derivatives
$\frac{D}{Dt} = \frac{\partial}{\partial t} + u_k \frac{\partial}{\partial X_k}$	substantial derivative
$\frac{\delta F[\varphi]}{\delta\varphi(X)}$	functional derivative
$\frac{\delta^N \Phi[\rho(X)]}{\delta\rho(X_1)\delta\rho(X_2)\ldots\delta\rho(X_N)}$	N-th order functional derivative
$(\boldsymbol{\vartheta}\cdot\boldsymbol{u}) = \int \boldsymbol{\vartheta}(M)\boldsymbol{u}(M)dM$	scalar product in the functional space
$D_k(M) = \frac{\delta}{\delta\vartheta_k(M)}$	functional derivative operator
i	imaginary unit
Im	imaginary part
$erf(z)$	probability integral

1
Basic Concepts of the Probability Theory

In order to formulate the theoretical concepts that will be crucial for the subsequent chapters, we must first mention some basic notions of the probability theory and their further ramifications. This chapter provides a brief outline. The interested reader will find a more detailed discussion of the relevant topics in textbooks and monographs on the probability theory and statistical physics (see the list of references at the end of the chapter).

1.1
Events, Set of Events, and Probability

Some of the typical situations considered in this chapter are as follows: a particle is in small volume (elementary volume) that includes a point X; N particles are in a small region of space; N_1 particles of type 1 and N_2 particles of type 2 are in a certain region. Speaking more generally, we shall be dealing with situations having to do with particle positions in space at different instants of time. Every such case can be thought of as a specific realization of some event. We shall define an event as an element of a certain space of events. In what follows, we will most often be using vector spaces defined by vectors $\mathbf{X}\,(X_1, X_2, \ldots, X_n)$, where X_i are real numbers. The probability theory introduces the notion of a set of events. Then the condition that an event ω belongs to a set of events A is written as $\omega \in A$.

Consider an event of finding a particle in some volume element ΔV centered at a point $\mathbf{X}$. Let all such events form an ensemble of events denoted as $A(\Delta V, \mathbf{X})$. It turns out that it is possible to determine whether the particle is in the vicinity of X, but it makes no sense to determine whether the particle is exactly at the point X. Physically, the probability of finding a particle exactly at some given point is zero. As an idealization one can consider a probability distribution given by a delta function, in which case the probability can be a finite number, and it is associated with an infininesimal interval around this point as the interval tends to zero. Let $\omega(\mathbf{X})$ denote the event that a particle is found in infinitesimal volume element centered at $\mathbf{X}$. We can then ascribe a certain probability to the condition $\omega(\mathbf{X}) \in A(\Delta V, \mathbf{X})$. This is just the probability for the particle to be in the volume element ΔV centered at $\mathbf{X}$.

Statistical Microhydrodynamics. Emmanuil G. Sinaiski and Leonid Zaichik

ISBN: 978-3-527-40656-2

The probability $P(A)$ of a set of events is defined as a function of A which satisfies the following probability axioms:

1. $P(A) \geq 0$ for all A;
2. $P(\Omega) = 1$;
3. if $\{A_i\}$ is a finite or countable sequence of non-overlapping sets, that is, $(A_i \cap A_j = \phi)$, then $P(\bigcup_i A_i) = \sum_i P(A_i)$.

The two consequences of these axioms are:

1. $P(\bar{A}) = 1 - P(A)$;
2. $P(\phi) = 0$.

Here $\cap$ and $\cup$ denote, respectively, the intersection and union operations on sets; Ω is the set of all events; ϕ – the empty set; $\bar{A}$ – the complement of a set A, that is set of all events not belonging to A.

Coexisting with the notion of probability is the notion of frequency of an event. To understand the difference between probability and frequency of events, consider an event ω selected at random from the full set Ω. The number of occurrences of $\omega \in A$ in N trials gives us the relative frequency of realization of the event $\omega \in A$. When N is increased, the relative frequency goes to the limit $P(A)$, which is defined as the probability of the event A. At $N \gg 1$, it is safe to assume the relative frequency of an event to be equal to the probability of occurrence of this event presuming that relative frequency has been normalized.

The axiom 3 is given for a finite or countable number of sets. But often it is necessary to deal with uncountable infinite number of sets. For instance, when studying the motion of particles under the action of external forces, one has to deal with sets of particle positions in spacetime. Let $\boldsymbol{X}$ is the position of a particle in space. The probability for the particle to be exactly at point $\boldsymbol{X}$ (it would then belong to a set consisting of only one element) is equal to zero, while the probability to find the particle in the vicinity of that point (that is, in the finite volume element ΔV centered at $\boldsymbol{X}$) is nonzero. The region ΔV can be visualized as a union of an infinite number of one-element sets of the type $\boldsymbol{X}$. A direct application of axiom 3 to this case would produce an uncertainty of the type $0 \cdot \infty$. Therefore the axiom 3 is unsuitable for infinitive sequences of sets, and the probability for the event to belong to the set ΔV cannot be obtained as the sum of such probabilities for the sets $\boldsymbol{X} \subset \Delta V$.

Axiom 3 is applicable only to incompatible events, that is, mutually exclusive events that belong to non-overlapping sets. Consider now the case of intersecting sets and overlapping events, that is, events belonging to two or more sets at the same time. Such events are called joint. Consider two sets A and B, whose intersection $A \cap B$ is not empty. We say that ω belongs to the intersection ($\omega \in A \cap B$) if $\omega \in A$ and $\omega \in B$. Then the probability of the joint event ω can be written as

$$P(A \cap B) = P(\omega \in A \quad \text{and} \quad \omega \in B). \tag{1.1}$$

As examples of joint events, consider two situations, which will prove to be of interest further on:

1. At a given time, the volume element ΔV centered at the space point $\boldsymbol{X}$ contains N_1 particles of type 1 (first event) and N_2 particles of type 2 (second event). The probability of this happening is given by the joint probability of both events.
2. A volume element ΔV centered at a space point $\boldsymbol{X}$ contains N_1 particles of type 1 and N_2 particles of type 2 at the time t_1 (first event) and n_1 particles of type 1 and n_2 particles type 2 at the time t_2 (second event). The probability of the joint event is the joint probability of both events at times t_1 and t_2.

Sometimes one is interested in the probability of an event given the occurrence of some other event. For example, we may want to know the probability of finding a particle in a volume element ΔV centered at the point $\boldsymbol{X}$ at the time t given that at the time $t_0 < t$, it was located in a volume element ΔV_0 centered at the point $\boldsymbol{X}_0 \neq \boldsymbol{X}$. Actually, we consider the set of all events C, where C denotes an event of finding the particle in the volume element ΔV at the time t. The particle could get into this element from any initial spatial position (with different probabilities), but we are interested only in some of those positions, that is, in a subset B of the set A. The probability of such an event is called a conditional probability. Conditional probability is defined as the probability of realization of an event $\omega \in A$ under the condition that $\omega \in B$ and is equal to

$$P(A|B) = P(A \cap B)/P(B). \tag{1.2}$$

The theory of stochastic processes is based (to a considerable degree) on the notion of joint probability. In this context, let us mention an important property of the joint probability. Suppose the full set Ω is divided into non-overlapping subsets B_i, that is,

$$B_i \cap B_j = \phi \quad \text{and} \quad \bigcup_i B_i = \Omega.$$

As far as

$$\bigcup_i (A \cap B_i) = A\mathbf{I}(\bigcup_i B_i) = A \cap W = A$$

and (see axiom 3)

$$\sum_i P(A \cap B_i) = P(\bigcup_i (A \bigcup_i B_i)) = P(A)$$

we find from (1.2):

$$\sum_i P(A|B_i)P(B_i) = P(A). \tag{1.3}$$

Thinking of the subset B_i as a variable, one can see from the last relation that summation over all mutually exclusive possibilities (i.e. over all sets B_i) eliminates this variable from the outcome.

Yet another important notion is the notion of independent events. Two sets of events A and B are called independent if the probability for an event to belong to set A and the probability to belong to set B are not correlated. Then

$$P(A \cap B) = P(A)P(B). \tag{1.4}$$

1.2
Random Variables, Probability Distribution Function, Average Value, and Variance

The concept of a random variable is of primary importance in stochastic processes. A random variable $F(X)$ is defined as a function of the element X of the space of probabilistic events $\mathbf{X}$. An event is specified by X, so X now stands for the event previously denoted by ω. The examples of random variables include position, momentum, and spatial orientation of a particle driven by random external forces (Brownian motion, motion in a turbulent flow). The introduction of a random variable notation simplifies operations with functions of random variables, calculations of random variable distributions, of averages and other statistical characteristics of distributions. Furthermore, the introduction of continuous random variables enables us to operate with stochastic differential equations and study the change of random variables in space and in time in the same way as we study deterministic systems by using differential equations.

The frequency (or probability) of realization of a definite event is equal to some value between zero and one. If the events are mutually exclusive, the sum of probabilities must be equal to one. This means that one of the events will realize.

Statistical mechanics is usually concerned with continuous random variables, that is, variables that can assume a continuous range of values. As far as the probability to get any given value from a continuum of possible values is zero, and the sum of all probabilities is one, it is necessary to look at the probability of realization of an event that is associated with an infinitesimal interval (set) of values rather than a single value. This probability is also an infinitely small quantity having the same order as the length of the interval (measure of event) and so is proportional to the measure of events, that is, to $d\mathbf{X}$. Thus the probability that a random variable is contained in the interval $(\mathbf{X}, \mathbf{X}+d\mathbf{X})$ can be represented as

$$P(X \in (\mathbf{X}, \mathbf{X} + d\mathbf{X})) = p(X)d\mathbf{X}. \tag{1.5}$$

The function $p(\mathbf{X})$ is called the probability density function (PDF) or simply the probability density. The condition that the sum of probabilities for a continuous random variable is equal to one can be written in the integral form:

$$\int_X p(\mathbf{X})d\mathbf{X} = 1, \tag{1.6}$$

where X is the domain of the n-dimensional space in which $\mathbf{X}$ varies. The relation (1.6) can be interpreted as the normalization condition for the PDF.

The introduction of a PDF enables us to find statistical characteristics of the distribution of a random variable X. The most important of them is the average value (aka mean value, or expectation value) of a random variable or random function:

$$\langle f \rangle = \int_X f(\mathbf{X})\, p(\mathbf{X}) d\mathbf{X}. \tag{1.7}$$

If X is a vector in an n-dimensional coordinate space, then (1.6) and (1.7) can be written in the coordinate form:

$$\int_{-\infty}^{\infty}\int_{-\infty}^{\infty} \cdots \int_{-\infty}^{\infty} f(X_1, X_2, \ldots, X_n)\, p(X_1, X_2, \ldots, X_n) dX_1, dX_2, \ldots, dX_n = 1.$$

$$\langle f \rangle = \int_{-\infty}^{\infty}\int_{-\infty}^{\infty} \cdots \int_{-\infty}^{\infty} f(X_1, X_2, \ldots, X_n)\, p(X_1, X_2, \ldots, X_n) dX_1, dX_2, \ldots, dX_n.$$

Another statistical characteristic is the variance σ^2. For a one-dimensional space, the variance is defined as

$$\sigma^2 = \int_X (f - \langle f \rangle)^2\, p(\mathbf{X}) d\mathbf{X}. \tag{1.8}$$

The square root σ of the variance is called the standard deviation. Sometimes the PDF has the form of a function with a sharp peak at the point $X = X_0$. In limiting case it is infinite at $X = X_0$ and zero at $X \neq X_0$. Such a case arises when we idealize a process. For example, we can choose to regard a mass that is continuously distributed in a small volume element centered at X_0 as localized at one space point X_0 (i.e. as "point mass"). Than the density of the substance differs from zero only at this point and the integral (1.7) has the meaning of the total mass. A similar reasoning leads to the concept of a point force – the net force with which we replace a force that is continuously distributed over a small volume element. To ensure the existence of integrals of such functions, we have to extend the notion of a function, what is achieved by the introduction of generalized functions.

1.3 Generalized Functions

The simplest and most extensively used generalized function is Diracs delta function $\delta(X - X_0)$, which can be defined as the limit of following sequence (sometimes referred to as "delta sequence"):

$$\delta(\boldsymbol{X}-\boldsymbol{X}_0) = \lim_{m\to\infty}\left(\frac{m}{\sqrt{\pi}}\right)^n \exp(-m^2(\boldsymbol{X}-\boldsymbol{X}_0)^2). \tag{1.9}$$

Here n is the number of dimensions and, accordingly, $\boldsymbol{X}$ is an n-dimensional vector with components $X_1, X_2, \ldots, X_n$. Eq. (1.9) can also be written for one-dimensional sequences of $X_i - X_i^0$. Then the following identity will hold:

$$\delta(X_1-X_1^0)\delta(X_2-X_2^0)\ldots\delta(X_n-X_n^0) = \delta(\boldsymbol{X}-\boldsymbol{X}_0). \tag{1.10}$$

The limit on the right-hand side of Eq. (1.9) is 0 at $\boldsymbol{X}\neq\boldsymbol{X}_0$ and $+\infty$ at $\boldsymbol{X}=\boldsymbol{X}_0$. Therefore Diracs delta function is not a function in the usual sense and should not be interpreted as giving the value of the dependent variable at each point. What is important, however, is that this function is still integrable, and behaves similarly to ordinary functions in its capacity as an integrand. In particular, the integral of the scalar product of Diracs delta function $\delta(\boldsymbol{X}-\boldsymbol{X}_0)$ and an ordinary function $\varphi(\boldsymbol{X})$ equals

$$\begin{aligned}(\delta(\boldsymbol{X}-\boldsymbol{X}_0), \varphi(\boldsymbol{X})) &= \int_X \delta(\boldsymbol{X}-\boldsymbol{X}_0)\varphi(\boldsymbol{X})d\boldsymbol{X} \\ &= \lim_{m\to\infty}\int_X\left(\frac{m}{\sqrt{\pi}}\right)^n \exp(-m^2(\boldsymbol{X}-\boldsymbol{X}_0)^2)\varphi(\boldsymbol{X})d\boldsymbol{X} = \varphi(\boldsymbol{X}_0)\end{aligned}$$

provided the domain contains the point X_0.

Thus, by its definition, the Delta function has two basic properties:

$$\delta(\boldsymbol{X}) = \begin{cases} 0, & \text{for} \quad \boldsymbol{X}\neq 0, \\ 1, & \text{for} \quad \boldsymbol{X} = 0, \end{cases} \tag{1.11a}$$

$$\int_X \delta(\boldsymbol{X}-\boldsymbol{X}_0)\varphi(\boldsymbol{X})d\boldsymbol{X} = \varphi(\boldsymbol{X}_0). \tag{1.11b}$$

In the particular cases $\varphi(\boldsymbol{X}) = 1$ and $\varphi(\boldsymbol{X}) = \boldsymbol{X}$ one gets:

$$\int_X \delta(\boldsymbol{X}-\boldsymbol{X}_0)d\boldsymbol{X} = 1. \tag{1.12}$$

and

$$\int_X \delta(\boldsymbol{X}-\boldsymbol{X}_0)\boldsymbol{X}d\boldsymbol{X} = \boldsymbol{X}_0. \tag{1.13}$$

Hence, according to Eq. (1.7), $\delta(\boldsymbol{X}-\boldsymbol{X}_0)$ can be taken as a PDF such that the random variable $\boldsymbol{X}$ has the average value $\boldsymbol{X}_0$. For the one-dimensional case, the following equality can be written:

$$(X-X_0)\delta(X-X_0) = 0$$

or

$$X\delta(X-X_0) = X_0\delta(X-X_0). \tag{1.14}$$

Taking $\varphi(\boldsymbol{X}) = (\boldsymbol{X}-\boldsymbol{X}_0)^2$, we can write

$$\int_X \delta(\boldsymbol{X}-\boldsymbol{X}_0)(\boldsymbol{X}-\boldsymbol{X}_0)^2 d\boldsymbol{X} = (\boldsymbol{X}_0-\boldsymbol{X}_0)^2 = 0. \tag{1.15}$$

The left-hand side of (1.23) coincides with the definition of the variance for the PDF $\delta(\boldsymbol{X}-\boldsymbol{X}_0)$. Thus, its variance is zero, and the delta function describes the case when one knows for sure that $\langle \boldsymbol{X} \rangle = \boldsymbol{X}_0$.

The (one-dimensional) Cauchy sequence is not the only sequence converging to the delta function. For example, the sequence

$$\delta(X-X_0) = \lim_{\varepsilon \to 0} \frac{\varepsilon}{\pi(X-X_0)^2 + \varepsilon^2} \tag{1.16}$$

can be used as an alternative representation of the delta function.

The delta function is an infinitely differentiable function. Its derivative can be defined by differentiating the integral

$$\int \varphi(X) \frac{d}{dX} \delta(X-X_0) dX$$

by parts and using the property (1.11):

$$\int_X \delta'(X-X_0)\varphi(X)dX = -\int_X \delta(X-X_0)\varphi'(X)dX = -\varphi'(X_0), \tag{1.17}$$

from which there follows a useful symbolic equality

$$\delta'(X) = -\frac{\delta(X)}{X}, \quad (X \neq 0),$$

In the more general case, one can write

$$\delta^{(r)}(X) = (-1)^r \frac{\delta(X)}{X^r}, \quad (X \neq 0, \quad r = 0, 1, \ldots). \tag{1.18}$$

If a function $Y = f(X)$ is single-valued, that is, if it can be solved with respect to X in a unique way, then $X = f^{-1}(Y)$ and

$$\delta(Y - f(X)) = \frac{\delta(X - f^{-1}(Y))}{|df/dX|}. \tag{1.19}$$

A similar relation takes place for a vector function $\boldsymbol{Y} = \boldsymbol{f}(\boldsymbol{X})$:

$$\delta(\boldsymbol{Y} - \boldsymbol{f}(\boldsymbol{X})) = \frac{\delta(\boldsymbol{X} - \boldsymbol{f}^{-1}(\boldsymbol{Y}))}{\Delta}. \tag{1.19$'$}$$

where Δ is the determinant of the Jacobian $||\partial f_i/\partial X_j||$.
The delta function can be connected with the unit step function (Heaviside function) defined as

$$H(X) = \begin{cases} 0, & \text{for} \quad X < 0, \\ 1, & \text{for} \quad X > 0 \end{cases}$$

through the symbolic relation

$$\delta(X) = \frac{d\mathrm{H}}{dX}. \tag{1.20}$$

If there is more then one independent variable, one has to use partial derivatives of the delta function. For example, if we take the delta function as a generalized vector function $d(\boldsymbol{X} - \boldsymbol{X}_0)$, its gradient is defined as

$$\nabla\delta = \frac{\partial\delta(\boldsymbol{X} - \boldsymbol{X}_0)}{\partial\boldsymbol{X}} = \left(\frac{\partial\delta}{\partial X_1}, \frac{\partial\delta}{\partial X_2}, \ldots, \frac{\partial\delta}{\partial X_n}\right). \tag{1.21}$$

1.4 Methods of Averaging

When looking at the hydrodynamic characteristics of a turbulent flow or at the motion of particles under the action of random external forces, we notice one distinguishing feature shared by these two types of motion: the presence of random fluctuations. Because of fluctuations, the dependences of hydrodynamic field parameters on spacetime coordinates, and the configuration of particles in space at different moments look irregular and have a confusing pattern. If a process is repeated multiple times under the identical set of initial and boundary conditions, the observed values of field parameters and particle positions will be different. This necessitates the use of averaging methods in any study of random motions. Averaging allows us to make a transition from irregular characteristics to much more smooth and regular mean values. In practice the mean value is determined by averaging over the time interval,

$$\langle f(X,t)\rangle_T = \frac{1}{T}\int_{-T/2}^{T/2} \omega(X,\tau)\, f(X, t+\tau)d\tau, \tag{1.22}$$

or by averaging over the considered spatial region,

$$\langle f(X,t)\rangle_V = \frac{1}{V}\int_V \omega(x,t)\, f(X+x, t)dx \tag{1.23}$$

or, most generally, by spacetime averaging,

$$\langle f(X,t)\rangle_{TV} = \frac{1}{VT}\int_V \int_{-T/2}^{T/2} \omega(x,\tau)\, f(X+x, t+\tau)dxd\tau, \tag{1.24}$$

where $\omega(x, t)$ is the weight function.

We can also introduce the autocorrelation function $\Psi(\tau)$, which is defined as follows: take a random function $f(t)$ at one and the same point of space but at instances of time t and $t+\tau$, form the product, and find its average value over the time interval $(0, T)$ for $T \to \infty$:

$$\Psi(\tau) = \lim_{T\to\infty} \frac{1}{T}\int_0^T f(t)\, f(t+\tau)dt. \tag{1.25}$$

This function plays an important role in many applications.

The three types of averaging mentioned above have one drawback, namely, they apply to only one instance of the process under consideration (turbulent velocity field, etc.). Another shortcoming is that one is faced with the problem of choosing the most convenient weight function.

If the process is repeated multiple times under the same initial conditions, we are dealing with many instances of the same process. In this case one can talk about a statistical set of identical processes (flows, particle motions etc.) taking place under fixed initial and external conditions. Let one and the same experiment be replicated N times under the same conditions, yielding different values u_i of one and the same parameter, for example, velocity u. By averaging the velocities u_i observed in a discrete set of similar tests, we obtain the mean value

$$\langle u(X,t)\rangle = \frac{1}{N}\sum_{i=1}^{N} u_i,$$

which is called the ensemble average. In many cases the ensemble average proves to be stable enough, in other words, the outcomes of a sufficiently large set of experiments show a very small variance.

Let a continuous random variable u ($-\infty < u < \infty$) be characterized by the PDF $p(u)$. If we are interested in the value of u at one and the same space point M, then $p(u)du$ signifies the probability for u to be found in interval $(u,\ u+du)$. Then the ensemble average of u is

$$\langle u(\mathbf{X}, t)\rangle = \int_{-\infty}^{\infty} u\, p(u)\, du. \tag{1.26}$$

By analogy, the ensemble average of any function F is equal to

$$\langle F(u(\mathbf{X}, t))\rangle = \int_{-\infty}^{\infty} F(u)\, p(u) du. \tag{1.27}$$

Now, let u be measured at different spacetime points $M_1 = (X_1,\ t_1)$, $M_2 = (X_2,\ t_2)$, ..., $M_N = (X_N,\ t_N)$. The resulting values of u are denoted by $u_1,\ u_2,\ \ldots,\ u_N$. We then introduce the N-dimensional PDF $p(u_1,\ u_2,\ \ldots,\ u_N)$, where $p(u_1,\ u_2,\ \ldots,\ u_N)du_1du_2\ldots du_N$ means the probability of finding u_i in the interval $(u_i,\ u_i+du_i)$. We have used a common convention where the index stands for all N variations, that is $(u_i, u_i + du_i) = (u_1 + du_1, \ldots, u_n + du_n)$. The average of any function will be written as

$$\langle F\rangle = \int_{-\infty}^{\infty}\int_{-\infty}^{\infty}\cdots\int_{-\infty}^{\infty} F(u_1, u_2, \ldots, u_N)\, p(u_1, u_2, \ldots, u_N) du_1 du_2 \ldots du_N. \tag{1.28}$$

By introducing an N-dimensional vector $\boldsymbol{u}(u_1,\ u_2,\ \ldots,\ u_N)$, we can rewrite the relation (1.28) in a more compact form:

$$\langle F\rangle = \int_{-\infty}^{\infty} F(\boldsymbol{u})\, p(\boldsymbol{u}) d\boldsymbol{u}. \tag{1.29}$$

Multidimensional PDFs are especially important for studying the behavior of an N-particle system in a random field of external forces. If u_i denotes the coordinate of the i-th particle, then the above-introduced PDF is called a multiparticle PDF. One-particle and two-particle PDFs are of particular interest in applications. Sometimes the two-particle PDF is also called "pair PDF" or "pair distribution". These PDFs can be derived from multidimensional PDFs by integrating them over all possible positions of the remaining particles. For instance, a single-particle PDF is obtained as

$$p(X_1) = \int_{-\infty}^{\infty}\int_{-8}^{\infty}\cdots\int_{-\infty}^{\infty} p(X_1, X_2, \ldots, X_N) dX_2 \ldots dX_N \tag{1.30}$$

Such PDFs are also called marginal PDFs.

If we consider spherical particles of different radii a_i, than the PDF $p(\boldsymbol{X}_1, \ldots, \boldsymbol{X}_N, a_1, \ldots, a_N)$ will be associated with the radius distribution in addition to the coordinate distribution, and $p(\boldsymbol{X}_1, \ldots, \boldsymbol{X}_N, a_1, \ldots, a_N)\, d\boldsymbol{X}_1 \ldots d\boldsymbol{X}_N\, da_1 \ldots da_N$ will have the meaning of probability to find the N-particle system in the volume element ($d\boldsymbol{X}_1 \ldots d\boldsymbol{X}_N$) with particle radii lying in the interval $(a_1 + da_1), \ldots, (a_N + da_N)$. The corresponding single-particle PDF is

$$p(\boldsymbol{X}_1, a_1) = \int_{-\infty}^{\infty}\int_{0}^{\infty} p(\boldsymbol{X}_1, \boldsymbol{X}_2, \ldots, \boldsymbol{X}_N, a_2, \ldots, a_N)d\boldsymbol{X}_2 \ldots d\boldsymbol{X}_N da_2 \ldots da_N. \quad (1.31)$$

If the radius a must be the same for all particles, then it is convenient to operate with particle distribution over the radius

$$n(\boldsymbol{X}, a) = N\, p(\boldsymbol{X}, a), \quad (1.32)$$

such that $n(\boldsymbol{X}, a)da$ is the probabilistic numerical concentration (aka number concentration) of particles with radius in the interval $(a + da)$ in the volume element $d\boldsymbol{X}$.

The multidimensional PDF should satisfy the following properties:

1. $p(\boldsymbol{u}) \geq 0$;
2. $\int_{-\infty}^{\infty} p(\boldsymbol{u})d\boldsymbol{u} = 1$;
3. $p(u_1, u_2, \ldots, u_N) = p(u_{i_1}, u_{i_2}, \ldots, u_{i_N})$, where the set $\{i_1, i_2, \ldots, i_N\}$ is formed from the set $\{1, 2, \ldots, N\}$ by changing the order.
4. $p(u_1, u_2, \ldots, u_n) = \int_{-\infty}^{\infty}\int_{-\infty}^{\infty} \ldots \int_{-\infty}^{\infty} p(u_1, u_2, \ldots, u_n, u_{n+1}, \ldots, u_N)du_{n+1} \ldots du_N \text{ for } n < N$;
5. For independent random variables $u_1, u_2, \ldots, u_N$, there holds:

$$p(u_1, u_2, \ldots, u_N) = p(u_1)\, p(u_2), \ldots, p(u_N) \quad (1.33)$$

Property 3 is known as the symmetry property and property 4 – as the consistency property.

It is now time to discuss the connection between different types of averaging. In practice, we use time or space averaging rather than ensemble averaging, because the latter requires a large number of experiments. In Statistical Mechanics, ensemble averaging, that is, averaging over the set of all possible states, is often replaced by time or space averaging, with the implicit assumption that by increasing the averaging interval we can always make the average values converge to the corresponding ensemble averages. This assumption is called the ergodic hypothesis, or, in those special cases when it can be rigorously proved, the ergodic theorem.

When studying such problems as the flow of a disperse medium or the filtration of a fluid through a porous medium, one often uses the so-called Saffman step function:

$$H(\boldsymbol{X}) = \begin{cases} 0, & \text{for } \boldsymbol{X} \text{ in rigid body,} \\ 1, & \text{for } \boldsymbol{X} \text{ in fluid.} \end{cases} \tag{1.34}$$

This function depends on statistical parameters of the distribution of moving particles of the disperse phase or fixed particles of the porous medium. After averaging over the particle ensemble, we get

$$\langle H(\boldsymbol{X})\rangle = 1-\varphi, \tag{1.35}$$

where φ is volume concentration of particles.

One can use the Saffman function to perform space averaging of hydrodynamic parameters. For example, the velocity of the fluid $\boldsymbol{u}$ will be averaged as

$$\bar{\boldsymbol{u}} = \langle H\boldsymbol{u}\rangle/\langle H\rangle = \langle \boldsymbol{u}\rangle/(1-\varphi), \tag{1.36}$$

where $\bar{\boldsymbol{u}}$ is the mean-flow-rate velocity through the microcapillaries of the porous medium. It should not be confused with the ensemble average $\langle \boldsymbol{u}\rangle$, although for a highly permeable medium ($\varphi \ll 1$), the two velocities are equal: $\langle u\rangle = \bar{u}$.

1.5 Characteristic Functions

Instead of using the PDF $p(u_1, u_2, \ldots, u_N)$, it is often convenient to use its Fourier transform:

$$\varphi(\rho_1, \rho_2, \ldots, \rho_N) = \int_{-\infty}^{\infty}\int_{-\infty}^{\infty}\cdots\int_{-\infty}^{\infty} \exp\left\{ i\sum_{k-1}^{N}\rho_k u_k \right\} p(u_1, u_2, \ldots, u_N)du_1 du_2 \ldots du_N$$

or, in the vector form,

$$\varphi(\rho) = \int_{-\infty}^{\infty} e^{i\rho\cdot\boldsymbol{u}} p(\boldsymbol{u})d\boldsymbol{u}. \tag{1.37}$$

Here ρ is an N-dimensional vector with components $(\rho_1, \rho_2, \ldots, \rho_N)$. The function $\varphi(\rho)$ is called the characteristic function or the moment-generating function. Because of Eq. (1.29), it can be represented as

$$\varphi(\rho) = \langle e^{i\rho\cdot\boldsymbol{u}}\rangle. \tag{1.38}$$

If the characteristic function is known, then the PDF is obtained as the inverse Fourier transform:

$$p(\boldsymbol{u}) = \frac{1}{(2\pi)^N} \int_{-\infty}^{\infty} e^{-i\rho \cdot \boldsymbol{u}} \varphi(\boldsymbol{\rho}) d\boldsymbol{\rho}. \tag{1.39}$$

So, the knowledge of the characteristic function is tantamount to the knowledge of the PDF. Hence the properties of the PDF are readily obtained from those of the characteristic function. The normalization condition for the PDF means that

$$\omega(\mathbf{0}) = \mathbf{1}. \tag{1.40}$$

For independent random variables we have, according to (1.33):

$$\varphi(\boldsymbol{\rho}) = \varphi(\rho_1)\varphi(\rho_2)\ldots\varphi(\rho_N). \tag{1.41}$$

The symmetry and consistency properties of characteristic function follow from properties 3 and 4 of the PDF:

$$\varphi(\rho_1, \rho_2, \ldots, \rho_N) = \varphi(\rho_{i_1}, \rho_{i_2}, \ldots, \rho_{i_N}), \tag{1.42}$$

$$\varphi(\rho_1, \rho_2, \ldots, \rho_n) = \varphi(\rho_1, \rho_2, \ldots, \rho_n, 0, 0, \ldots, 0), \tag{1.43}$$

where $i_1, i_2, \ldots, i_N$ is any combination of non-repeating numbers $1, 2, \ldots, N$. In the last relation, $n < N$ and the number of zeros is equal to $N - n$. The property (1.57) allows us to obtain the characteristic function for a smaller number of dimensions (smaller number of particles) from the N-dimensional (N-particle) characteristic function, and then to get the corresponding marginal PDF by using the inverse Fourier transform (1.39). Therefore one can specify all PDFs, describing random variables at all possible points, through a single characteristic function known as the characteristic functional. In particular, for one-dimensional random function $u(X)$ defined on a finite interval $a \leq X \leq b$, the characteristic functional is

$$\Phi(\rho(X)) = \left\langle \exp\left\{ i \int_a^b \rho(X) u(X) dX \right\} \right\rangle, \tag{1.44}$$

where $\rho(X)$ is a function selected in such a way that the integral in the exponent converges. The left-hand side is a function of a function, which is why it is called a functional.

If

$$\rho(X) = \sum_{i=1}^{N} \rho_i \delta(X - X_i),$$

then Eq. (1.44) gives us:

$$\Phi(\rho(X)) = \left\langle \exp\left\{ i\sum_{k=1}^{N} \rho_k X_k \right\} \right\rangle = \varphi(\rho_1, \rho_2, \ldots, \rho_N), \tag{1.45}$$

Thus the characteristic functional turns into a characteristic function of the multidimensional PDF for $u(X_1)$, $u(X_2)$, ..., $u(X_N)$. Additional information about the characteristic functional can be found in Section 1.15.

1.6
Moments and Cumulants of Random Variables

To solve a specific problem in a rigourous way, one has to specify a multidimensional (multiparticle) PDF. However, one run into difficulties with this approach because the PDF cannot be determined with a sufficient accuracy. Furthermore, it is inconvenient to use because it results in cumbersome expressions. In practice, when solving applied problems, one usually considers only the more simple parameters that characterize specific statistical properties of the process. The most important of these parameters are moments.

Let us consider a set of N random variables with N-dimensional PDF $p(u_1, u_2, \ldots, u_N)$. The moments are defined as follows:

$$\begin{aligned} B_{k_1 k_2 \ldots k_N} &= \left\langle u_1^{k_1} u_2^{k_2} \ldots u_N^{k_N} \right\rangle \\ &= \int_{-\infty}^{\infty}\int_{-\infty}^{\infty} \ldots \int_{-\infty}^{\infty} u_1^{k_1} u_2^{k_2} \ldots u_N^{k_N}\, p(u_1, u_2, \ldots, u_N) du_1 du_2 \ldots du_N, \end{aligned} \tag{1.46}$$

where k_1, k_2, ..., k_N are non-negative integers, whose sum $K = k_1 + k_2 + \ldots + k_N$ is called the moments order. It is evident that moments of the first order are simply the mean values of random variables u_1, u_2, ..., u_N.

In addition to ordinary moments (1.61) one often uses some combinations of moments. In particular, central moments are defined as moments of fluctuations (deviations of random variables u_1, u_2, ..., u_N from their mean values):

$$b_{k_1 k_2 \ldots k_N} = \left\langle (u_1 - \bar{u}_1)^{k_1} (u_2 - \bar{u}_2)^{k_2} \ldots (u_N - \bar{u}_N)^{k_N} \right\rangle. \tag{1.47}$$

For $N = 1$ and $k_1 = 2$, we get the second-order central moment $b_2 = \sigma^2$.

If u_i are the velocities of a turbulent flow at spatial points x_i, then the differences $u_i - \langle u_i \rangle$ have the meaning of velocity fluctuations at these points. Thus central moments characterize statistical properties of random variables – velocity fluctuations. In the case when u_i are the random positions of particles driven by a random external force, central moments characterize the statistical properties of the disperse phase in the suspension.

By removing brackets in Eq. (1.47) and making use of Eq. (1.46), we can obtain connections between central and ordinary moments. The case of $N = 1$ yields

$$b_1 = 0; \quad b_2 = B_2 - B_1^2; \quad b_3 = B_3 - 3B_1B_2 + 2B_1^3; \\ b_4 = B_4 - 4B_1B_3 + 6B_1^2B_2 - 3B_2^4 \quad \text{etc.} \tag{1.48}$$

When $\bar{u}_i = 0$, the central and ordinary moments coincide. The two combinations of central moments,

$$\sqrt{\beta_1} = \frac{b_3}{b_2^{3/2}}, \quad \beta_2 = \frac{b_4}{b_2^2} - 3 \tag{1.49}$$

serve as statistical characteristics of a random quantity and are called, repectively, the asymmetry and the excess.

The moments of random variables $u_1, u_2, \ldots, u_N$ can be expressed through a corresponding characteristic function $\varphi(\rho_1, \rho_2, \ldots, \rho_N)$ by comparing the relations (1.46) and (1.37):

$$B_{k_1k_2\ldots k_N} = (-i)^K \frac{\partial^K \varphi(\rho_1, \rho_2, \ldots, \rho_N)}{\partial \rho_1^{k_1} \partial \rho_2^{k_2} \ldots \partial \rho_N^{k_N}}\bigg|_{\rho_1 = \rho_2 = \ldots = \rho_N = 0}, \tag{1.50}$$

from which one can see that moments can also be thought of as coefficients in the Taylor expansion of the characteristic function:

$$\varphi(\rho_1, \rho_2, \ldots, \rho_N) = \sum_{k_1, k_2, \ldots, k_N} i^K \frac{B_{k_1k_2\ldots k_N}}{k_1!k_2!\ldots k_N!} \rho_1^{k_1} \rho_2^{k_2} \ldots \rho_N^{k_N}. \tag{1.51}$$

Thus, if the moments are known, Eq. (1.51) gives us the characteristic function, and then the PDF follows from Eq. (1.39). It means that the PDF is uniquely defined by the moments of the distribution.

The other category of combinations of central moments are the so-called cumulants (aka semi-invariants) $S_{k_1k_2k_N}$. Let us introduce the logarithm of the characteristic function

$$\psi(\rho_1, \rho_2, \ldots, \rho_N) = \ln \varphi(\rho_1, \rho_2, \ldots, \rho_N), \tag{1.52}$$

which is called the generating function of cumulants. Cumulants can then be defined as the coefficients in the Taylor expansion of ψ in just as the moments were defined as the coefficients in the expansion (1.66):

$$\psi(\rho_1, \rho_2, \ldots, \rho_N) = \sum_{k_1, k_2, \ldots, k_N} i^K \frac{S_{k_1k_2\ldots k_N}}{k_1!k_2!\ldots k_N!} \rho_1^{k_1} \rho_2^{k_2} \ldots \rho_N^{k_N}, \tag{1.53}$$

$$S_{k_1k_2\ldots k_N} = (-i)^K \frac{\partial^K \varphi(\rho_1, \rho_2, \ldots, \rho_N)}{\partial \rho_1^{k_1} \partial \rho_2^{k_2} \ldots \partial \rho_N^{k_N}}\bigg|_{\rho_1 = \rho_2 = \ldots = \rho_N = 0}, \tag{1.54}$$

Recalling that $\varphi(0, \ldots, 0) = 1$ and taking $N = 1$, we can express cumulants in terms of ordinary and central moments:

$$\begin{aligned} S_1 &= B_1; \quad S_2 = B_2 - B_1^2 = b_2; \quad S_3 = B_3 - 3B_1B_2 + 2B_1^3 = b_3; \\ S_4 &= B_4 - 4B_1B_3 - 3B_2^2 + 12B_1^2B_2 - 6B_1^4 = b_4 - 3b_2^2; \\ S_5 &= b_5 - 10b_2b_3 \quad \text{etc.} \end{aligned} \tag{1.55}$$

By the same token, moments could be expressed in terms of cumulants:

$$B_1 = S_1; \quad B_2 = S_2 + S_1^2; \quad B_3 = S_3 + 3S_2S_1 + S_1^2; \quad \text{etc.} \tag{1.56}$$

In the case of one-dimensional PDF $p(u)$ we have the following expressions for moments and cumulants:

$$B_n = \int_{-\infty}^{\infty} p(u)u^n du = \left(\frac{1}{i}\frac{d}{d\rho}\right)^n \varphi(\rho)|_{\rho=0}, \tag{1.57}$$

$$S_n = \left(\frac{1}{i}\frac{d}{d\rho}\right)^n \psi(\rho)|_{\rho=0}. \tag{1.58}$$

There is a recurrent relation between moments and cumulants. In one-dimensional case it has the following form:

$$B_0 = 1; \quad B_n = \sum_{k=0}^{n} \frac{(n-1)!}{(k-1)!(n-k)!} S_k B_{n-k}; \quad (n = 1, 2, \ldots). \tag{1.59}$$

From this relation, one can derive Eq. (1.56).

1.7
Correlation Functions

In statistical mechanics of disperse media as well as in the turbulence theory, one often encounters random fields described by a random function $u(M)$ of a spacetime point M. Following the definition (1.46), let us call expressions

$$B_{uu\ldots u}(M_1, M_2, \ldots, M_k) = \langle u(M_1)u(M_2)\ldots u(M_k)\rangle \tag{1.60}$$

the k-th order moments of such a field. Generally speaking, some points may coincide. The number of different points is called the type of the moment. In this context, one can talk about single-point moments, two-point moments, and so on. The average values of products of several correleted random functions of different fields are called mixed moments.

Consider, for example, the field of velocities in a turbulent flow given by the velocity vector $\boldsymbol{u}(u_1, u_2, u_3)$. Components of this vector can be regarded as different mutually correlated random functions. Since the values of velocity components could be taken at the same point or at different points, there are moments of various types and orders. Of primary importance in statistical mechanics are two-point second-order moments known as correlation functions (or simple correlations):

$$B_{ij}(M_1, M_2) = \langle u_i(M_1)u_j(M_2)\rangle. \tag{1.61}$$

B_{ij} are the components of a second rank tensor $\boldsymbol{B}$. It is obvious that the relation (1.61) could be written in a matrix form:

$$\boldsymbol{B}(M_1, M_2) = \langle \boldsymbol{u}(M_1)\boldsymbol{u}^T(M_2)\rangle,$$

where the superscript T stands for transpose.

When $M_1, M_2, \ldots, M_K$ are points in spacetime, the corresponding moments and correlations are called spacetime moments/correlations. In statistical mechanics one usually has to deal with correlations of random functions at different points at one and the same instant of time or with correlations at one and the same point but at different instants of time. The former correlations are called spatial correlations and the latter – time correlations.

Let us metion some important properties of correlation functions. A correlation function $B_{uu}(M_1, M_2) = \langle u(M_1)u(M_2)\rangle$ is symmetric with respect to the pair of points M_1, M_2:

$$B_{uu}(M_1, M_2) = B_{uu}(M_2, M_1). \tag{1.62}$$

A quadratic form with coefficients B_{uu} (M_i, M_j) is always non-negative, that is,

$$\sum_{i=1}^{n}\sum_{j=1}^{n} B_{uu}(M_i, M_j)X_iX_j \geqslant 0, \tag{1.63}$$

for all real X_i, non-negative integer n and any selection of points $M_1, M_2, \ldots, M_n$. In particular, at $n = 2$ the expression (1.63) becomes

$$|B_{uu}(M_1, M_2)| \leq |B_{uu}(M_1, M_1)|^{1/2}|B_{uu}(M_2, M_2)|^{1/2}. \tag{1.64}$$

In addition to the above-mentioned two-point moments of one random function at different points, $u(M_1)$ and $u(M_2)$, one can consider two-point moments of different random functions, $u(M_1)$ and $v(M_2)$. A mixed two-point moment B_{uv} (M_1, M_2) is called the mutual correlation function. Its properties are similar to those of "ordinary" moments. For example, the symmetry property still holds:

$$B_{uv}(M_1, M_2) = B_{vu}(M_2, M_1). \tag{1.65}$$

Two-point moments of orders higher than two are referred to as correlation functions of higher order.

By analogy, one can define two-point central moments of the second order:

$$\begin{aligned} b_{uu} &= \langle (u(M_1)-\langle u(M_1)\rangle)(u(M_2)-\langle u(M_2)\rangle)\rangle \\ &= B_{uu}(M_1, M_2)-\langle u(M_1)\rangle\langle u(M_2)\rangle, \end{aligned} \tag{1.66}$$

$$\begin{aligned} b_{uv} &= \langle (u(M_1)-\langle u(M_1)\rangle)(v(M_2)-\langle v(M_2)\rangle)\rangle \\ &= B_{uv}(M_1, M_2)-\langle u(M_1)\rangle\langle v(M_2)\rangle \end{aligned} \tag{1.67}$$

The variances of distributions of random variables u and v can be expressed through central moments:

$$\sigma_u^2(M) = b_{uu}(M, M), \quad \sigma_v^2(M) = b_{vv}(M, M). \tag{1.68}$$

Two-point central moments of the second order relate the deviations of random functions from their mean values (i.e. fluctuations) at two different points. This is why they are also called correlation functions of fluctuations.

Another important statistical parameter is the correlation coefficient, defined as

$$\Psi_{uu}(M) = \frac{b_{uu}(M_1, M_2)}{\sigma_u(M_1)\sigma_u(M_2)}, \quad \Psi_{uv}(M) = \frac{b_{uv}(M_1, M_2)}{\sigma_u(M_1)\sigma_v(M_2)}. \tag{1.69}$$

As a consequence of the Schwartz inequality, these coefficients satisfy $|\psi_{uu}| \leq 1$ and $|\psi_{uv}| \leq 1$. If the correlation coefficient vanishes, the correlation between fluctuations at different spatial points is absent.

An important property that follows from physical considerations is the damping of correlation between random variables at different spacetime points as we increase the distance between the points. As the distance goes to infinity, the correlation function will tend to zero. Of course, the "distance" that goes to infinity can be the geometrical distance, $|\boldsymbol{X}_2 - \boldsymbol{X}_1| \to \infty$ at $t_2 = t_1$, the temporal distance (time interval), $|t_2 - t_1| \to \infty$ at $\boldsymbol{X}_2 = \boldsymbol{X}_1$, or both.

1.8 Bernoulli, Poisson, and Gaussian Distributions

Let us consider the three distributions that are most frequently used in physical applications – the Bernoulli, Poisson, and Gaussian (normal) distributions.

Bernoulli distribution To begin, let us formulate the random walk problem, which will be considered in more detail farther on, in the chapter dedicated to Brownian motion. A particle undergoes a sequence of random displacements along a straight line. Each displacement is a step of the same length 1, and each step can be directed either forward or backward with the same probability of 0.5. The origin of the reference system is coincident with the initial position of the particle. Then the particles coordinate can assume only integer values $\ldots -N, -N+1, \ldots, 0, 1, \ldots,$ $N-1, N\ldots$. The probability of finding the particle at a point m after N steps is given by the Bernoulli distribution,

$$P(m, N) = C^N_{(N+m)/2}\left(\frac{1}{2}\right)^N, \tag{1.70}$$

where $C^N_{(N+m)/2} = \frac{N!}{\left(\frac{1}{2}(N+m)\right)!\left(\frac{1}{2}(N-m)\right)!}$ are binomial coefficients.

The mean and root-mean-square displacements of the particle are, respectively,

$$\langle m\rangle = 0, \quad \sqrt{\langle m^2\rangle} = \sqrt{N}.$$

In the limiting case where $N \gg 1$ and $m \ll N$ this reduces to an asymptotic formula

$$P(m, N) \approx \left(\frac{2}{\pi N}\right)^{1/2} \exp\left(-\frac{m^2}{2N}\right). \tag{1.71}$$

Poisson distribution Let N particles be randomly distributed in a volume V. Then the probability of finding n particles in a volume element v, where v is a small part of V, is given by the Bernoulli distribution

$$P_N(n) = \frac{N!}{n!(N-n)!}\left(\frac{v}{V}\right)^n\left(1-\frac{v}{V}\right)^{N-n}. \tag{1.72}$$

For a given N, V and v, the mean value of n equals

$$\langle n\rangle = N\left(\frac{v}{V}\right) \equiv \nu.$$

In the limiting case where $N \to \infty$ and $V \to \infty$ but v remains finite, the distribution (1.72) tends asymptotically to the Poisson distribution

$$P(n) = \frac{\nu^n e^{-\nu}}{n!}. \tag{1.73}$$

If ν is large and ν is of the same order as $\boldsymbol{n}$, the Poisson distribution is close to the distribution

$$P(n) = \left(\frac{1}{2\pi\nu}\right)^{1/2} \exp\left(-\frac{(n-\nu)^2}{2\nu}\right). \tag{1.74}$$

Gaussian (normal) distribution The distributions (1.88) and (1.92) are both special cases of the Gaussian (aka normal) distribution. In the general N-dimensional case this distribution has the following normalized form:

$$p(u_1, u_2, \ldots, u_N) = C \exp\left\{ -\frac{1}{2}\sum_{j=1}^{N}\sum_{k=1}^{N} g_{jk}(u_j - a_j)(u_k - a_k) \right\}, \tag{1.75}$$

Here a_j are real numbers; g_{jk} are the elements of the positive definite matrix $||g_{jk}||$; $C = g^{1/2}/(2\pi)^{N/2}$ is a constant that is given by the normalization condition for probability density (see Eq. (1.6); $g = |g_{jk}|$ is the determinant of the matrix $||g_{jk}||$. The constants a_j and g_{jk} are related to the first and second moments of the distribution (1.75) (see Eq. (1.46) and Eq. (1.47):

$$\langle u_j \rangle = a_j; \quad b_{jk} = \langle (u_j - \langle u_j \rangle)(u_k - \langle u_k \rangle) \rangle = \frac{g_{jk}}{g} \tag{1.76}$$

Here $G_{jk} = \partial g / \partial g_{jk}$ is the algebraic complement of the element g_{jk} in the determinant g. It means that the matrices $||g_{jk}||$ and $||b_{jk}||$ are mutually inverse.

The Gaussian distribution can also represented in the matrix form:

$$p(\boldsymbol{u}) = \frac{1}{(2\pi)^{N/2} b^{1/2}}\left\{ -\frac{1}{2}(\boldsymbol{u} - \langle \boldsymbol{u} \rangle)^T \boldsymbol{b}^{-1}(\boldsymbol{u} - \langle \boldsymbol{u} \rangle) \right\}, \tag{1.77}$$

where $\boldsymbol{b} = ||b_{jk}||$ and $b = |b_{jk}|$.

The ordinary second-order moments B_{jk} can be expressed in terms of the normal distribution parameters according to Eq. (1.76):

$$B_{jk} = \langle u_j u_k \rangle = \frac{G_{jk}}{G} + a_j a_k. \tag{1.78}$$

We see from Eq. (1.76) and Eq. (1.77) that the first two moments completely determine the PDF, and thereby the entire statistics of random variables, for a normal distribution. Hence the knowledge of mean values and correlation functions provides a complete statistical description of a random Gaussian field $\boldsymbol{u}(M) = [u_1(M), u_2(M), \ldots, u_N(M)]$. Central moments can be obtained from the property of normal distributions, which states that all central moments of an odd order are zero, whereas central moments of an even order are expressed through central moments of the second order:

$$b_{k_1 k_2 \ldots k_N} = \left\langle (u_1 - \langle u_1 \rangle)^{k_1} (u_2 - \langle u_2 \rangle)^{k_2} \ldots (u_N - \langle u_N \rangle)^{k_N} \right\rangle = \sum b_{i_1 i_2} b_{i_3 i_4} \ldots b_{i_{2K-1}} b_{i_{2K}}, \tag{1.79}$$

where $k_1 + k_2 + \ldots + k_N = 2K$ and subscript pairs are formed from numbers 1, 2, … $2K$ so that the first index is less than the second, for example,

$$b_{1111} = \langle (u_1 - \langle u_1 \rangle)(u_2 - \langle u_2 \rangle)(u_3 - \langle u_3 \rangle)(u_4 - \langle u_4 \rangle) \rangle \\ = b_{12}b_{34} + b_{13}b_{24} + b_{14}b_{23}.$$

When studying random variables described by a normal distribution it is convenient to use characteristic functions because of their simple form (see Eq. (1.37)):

$$\varphi(\rho_1, \rho_2, \ldots, \rho_N) = \int_{-\infty}^{\infty}\int_{-\infty}^{\infty} \ldots \int_{-\infty}^{\infty} \exp\left\{ i \sum_{k=1}^{N} \rho_k u_k \right\} p(u_1, u_2, \ldots, u_N) du_1 du_2 \ldots du_N \\ = \exp\left\{ i \sum_{k=1}^{N} a_k \rho_k - \frac{1}{2} \sum_{j=1}^{N} \sum_{k=1}^{N} b_{jk} \rho_j \rho_k \right\}. \quad (1.80)$$

Cumulants can be obtained from Eq. (1.53), using Eq. (1.80). For a Gaussian distribution, the cumulants of the first and second orders are respectively equal to a_j and b_{jk} whereas cumulants of higher orders are identically equal to zero. By using characteristic functions, one can prove that any linear combination of Gaussian random variables will also result in a Gaussian distribution.

Gaussian distributions are of great importance in applications due to a number of reasons. First, the behavior of many random variables is well approximated by a Gaussian distribution. Secondly, according to the central limit theorem, a random variable that is a sum of a large number of independent components with arbitrary distributions (which is the most common situation in statistical mechanics) is Gaussian.

Let us consider a one-dimensional Gaussian distribution

$$p(u) = \frac{1}{\sqrt{2\pi}\sigma} \exp\left\{ -\frac{u^2}{2\sigma^2} \right\}, \quad (\sigma^2 = \langle u^2 \rangle). \quad (1.81)$$

The characteristic function follows from the relations (1.37) and (1.52):

$$\varphi(\rho) = \exp\left\{ -\frac{\rho^2 \sigma^2}{2} \right\}, \quad \psi(\rho) = -\frac{\rho^2 \sigma^2}{2} \quad (1.82)$$

Then Eqs. (1.57)–(1.58) yield

$$B_1 = S_1 = 0; \quad B_2 = S_2 = \sigma^2; \quad S_{n>2} = 0. \quad (1.83)$$

The recurrent relation (1.59) takes the form

$$B_n = (n-1)\sigma^2 B_{n-2}, \quad (1.84)$$

from which there follows

$$B_{2n+1} = 0; \quad B_{2n} = (2n-1)!!\sigma^{2n}. \quad (1.85)$$

Consider yet another average value $\langle Xf(X)\rangle$, which is helpful in many applications. Here X is a Gaussian random variable given by Eq. (1.81) and $f(X)$ is an arbitrary deterministic function. We make a further assumption that $f(X)\exp(-X^2/2\sigma^2)\to 0$ for $X\to\pm\infty$ (i.e. the exponent dominates at large values of X). Then

$$\langle X f(X)\rangle = \frac{1}{\sqrt{2\pi}\sigma}\int_{-\infty}^{\infty} X f(X)\exp\left\{-\frac{X^2}{2\sigma^2}\right\}dX.$$

Integrating by parts, we arrive at the following expression:

$$\langle X f(X)\rangle = \frac{\sigma}{\sqrt{2\pi}}\int_{-\infty}^{\infty} \frac{d f(X)}{dX}\exp\left\{-\frac{X^2}{2\sigma^2}\right\}dX = \sigma^2\left\langle\frac{d f(X)}{dX}\right\rangle. \tag{1.86}$$

A similar expression can be obtained for a Gaussian random vector $\boldsymbol{X} = (X_1, X_2, \ldots, X_N)$ with a multidimensional distribution given by Eq. 1.75:

$$\langle X_i f(\boldsymbol{X})\rangle = B_{ij}\left\langle\frac{d f(\boldsymbol{X})}{d\boldsymbol{X}}\right\rangle, \tag{1.87}$$

where $B_{ij} = \langle X_i X_j\rangle$ are components of the correlation matrix.

1.9
Stationary Random Functions, Homogeneous Random Fields

When discussing the problem of random variable averaging in Section 4, we mentioned the ergodic hypothesis, which states that as we increase the temporal or spatial averaging interval to infinity, the corresponding mean values tend to the ensemble average. For the ergodic hypothesis to be valid, some necessary conditions should be satisfied. We thus arrive to a special class of random fields $u(\boldsymbol{X}, t)$ satisfying the ergodicity conditions. These fields are frequently encountered in Statistical Mechanics, in particular, in problems that involve Brownian motion and turbulence.

Consider first the time averaging of a function $u(t)$, written for simplicity as a function of one variable because its dependence on space coordinates $\boldsymbol{X}$ is of no relevance to the problem. The time average will be denoted by $\langle u\rangle_T$. Then, in accordance with Eq. (1.22),

$$\langle u(t)\rangle_T = \frac{1}{T}\int_{-T/2}^{T/2} u(t+\tau)d\tau. \tag{1.88}$$

According to the ergodic hypothesis, $\langle u(t)\rangle_T$ should tend to the ensemble average $\langle u(t)\rangle$ at $T \to \infty$. For this to happen, the following simple relation must take place:

$$\langle u(t)\rangle_T = U = const. \tag{1.89}$$

This condition can be derived by considering the difference between the average values of the random variable calculated at different moments, t and t_1, where $t_1 > t$:

$$\begin{aligned}\langle u(t)\rangle_T - \langle u(t_1)\rangle_T &= \frac{1}{T}\left\{\int_{-T/2}^{T/2} u(t+\tau)d\tau - \int_{-T/2}^{T/2} u(t_1+\tau)d\tau\right\}\\ &= \frac{1}{T}\left\{\int_{-T/2+t}^{-T/2+t_1} u(s)ds - \int_{-T/2+t}^{T/2+t_1} u(s)ds\right\}.\end{aligned} \tag{1.90}$$

At $T \to \infty$ the right-hand side of Eq. (1.90) goes to zero, thus giving rise to the condition (1.89).

Similarly, by time-averaging the product $u(t)u(t_1) = u(t)u(t+s)$, where $s = t_1 - t$, and letting T go to ∞, we conclude that the time average of the correlation function $(B_{uu}(t, t_1))_T$ can be equal to its ensemle average $(B_{uu}(t, t_1)) = \langle u(t)u(t_1)\rangle$ only if for any two instants of time t_1 and t_2, where $t_2 > t_1$, the following condition is satisfied:

$$B_{uu}(t_2, t_1) = B_{uu}(t_2 - t_1). \tag{1.91}$$

For a moment of N-th order, this condition takes the form

$$B_{uu\ldots u}(t_1, t_2, \ldots, t_N) = B_{uu}(t_2 - t_1, \ldots, t_N - t_1). \tag{1.92}$$

In order for the ensemble averages of random values $u(t_1)$, $u(t_2)$, ..., $u(t_N)$ to be obtainable by time averaging, it is necessary to consider only those random functions $u(t)$ for which the N-dimensional PDF (at any N and t_1, t_2, ..., t_N) will depend on $N-1$ parameters $t_2 - t_1$, $t_3 - t_1$, ..., $t_N - t_1$, rather than on N parameters t_1, t_2, ..., t_N. In other words, the PDF must satisfy the condition

$$p_{i_1,\ldots,i_N}(u_1, u_2, \ldots, u_N) = p_{t_2-t_1,\ldots,t_N-t_1}(u_1, u_2, \ldots, u_N). \tag{1.93}$$

It should be noted that the condition (1.93) leads to the conditions (1.98), (1.91) and (1.92), and if the random function is Gaussian, then from Eq. (1.89) and Eq. (1.91) one can derive the properties (1.92) and (1.93).

The condition (1.93) describes a class of random functions whose PDF does not vary as we shift the time t_i by any time interval. Such functions are called stationary random functions or stationary random processes. One example is a steady turbulent flow, whose average characteristics (velocity, pressure, temperature, etc.) do not change with time. Any hydrodynamic parameter $\boldsymbol{u}(u_1(t), u_2(t), \ldots, u_N(t))$, for example, flow velocity at different space points, whose PDF for any set of $u_{i_1}(t_1), u_{i_2}(t_2), \ldots, u_{i_N}(t_N)$ does not vary as we shift all the instants of time $t_1, t_2, \ldots, t_N$ by one and the same value, represents a multidimensional stationary random process. Then all the mixed moments of functions $u(t)$ will also depend only on the differences between the corresponding instants of time. For example, all mutual correlation functions $B_{jk}(t_1, t_2) = \langle u_j(t_1)u_k(t_2)\rangle$ depend only on the time difference $\tau = t_2 - t_1$.

Consider now the space averaging of a random function $u(\boldsymbol{X})$, where $\boldsymbol{X}(X_1, X_2, X_3)$ is a space point. The space average (recall the definition (1.23)) is equal to

$$\langle u(\boldsymbol{X})\rangle_{ABC} = \frac{1}{ABC}\int_{-A/2}^{A/2}\int_{-B/2}^{B/2}\int_{-C/2}^{C/2} u(X_1+\xi_1, X_2+\xi_2, X_3+\xi_3)d\xi_1 d\xi_2 d\xi_3. \tag{1.94}$$

By analogy with time averaging, we can find the conditions that must hold in order for $\langle u(\boldsymbol{X})\rangle_{ABC}$ to coinside with the ensemble average $\langle u(\boldsymbol{X})\rangle$ at $A\to\infty$, $B\to\infty$, $C\to\infty$ (or when at least one of the intervals A,B,C goes to the limit). It is evident that the necessary conditions would be relations similar to (1.89), (1.91)–(1.93) with t replaced by $\boldsymbol{X}$:

$$\langle u(\boldsymbol{X})\rangle = U = const, \tag{1.95}$$

$$B_{uu}(\boldsymbol{X}_1, \boldsymbol{X}_2) = B_{uu}(\boldsymbol{X}_2 - \boldsymbol{X}_1), \tag{1.96}$$

$$p_{X_1,X_2,\ldots,X_N}(u_1, u_2, \ldots, u_N) = p_{X_2-X_1,\ldots,X_N-X_1}(u_1, u_2, \ldots, u_N). \tag{1.97}$$

where $B_{uu}(\boldsymbol{X}_1, \boldsymbol{X}_2) = \langle u(\boldsymbol{X}_1)u(\boldsymbol{X}_2)\rangle$.

A random field $u(\boldsymbol{X})$ satisfying the conditions (1.95)–(1.97) is called a statistically homogeneous field.

Thus, in order for the space averaging of a function of random variables to produce the same results as ensemble averaging, it is necessary for the field $u(\boldsymbol{X})$ to be homogeneous. Parameters of a homogeneous turbulent flow (velocity, pressure, temperature, etc.), which do not depend on spatial coordinates, are good examples. It is clear that homogeneity of the flow cannot be realized in the entire flow region, because any flow is always restricted by boundaries, and the flow near the boundary is essentially inhomogeneous. In reality, the property of homogeneity can be realized only far enough from the boundary.

It should be mentioned that generally speaking, the conditions of stationarity and homogeneity are not sufficient for the convergence of time and space averages to ensemble averages. The necessary and sufficient conditions are formulated by ergodic theorem. Namely, it is necessary and sufficient to ensure the fulfillment of the following condition for the correlation function of fluctuations $b_{uu}(\tau)$:

$$\lim_{T\to\infty}\frac{1}{T}\int_0^T b_{uu}(\tau)d\tau = 0. \tag{1.98}$$

The necessary averaging interval T can be estimated from the corresponding correlation time T_1, which is given by

$$T_1 = \frac{1}{b_{uu}(0)} \int_0^T b_{uu}(\tau) d\tau. \tag{1.99}$$

For sufficiently large T, the following asymptotic formula for root-mean-square deviation of the time average from the ensemble average is valid:

$$\left\langle \left| \langle u \rangle_T - \langle u \rangle \right|^2 \right\rangle \approx 2 \frac{T_1}{T} b_{uu}(0). \tag{1.100}$$

Eq. (1.99) allows us to determine the minimum averaging time for a given deviation of $\langle u \rangle_T$ from $\langle u \rangle$. In the case of spatial averaging, a similar estimation for the root-mean-square deviation of $\langle u \rangle_V$ from $\langle u \rangle$ gives

$$\left\langle \left| \langle u \rangle_V - \langle u \rangle \right|^2 \right\rangle \approx 2 \frac{V_1}{V} b_{uu}(0). \tag{1.101}$$

Here $\langle u \rangle_V$ is the spatial (volume) average, and V_1 is the correlation volume equal to

$$V_1 = \frac{1}{b_{uu}(0)} \int_{-\infty}^{\infty} \int_{-\infty}^{\infty} \int_{-\infty}^{\infty} b_{uu}(\boldsymbol{r}) dr_1 dr_2 dr_3. \tag{1.102}$$

1.10 Isotropic Random Fields. Spectral Representation

A scalar random field $u(\boldsymbol{X})$ is called isotropic when all finite-dimensional PDFs $p_{X_1,X_1,\ldots,X_N}(u_1, u_2, \ldots, u_N)$ corresponding to this field are invariant under rotations of points $\boldsymbol{X}_1, \boldsymbol{X}_2, \ldots, \boldsymbol{X}_N$ around the axis passing through the origin of the coordinate system and under mirror reflections of this set of points relative to planes passing through the origin. In applications, random fields that are both homogeneous and isotropic present the greatest interest. Henceforth these fields will be called simply isotropic. Thus the term "isotropic field" will imply a field whose PDF $p_{X_1,X_2,\ldots,X_N}(u_1, u_2, \ldots, u_N)$ is invariant under parallel translations, rotations, and specular reflections of the set of points $\boldsymbol{X}_1, \boldsymbol{X}_2, \ldots, \boldsymbol{X}_N$.

The homogeneity condition (1.95) for the field $u(\boldsymbol{X})$ means that its average value $\langle u(\boldsymbol{X}) \rangle$ should be constant. This constant is often made equal to zero by replacing the initial field $u(\boldsymbol{X})$ with the field $u'(\boldsymbol{X}) = u(\boldsymbol{X}) - \langle u(\boldsymbol{X}) \rangle$.

The correlation function $B(\boldsymbol{X}, \boldsymbol{X}') = \langle u(\boldsymbol{X}) u(\boldsymbol{X}') \rangle$ of an isotropic field has the same values at any pair of points $(\boldsymbol{X}, \boldsymbol{X}')$ and $(\boldsymbol{X}_1, \boldsymbol{X}_1')$ that would coincide after a combination of parallel translation and rotation. If the distance between the points $\boldsymbol{X}$ and

X' is the same as the distance between X_1 and X'_1, then $B(X, X') = B(X_1, X'_1)$. Hence the correlation function $B(X, X')$ depends only on the distance r between the points X and $X' = X + r$. Here $r = |X' - X| = |r|$, and correlation function can be written as

$$\langle u(X)u(X')\rangle = B(r). \tag{1.103}$$

Application of harmonic (Fourier) analysis to random processes and random fields, that is, expansion of random functions as Fourier series (for functions defined on a finite domain) or Fourier integrals (for functions defined on an infinite domain) has proved to be a very successful approach. For any stationary random functions or homogeneous random fields, which by their definition cannot decay on the infinity, it is possible to carry out Fourier expansion (another common term is "spectral representation" or "spectral expansion"). It has a clear physical meaning: superposition of harmonic oscillations (for stationary random processes) or plane waves (for homogeneous random fields). The integral representation of the correlation function of a homogeneous random field is

$$B(\mathbf{r}) = \int e^{i\mathbf{kr}} F(\mathbf{k}) d\mathbf{k}, \tag{1.104}$$

$$F(\mathbf{k}) = \frac{1}{8\pi^3} \int e^{-i\mathbf{kr}} B(\mathbf{r}) d\mathbf{r}, \tag{1.105}$$

where $F(\mathbf{k})$ is called the spectrum of the homogeneous field, and $\mathbf{k}$ is the wave vector.

For an isotropic field, the condition (1.103) holds, so the spectrum depends on $k = |\mathbf{k}|$ rather than on $\mathbf{k}$. If we represent x, y, z in terms of spherical coordinates as $x = r \sin\theta \cos\Phi$, $y = r \sin\theta \sin\Phi$, $z = r\cos\theta$, the relation (1.105) will take the following form:

$$\begin{aligned} F(\mathbf{k}) &= \frac{1}{8\pi^3} \int e^{-i\mathbf{kr}} B(\mathbf{r}) d\mathbf{r} = \frac{1}{8\pi^3} \int_0^{\infty}\int_{-\pi}^{\pi}\int_0^{\pi} e^{-ikr\cos\theta} B(r) r^2 \sin\theta \, d\theta \, d\Phi \, dr \\ &= \frac{1}{2\pi^2} \int_{-\infty}^{\infty} \frac{\sin(kr)}{kr} B(r) r^2 dr = F(k). \end{aligned} \tag{1.106}$$

Similarly,

$$B(r) = 4\pi \int_{-\infty}^{\infty} \frac{\sin(kr)}{kr} F(k) k^2 dk. \tag{1.107}$$

Instead of looking at $F(\mathbf{k})$, we can consider the following statistical characteristic:

$$E(\mathbf{k}) = \int\int_{|\mathbf{k}|=k} F(\mathbf{k}) dS(\mathbf{k}), \tag{1.108}$$

where $S(\mathbf{k})$ is a surface element of the sphere $|\mathbf{k}| = k$.

Putting $\boldsymbol{r} = 0$ into Eq. (1.104) and recalling Eq. (1.103), we get

$$B(0) = \left\langle [u(\boldsymbol{X})]^2 \right\rangle = \int_0^\infty E(k)dk. \tag{1.109}$$

If u is the velocity (for instance, the velocity of a turbulent flow), then $B(0)$ stands for the total energy of the field $u(\boldsymbol{X})$. Therefore $E(k)dk$ has the meaning of the energy of plane waves with wave numbers in the interval $(k,\ k+dk)$.

The derivations above can be generalized for the case of an isotropic multidimensional random field $\boldsymbol{u}(\boldsymbol{X}) = (u_1(\boldsymbol{X}), u_2(\boldsymbol{X}), \ldots, u_N(\boldsymbol{X}))$ characterized by the correlation matrix

$$||B_{ij}|| = \langle u_i(\mathrm{X})u_j(\mathrm{X} + \mathbf{r})\rangle \tag{1.110}$$

The components of such a matrix are functions of $r = |\boldsymbol{r}|$. Hence the spectral representation of this field will be written as

$$B_{ij}(r) = 4\pi \int_0^\infty \frac{\sin(kr)}{kr} F_{ij}(k)k^2 dk, \tag{1.111}$$

$$F_{ij}(k) = \frac{1}{2\pi^2} \int_0^\infty \frac{\sin(kr)}{kr} B_{ij}(r)r^2 dr.$$

The above-formulated definition of an isotropic random field is valid for scalar random functions, for example, pressure $p(\boldsymbol{X})$, temperature $\vartheta(\boldsymbol{X})$, one-dimensional velocity $u(\boldsymbol{X})$ and so on. In the case of vector random fields such as three-dimensional velocity, as well as for the fields given by a set of vector and scalar hydrodynamic parameters (for example, a field of three-dimensional velocity, pressure, and temperature), isotropy is defined in the following way. A random vector field $\boldsymbol{u}(\boldsymbol{X})$ is called isotropic if the PDF of the components of the vector $\boldsymbol{u}(\boldsymbol{X})$ taken at an arbitrary set of points $\boldsymbol{X}_1, \boldsymbol{X}_2, \ldots, \boldsymbol{X}_N$ is invariant under parallel translations, rotations, and mirror reflections of this set of points accompanied by rotation or mirror reflection of the coordinate system. Using the theory of invariants of the rotation-reflection group, we may conclude from this definition that the correlation tensor $B_{ij}(\boldsymbol{r})$ should be a linear combination of the constant invariant tensor δ_{ij} ("Kroneckers delta function") and the tensor $r_i r_j$. The coefficients in this linear combination will depend on the only invariant that can be built from components of the vector $\boldsymbol{r}$, that is, on the length $r = |\boldsymbol{r}|$:

$$B_{ij}(\boldsymbol{r}) = A_1(r)r_i r_j + A_2(r)\delta_{ij}. \tag{1.112}$$

1.11
Stochastic Processes. Markovian Processes. The Chapman–Kolmogorov Integral Equation

The term "stochastic process" implies that the time evolution of a system is described probabilistically. This means that there is a certain time-dependent random variable. Examples of stochastic processes include Brownian motion of a particle driven by a random force and the motion of particles suspended in a turbulent flow. The random variable is the spatial position $\boldsymbol{X}$ of the particle at different instants of time. One can measure the values $\boldsymbol{X}_1, \boldsymbol{X}_2, \boldsymbol{X}_3, \ldots$ at the instants of time $t_1, t_2, t_3, \ldots$ and assume that there should exist a joint PDF $p(\boldsymbol{X}_1, t_1; \boldsymbol{X}_2, t_2; \boldsymbol{X}_3, t_3; \ldots)$ such that $p(\boldsymbol{X}_1, t_1; \boldsymbol{X}_2, t_2; \boldsymbol{X}_3, t_3; \ldots)d\boldsymbol{X}_1, d\boldsymbol{X}_2, \ldots$ would give the probability for the particle to be located in the interval $(\boldsymbol{X}_1 + d\boldsymbol{X}_1)$ at the instant of time t_1, in the interval $(\boldsymbol{X}_2 + d\boldsymbol{X}_2)$ at the instant t_2, and so on. When the particle moves under the action of a rapidly fluctuating random force (in the case of Brownian motion this random force is the sum of interaction forces between the particle and the molecules of the surrounding fluid, or, to use a more casual term, the sum of collision forces), it can change its direction millions times per second. In this context, when considering two successive particle positions $\boldsymbol{X}_i$ and $\boldsymbol{X}_{i+1}$ at the instants t_i and t_{i+1} such that the time increment $\Delta t_i = t_{i+1} - t_i$ is much smaller than the characteristic time of the process but large as compared to the time between successive collisions of the particle with the surrounding molecules, it is natural to suggest a model where the particles position $\boldsymbol{X}_{i+1}$ at the instant t_{i+1} is determined by its position $\boldsymbol{X}_i$ at the previous instant t_i and does not depend on the earlier instants of time $\ldots, t_{i-2}, t_{i-1}$. In other words, in the proces of a chaotic small-scale random walk, the particle forgets its past very quickly. Such processes are known as Markovian processes.

Hence a Markovian process is a stochastic process characterized by the independence of the future from the past, where the past is defined as the set of all events observed up to the present instant of time t. In other words, one has to deal with random functions whose variations are statistically independent from one another.

A Markovian process can be described by using the concept of conditional probability. Let us consider an ordered sequence of times $t_1 \geq t_2 \geq t_3 \geq \ldots \geq \tau_1 \geq \tau_2 \geq \ldots$, where $\tau_1, \tau_2, \ldots$ belong to the past and $\ldots t_3, t_2, t_1$ – to the future. Let $\boldsymbol{Y}_1, \boldsymbol{Y}_2, \ldots$ denote the values of a random variable at the past instants of time $\tau_1, \tau_2, \ldots$ and $\ldots \boldsymbol{X}_3, \boldsymbol{X}_2, \boldsymbol{X}_1$ denote its values at the future instants of time $\ldots t_3, t_2, t_1$. The PDF of the events $\boldsymbol{X}_1, \boldsymbol{X}_2, \ldots$ under the condition that the events $\boldsymbol{Y}_1, \boldsymbol{Y}_2, \ldots$ have already occurred (the conditional PDF) is then written as

$$p(\boldsymbol{X}_1, t_1; \boldsymbol{X}_2, t_2; \boldsymbol{X}_3, t_3; \ldots | Y_1, \tau_1; Y_2, \tau_2; \ldots).$$

In accordance with the Markovian principle, we demand that the conditional probability must be completely determined by the state of the system at the most recent instant of time, that is, by the knowledge of the random variable at τ_1. Then the following equality must be valid:

$$p(\boldsymbol{X}_1, t_1; \boldsymbol{X}_2, t_2; \boldsymbol{X}_3, t_3; \ldots | Y_1, \tau_1; Y_2, \tau_2; \ldots) \\ = p(\boldsymbol{X}_1, t_1; \boldsymbol{X}_2, t_2; \boldsymbol{X}_3, t_3; \ldots | Y_1, \tau_1). \quad (1.113)$$

This relation means that any conditional probability can be expressed through an ordinary conditional probability of the type $p(\boldsymbol{X}_1, t_1 | Y_1, \tau_1)$. Indeed, from the definition (1.2) of conditional probability we have:

$$p(\boldsymbol{X}_1, t_1; \boldsymbol{X}_2, t_2 | Y_1, \tau_1) = p(\boldsymbol{X}_1, t_1 | \boldsymbol{X}_2, t_2; Y_1, \tau_1)\, p(\boldsymbol{X}_2, t_2 | Y_1, \tau_1).$$

Applying the postulate (1.113) to the first factor on the right-hand side, we express the joint PDF through ordinary conditional PDFs:

$$p(\boldsymbol{X}_1, t_1; \boldsymbol{X}_2, t_2; | Y_1, \tau_1) = p(\boldsymbol{X}_1, t_1 | \boldsymbol{X}_2, t_2)\, p(\boldsymbol{X}_2, t_2 | Y_1, \tau_1). \quad (1.114)$$

Continuing this procedure, we obtain for N successive events:

$$p(\boldsymbol{X}_1, t_1; \boldsymbol{X}_2, t_2; \ldots; \boldsymbol{X}_N, t_N) = p(\boldsymbol{X}_1, t_1 | \boldsymbol{X}_2, t_2)\, p(\boldsymbol{X}_2, t_2 | \boldsymbol{X}_3, t_3) \times \\ \ldots \times p(\boldsymbol{X}_{N-1}, t_{N-1} | \boldsymbol{X}_N, t_N). \quad (1.115)$$

As one could expect, the Markovian principle results in the independence of conditional pairs of successive events.

From the consistency property (see property 4 in Section 4) of the PDF for two successive events $\boldsymbol{X}_2$ and $\boldsymbol{X}_1$ and from the relation (1.2) for the conditional probability there follows:

$$p(\boldsymbol{X}_1, t_1) = \int p(\boldsymbol{X}_1, t_1; \boldsymbol{X}_2, t_2) d\boldsymbol{X}_2 = \int p(\boldsymbol{X}_1, t_1 | \boldsymbol{X}_2, t_2) p(\mathbf{X}_2, t_2) d\boldsymbol{X}_2. \quad (1.116)$$

A similar relation can be written for the conditional probability:

$$p(\boldsymbol{X}_1, t_1 | \boldsymbol{X}_3, t_3) = \int p(\boldsymbol{X}_1, t_1; \boldsymbol{X}_2, t_2 | \boldsymbol{X}_3, t_3) d\boldsymbol{X}_2 \\ = \int p(\boldsymbol{X}_1, t_1 | \boldsymbol{X}_2, t_2; \boldsymbol{X}_3, t_3)\, p(\boldsymbol{X}_2, t_2 | \boldsymbol{X}_3, t_3) d\boldsymbol{X}_2.$$

As far as $t_1 \geq t_2 \geq t_3$, the Markovian principle allows to drop the dependence on $\boldsymbol{X}_3$ in the first factor of the integrand:

$$p(\boldsymbol{X}_1, t_1 | \boldsymbol{X}_3, t_3) = \int p(\boldsymbol{X}_1, t_1 | \boldsymbol{X}_2, t_2)\, p(\boldsymbol{X}_2, t_2 | \boldsymbol{X}_3, t_3) d\boldsymbol{X}_2. \quad (1.117)$$

The integral equation (1.117) is called the Chapman–Kolmogorov equation. This equation forms the basis of the theory of stochastic processes.

When considering a Markovian process, it is important to know whether the range of the random value is continuous or discrete and whether the trajectory $X(t)$ is a continuous function of t. As an example, consider rarefied gas molecules characterized by the velocity $V(t)$ and by the position $X(t)$. In this example, the velocity range is obviously continuous, but the function $V(t)$ can be discontinuious, which happens when the interactions between molecules are modeled by the elastic collisions of rigid spheres. However, even in such a model, the position of a gas molecule $X(t)$ remains a continuous function. In reality, molecules do not interact as rigid spheres. There exists a molecular interaction between them that is characterized by some interaction potential (for example, the Lennard–Jones potential). If we account for this potential, we will find that the molecules trajectory deflects continuously in the process of collision. The characteristic time of molecular collisions is extremely short. It is much shorter then the time intervals that make up a Markovian chain. It can be said that the Markovian method circumnavigates the issue of continuity of a random variable by approximating the real process on a large-scale time grid. Hence, irrespective of how the collision process is modeled, on large-scale time grid, the collision will always be marked by a velocity jump. By the same token, the trajectories are not necessarily continuous on this time grid. Another example is a chemical reaction that involves production consumption of molecules of a certain substance. The characteristic time of a chemical reaction is also very short as a rule. Therefore the random value, for example, molecular concentration, changes discontinuously on the large-scale time grid during the reaction.

In this context, the following continuity condition looks quite self-intuitive: if for any $\varepsilon > 0$ uniformly in Z, t and Δt there holds

$$\lim_{\Delta t \to 0} \frac{1}{\Delta t} \int\limits_{|X-Z|>\varepsilon} p(X, t+\Delta t|Z, t)dX = 0, \tag{1.118}$$

then the realization of $X(t)$ is continuous function of t, with the probability 1. It means that the probability that the position X differs from Z by a finite amount at $\Delta t \to 0$ goes to zero faster than Δt. This is known as the Lindenberg continuity condition for a random function $X(t)$.

One can show that Einsteins solution of the Brownian motion problem, which is a Gaussian PDF written as

$$p(X, t+\Delta t|Z, t) = \frac{1}{(4\pi D\Delta t)^{1/2}} \exp\left\{-\frac{(X-Z)^2}{4D\Delta t}\right\}, \tag{1.119}$$

satisfies the condition (1.118). On the other hand, the PDF

$$p(X, t+\Delta t|Z, t) = \frac{\Delta t}{\pi[(X-Z)^2 + \Delta t^2]}, \tag{1.120}$$

which describes a Cauchy process, does not satisfy this condition. Both distributions tend to the delta function $\delta(X-Z)$ at $\Delta t \to 0$ (see Eq. (1.9) and Eq. (1.16) and satisfy

Eq. (1.117). So the Chapman–Kolmogorov equation allows for both continuous and discontinuous solutions (PDFs).

1.12 The Chapman–Kolmogorov, Chapman–Feller, Fokker–Planck, and Liouville Differential Equations

1.12.1 Derivation of the Differential Chapman–Kolmogorov Equation

When solving concrete problems, one uses the differential form of the Chapman–Kolmogorov equation, which can be derived from the integral equation (1.117) under some additional assumptions. An additional assumption of continuity of the random process leads us to the Fokker–Planck equation. But discontinuous processes can also take place, as was mentioned in the previous section. Thus the Chapman–Kolmogorov differential equation should be able to describe both continuous and discontinuous processes. To satisfy this general requirement, we shall demand realization of the following conditions:

1. $$\lim_{\Delta t \to 0} \frac{1}{\Delta t} p(\boldsymbol{X}, t + \Delta t | \boldsymbol{Z}, t) = \boldsymbol{W}(\boldsymbol{X}|\boldsymbol{Z}, t) \tag{1.121}$$

   should take place in the region $|\boldsymbol{X} - \boldsymbol{Z}| \geq \varepsilon$ uniformly for all $\boldsymbol{X}$, $\boldsymbol{Z}$, t, and the limit should not depend on ε;

2. $$\lim_{\Delta t \to 0} \frac{1}{\Delta t} \int_{|\boldsymbol{X}-\boldsymbol{Z}|<\varepsilon} (X_i - Z_i)\, p(\boldsymbol{X}, t + \Delta t | \boldsymbol{Z}, t) d\boldsymbol{X} = A_i(\boldsymbol{Z}, t) + 0(\varepsilon); \tag{1.122}$$

3. $$\lim_{\Delta t \to 0} \frac{1}{\Delta t} \int_{|\boldsymbol{X}-\boldsymbol{Z}|<\varepsilon} (X_i - Z_i)(X_j - Z_j)\, p(\boldsymbol{X}, t + \Delta t | \boldsymbol{Z}, t) d\boldsymbol{X} = D_{ij}(\boldsymbol{Z}, t) + 0(\varepsilon). \tag{1.123}$$

The conditions 2 and 3 assume a uniform convergence with respect to $\boldsymbol{Z}$, ε, and t. Condition 1 is responsible for the continuity of the process. If $\boldsymbol{W}(\boldsymbol{X}|\boldsymbol{Z}, t) = 0$, the process can be described by continuous trajectories; otherwise the trajectories are discontinuous.

To derive the differential equation, let us consider how the average value of some (twice differentiable) function $f(\boldsymbol{X})$ varies with time. According to Eq. (1.7), the average is written as

$$\langle f(\boldsymbol{X}) \rangle = \int f(\boldsymbol{X})\, p(\boldsymbol{X}, t | \boldsymbol{Y}, t') d\boldsymbol{X}.$$

Then

$$\frac{\partial\langle f\rangle}{\partial t} = \left\{\lim_{\Delta t\to 0}\frac{1}{\Delta t}\int f(\mathbf{X})[p(\mathbf{X},t+\Delta t|\mathbf{Y},t')-p(\mathbf{X},t|\mathbf{Y},t')]d\mathbf{X}\right\}.$$

Let us now put $\mathbf{X}_1=\mathbf{X}$, $t_1=t+\Delta t$, $\mathbf{X}_2=\mathbf{Z}$, $t_2=t$, $\mathbf{X}_3=\mathbf{Y}$, $t_3=t'$ into the Chapman–Kolmogorov integral equation (1.110). It means that in addition to the point $\mathbf{Y}$ at the time t' and the point $\mathbf{X}$ at the time $t+\Delta t$, we take yet another point $\mathbf{Z}$ (between these two) on the trajectory at the intermediate time t ($t'<t<t+\Delta t$). Then the last relation transforms into

$$\frac{\partial}{\partial t}\left\{\int f(\mathbf{X})\,p(\mathbf{X},t|\mathbf{Y},t')d\mathbf{X}\right\} = \lim_{\Delta t\to 0}\frac{1}{\Delta t}\left\{\int d\mathbf{X}\int f(\mathbf{X})\right.$$
$$\left. p(\mathbf{X},t+\Delta t|\mathbf{Z},t)p(\mathbf{Z},t|\mathbf{Y},t')d\mathbf{Z}-\int f(\mathbf{Z})\,p(\mathbf{Z},t|\mathbf{Y},t')d\mathbf{Z}\right\}. \qquad (1.124)$$

We replaced the integration variable $\mathbf{X}$ by $\mathbf{Z}$ in the last term of the right-hand side. Let us now subdivide the region of integration over $\mathbf{X}$ into two regions: $|\mathbf{X}-\mathbf{Z}|\geq\varepsilon$ and $|\mathbf{X}-\mathbf{Z}|<\varepsilon$. In the latter region, we shall perform a Taylor series expansion of the function $f(\mathbf{X})$ that appears in the integrand:

$$f(\mathbf{X}) = f(\mathbf{Z})+\sum_i\frac{\partial f(\mathbf{Z})}{\partial Z_i}(X_i-Z_i)$$
$$+\sum_i\sum_j\frac{1}{2}\frac{\partial^2 f(\mathbf{Z})}{\partial Z_i\partial Z_j}(X_i-Z_i)(X_j-Z_j)+|\mathbf{X}-\mathbf{Z}|^2R(\mathbf{X},\mathbf{Z}), \qquad (1.125)$$

where the last term on the right-hand side of (1.125) is the residual term that obeys the condition $R(\mathbf{X},\mathbf{Z})\to 0$ at $|\mathbf{X}-\mathbf{Z}|\to 0$. Now, substituting Eq. (1.125) into the right-hand side of Eq. (1.124) and grouping the terms, we get:

$$\frac{\partial}{\partial t}\left\{\int f(\mathbf{X})\,p(\mathbf{X},t|\mathbf{Y},t')d\mathbf{X}\right\}$$
$$= \lim_{\Delta t\to 0}\frac{1}{\Delta t}\left\{\iint\limits_{|\mathbf{X}-\mathbf{Z}<\varepsilon}\left[\sum_i\frac{\partial f(\mathbf{Z})}{\partial Z_i}(X_i-Z_i)+\sum_i\sum_j\frac{1}{2}\frac{\partial^2 f(\mathbf{Z})}{\partial Z_i\partial Z_j}(X_i-Z_i)(X_j-Z_j)\right]\right.$$
$$\times\, p(\mathbf{X},t+\Delta t|\mathbf{Z},t)\,p(\mathbf{Z},t|Y,t')d\mathbf{X}d\mathbf{Z}$$
$$+\iint\limits_{|\mathbf{X}-\mathbf{Z}|<\varepsilon}|\mathbf{X}-\mathbf{Z}|^2R(\mathbf{X},\mathbf{Z})\,p(\mathbf{X},t+\Delta t|\mathbf{Z},t)\,p(\mathbf{Z},t|\mathbf{Y},t')d\mathbf{X}d\mathbf{Z}$$
$$+\iint\limits_{|\mathbf{X}-\mathbf{Z}|\geqslant\varepsilon}f(\mathbf{X})\,p(\mathbf{X},t+\Delta t|\mathbf{Z},t)\,p(\mathbf{Z},t|\mathbf{Y},t')d\mathbf{X}d\mathbf{Z}$$
$$\left.+\iint\limits_{|\mathbf{X}-\mathbf{Z}|<\varepsilon}f(\mathbf{Z})\,p(\mathbf{X},t+\Delta t|\mathbf{Z},t)\,p(\mathbf{Z},t|\mathbf{Y},t')d\mathbf{X}d\mathbf{Z}-\int f(\mathbf{Z})\,p(\mathbf{Z},t|\mathbf{Y},t')d\mathbf{Z}\right\}.$$
$$(1.126)$$

Consider the terms in the left-hand side of Eq. (1.126) in the consecutive order. Since $p(\boldsymbol{X}, t+\Delta t|\boldsymbol{Z}, t)$ is the PDF, we note that

$$\int p(\boldsymbol{X}, t+\Delta t|\boldsymbol{Z}, t)d\boldsymbol{X} = 1.$$

With this in mind, and using the condition of uniform convergence that allows us to take the limit of the integrand, the last term in Eq. (1.126) can be rewritten as

$$\begin{aligned}&\int f(\boldsymbol{Z})\, p(\boldsymbol{Z}, t|\boldsymbol{Y}, t')d\boldsymbol{Z}\\ &= \int f(\boldsymbol{Z})\, p(\boldsymbol{Z}, t|\boldsymbol{Y}, t')d\boldsymbol{Z} \int p(\boldsymbol{X}, t+\Delta t|\boldsymbol{Z}, t)d\boldsymbol{X}\\ &= \iint f(\boldsymbol{Z})\, p(\boldsymbol{X}, t+\Delta t|\boldsymbol{Z}, t)\, p(\boldsymbol{Z}, t|\boldsymbol{Y}, t')d\boldsymbol{X}d\boldsymbol{Z}.\end{aligned}$$

To transform the first term, it is necessary to exploit the conditions 2 and 3:

$$\begin{aligned}&\lim_{\Delta t\to 0}\frac{1}{\Delta t}\Bigg\{\iint\limits_{|X-Z|<\varepsilon}\left[\sum_i \frac{\partial f(\boldsymbol{Z})}{\partial Z_i}(X_i - Z_i) + \sum_i\sum_j \frac{1}{2}\frac{\partial^2 f(\boldsymbol{Z})}{\partial Z_i \partial Z_j}(X_i - Z_i)(X_j - Z_j)\right]\\ &\times p(\boldsymbol{X}, t+\Delta t|\boldsymbol{Z}, t)\, p(\boldsymbol{Z}, t|\boldsymbol{Y}, t')d\boldsymbol{X}d\boldsymbol{Z}\Bigg\}\\ &= \int\left[\sum_i A_i(\boldsymbol{Z}, t)\frac{\partial f}{\partial Z_i} + \sum_i\sum_j \frac{1}{2} D_{ij}(\boldsymbol{Z}, t)\frac{\partial^2 f}{\partial Z_i \partial Z_j}\right] p(\boldsymbol{Z}, t|Y, t')d\boldsymbol{Z} + 0(\varepsilon).\end{aligned}$$

The second term tends to zero because of the condition $\boldsymbol{R}(\boldsymbol{X}, \boldsymbol{Z}) \to 0$ at $\varepsilon \to 0$, since $|\boldsymbol{X} - \boldsymbol{Z}| \to 0$.

Carrying out integration in the last term over two subdomains, $|\boldsymbol{X} - \boldsymbol{Z}| \geq \varepsilon$ and $|\boldsymbol{X} - \boldsymbol{Z}| < \varepsilon$ and taking into account Property 1, one can reduce the last three terms to

$$\begin{aligned}&\lim_{\Delta t\to 0}\frac{1}{\Delta t}\Bigg\{\iint\limits_{|X-Z|\geqslant\varepsilon} f(\boldsymbol{X})\, p(\boldsymbol{X}, t+\Delta t|\boldsymbol{Z}, t)\, p(\boldsymbol{Z}, t|\boldsymbol{Y}, t')d\boldsymbol{X}d\boldsymbol{Z}\\ &+ \iint\limits_{|X-Z|<\varepsilon} f(\boldsymbol{Z})\, p(\boldsymbol{X}, t+\Delta t|\boldsymbol{Z}, t)\, p(\boldsymbol{Z}, t|\boldsymbol{Y}, t')d\boldsymbol{X}d\boldsymbol{Z}\\ &- \iint\limits_{|X-Z|\geqslant\varepsilon} f(\boldsymbol{Z})\, p(\boldsymbol{X}, t+\Delta t|\boldsymbol{Z}, t)\, p(\boldsymbol{Z}, t|\boldsymbol{Y}, t')d\boldsymbol{X}d\boldsymbol{Z}\\ &- \iint\limits_{|X-Z|<\varepsilon} f(\boldsymbol{Z})\, p(\boldsymbol{X}, t+\Delta t|\boldsymbol{Z}, t)\, p(\boldsymbol{Z}, t|\boldsymbol{Y}, t')d\boldsymbol{X}d\boldsymbol{Z}\Bigg\}\\ &= \iint\limits_{|X-Z|\geqslant\varepsilon}[f(\boldsymbol{X})\, W(\boldsymbol{X}|\boldsymbol{Z}, t)\, p(\boldsymbol{Z}, t|\boldsymbol{Y}, t') - f(\boldsymbol{Z})\, W(\boldsymbol{X}|\boldsymbol{Z}, t)\, p(\boldsymbol{Z}, t|\boldsymbol{Y}, t')]d\boldsymbol{X}d\boldsymbol{Z}\end{aligned}$$

Nothing will be changed if we swap the variables X and Z in the first term:

$$\iint\limits_{|X-Z|\geqslant\varepsilon} f(Z)[W(Z|X,t)\,p(X,t|Y,t')-W(X|Z,t)\,p(Z,t|Y,t')]dX\,dZ.$$

Going to the limit $\varepsilon\to 0$ in Eq. (1.126), one obtains the following relation:

$$\begin{aligned}
&\frac{\partial}{\partial t}\left\{\int f(X)\,p(X,t|Y,t')dX\right\}\\
&=\int\left[\sum_i A_i(Z,t)\frac{\partial f}{\partial Z_i}+\sum_i\sum_j\frac{1}{2}D_{ij}(Z,t)\frac{\partial^2 f}{\partial Z_i\partial Z_j}\right]p(Z,t|Y,t')dZ\\
&\quad+\int f(Z)dZ\int[W(Z|X,t)\,p(X,t|Y,t')-W(X|Z,t)\,p(Z,t|Y,t')]dX,
\end{aligned}\tag{1.127}$$

which is valid when the integral

$$\int W(Z|X,t)\,p(X,t|Y,t')dX \tag{1.128}$$

exists.

The condition (1.121) determines the function $W(Z|X, t)$ only if $X\neq Z$. But there are situations when $X=Z$, for example, a Cauchy process given by Eq. (1.120). In this case the value of the integral (1.128) should be interpreted as the principal integral value. Because such singular cases are rare, we will not be using the principal integral value symbol in the discussion below.

Integration by parts of the first term on the right-hand side of Eq. (1.127) yields

$$\begin{aligned}
&\int f(Z)\frac{\partial p(Z,t|Y,t')}{\partial t}dZ\\
&=\int f(Z)\left\{-\sum_i\frac{\partial}{\partial Z_i}[A_i(Z,t)\,p(Z,t|Y,t')]dZ\right.\\
&\quad+\sum_i\sum_j\frac{1}{2}\frac{\partial^2 f}{\partial Z_i\partial Z_j}[D_{ij}(Z,t)\,p(Z,t|Y,t')]\\
&\quad\left.+\int[W(Z|X,t)\,p(X,t|Y,t')-W(X|Z,t)\,p(Z,t|Y,t')]dX\right\}+\ldots,
\end{aligned}$$

where the dots denote the surface integrals over the boundary enclosing the considered region. As far as the function $f(Z)$ was chosen arbitrarily with the only requirement that it should be at least twice differentiable, we can impose an additional requirement that this function should vanish on the regions boundary. Then all surface integrals also vanish, and finally, we obtain the Chapman–Kolmogorov differential equation:

$$\frac{\partial p(\boldsymbol{Z},t|\boldsymbol{Y},t')}{\partial t} = -\sum_i \frac{\partial}{\partial Z_i}[A_i(\boldsymbol{Z},t)\,p(\boldsymbol{Z},t|\boldsymbol{Y},t')] + \sum_i\sum_j \frac{1}{2}\frac{\partial^2 f}{\partial Z_i \partial Z_j}[D_{ij}(\boldsymbol{Z},t)\,p(\boldsymbol{Z},t|\boldsymbol{Y},t')] + \int [W(\boldsymbol{Z}|\boldsymbol{X},t)\,p(\boldsymbol{X},t|\boldsymbol{Y},t') - W(\boldsymbol{X}|\boldsymbol{Z},t)\,p(\boldsymbol{Z},t|\boldsymbol{Y},t')]d\boldsymbol{X}. \tag{1.129}$$

Consider now some particular cases of the Chapman–Kolmogorov equation.

1.12.2
Discontinuous ("Jump") Processes. The Kolmogorov–Feller Equation

This equation follows from the Chapman–Kolmogorov equation at $A_i = D_{ij} = 0$:

$$\frac{\partial p(\boldsymbol{Z},t|\boldsymbol{Y},t')}{\partial t} = \int [W(\boldsymbol{Z}|\boldsymbol{X},t)\,p(\boldsymbol{X},t|\boldsymbol{Y},t') - W(\boldsymbol{X}|\boldsymbol{Z},t)\,p(\boldsymbol{Z},t|\boldsymbol{Y},t')]d\boldsymbol{X}. \tag{1.130}$$

If we take $p(Z, t|Y, t') = \delta(Y - Z)$ at $t = t'$ as the initial condition, then for small values of Δt the solution will be approximately equal to

$$p(\boldsymbol{Z}, t+\Delta t|Y,t') \approx \delta(\boldsymbol{Y}-\boldsymbol{Z})[1 - \int W(\boldsymbol{X}|\boldsymbol{Z},t)\Delta t d\boldsymbol{X}] + W(\boldsymbol{Z}|\boldsymbol{X},t)\Delta t.$$

It is implied by this solution that for any Δt, there is a finite probability

$$1 - \int W(\boldsymbol{X}|\boldsymbol{Z},t)\Delta t\, d\boldsymbol{X}$$

to find the particle at the initial position Y, and the distribution of particles leaving $\boldsymbol{Y}$ is given by the function $W(\boldsymbol{Z}|Y, t)$. Hence, the trajectory $\boldsymbol{X}(t)$ consists of linear segments $\boldsymbol{X} = const$ alternating with jumps whose distribution is given by the function $W(\boldsymbol{Z}|Y, t)$. That is why the process has discontinuous character and the trajectories have discontinuities in a discrete set of points.

1.12.3
Diffusion Processes. The Fokker–Planck Equation

If the process is continuous, then $W(\boldsymbol{Z}|Y, t) = 0$, and the Chapman–Kolmogorov equation is reduced to the Fokker–Planck equation:

$$\frac{\partial p(\mathbf{Z},t|\mathbf{Y},t')}{\partial t} = -\sum_i \frac{\partial}{\partial Z_i}[A_i(\mathbf{Z},t)\,p(\mathbf{Z},t|\mathbf{Y},t')] + \sum_i\sum_j \frac{1}{2}\frac{\partial^2}{\partial Z_i \partial Z_j}[D_{ij}(\mathbf{Z},t)\,p(\mathbf{Z},t|\mathbf{Y},t')] \qquad (1.131)$$

Such a process is called the diffusion process. The vector $\boldsymbol{A}(\boldsymbol{Z}, t)$ is called the drift vector. It is similar to the velocity vector in the convective term of the transport equation. The matrix $\mathbf{D}(\boldsymbol{Z}, t) = ||D_{ij}(\boldsymbol{Z}, t)||$ is called the dispersion matrix. According to its definition (see Eq. (1.123)), it is non-negative, definite and symmetric. It can be shown that $\boldsymbol{D}$ is a tensor. It is known as the dispersion tensor.

To understand the physical meaning of $\boldsymbol{A}$ and $\boldsymbol{D}$, consider the initial phase of the process in the same manner as we just did for the Kolmogorov–Feller equation, with the same initial condition $p(\boldsymbol{Z}, t|\boldsymbol{Y}, t') = \delta(\boldsymbol{Y} - \boldsymbol{Z})$ at $t = t'$.

Assuming that during a small time $\Delta t \ll 1$ the values of A_j and D_{ij}. will not change much as compared to p, the equation transforms to

$$\frac{\partial p(\mathbf{Z},t|\mathbf{Y},t')}{\partial t} = -\sum_i A_i(\mathbf{Z},t)\frac{\partial}{\partial Z_i}[p(\mathbf{Z},t|\mathbf{Y},t')] + \sum_i\sum_j \frac{1}{2}D_{ij}(\mathbf{Z},t)\frac{\partial^2}{\partial Z_i \partial Z_j}[p(\mathbf{Z},t|\mathbf{Y},t')], \qquad (1.132)$$

where $t - t' = \Delta t \ll 1$. On this small time interval, $A_i(\mathbf{Y},\boldsymbol{t})$ and $D_{ij}(\mathbf{Y},\boldsymbol{t})$ are regarded as dependent on the initial position $\mathbf{Y}$ but independent of time t. The solution of Eq. (1.132) has the form

$$p(\mathbf{Z},t+\Delta t|\mathbf{Y},t) = \frac{1}{(2\pi)^{N/2}|\mathbf{D}(\mathbf{Y},t)|^{1/2}\Delta t^{1/2}} \times \exp\left\{-\frac{1}{2}\frac{[\mathbf{Z}-\mathbf{Y}-\boldsymbol{A}(\mathbf{Y},t)\Delta t]^T[\mathbf{D}(\mathbf{Y},t)]^{-1}[\mathbf{Z}-\mathbf{Y}-\boldsymbol{A}(\mathbf{Y},t)\Delta t]}{\Delta t}\right\} \qquad (1.133)$$

where $D = |\boldsymbol{D}|$ is the determinant of the matrix $\boldsymbol{D}$.

Eq. (1.333) indicates that at the initial stage, the diffusion process is described by the Gaussian law (see Eq. 1.77) and that fluctuations with the correlation matrix $\boldsymbol{D}(\boldsymbol{Y}, t)\Delta t$ are superimposed on the regular drift with the velocity $\boldsymbol{A}(\boldsymbol{Y}, t)$. It means that at the initial stage, the systems trajectory can be represented as

$$\mathbf{Z}(t+\Delta t) = \mathbf{Y}(t) + \boldsymbol{A}(\mathbf{Y}(t),t)\Delta t + \boldsymbol{\eta}(t)(\Delta t)^{1/2}, \qquad (1.134)$$

where $\boldsymbol{\eta}(t)$ is a random vector with the mean value and the correlation matrix given by

$$\langle\boldsymbol{\eta}\rangle = 0, \quad \langle\boldsymbol{\eta}(t)\boldsymbol{\eta}^T(t)\rangle = \mathbf{D}(\mathbf{Y},t). \qquad (1.135)$$

In a diffusion process, trajectories are continuous everywhere because $\mathbf{Z}(t+\Delta t) \rightarrow \mathbf{Z}(t)$ at $\Delta t \rightarrow 0$. They are also non-differentiable at any point because of the term proportional to $(\Delta t)^{1/2}$. Since $\mathbf{Z}(t+\Delta t) - \mathbf{Y}(t) = \Delta \mathbf{Z}$ is a random increment of the particles position, we can divide both parts of Eq. (1.134) by Δt, obtaining

$$\frac{\Delta Z}{\Delta t} = \boldsymbol{A}(\mathbf{Y}(t), t) + \boldsymbol{\eta}(t)(\Delta t)^{-1/2}. \tag{1.136}$$

Eq. (1.136) is a stochastic differential equation that has a fundamental role in describing the motion of particles driven by an external random force.

In a three-dimensional case, the Gaussian distribution (1.133) can be written in a simpler form. Let us direct the Cartesian axes Z_1, Z_2, Z_3, so that they would coincide with the principal directions of the dispersion tensor $\boldsymbol{D}$. Let D_{ij} be the principal values of the dispersion tensor matrix $\|D_{ij}\|$. In this coordinate system, the distribution (1.133) transforms into

$$p(\mathbf{Z}, t+\Delta t|\mathbf{Y}, t) = \frac{1}{(2\pi)^{3/2}[D_{11}(t)D_{22}(t)D_{33}(t)]^{1/2}} \times \exp\left\{-\frac{(Z_1-Y_1)^2}{2D_{11}(t)} - \frac{(Z_2-Y_2)^2}{2D_{22}(t)} - \frac{(Z_3-Y_3)^2}{2D_{33}(t)}\right\}. \tag{1.137}$$

Let us now introduce the probability flux with components

$$J_i(\mathbf{Z}, t) = A_i(\mathbf{Z}, t)\, p(\mathbf{Z}, t+\Delta t|\mathbf{Y}, t') - \frac{1}{2}\sum_j \frac{\partial}{\partial Z_j}[D_{ij}(\mathbf{Z}, t)\, p(\mathbf{Z}, t|\mathbf{Y}, t')].$$

Then the Fokker–Planck equation can be written in a compact, universally accepted form of a conservation equation:

$$\frac{\partial p(\mathbf{Z}, t|\mathbf{Y}, t')}{\partial t} + \sum_i \frac{\partial J_i(\mathbf{Z}, t)}{\partial Z_i} = 0. \tag{1.138}$$

Introduction of the probability flux allows us to formulate the boundary conditions for the Fokker–Planck equation. Consider the process in a region R bounded by the surface S. One can see the following possible types of boundary conditions.

a) Absorbing boundary

It is assumed that as soon as a particle reaches the boundary, it vanishes, that is, it leaves the system, for example, adheres to the surface or reacts with the boundary surface. Hence the probability to find the particle at the boundary is equal to zero, and the boundary condition for the PDF is

$$p(\mathbf{Z}, t|\mathbf{Y}, t') = 0 \quad \text{at} \quad \mathbf{Z} \in S. \tag{1.139}$$

b) Reflecting boundary

If the particle cannot leave the region R, then the probability flux in the $\boldsymbol{n}$ direction at the boundary surface should be equal to zero, that is,

$$\boldsymbol{n}\cdot\boldsymbol{J}(Z,t)=0 \quad \text{at} \quad Z\in S, \tag{1.140}$$

where $\boldsymbol{n}$ is the normal to the boundary.

c) Surface of discontinuity

Suppose that the coefficients A_i and D_{ij} experience a jump at the surface S, but particles can cross the surface freely. Such a behavior is possible, when the surface is an interface between two media with different properties. At such a surface, the probabilities and the normal components of probability fluxes should be equal at both sides of the boundary surface:

$$p(\mathbf{Z},t|\mathbf{Y},t')|_{S+}=p(\mathbf{Z},t|\mathbf{Y},t')|_{S-}, \quad \boldsymbol{n}\cdot\boldsymbol{J}(Z,t)|_{S+}=\boldsymbol{n}\cdot\boldsymbol{J}(Z,t)|_{S-}. \tag{1.141}$$

d) Conditions at infinity

If the process is considered in an infinite region, then, depending on the problem under consideration, one of the following two boundary conditions must be valid:

$$p(Z,t)\to 0 \quad \text{or} \quad 1 \quad \text{and} \quad \frac{\partial p(Z,t)}{\partial Z}\to 0 \quad \text{at} \quad |Z|\to\infty. \tag{1.142}$$

Of special interest is the one-dimensional case of the Fokker–Planck equation, in which the drift and dispersion coefficients are become scalar quantities A and D:

$$\frac{\partial p(Z,t|Y,t')}{\partial t}=-\frac{\partial}{\partial Z}[A(Z,t)\,p(Z,t|Y,t')]+\frac{1}{2}\frac{\partial^2}{\partial Z^2}[D(Z,t)\,p(Z,t|Y,t')].$$

This equation can be rewritten as

$$\begin{aligned}\frac{\partial p(Z,t|Y,t')}{\partial t}&=\frac{\partial}{\partial Z}\left(\frac{1}{2}D(Z,t)\frac{\partial}{\partial Z}p(Z,t|Y,t')\right)\\ &-\frac{\partial}{\partial Z}\left(\left(A(Z,t)-\frac{1}{2}\frac{\partial D}{\partial Z}\right)p(Z,t|Y,t')\right).\end{aligned} \tag{1.143}$$

Now it is possible to compare it with the molecular diffusion equation, which describes the change of concentration C of a substance in the solution due to the thermal motion of solvent molecules:

$$\frac{\partial \boldsymbol{C}}{\partial \boldsymbol{t}} = \frac{\partial(\boldsymbol{\nu}\boldsymbol{C})}{\partial \boldsymbol{Z}} + \frac{\partial}{\partial \boldsymbol{Z}}\left(\boldsymbol{D}\frac{\partial \boldsymbol{C}}{\partial \boldsymbol{Z}}\right), \tag{1.144}$$

where ν is velocity of the substance under the action of external force. A comparison of equations (1.143) and (1.144) shows that $D/2$ has the meaning of diffusion coefficient D and the drift A

$$A = \nu + \frac{1}{2}\frac{\partial D}{\partial Z}$$

has the meaning of average velocity of particle displacement. The latter consists of two terms. The first term is the drift caused by external forces, and the second term is the drift caused by the inhomogeneneity of the medium. Another difference between equations (1.143) and (1.144) is the difference between the unknown variables: probability density p in (1.143) and concentration C in (1.144). But concentration can be obtained from the probability density by multiplying p by the number of particles N in a unit volume. Therefore if C/N is taken instead of C, then we can also take C/N instead of p under the condition that $N = const.$

A process described by the one-dimensional Fokker–Planck equation with $A = 0$ and $D = 1$,

$$\frac{\partial p(Z, t|Y, t')}{\partial t} = \frac{1}{2}\frac{\partial^2}{\partial Z^2}[p(Z, t|Y, t')]. \tag{1.145}$$

is called the Wiener process. Under the initial condition

$$p(\boldsymbol{Z}, t|\boldsymbol{Y}, t) = \delta(\boldsymbol{Y} - \boldsymbol{Z}) \quad \text{at} \quad t = t'$$

its solution is

$$p(Z, t|Y, t') = \frac{1}{(2\pi)^{1/2}}\exp\left\{-\frac{(Z-Y)^2}{2(t-t')}\right\}. \tag{1.146}$$

A multidimensional Wiener process is described by the multidimensional Fokker–Planck equation

$$\frac{\partial p(\boldsymbol{Z}, t|\boldsymbol{Y}, t')}{\partial t} = \frac{1}{2}\sum_i \frac{\partial^2}{\partial Z_i^2}[p(\boldsymbol{Z}, t|\boldsymbol{Y}, t')], \tag{1.147}$$

whose solution is

$$p(\boldsymbol{Z}, t|\boldsymbol{Y}, t') = \frac{1}{(2\pi)^{n/2}}\exp\left\{-\frac{(Z-Y)^2}{2(t-t')}\right\}. \tag{1.148}$$

Sometimes the Wiener process is causally referred to as "Brownian motion".

1.12.4
Deterministic Processes. The Liouville Equation

The Liouville equation follows from the Chapman–Kolmogorov equation at $W(Z|Y, t)=0$ and $D_{ij}=0$:

$$\frac{\partial p(Z,t|Y,t')}{\partial t} = -\sum_i \frac{\partial}{\partial Z_i}[A_i(Z,t)\,p(Z,t|Y,t')]. \tag{1.149}$$

Eq. (1.149) describes deterministic motion, which is given by a single-valued function and is determined by the initial conditions. Indeed, consider a trajectory $\boldsymbol{X}(t)$ that is a solution of the characteristic equation

$$\frac{d\boldsymbol{X}(t)}{dt} = \boldsymbol{A}(\boldsymbol{X}(t),t)$$

with the initial condition $\boldsymbol{X}(Y, t') = Y$. Here $\boldsymbol{A}$ is a vector with components A_i. Let us show that

$$p(Z,t|Y,t') = \delta(Z-\boldsymbol{X}(Y,t))$$

is the solution of Eq. (1.149) with the initial condition

$$p(Z,t'|\boldsymbol{X},t') = (Z-\boldsymbol{X}).$$

Substituting it into Eq. (1.149), one obtains:

$$\begin{aligned}&\frac{\partial\delta(Z-\boldsymbol{X}(Y,t))}{\partial t}\\ &= -\sum_i\left\{\frac{\partial\delta(Z-\boldsymbol{X}(Y,t))}{\partial Z_i}\frac{dX_i(Y,t)}{dt}\right\} = -\sum_i\left\{\frac{\partial\delta(Z-\boldsymbol{X}(Y,t))}{\partial Z_i}A_i(\boldsymbol{X}(Y,t),t)\right\}\\ &= -\sum_i\frac{\partial}{\partial Z_i}\{A_i(\boldsymbol{X}(Y,t),t)\delta(Z-\boldsymbol{X}(Y,t))\} = -\sum_i\frac{\partial}{\partial Z_i}\{A_i(Z,t)\delta(Z-\boldsymbol{X}(Y,t))\}.\end{aligned}$$

The example above considers the motion of a single particle. In the case of many particles, the Liouville equation has a somewhat different form. Statistical Mechanics uses the Liouville equation to describe the motion of a particle ensemble as a set of mass points moving in accordance with Newtons second law,

$$\ddot{\boldsymbol{X}}_i = \frac{d^2\boldsymbol{X}}{dt^2} = \boldsymbol{F}_i \quad \text{or} \quad \dot{\boldsymbol{V}}_i = \frac{d\boldsymbol{V}_i}{dt} = \boldsymbol{F}_i; \quad \dot{\boldsymbol{X}}_i = \frac{d\boldsymbol{X}_i}{dt} = \boldsymbol{V}_i, \tag{1.150}$$

where $\boldsymbol{X}_i$, $\boldsymbol{V}_i$, $\boldsymbol{F}_i$ are, respectively, the radius vector, the velocity, and the force exerted on a unit mass of i-th particle by other particles and the environment. Let

$$\boldsymbol{X}_i(0) = \boldsymbol{X}_i^0, \quad \boldsymbol{V}_i(0) = \boldsymbol{V}_i^0. \tag{1.151}$$

be the initial positions and velocities of the points. Then in order to describe the time evolution of the state of an N-particle system, one has to integrate the system of equations (1.150) with the initial conditions (1.151). For $N \gg 1$, this is practically impossible, so one has to use statistical methods. Statistical Mechanics usually studies the dynamics of a mass point system by introducing generalized coordinates, which are either the ordinary coordinates $\boldsymbol{X}_i$ and velocities $\boldsymbol{V}_i$ or the ordinary coordinates $\boldsymbol{X}_i$ and momenta $m_i\boldsymbol{V}_i$ of particles.

The state of an N-particle system is characterized by the joint probability density taken at one given instant of time, which determines the chance to find particle 1 in the generalized coordinate interval $(\boldsymbol{X}_1 + d\boldsymbol{X}_1,\ \boldsymbol{V}_1 + d\boldsymbol{V}_1)$, AND to find particle 2 in the interval $(\boldsymbol{X}_2 + d\boldsymbol{X}_2,\ \boldsymbol{V}_2 + d\boldsymbol{V}_2)$, ..., and to find particle N in the interval $(\boldsymbol{X}_N + d\boldsymbol{X}_N,\ \boldsymbol{V}_N + d\boldsymbol{V}_N)$. If particle trajectories are known, that is, the functions

$$\boldsymbol{X}_i = \boldsymbol{X}_i(t), \quad \boldsymbol{V}_i = \boldsymbol{V}_i(t)$$

are given, then the probability density is equal to zero if $\boldsymbol{X}_i \neq \boldsymbol{X}_i(t)$ or $\boldsymbol{V}_i \neq \boldsymbol{V}_i(t)$ for even one value of i. It means that probability reduces to certainty and the probability density is given by a product of delta functions:

$$p(\boldsymbol{X}, \boldsymbol{V}, t) = \prod_{i=1}^{N} \delta(\boldsymbol{X}_i - \boldsymbol{X}_i(t))\delta(\boldsymbol{V}_i - \boldsymbol{V}_i(t)). \tag{1.152}$$

It is easy to show that this PDF satisfies the Liouville equation. Taking the derivative of the product of functions and using properties (1.10), (1.18), (1.21) of the delta function, we write:

$$\begin{aligned}
&\frac{\partial p(\boldsymbol{X}, \boldsymbol{V}, t)}{\partial t} \\
&= -\sum_{j=1}^{N} \left(\prod_{\substack{k=1 \\ k \neq j}}^{N} \delta(\boldsymbol{X}_k - \boldsymbol{X}_k(t))\delta(\boldsymbol{V}_k - \dot{\boldsymbol{X}}_k(t)) \right) \dot{\boldsymbol{X}}_j \\
&\times \frac{\partial \delta(\boldsymbol{X}_j - \boldsymbol{X}_j(t))}{\partial \boldsymbol{X}_j} \delta(\boldsymbol{V}_j - \dot{\boldsymbol{X}}_j(t)) - \sum_{j=1}^{N} \left(\prod_{\substack{k=1 \\ k \neq j}}^{N} \delta(\boldsymbol{X}_k - \boldsymbol{X}_k(t))\delta(\boldsymbol{V}_k - \dot{\boldsymbol{X}}_k(t)) \right) \\
&\times \delta(\boldsymbol{X}_j - \boldsymbol{X}_j(t))\ddot{\boldsymbol{X}}_j \cdot \frac{\partial \delta(\boldsymbol{V}_j - \dot{\boldsymbol{X}}_j(t))}{\partial \boldsymbol{V}_j}.
\end{aligned}$$

Now, using the relation (1.14) and the first equation (1.150), we get

$$\frac{\partial p(\boldsymbol{X}, \boldsymbol{V}, t)}{\partial t} = -\sum_{j=1}^{N} \boldsymbol{V}_j \cdot \frac{\partial \delta(\boldsymbol{X}_j - \boldsymbol{X}_j(t))}{\partial \boldsymbol{X}_j} \delta(\boldsymbol{V}_k - \dot{\boldsymbol{X}}_k(t))$$

$$\times \left(\prod_{\substack{k=1 \\ k \neq j}}^{N} \delta(\boldsymbol{X}_k - \boldsymbol{X}_k(t)) \delta(\boldsymbol{V}_k - \dot{\boldsymbol{X}}_k(t)) \right) - \sum_{j=1}^{N} \boldsymbol{F}_j \cdot \frac{\partial \delta(\boldsymbol{V}_j - \dot{\boldsymbol{X}}_j(t))}{\partial \boldsymbol{V}_j}$$

$$\times \delta(\boldsymbol{X}_k - \boldsymbol{X}_k(t)) \left(\prod_{\substack{k=1 \\ k \neq j}}^{N} \delta(\boldsymbol{X}_k - \boldsymbol{X}_k(t)) \delta(\boldsymbol{V}_k - \dot{\boldsymbol{X}}_k(t)) \right).$$

It is easy to verify that

$$\frac{\partial p(\boldsymbol{X}, \boldsymbol{V}, t)}{\partial \boldsymbol{X}_j} = \frac{\partial \delta(\boldsymbol{X}_j - \boldsymbol{X}_j(t))}{\partial \boldsymbol{X}_j} \delta(\boldsymbol{V}_k - \dot{\boldsymbol{X}}_k(t)) \times \left(\prod_{\substack{k=1 \\ k \neq j}}^{N} \delta(\boldsymbol{X}_k - \boldsymbol{X}_k(t)) \delta(\boldsymbol{V}_k - \dot{\boldsymbol{X}}_k(t)) \right),$$

$$\frac{\partial p(\boldsymbol{X}, \boldsymbol{V}, t)}{\partial \boldsymbol{V}_j} = \frac{\partial \delta(\boldsymbol{V}_j - \dot{\boldsymbol{X}}_j(t))}{\partial \boldsymbol{V}_j} \times \delta(\boldsymbol{X}_k - \boldsymbol{X}_k(t)) \times \left(\prod_{\substack{k=1 \\ k \neq j}}^{N} \delta(\boldsymbol{X}_k - \boldsymbol{X}_k(t)) \delta(\boldsymbol{V}_k - \dot{\boldsymbol{X}}_k(t)) \right)$$

The Liouville equation finally reduces to

$$\frac{\partial p(\boldsymbol{X}, \boldsymbol{V}, t)}{\partial t} = -\sum_{i=1}^{N} \boldsymbol{V}_i \cdot \frac{\partial p(\boldsymbol{X}, \boldsymbol{V}, t)}{\partial \boldsymbol{X}_i} - \sum_{i=1}^{N} \boldsymbol{F}_i \cdot \frac{\partial p(\boldsymbol{X}, \boldsymbol{V}, t)}{\partial \boldsymbol{V}_i}. \tag{1.153}$$

If the generalized coordinates $\boldsymbol{Z}_i$ are defined as coordinates $\boldsymbol{X}_i$ and momenta per unit particle mass $\boldsymbol{V}_i$, then Eq. (1.153) will assume the following compact form:

$$\frac{\partial p(\boldsymbol{Z}, t)}{\partial t} = -\sum_{i=1}^{N} \boldsymbol{A}_i \cdot \frac{\partial p(\boldsymbol{Z}, t)}{\partial \boldsymbol{Z}_i}, \tag{1.154}$$

where $\boldsymbol{A}$ is the generalized vector that includes $\boldsymbol{V}_i$ and $\boldsymbol{F}_i$.

1.13 Stochastic Differential Equations. The Langevin Equation

When a randomly and rapidly fluctuating function of time and (or) spatial coordinates appears in a differential equation, this equation is called stochastic. The presence of a random component in the equation means that the solution will also be a random function. One example of a stochastic differential equation is the Langevin equation describing random trajectories of a particle driven by a random force. Another example is the equation of diffusion, which takes into account chemical reactions that are responsible for fluctuations. Let us consider these equations in more detail.

1.13.1 The Langevin Equation

In the theory of Brownian motion one frequently encounters the following Langevin equation:

$$\frac{d\mathbf{X}}{dt} = \mathbf{a}(\mathbf{X}, t) + b(\mathbf{X}, t)\boldsymbol{\xi}(t), \tag{1.155}$$

where $\mathbf{a}(\mathbf{X}, t)$ and $b(\mathbf{X}, t)$ are known functions and $\boldsymbol{\xi}(t)$ is a random fluctuating function.

The problem of Brownian motion of a particle driven by a fluctuating external force (that is, the force resulting from collisions with molecules of the surrounding fluid) reduces to this equation. Every second, the particle experiences millions of collisions, each collision resulting in a microscopic motion in the direction of impact. Therefore the particles position at the time t is determined only by its position at previous instant of time and does not depend on the motion prehistory. The particles motion can be considered as a Markowian process that is taking place under the action of a random force $\boldsymbol{\xi}(t)$ with the average value of zero:

$$\langle \boldsymbol{\xi}(t) \rangle = 0 \tag{1.156}$$

The force $\boldsymbol{\xi}(t)$ is a "pseudoforce" that models the real impact forces. There is no correlation between the forces at different instants of time, in other words, $\boldsymbol{\xi}(t)$ and $\boldsymbol{\xi}(t')$ are mutually independent if $t \neq t'$. Or, to use another term, they are delta-correlated:

$$\langle \boldsymbol{\xi}(t)\boldsymbol{\xi}(t') \rangle = \boldsymbol{\xi}\delta(t-t'). \tag{1.157}$$

Equations (1.155) and (1.157) serve as idealizations of Brownian motion. The actual equation of motion of a particle is given by Newtons second law:

$$m\frac{d^2\mathbf{X}}{dt^2} = -h\frac{d\mathbf{X}}{dt} + \mathbf{F} + \mathbf{f}, \tag{1.158}$$

where $-h dX/dt$ is the drag force exerted on the particle by the surrounding fluid (see Example 4 for more details), F is a systematic force such as gravitational, centrifugal, or electrostatic force (in other words, an external force), and f is the stochastic force. The small size of the particle allows us to neglect its inertia in the first approximation, in other words, the left-hand side of Eq. (1.158) is set to zero. This immediately results in Eq. (1.155). The condition (1.157) is also an idealization, because at $t = t'$ it gives an infinitely large dispersion, which is an obvious impossibility. The assumption (1.157) is similar to the idealization inherent in the concept of white noise in electrical engineering. The delta function representation of the correlation function is a natural idealization that helps us make the transition from a small time scale to a large scale typical for Brownian motion. Note, however, that differential equations that include perturbating terms given by random delta-correlated functions must be treated carefully, because the usual calculation rules will not always apply.

1.13.2 The Diffusion Equation

A deterministic description of processes in continuous media requires the use of conservation equations (conservation of mass, momentum, energy, and so on). But in order to get concrete solutions, the system of conservation equations must be complemented by constitutive relations or equations. As examples of such relations, we can mention Ficks law for diffusion (mass transport) processes, Fouriers law for heat transport processes, and the Navier–Stokes law describing the hydromechanics of viscous fluids. You may ask how these equations account for fluctuations of the relevant quantities. We shall answer this question by taking the diffusion equation as an example.

According to Ficks law, the diffusive flux of the substance $\boldsymbol{j}(\boldsymbol{X}, t)$ is proportional to the concentration gradient:

$$\boldsymbol{j}(\boldsymbol{X}, t) = -D\nabla C(\boldsymbol{X}, t). \tag{1.159}$$

On the other hand, we have the equation of conservation of mass (the continuity equation):

$$\frac{\partial C}{\partial t} + \nabla\cdot\boldsymbol{j}(\boldsymbol{X}, t) = 0. \tag{1.160}$$

Substitution of $\boldsymbol{j}(\boldsymbol{X}, t)$ from (1.159) results in the standard diffusion equation:

$$\frac{\partial C}{\partial t} = \nabla\cdot[D\nabla C(\boldsymbol{X}, t)]. \tag{1.161}$$

Fluctuations could be taken into account by adding a fluctuating term to the right-hand side Eq. (1.159):

$$\boldsymbol{j}(\boldsymbol{X}, t) = -D\nabla C(\boldsymbol{X}, t) + \boldsymbol{j}^{fl}(\boldsymbol{X}, t). \tag{1.162}$$

Just as we did for the fluctuating term in the Langevin equation, we assume that this term has the following statistical properties:

$$\left\langle j^{fl}(X,t)\right\rangle = 0, \left\langle j_i^{fl}(X,t)j_k^{fl}(X',t')\right\rangle = K(X,t)\delta_{jk}\delta(X-X')\delta(t-t'). \tag{1.163}$$

The second property says that different components of the fluctuating flux vector j^{fl} taken at one and the same spatial point, as well as the values of one and the same component taken at different instants of time and/or at different points are assumed to be statistically independent. Or, to say it in fewer words, this property states that fluctuations have local behavior.

Eqs. (1.160) and (1.162) give us a diffusion equation where the additional term is expressed as the divergence of a vector:

$$\frac{\partial C}{\partial t} = \nabla\cdot[D\nabla C(X,t)] - \nabla\cdot j^{fl}(X,t) \tag{1.164}$$

It can be shown that this term possesses the following statistical properties:

$$\begin{gathered}\left\langle \nabla\cdot j^{fl}(X,t)\right\rangle = 0,\\ \left\langle \nabla\cdot j^{fl}(X,t)\nabla'\cdot j^{fl}(X',t')\right\rangle = \nabla\cdot\nabla'[K_1(X,t)\delta(X-X')]\delta(t-t').\end{gathered} \tag{1.165}$$

1.13.2.1 *The Diffusion Equation with Chemical Reactions Taken into Account*

The deterministic diffusion equation with a source/sink term arising due to chemical reactions has the form

$$\frac{\partial C}{\partial t} + \nabla\cdot j(X,t) = F[C(X,t)], \tag{1.166}$$

where the rate of substance production/consumption in the course of the chemical reaction appears in the right-hand side. This term usually depends on the substance concentrationl; depending on the reaction kinetics, it could be a linear or a nonlinear function of concentration. Production/consumption of matter gives rise to fluctuations, which can be taken into consideration by adding a fluctuating term $g^{fl}(X, t)$ to the systematic term $F[C(X, t)]$. The new term must satisfy the following statistical conditions:

$$\langle g^{fl}(X,t)\rangle = 0, \langle g^{fl}(X,t)g^{fl}(X',t')\rangle = K_2(X,t)\delta(X-X')\delta(t-t'). \tag{1.167}$$

The second condition expresses locality (lack of correlation between fluctuations at different points) as well as the Markovian character of chemical reactions. Eqs. (1.162), (1.166), and (1.167) give us the stochastic diffusion equation that accounts for chemical reactions:

$$\frac{\partial C}{\partial t} + \nabla \cdot \boldsymbol{j}(\boldsymbol{X}, t) = F[C(\boldsymbol{X}, t)] + G^{fl}(\boldsymbol{X}, t), \tag{1.168}$$

where

$$G^{fl}(\boldsymbol{X}, t) = -\nabla \cdot \boldsymbol{j}^{fl}(\boldsymbol{X}, t) + g^{fl}(\boldsymbol{X}, t).$$

The fluctuating term has the following properties:

$$\begin{aligned}&\langle G^{fl}(\boldsymbol{X}, t) G^{fl}(\boldsymbol{X}, t)\rangle = 0; \quad \langle G^{fl}(\boldsymbol{X}, t) G^{fl}(\boldsymbol{X}', t')\rangle \\ &= \{K_2(\boldsymbol{X}-\boldsymbol{X}')\delta(\boldsymbol{X}-\boldsymbol{X}') + \nabla \cdot \nabla' [K_1(\boldsymbol{X}, t)\delta(\boldsymbol{X}-\boldsymbol{X}')]\}\delta(t-t').\end{aligned}$$

Eqs. (1.164) and (1.168) are both Langevin equations. The main shortcoming of these equations is that the functions $K_1(\boldsymbol{X}, t)$ and $K_2(\boldsymbol{X}, t)$ are not known in advance. Information about these functions can be obtained by studying the process on the microscopic level, using the same approach that helped us derive the Chapman–Kolmogorov equation. This approach allows to interpret diffusion coefficients as components of a dispersion tensor or a correlation matrix. The difference between equations (1.168) and (1.164) is that $F[C(\boldsymbol{X}, t)]$ that appears in Eq. (1.168) is not a function but a functional.

1.13.2.2 *Brownian Motion of a Particle in a Hydrodynamic Medium*

Slow motion of a particle in a fluctuating hydrodynamic medium (the medium is assumed to be at rest at the infinity) under the action of the average viscous force exerted by the surrounding fluid $\langle \boldsymbol{F}(t)\rangle$ and the fluctuating force $\boldsymbol{F}^{fl}(t)$ (a force induced by thermal hydrodynamic fluctuations in the fluid or by some other source of random forces) is described by the Langevin equation (a stochastic analogue of Newtons second law):

$$m\frac{d\boldsymbol{U}}{dt} = -\langle \boldsymbol{F}(t)\rangle + \boldsymbol{F}^{fl}, \quad \boldsymbol{U} = \frac{d\boldsymbol{X}}{dt}. \tag{1.169}$$

To find the statistical properties of the random force $\boldsymbol{F}^{fl}$, it is necessary to use the hydrodynamic equations describing the fields of velocity $\boldsymbol{u}$ and pressure p in the fluid. The particle motion occurs at low Reynolds numbers, so one can use Navier–Stokes equations in the inertialess approximation, (i.e., Stokes equations):

$$\nabla \cdot \boldsymbol{u} = 0; \quad \nabla \cdot \boldsymbol{T} = 0, \tag{1.170}$$

where $\boldsymbol{T}(\boldsymbol{r}, t) = \boldsymbol{T}^s + \boldsymbol{T}^{fl}$ is the stress tensor in the fluid, which consists of a systematic component $\boldsymbol{T}^s = -p\boldsymbol{I} + 2\mu_e \boldsymbol{E}$ and a fluctuating component $\boldsymbol{T}^{fl}$; $-p\boldsymbol{I}$ is the spherical tensor (the isotropic part of the stress tensor); $2\mu_e\boldsymbol{E}$ is the deviator (the deviatoric part of the stress tensor); $\boldsymbol{E} = 0.5(\nabla\boldsymbol{u} + \nabla\boldsymbol{u}^T)$ is the symmetric rate-of-strain tensor, μ_e is the coefficient of dynamic viscosity of the carrier (external) fluid.

In a statistically homogeneous medium, the fluctuating stress tensor T^{fl} is symmetric and its trace is equal to zero, that is, $T^{fl}_{ii} = 0$. The components of T^{fl} are usually assumed to have a Gaussian distribution with the average value $\left\langle T^{fl}_{ik} \right\rangle = 0$ and the correlation matrix

$$\left\langle \boldsymbol{T}^{fl}_{ik}(\boldsymbol{r}_1, \boldsymbol{t}_1)\boldsymbol{T}^{fl}_{lm}(\boldsymbol{r}_2, t_2)\right\rangle = 2k\theta\mu_e(\delta_{il}\delta_{km} + \delta_{im}\delta_{kl} - \frac{2}{3}\delta_{ik}\delta_{lm}) \times \delta(\boldsymbol{r}_1 - \boldsymbol{r}_2)\delta(t_1 - t_2). \tag{1.171}$$

Equations (1.170) can then be rewritten in the coordinate form:

$$\frac{\partial u_i}{\partial X_i} = 0; \quad \frac{\partial T^s_{ij}}{\partial X_j} = -\frac{\partial T^{fl}_{ij}}{\partial X_j}. \tag{1.172}$$

Following the method of small perturbations, we shall be looking for the solutions in the form

$$u_i = \langle u_i \rangle + \tilde{u}_i; \quad T^s_{ij} = \left\langle T^s_{ij} \right\rangle + \tilde{T}^s_{ij}, \tag{1.173}$$

where $\langle \boldsymbol{u} \rangle$, $\langle \boldsymbol{T}^s \rangle$ are the average values and $\tilde{\boldsymbol{u}}, \tilde{\boldsymbol{T}}^s$ are small fluctuating "additions" (perturbations).

Substituting (1.173) into (1.172) and neglecting small terms of higher orders, we get the equations for the average values and for the fluctuating terms:

$$\frac{\partial \langle u_i \rangle}{\partial X_i} = 0; \quad \frac{\partial \left\langle T^s_{ij} \right\rangle}{\partial X_j} = 0, \quad \frac{\partial \tilde{u}_i}{\partial X_i} = 0; \quad \frac{\partial \tilde{T}^s_{ij}}{\partial X_j} = -\frac{\partial \tilde{T}^{fl}_{ij}}{\partial X_j}. \tag{1.174}$$

Now, we must introduce two boundary conditions. First, the relative velocity at the interface between the fluid and the solid particle should be zero: $\langle u_i \rangle = U_i$, $\tilde{u}_i = 0$. Secondly, the fluid must be at rest at the infinity: $\langle u_i \rangle = \tilde{u}_i = 0$.

As a rule, the main purpose of hydrodynamic calculations is to find the force the surrounding fluid exerts on a moving particle. This force has systematic and fluctuating parts $\boldsymbol{F}$ and $\boldsymbol{F}^{fl}$, whose components are

$$\langle F_i \rangle = \int_S \left\langle T^s_{ij} \right\rangle n_j ds; \quad F^{fl}_i = \int_S \tilde{T}^s_{ij} n_j ds, \tag{1.175}$$

where n_i are components of the outer normal vector and S is the surface of the particle.

Solving the first two equations (1.174), we get the relation between the particles velocity and the drag force:

$$\langle F_i \rangle = R_{ij} U_j, \tag{1.176}$$

where R_{ij} are components of the resistance tensor $\boldsymbol{R}$.

In order to determine statistical characteristics of the random force $\tilde{F}_i$, let us use the relation that follows in a self-obvious way from Gausss theorem, equations (1.174), and the above-mentioned boundary conditions:

$$\int_V \left(\langle u_i \rangle \frac{\partial \tilde{T}^s_{ij}}{\partial X_j} - \tilde{u}_i \frac{\partial \left\langle T^s_{ij} \right\rangle}{\partial X_j} \right) dV = \int_V \frac{\partial}{\partial X_j} (\langle u_i \rangle \tilde{T}^s_{ij} - \tilde{u}_i \left\langle T^s_{ij} \right\rangle) dV$$

$$= \int_S (\langle u_i \rangle \tilde{T}^s_{ij} - \tilde{u}_i \left\langle \tilde{T}^s_{ij} \right\rangle) n_j ds = - \int_S \langle u_i \rangle \frac{\partial \tilde{T}^s_{ij}}{\partial X_j} ds = \int_S U_i \tilde{T}^s_{ij} n_j ds = U_i \tilde{F}_i,$$

where integration is carried out over the volume V occupied by the fluid. Using this relation together with the property (1.171), we finally get the two-time correlation between components of the fluctuating force:

$$\left\langle \tilde{F}_i(t) \tilde{F}_j(t') \right\rangle = \frac{1}{U_i U_j} 2kT\delta(t-t') \int_S \langle u_i \rangle \langle T_{ij} \rangle n_j ds = 2kTR_{ij}\delta(t-t'). \qquad (1.177)$$

1.14
Variational (Functional) Derivatives

When considering random processes and random fields, one can see that the PDF depends on random variables, which, in turn, are functions of time and (or) spatial coordinates. Therefore, the variables in the PDF are random functions rather than random variables.

It was shown in Sections 1.5 and 1.6 that knowing the characteristic function φ and the characteristic functional, one can determine statistical parameters of random quantities such as moments and cumulants from the formulas (1.50) and (1.54). These formulas contain derivatives of φ and $\ln\varphi$. When considering random processes and random fields, we treat these derivatives as derivatives with respect to functions and not as derivatives with respect to random variables. Hence, instead of ordinary differentiation we have to perform functional, or, to use another term, variational, differentiation. The corresponding derivatives are called functional (variational) derivatives.

Let us recal the general definition of a functional. We say that a functional is given when there exists a rule that assigns a definite number to each function belonging to some set of functions. Some examples are given below:

- *linear functional*

$$F[\varphi(X)] = \int_{X_1}^{X_2} a(X)\varphi(X)dX,$$

where $a(X)$ is a given function and the limits X_1 and X_2 can be either finite or infinite;

- *quadratic functional*

$$F[\varphi(X)] = \int_{X_1}^{X_2}\int_{X_1}^{X_2} B(\tau_1, \tau_2)\varphi(\tau_1)\varphi(\tau_2)d\tau_1 d\tau_2,$$

where $B(\tau_1, \tau_2)$ is a given function;
- *function of functional*

$$F[\varphi(X)] = f(\Phi[\varphi])$$

where $f(X)$ is a given function and $\Phi[\varphi(X)]$ is a functional.

Consider the difference in values of one and the same functional taken for two functions $\varphi(\tau)$ and $\varphi(\tau) + \delta\varphi(\tau)$, where t lies in the interval $\tau \in (X - 0.5\Delta X; X + 0.5\Delta X)$. The difference (more strictly, the linear (with respect to $\delta\varphi(\tau)$) part of that difference) is called variation of the functional:

$$\delta F[\varphi] = \{F[\varphi + \delta\varphi] - F[\varphi]\}.$$

The variational (functional) derivative is defined as

$$\frac{\delta F[\varphi]}{\delta\varphi(X)} = \lim_{\Delta X \to 0} \frac{\delta F[\varphi]}{\int_{\Delta X} \varphi(\tau)d\tau}. \tag{1.178}$$

The variational derivative of a functional $F[\varphi]$ is itself a functional of $\varphi(\tau)$ that also depends on the point X as a parameter. Thus the variational derivative itself has two different derivatives. One can differentiate it in the ordinary way with respect to the parameter X; or one can take the variational derivative with respect to $\varphi(\tau)$ at the point $\tau = \tilde{X}$. The latter would be the second variational derivative of the initial functional $F[\varphi]$:

$$\frac{\delta}{\delta\varphi(\tilde{X})}\left[\frac{\delta F[\varphi]}{\delta\varphi(X)}\right] = \frac{\delta^2 F[\varphi]}{\delta\varphi(\tilde{X})\delta\varphi(X)}. \tag{1.179}$$

The second variational derivative is also a functional of $\varphi(\tau)$, but, in contrast to the first variational derivative, it now depends on two points: X and $\tilde{X}$. Variational derivatives of higher orders can be defined in a similar way. As examples, consider variational derivatives of the above-mentioned functionals.

$$\delta F = F[\varphi + \delta\varphi] - F[\varphi] = \int_{X_1}^{X_2} a(\tau)\delta\varphi(\tau)d\tau = \int_{X-0.5\Delta X}^{X+0.5\Delta X} a(\tau)\delta\varphi(\tau)d\tau.$$

If the function $a(\tau)$ is continuous on the interval ΔX, then, according to the mean value theorem,

$$\delta F[\varphi] = a(X') \int\limits_{\Delta X} \delta\varphi(\tau)d\tau,$$

where $X' \in (X - 0.5\Delta X;\ X + 0.5\Delta X)$. The definition (1.178) yields

$$\frac{\delta F[\varphi]}{\delta\varphi(X)} = \lim_{\Delta X \to 0} a(X') \left(\int\limits_{\Delta X} \delta\varphi(\tau)d\tau \Big/ \int\limits_{\Delta X} \delta\varphi(\tau)d\tau \right) = a(X). \tag{1.180}$$

Applying the same approach to the quadratic functional, we get

$$\frac{\delta F[\varphi]}{\delta\varphi(X)} = \int\limits_{X_1}^{X_2} (B(\tau, X) + B(X, \tau))\varphi(\tau)d\tau; \quad (X_1 < \tau < X_2). \tag{1.181}$$

Note that in many cases the function $B(\tau, X)$ is a symmetric function of its arguments, that is, $B(\tau, X) = B(X, \tau)$.

Finally, for a function of a functional,

$$\begin{aligned} F[\varphi + \delta\varphi] &= f(\Phi[\varphi + \delta\varphi]) = f(\Phi[\varphi] + \delta\Phi) \\ &= f(\Phi[\varphi]) + \frac{\partial f(\Phi[\varphi])}{\partial\Phi}\delta\Phi + \ldots = F[\varphi] + \frac{\partial f(\Phi[\varphi])}{\partial\Phi}\delta\Phi + \ldots \end{aligned}$$

and

$$\frac{\delta}{\delta\varphi(X)} f(\Phi[\varphi]) = \frac{\partial f(\Phi[\varphi])}{\partial\Phi} \frac{\delta\Phi[\varphi]}{\delta\varphi(X)}. \tag{1.182}$$

Consider now some properties of variational derivatives. Let a functional be a product of two functionals: $\Phi[\varphi] = F_1[\varphi]F_2[\varphi]$. The variation and variational derivative of this functional are:

$$\begin{aligned} \delta\Phi = \Phi[\varphi + \delta\varphi] - \Phi[\varphi] &= F_1[\varphi + \delta\varphi]F_2[\varphi + \delta\varphi] - F_1[\varphi]F_2[\varphi] \\ &= F_1[\varphi]\delta F_2[\varphi] + F_2[\varphi]\delta F_1[\varphi], \end{aligned}$$

$$\frac{\delta}{\delta\varphi(X)}[F_1[\varphi]F_2[\varphi]] = F_1[\varphi]\frac{\delta F_2[\varphi]}{\delta\varphi(X)} + F_2[\varphi]\frac{\delta F_1[\varphi]}{\delta\varphi(X)}. \tag{1.183}$$

Of special interest is the functional of a Gaussian distribution,

$$F[\varphi] = \int\limits_0^\infty \frac{1}{\sqrt{2\pi}\sigma} \exp\left\{ -\frac{(\tau - \tau_0)^2}{2\sigma^2} \right\} \varphi(\tau)d\tau. \tag{1.184}$$

Eq. (1.180) gives us its variatlional derivative:

$$\frac{\delta F[\varphi]}{\delta\varphi(\tau)} = \frac{1}{\sqrt{2\pi}\sigma}\exp\left\{-\frac{(\tau-\tau_0)^2}{2\sigma^2}\right\}. \tag{1.185}$$

Going to the limit $\sigma \to 0$ in Eqs. (1.183), (1.184) and using the definition of the delta function Eq. (1.9) and its property (1.11), we obtain:

$$\lim_{\sigma\to 0} F[\varphi] = \varphi(\tau_0), \quad \lim_{\sigma\to 0} = \frac{\delta F[\varphi]}{\delta\varphi(\tau)} = \delta(\tau-\tau_0).$$

Therefore,

$$\frac{\delta\varphi[\tau_0]}{\delta\varphi(\tau)} = \delta(\tau-\tau_0). \tag{1.186}$$

The relation (1.186) facilitates the derivation of formulas for some variational derivatives. As an example, let us derive the formula (1.181) for the quadratic functional:

$$\begin{aligned}
&\frac{\delta}{\delta\varphi(X)}\left\{\int_{X_1}^{X_2}\int_{X_1}^{X_2} B(\tau_1,\tau_2)\varphi(\tau_1)\varphi(\tau_2)d\tau_1 d\tau_2\right\}\\
&= \int_{X_1}^{X_2}\int_{X_1}^{X_2} B(\tau_1,\tau_2)\frac{\delta}{\delta\varphi(X)}[\varphi(\tau_1)\varphi(\tau_2)]d\tau_1 d\tau_2\\
&= \int_{X_1}^{X_2}\int_{X_1}^{X_2} B(\tau_1,\tau_2)\left[\frac{\delta\varphi(\tau_1)}{\delta\varphi(X)}\varphi(\tau_2) + \varphi(\tau_1)\frac{\delta\varphi(\tau_2)}{\delta\varphi(X)}\right]d\tau_1 d\tau_2\\
&= \int_{X_1}^{X_2}\int_{X_1}^{X_2} B(\tau_1,\tau_2)[\delta(\tau_1-X)\varphi(\tau_2) + \varphi(\tau_1)\delta(\tau_2-X)]d\tau_1 d\tau_2\\
&= \int_{X_1}^{X_2} [B(\tau,X) + B(X,\tau)]\varphi(\tau)d\tau, \quad (X_1<\tau<X_2).
\end{aligned}$$

Another example is the variational derivative of the functional

$$F[\varphi(\tau)] = \int_{X_1}^{X_2} L[(\tau,\varphi(\tau),\dot{\varphi}(\tau)]d\tau, \quad \dot{\varphi}(\tau) = \frac{d\varphi(\tau)}{d\tau}.$$

We have:

$$\frac{\delta F[\varphi]}{\delta\varphi(X)} = \int_{X_1}^{X_2}\left[\frac{\partial L}{\partial\varphi}+\frac{\partial L}{\partial\dot\varphi}\frac{d}{d\tau}\right]\frac{\delta\varphi(\tau_0)}{\delta\varphi(X)}d\tau = \int_{X_1}^{X_2}\left[\frac{\partial L}{\partial\varphi}+\frac{\partial L}{\partial\dot\varphi}\frac{d}{d\tau}\right]\delta(\tau-X)d\tau$$
$$= \left[-\frac{d}{dX}\frac{\partial}{\partial\dot\varphi}+\frac{\partial}{\partial\varphi}\right]L[(\tau,\varphi(\tau),\dot\varphi(\tau)], \quad X\in(X_1,X_2).$$

The functional $F[\varphi(\tau)+\eta(\tau)]$ can be expanded as a functional Taylor series in the vicinity of the point $\eta\sim 0$:

$$F[\varphi(\tau)+\eta(\tau)] = F[\varphi(\tau)] + \int\frac{\delta F[\varphi]}{\delta\varphi(X)}\eta(X)dX$$
$$+\frac{1}{2!}\int\int\frac{\delta F[\varphi]}{\delta\varphi(X_1)\delta\varphi(X_2)}\eta(X_1)\eta(X_2)dX_1dX_2+\ldots. \qquad (1.187)$$

Here and later, when the range of integration is not pointed out, it is assumed to be infinite.

The Taylor series (1.187) could be written in the compact form

$$F[\varphi(\tau)+\eta(\tau)] = \exp\left\{\int dX\eta(X)\frac{\delta}{\delta\varphi(X)}\right\}F[\varphi(\tau)]$$

using the following operator notation:

$$\exp\left\{\int dX\eta(X)\frac{\delta}{\delta\varphi(X)}\right\} = 1+\int dX\eta(X)\frac{\delta}{\delta\varphi(X)}$$
$$+\frac{1}{2!}\int\int dX_1dX_2\eta(X_1)\eta(X_2)\frac{\delta^2}{\delta\varphi(X_1)\delta\varphi(X_2)}+\ldots$$
$$=1+\int dX\eta(X)\frac{\delta}{\delta\varphi(X)}+\frac{1}{2!}\left\{\int dX\eta(X)\frac{\delta}{\delta\varphi(X)}\right\}^2+\ldots.$$

Consider the transformation of variational derivatives as we change the functional variables. Let us replace the function $\varphi(t)$ by a new function $\psi(t)$ given by the equality $\varphi(t)=\Psi[\psi(t);t]$, where ψ is the functional of a function $\psi(t)$, which also depends on t. Then the functional $F[\varphi(\tau)]$ is a composite functional of $\psi(\tau)$:

$$F[\varphi(\tau)] = F[\Psi[\psi(\tau);\tau]]\equiv F_1[\psi(\tau)].$$

For such functional, there exists the following expression for the variational derivative:

$$\frac{\delta F_1[\psi(\tau)]}{\delta\varphi(t)} = \int dt'\frac{\delta F[\varphi(\tau)]}{\delta\varphi(t')}\frac{\delta\Psi[\psi(\tau);X']}{\delta\psi(t)}. \qquad (1.188)$$

A functional change of variables plays an important role in Fourier transforms:

$$\varphi(X) = \int \psi(\omega) e^{i\omega X} d\omega = \varphi[\psi(\omega); X].$$

According to the formula (1.173), we have

$$\frac{\delta\varphi[\psi(\omega); X]}{\delta\psi(\omega')} = e^{i\omega' X}$$

and from (1.180), there follows

$$\frac{\delta F_1[\varphi[\psi(\omega); X]]}{\delta\varphi(\omega')} = \int \frac{\delta F[\varphi(X)]}{\delta\varphi(X')} e^{i\omega' X'} dX'. \tag{1.189}$$

1.15
The Characteristic Functional

It was shown in Section 1.5 that a random value u is completely determined by its characteristic function $\varphi(\rho) = \langle \exp(i\rho u)\rangle$, which allows us to use the inverse Fourier transform to get the PDF,

$$p(u) = \frac{1}{2\pi} \int e^{-i\rho u} d\rho$$

the moments,

$$B_n = \langle u^n \rangle = \left[\frac{1}{i}\frac{d}{d\rho}\right]^n \varphi(\rho)|_{\rho=0},$$

the cumulants,

$$S_n = \left[\frac{1}{i}\frac{d}{d\rho}\right]^n [\ln\varphi(\rho)]_{\rho=0}$$

and other statistical parameters of the PDF.

For a multidimensional random quantity $\boldsymbol{u} = (u_1, u_2, \ldots, u_N)$, the complete description is contained in the characteristic function

$$\varphi(\boldsymbol{\rho}) = \varphi(\rho_1, \rho_2, \ldots, \rho_N) = \langle \exp(i\boldsymbol{\rho}\cdot\mathbf{u})\rangle = \left\langle \exp(i\sum_k \rho_k u_k)\right\rangle. \tag{1.190}$$

The corresponding joint PDF of the random values $u_1, u_2, \ldots, u_N$ is the Fourier transform of the characteristic function $\varphi(\rho_1, \rho_2, \ldots, \rho_N)$:

$$
\begin{aligned}
p(X_1, X_2, \ldots, X_N) &= \frac{1}{(2\pi)^N} \int e^{-i\boldsymbol{\rho}\cdot X} \varphi(\boldsymbol{\rho}) d\boldsymbol{\rho} \\
&= \frac{1}{(2\pi)^N} \int \exp(-i \sum_k \rho_k X_k) \varphi(\rho_1, \rho_2, \ldots, \rho_N) d\rho_1 d\rho_2 \ldots d\rho_N \\
&= \delta(u_1 - X_1)\delta(u_2 - X_2) \ldots \delta(u_N - X_N).
\end{aligned}
\tag{1.191}
$$

Consider now a random function $u(X)$. For its complete statistical description, it is sufficient to know the characteristic functional

$$
\Phi[\rho] = \left\langle \exp\left\{ i \int \rho(\tau) u(\tau) d\tau \right\} \right\rangle, \tag{1.192}
$$

where the function $\rho(\tau)$ is an arbitrary function replacing the set of numbers ρ_1, ρ_2, $\ldots$, ρ_N in (1.190).

Given $\Phi[\rho]$, we can find statistical characteristics of the random function $u(X)$, for example, the average value $\langle u(X) \rangle$, N-point moments $B_{uu\ldots u} = \langle u(X_1) \ldots u(X_N) \rangle$, etc.

To find the variational derivative, let us use the results obtained in Section 1.14, keeping in mind that the averaging operator commutes with the operator $\delta/\delta\rho(X)$:

$$
\begin{aligned}
\frac{\delta \Phi[\rho]}{\delta \rho(X)} = \frac{\delta}{\delta \varphi(X)} \exp\left\langle \left\{ i \int \rho(\tau) u(\tau) d\tau \right\} \right\rangle &= \left\langle \frac{\delta}{\delta \varphi(X)} \exp\left\{ i \int \rho(\tau) u(\tau) d\tau \right\} \right\rangle \\
&= i \left\langle u(X) \exp\left\{ i \int \rho(\tau) u(\tau) d\tau \right\} \right\rangle.
\end{aligned}
$$

Similarly, we write

$$
\left(\frac{1}{i} \frac{\delta}{\delta \rho(X_1)} \right) \cdots \left(\frac{1}{i} \frac{\delta}{\delta \rho(X_n)} \right) \Phi = \left\langle u(X_1) \ldots u(X_n) \exp\left\{ i \int \rho(\tau) u(\tau) d\tau \right\} \right\rangle.
$$

Setting $\rho = 0$ in the last relation, we obtain:

$$
\frac{1}{i^n} \frac{\delta^n}{\delta \rho(X_1) \ldots \delta \rho(X_n)} \Phi \bigg|_{\rho=0} = \langle u(X_1) \ldots u(X_n) \rangle = B_{uu\ldots u}. \tag{1.193}
$$

So, it is possible to find the multi-point moments for a given characteristic functional. Expanding the functional $\Phi[\rho]$ into a functional Taylor series according to Eq. (1.187) and taking into account Eq. (1.193), one can express the characteristic functional in terms of moments:

$$
\begin{aligned}
\Phi[\rho] = \sum_{n=0}^{\infty} \frac{i^n}{n!} \int \cdots \int & B_{uu\ldots u}(X_1, X_2, \ldots, X_n) \\
& \times \rho(X_1)\rho(X_2) \ldots \rho(X_n) dX_1 dX_2 \ldots dX_n.
\end{aligned}
\tag{1.194}
$$

If the functional has the form

$$\Phi[\rho] = \exp(\psi[\rho]),$$

then ψ is called the cumulant-generating function (see Section 1.6), and the cumulants themselves given by

$$S_{11\ldots1}(X_1, X_2, \ldots, X_n) = (-i)^k \frac{\delta^n}{\delta\rho(X_1)\ldots\delta\rho(X_n)}\bigg|_{\rho=0}. \tag{1.195}$$

As an example, let us find the characteristic functional for a Gaussian random process $u(X)$ with $\langle u(X)\rangle = 0$, under the additional assumption that the joint distribution for any two given values of X is also Gaussian. Let the random variable

$$\langle A\rangle = \int_{-\infty}^{\infty} \rho(\tau)u(\tau)d\tau, (\rho(\pm\infty) = 0,$$

have a Gaussian distribution

$$p(A) = \frac{1}{\sqrt{2\pi}\sigma_A}\exp\left\{-\frac{(A-\langle A\rangle)^2}{2\sigma_A^2}\right\}$$

with the parameters

$$A = \int_{-\infty}^{\infty} \rho(\tau)\langle u\rangle(\tau)d\tau = 0,$$

$$\sigma_A^2 = \langle A^2\rangle - (\langle A\rangle)^2 = \int_{-\infty}^{\infty}\int_{-\infty}^{\infty} B(\tau_1, \tau_2)\rho(\tau_1)\rho(\tau_2)d\tau_1 d\tau_2,$$

where $B(\tau_1, \tau_2) = \langle u(\tau_1)u(\tau_2)\rangle$ is a two-point correlation function. The characteristic functional of the random quantity A is

$$\langle\exp(e^{iA})\rangle = \exp\left(-\frac{1}{2}\sigma_A^2\right).$$

Therefore the characteristic functional of a Gaussian random process is

$$\Phi[\rho] = \exp\left\{-\frac{1}{2}\int_{-\infty}^{\infty}\int_{-\infty}^{\infty} B(\tau_1, \tau_2)\rho(\tau_1)\rho(\tau_2)d\tau_1 d\tau_2\right\}. \tag{1.196}$$

Characteristic functionals and variational derivatives are widely used in problems pertaining to particle motion under the action of random fluctuating forces, for example, in the theories of Brownian motion, diffusion, turbulence, etc. Such problems are the subject of the statistical theory of dynamic systems with fluctuating parameters. These parameters (e.g., coordinates and velocities of particles suspended in a fluid), are described by ordinary or partial differential equations (stochastic equations). The main challenge is to obtain and then solve a closed system of equations. It turns out that many processes can be treated as Markowian processes. Also, the distribution of fluctuating parameters (random variables) has proved to be Gaussian.

As an example, consider a simplified form of the Langevin equation (see Section 1.13) that describes the time rate of change of velocity $V(t)$ of a particle driven by a fluctuating force $\xi(t)$:

$$\frac{dV}{dt} = -hV + \xi(t); \quad X(0) = 0, \tag{1.197}$$

where h is a constant and $\xi(t)$ is a random function.

For a given $\xi(t)$, the solution of Eq. (1.197) has the form

$$V(t) = \int_0^i \xi(t)\exp(-h(t-\tau))d\tau. \tag{1.198}$$

From Eq. (1.198), one can derive all statistical characteristics of the random process $V(t)$. On the other hand, all statistical characteristics are contained in the characteristic function

$$\varphi(\rho) = (\exp(i\rho V(t)) = \left\langle \exp\left\{ i\rho \int_0^t \xi(t)\exp(-h(t-\tau))d\tau \right\}\right\rangle. \tag{1.199}$$

Taking into account the relation (1.192), the last equation can be rewritten as

$$\varphi(\rho) = \Phi\{\rho\exp(-h(t-\tau))\}. \tag{1.200}$$

Finally, we should mention the Furutsu–Donsker–Novikov correlation formula for the product of a Gaussian random function $X(t)$ and a functional $R[X(\tau)]$ that may depend (either explicitly or implicitly) on $X(\tau)$:

$$\langle X(t)R[X(\tau)]\rangle = \int_{-\infty}^{\infty} B(t,\tau_1)\left\langle \frac{\delta R[X(\tau)]}{\delta X(\tau_1)} \right\rangle d\tau_1. \tag{1.201}$$

This formula is the functional analogue of Eq. (1.86).

References

1 Gardiner, C.W. (1985) *Handbook of Stochastic Methods*, 2nd ed., Springer–Verlag.

2 Klyatskin, V.I. (1975) *Statistical Description of Dynamical Systems with Fluctuating Parameters*, Nauka, Moscow, (in Russian).

3 Klyatskin, V.I. (1980) *Stochastic Equations and Waves in Random Inhomogeneous Media*, Nauka, Moscow, (in Russian).

4 Landau, L.D. and Lifshitz E.M. (1964) *Statistical Physics*, Nauka, Moscow, (in Russian).

5 Leontovitch, M.A. (1983) *Introduction to Thermodynamics. Statistical Physics*, Nauka, Moscow, (in Russian).

6 Monin, A.C. and Yaglom, A.M. (1971) *Statistical Fluid Mechanics: Mechanics of Turbulence*, Vol. **1**, MIT Press, Cambridge, MA.

7 Monin, A.C. and Yaglom, A.M. (1975) *Statistical Fluid Mechanics: Mechanics of Turbulence*, Vol. **2**, MIT Press, Cambridge, MA.

7 Feller, W. (1974) *An Introduction to Probability Theory and Its Applications*, Vol. **I**, **II**, Wiley.

8 Chandrasekhar, S. (1943) Stochastic Problems in Physics and Astronomy, *Rev. Mod. Phys.*, **15**, 1–89.

9 Cercignani, C. (1969) *Mathematical Methods in Kinetic Theory*, Macmillan.

10 Einstein, A. and Smoluchowski, M. (1936) *Brownian Motion*, ONTI, Moscow, (in Russian).

11 Pope, S.B. (2000) *Turbulent Flows*, Cambridge Univ. Press.

12 Saffman, P.G. (1971) On the Boundary Condition at the Surface of a Porous Medium, *Studies in Appl. Math.*, **50** (2), 93–101.

2
Elements of Microhydrodynamics

The branch of hydrodynamics studying the motion of microparticles in liquid at low Reynolds numbers is called microhydrodynamics. The present chapter is devoted to the fundamentals of low Reynolds number hydrodynamics which will be a prerequisite for the subsequent discussion.

Low Reynolds number hydrodynamics is based on Stokes equations [1–5]

$$\nabla \cdot \boldsymbol{v} = 0, \quad \nabla^2 \boldsymbol{v} = \frac{1}{\mu_e} \nabla p. \tag{2.1}$$

The boundary conditions for these equations follow from the restrictions imposed on velocities at particle surfaces, at flow region boundaries (if any), and far away from the particles ("at infinity") if the fluid is unbounded. Exact solution of concrete boundary value problems presents considerable mathematical difficulties even for the single particle case. Still, there are some general properties common for this type of boundary value problems that prove to be very useful for derivation of solutions. For simplicity, we consider the behavior of particles in an unbounded fluid.

The first property is called the reciprocity theorem. Let $(\boldsymbol{v}',\boldsymbol{T}')$ and $(\boldsymbol{v}'',\boldsymbol{T}'')$ be the velocity $\boldsymbol{v}$ and stress $\boldsymbol{T}$ fields corresponding to the flows of two fluids having different viscosities μ_e' and μ_e''. Both flows are described by the same equations (2.1) and by the same boundary conditions. Then for any closed surface S bounding the fluid volume V (this surface could consists of several different surfaces, including surfaces of particles contained in the considered volume V), there should hold

$$\mu_e'' \int_S d\boldsymbol{S} \cdot \boldsymbol{T}' \cdot \boldsymbol{V}'' = \mu_e' \int_S d\boldsymbol{S} \cdot \boldsymbol{T}'' \cdot \boldsymbol{V}'. \tag{2.2}$$

Another important property, which is often used to determine the hydrodynamic force $\boldsymbol{F}$ and the torque $\boldsymbol{L}$ acting on a spherical particle that has translational velocity $\boldsymbol{U}$ and rotates with angular velocity $\boldsymbol{\Omega}$, is given by Faxen's laws:

$$\boldsymbol{F} = 6\,\pi\mu_e a(\boldsymbol{v}^\infty|_0 - \boldsymbol{U}) + \mu_e \pi a^3 (\nabla \boldsymbol{v}^\infty)_0,$$

$$\boldsymbol{L}_0 = 8\pi\mu_e a^3 \left(\frac{1}{2}(\nabla \times \boldsymbol{v}^\infty)_0 - \boldsymbol{\Omega}\right). \tag{2.3}$$

Statistical Microhydrodynamics. Emmanuil G. Sinaiski and Leonid Zaichik

ISBN: 978-3-527-40656-2

Here $\boldsymbol{v}^{\infty}$ is the (translational) velocity of the liquid at the infinity. The subscript 0 indicates that the corresponding quantity is measured at the point occupied by the center of the sphere.

When solving problems on slow motion of particles, it is important to estimate hydrodynamic field perturbations in the fluid that are induced by the moving particle. If the considered suspension is low-concentrated (dilute), then the effect of the particle on the fluid can be approximated by a point force applied at the particle center and equal to the drag force on the particle, but pointing in the opposite direction. The resultant problem is handled by writing Stokes equations and obtaining the fundamental solution. If we assume the volume concentration of particles φ to be small, then the average distance between particles is large compared to their radius, and we need to consider only pair interactions between particles. Since the particles are far enough from each other, the hydrodynamic influence of particle B on particle A can be accounted for by replacing particle B with a point force $\boldsymbol{F}_B$ applied at the particle center and equal to the drag force on particle B but having the opposite sign. The translational velocity of the fluid at the infinity is assumed to be constant: $\boldsymbol{u}^{\infty}$ $U_0\boldsymbol{e}_x$; we also assume that the line of centers of the particle pair is parallel to $\boldsymbol{U}_0$ and coincides with the X-axis. Then in order to find the drag force of particle A, it is necessary to solve the Stokes equations with the source term

$$\mu_e \Delta \boldsymbol{u} - \nabla p = F_B \delta(X - X_B)\delta(Y)\delta(Z)\boldsymbol{e}_x, \quad \nabla \cdot \boldsymbol{u} = 0 \tag{2.4}$$

and the boundary conditions

$$\begin{aligned} &\boldsymbol{u} = 0 \quad \text{on} \quad S_A, \\ &\boldsymbol{u} \to \boldsymbol{U}_0 \quad \text{at} \quad X \to \infty. \end{aligned} \tag{2.5}$$

The fundamental solution is

$$\begin{aligned} \boldsymbol{u} &= \frac{F_B}{8\pi\mu_e}\left(\frac{1}{|\boldsymbol{X} - \boldsymbol{X}_B|}\boldsymbol{e}_x + \frac{(X - X_B)(\boldsymbol{X} - \boldsymbol{X}_B)}{|\boldsymbol{X} - \boldsymbol{X}_B|^3}\right), \\ p &= \frac{F_B}{4\pi}\left(\frac{(X - X_B)}{|\boldsymbol{X} - \boldsymbol{X}_B|^3}\right). \end{aligned} \tag{2.6}$$

Owing to the linearity of the inhomogeneous boundary value problem, its solution can be obtained as a superposition of solutions of the corresponding homogeneous problems: Eqs. (2.4) at $F_B = 0$ with the boundary conditions (2.5); and Eqs. (2.4) with the boundary conditions (2.5) when $U_0 = 0$. The solutions of these particular problems are found as sums of spherical functions, while keeping in mind the axial symmetry of the problem. To determine the drag force acting on a particle of finite radius, one should average the velocity over the particle surface. The final expression for the force exerted on a single particle by the surrounding fluid will be given by the Stokes formula:

$$F_B = 6\pi\mu_e b U_0. \tag{2.7}$$

For a pair of particles, A and B, the resulting drag forces have the form

$$D_A = 6\pi\mu_e a U_0 + \frac{3}{2}\left(\frac{a^3}{3r^3} - \frac{a}{r}\right)F_B, \quad D_B = 6\pi\mu_e b U_0 + \frac{3}{2}\left(\frac{b^3}{3r^3} - \frac{b}{r}\right)F_A, \quad (2.8)$$

where a, b are the particle radii and r is the distance between their centers. For identical particles ($a = b$),

$$D_A = D_B = 6\pi\mu_e u_0 \lambda(a/r), \quad (2.9)$$

where λ is a dimensionless resistance coefficient. Up to the order a/r, it is equal to

$$\lambda = 1 - \frac{3}{2}\frac{a}{r}. \quad (2.10)$$

With further applications in mind, we shall focus our attention on the motion of one particle, two particles, and, finally, many (i.e., more then two) particles in a quiescent fluid or in a shear flow of viscous incompressible fluid.

2.1
Motion of an Isolated Particle in a Quiescent Fluid

Consider a solid particle of an arbitrary shape moving with translational velocity $\boldsymbol{U}_0$ and rotating with angular velocity $\boldsymbol{\Omega}$. By its nature, such a motion is unstable. But if the Reynolds numbers derived from translational and rotational velocities U_0 and Ω are small, that is, if $\mathrm{Re}_t = a U_0 \rho_e/\mu_e \ll 1$ and $\mathrm{Re}_r = a^2 \Omega \rho_e/\mu_e \ll 1$ where a is the characteristic linear size of the particle and ρ_e and μ_e are the density and viscosity of the fluid, then it is legitimate to derive the velocity and pressure fields $\boldsymbol{v}$ and p induced in the fluid by the moving particle by using quasi-stationary Stokes equations

$$\nabla \cdot \boldsymbol{v} = 0, \quad \nabla^2 \boldsymbol{v} = \frac{1}{\mu_e} \nabla p. \quad (2.11)$$

Let 0 be a point attached to the particle, and let $\boldsymbol{U}_0$ be the instantaneous translational velocity of this point, and $\boldsymbol{\Omega}$ – the instantaneous angular velocity of the particle. The particle is assumed to be solid, and thus the condition of zero velocity at the particle surface $\boldsymbol{S}_a$ yields

$$\boldsymbol{v} = \boldsymbol{U}_0 + \boldsymbol{\Omega} \times \boldsymbol{r}_0 \quad \text{at} \quad S_a, \quad (2.12)$$

where $\boldsymbol{r}_0$ is the radius vector measured with respect to 0.

The second boundary condition is that fluid should be at rest at the infinity:

$$\boldsymbol{v} \to 0 \quad \text{at} \quad r_0 \to \infty. \tag{2.13}$$

The solution of thusly formulated problem gives the hydrodynamic force $\boldsymbol{F}$ and the torque $\boldsymbol{L}$ acting on the particle [6]:

$$\boldsymbol{F} = \int_{S_a} d\boldsymbol{S} \cdot \boldsymbol{T}; \qquad \boldsymbol{L} = \int_{S_a} \boldsymbol{r}_0 \times (d\boldsymbol{S} \cdot \boldsymbol{T}). \tag{2.14}$$

Here $d\boldsymbol{S} = \boldsymbol{n}\, dS$ is a particle surface element (directed towards the fluid), $\boldsymbol{T}$ is the stress tensor of an incompressible Newtonian fluid:

$$\boldsymbol{T} = -p\boldsymbol{I} + \mu_e(\nabla \boldsymbol{v} + \nabla \boldsymbol{v}^T), \tag{2.15}$$

$\boldsymbol{I}$ is the unit tensor with components δ_{ij}, and $\nabla \boldsymbol{v}$ is the velocity gradient tensor with components $\nabla_i v_j$. Due to the linearity of the boundary value problem defined by Eqs. (2.11)–(2.13), it is possible to represent the velocity, the pressure, and the stress tensor as sums of translational and rotational terms, $(\boldsymbol{v}_0^t, p_0^t, \boldsymbol{T}_0^t)$ and $(\boldsymbol{v}_0^r, p_0^r, \boldsymbol{T}_0^r)$, each term describing its corresponding type of particle motion:

$$\boldsymbol{v} = \boldsymbol{v}_0^t + \boldsymbol{v}_0^r, \quad p = p_0^t + p_0^r, \quad \boldsymbol{T} = \boldsymbol{T}^t + \boldsymbol{T}^r. \tag{2.16}$$

As a result, the boundary value problem separates into the translational problem defined by

$$\nabla \cdot \boldsymbol{v}_0^t = 0, \quad \nabla^2 \boldsymbol{v}_0^t = \frac{1}{\mu_e} \nabla p_0^t,$$
$$\boldsymbol{v}_0^t = \boldsymbol{U}_0 \quad \text{on} \quad S_a, \quad \boldsymbol{v}_0^t \to 0 \quad \text{at} \quad r_0 \to \infty \tag{2.17}$$

and the rotational problem defined by

$$\begin{gathered} \nabla \cdot \boldsymbol{v}_0^r = 0, \quad \nabla^2 \boldsymbol{v}_0^r = \frac{1}{\mu_e} \nabla p_0^r, \\ \boldsymbol{v}_0^r = \boldsymbol{\Omega} \times \boldsymbol{r}_0 \quad \text{on} \quad S_a, \quad \boldsymbol{v}_0^r \to 0 \quad \text{at} \quad r_0 \to \infty. \end{gathered} \tag{2.18}$$

Using the formulas (2.14), we can obtained the hydrodynamic forces $\boldsymbol{F}_0^t, \boldsymbol{F}_0^r$ and torques $\boldsymbol{L}_0^t, \boldsymbol{L}_0^r$ induced by translational and rotational motions of the particle:

$$\boldsymbol{F}_0^{t,r} = \int_{S_a} d\boldsymbol{S} \cdot \boldsymbol{T}_0^{t,r}, \quad \boldsymbol{L}_0^{t,r} = \int_{S_a} \boldsymbol{r}_0 \times (d\boldsymbol{S} \cdot \boldsymbol{T}_0^{t,r}). \tag{2.19}$$

where $\boldsymbol{T}_0^{t,r}$ are stress tensors corresponding to the translational and rotational particle motions:

$$\boldsymbol{T}_0^{t,r} = -p_0^{t,r}\boldsymbol{I} + (\nabla \boldsymbol{v}_0^{t,r} + (\nabla \boldsymbol{v}_0^{t,r})^T). \tag{2.20}$$

The resulting values of the hydrodynamic force $\boldsymbol{F}$ and torque $\boldsymbol{L}$ acting on the particle are

$$\boldsymbol{F} = \boldsymbol{F}_0^t + \boldsymbol{F}_0^r, \quad \boldsymbol{L} = \boldsymbol{L}_0^t + \boldsymbol{L}_0^r. \tag{2.21}$$

Both translational and rotational fields, as well as forces and torques depend on the fluid's viscosity μ_e, on the length and orientation of vectors $\boldsymbol{U}_0$ and $\boldsymbol{\Omega}$ relative to the particle's coordinate system, and on the particle's geometry. But such complex dependences are inconvenient to use. Let us try to separate these dependences. To this end, for each type of motion we shall introduce quantities that characterize only the geometrical properties of the particle, thus determining the geometrical dependence once and for all for any given class of geometrical shapes.

For translational motion, there exists a velocity tensor $\boldsymbol{V}^t$ (asymmetrical in the general case) and an associative dynamic pressure vector $\boldsymbol{P}^t$ satisfying the relations

$$\boldsymbol{v}_0^t = \boldsymbol{V}^t \cdot \boldsymbol{U}_0, \quad p_0 = \mu_e \boldsymbol{P}^t \cdot \boldsymbol{U}_0 \tag{2.22}$$

and the equations

$$\begin{aligned} &\nabla \cdot \boldsymbol{V}^t = 0, \quad \nabla^2 \boldsymbol{V}^t - \nabla \boldsymbol{P}^t = 0, \\ &\boldsymbol{V}^t = \boldsymbol{I} \quad \text{on} \quad S_a, \quad \boldsymbol{V}^t \to 0 \quad \text{at} \quad r_0 \to \infty, \end{aligned} \tag{2.23}$$

and independent of the fluid's viscosity (at any point in the fluid), of the choice of reference frame (i.e., of the origin 0), and of the length and orientation of the vector $\boldsymbol{U}_0$. They depend only on the geometry of the particle's surface and on the radius vector (which can be taken relative to any origin of coordinates attached to the particle).

Consider a translational Stokesian flow that goes around a spherical solid particle of radius a when the fluid's velocity at the infinity is $U = const$. The stream function for this flow is expressed in spherical coordinates as

$$\psi = \frac{1}{4} U r^2 \sin^2\theta \left[\left(\frac{a}{r}\right)^3 - 3\frac{a}{r} \right].$$

The velocity components and the dynamic pressure are given by

$$\begin{aligned} v_r &= -\frac{1}{r^2 \sin\theta} \frac{\partial \psi}{\partial \theta} = -\frac{1}{2} U \cos\theta \left(\frac{a}{r}\right)^2 \left(\frac{a}{r} - 3\frac{r}{a}\right), \\ v_\theta &= -\frac{1}{r \sin\theta} \frac{\partial \psi}{\partial r} = -\frac{1}{4} U \sin\theta \left(\frac{a}{r}\right) \left[\left(\frac{a}{r}\right)^2 + 3 \right], \\ p - p^\infty &= \frac{3}{2} a \mu_e U \frac{\cos\theta}{r^2}, \end{aligned}$$

where p^∞ is the static pressure at the infinity.

The velocity tensor and the pressure vector for such a flow are equal to

$$V^r = \frac{3}{4}\frac{a}{r}\left(\frac{\boldsymbol{rr}}{r} + \boldsymbol{I}\right) - \frac{1}{4}\left(\frac{a}{r}\right)^3\left(\frac{3\boldsymbol{rr}}{r^2} - \boldsymbol{I}\right), \quad \boldsymbol{P}^r = \frac{3}{2}a\frac{\boldsymbol{r}}{r^3}, \tag{2.24}$$

where $\boldsymbol{r}$ is measured from the center of the sphere.

We see from Eq. (2.15) and Eq. (2.22) that the stress tensor can be written as

$$\boldsymbol{T}^t = \mu_e \boldsymbol{\Pi}^t \boldsymbol{U}_0, \tag{2.25}$$

where we have just introduced a tensor of the third rank (a triadic tensor)

$$\Pi^t = -\boldsymbol{I}\cdot\boldsymbol{P}^t + (\nabla \boldsymbol{V}^t + {}^T(\nabla \boldsymbol{V}^t)).$$

The superscript T to the left of the tensor (it is known as a polyadic) means that ${}^T(\boldsymbol{abc}\ldots) = \boldsymbol{bac}\ldots$.

Let us introduce the tensors $\boldsymbol{K}^t$ and $\boldsymbol{K}_0^c$ characterizing the given particle:

$$\boldsymbol{K}^t = -\int_{S_a} d\boldsymbol{S}\cdot\boldsymbol{\Pi}^t; \qquad \boldsymbol{K}_0^c = -\int_{S_a} \boldsymbol{r}_0 \times (d\boldsymbol{S}\cdot\boldsymbol{\Pi}^t).$$

Then expressons for the force and torque acting on the particle undergoing a translational motion can be written in a compact form:

$$\boldsymbol{F}_0 = -\mu_e \boldsymbol{K}^t\cdot\boldsymbol{U}_0, \quad \boldsymbol{L} = -\mu_e \boldsymbol{K}_0^c\cdot\boldsymbol{U}_0. \tag{2.26}$$

The tensor $\boldsymbol{K}^t$ is called the tensor of translational resistance. It depends only on the size and shape of the particle and does not depend on the particle's velocity and orientation in space or on the properties of the fluid. The tensor of translational resistance has the dimensionality of length and characterizes resistance for the particle's motion at low Reynolds numbers. Let us show that it is a symmetric tensor.

Consider two motions of one and the same particle, in the same fluid, but with different velocities $\boldsymbol{U}'$ and $\boldsymbol{U}''$. For these two motions we have the corresponding fields of velocities and stresses ($\boldsymbol{v}'$, $\boldsymbol{T}'$) and ($\boldsymbol{v}''$, $\boldsymbol{T}''$) in the surrounding fluid. As a consequence of the reciprocity theorem (2.2), we have for $\mu_e' = \mu_e''$:

$$\int_S d\boldsymbol{S}\cdot\boldsymbol{T}'\cdot\boldsymbol{v}'' = \int_S d\boldsymbol{S}\cdot\boldsymbol{T}''\cdot\boldsymbol{v}'. \tag{2.27}$$

Suppose the surface of integration consists of a spherical surface S^∞ of large radius (the whole sphere is still positioned inside the fluid) and the particle surface S_a which lies wholly inside the volume bounded by S^∞. Then the integral (2.27) can be written as

$$\int_S F\,dS = \int_{S^\infty} F\,dS + \int_{S_a} F\,dS.$$

Since $dS \sim r^2$, and for large r, the fluid velocity is $v \sim r^{-1}$ and the stress is $T \sim r^{-2}$, we have $vTdS \sim r^{-1}$ for large r. Therefore at $r \to \infty \int_S \sim \int_{S_a}$, and Eq. (2.27) converts into

$$\int_{S_a} d\boldsymbol{S} \cdot \boldsymbol{T}' \cdot \boldsymbol{v}'' = \int_{S_a} d\boldsymbol{S} \cdot \boldsymbol{T}'' \cdot \boldsymbol{v}'. \tag{2.28}$$

On the particle surface S, the boundary conditions $\boldsymbol{v}'_a = \boldsymbol{U}'$ and $\boldsymbol{v}''_a = \boldsymbol{U}''$ must be satisfied. A substitution of these relations into Eq. (2.28) yields

$$\boldsymbol{U}'' \int_{S_a} d\boldsymbol{S} \cdot \boldsymbol{T}' = \boldsymbol{U}' \int_{S_a} d\boldsymbol{S} \cdot \boldsymbol{T}''. \tag{2.29}$$

Due to Eq. (2.14), we have

$$\int_{S_a} d\boldsymbol{S} \cdot \boldsymbol{T}' = \boldsymbol{F}', \quad \int_{S_a} d\boldsymbol{S} \cdot \boldsymbol{T}'' = \boldsymbol{F}''$$

and Eq. (2.29) reduces to

$$\boldsymbol{U}'' \cdot \boldsymbol{F}' = \boldsymbol{U}' \cdot \boldsymbol{F}''. \tag{2.30}$$

Let us now substitute (2.30) into (2.26). Then the forces acting on the particle in our two cases are equal to

$$\boldsymbol{F}' = -\mu_e \boldsymbol{K}^t \cdot \boldsymbol{U}', \quad \boldsymbol{F}'' = \mu_e \boldsymbol{K}^t \cdot \boldsymbol{U}''. \tag{2.31}$$

Note that one and the same tensor $\boldsymbol{K}^t$ enters the right-hand sides of both formulas, since it depends only on the size and shape of the particle.

From Eq. (2.30) and Eq. (2.31) there follows

$$\boldsymbol{U}'' \cdot \boldsymbol{K}^t \cdot \boldsymbol{U}' = \boldsymbol{U}' \cdot \boldsymbol{K}^t \cdot \boldsymbol{U}''. \tag{2.32}$$

From the properties of scalar products of tensors and vectors we find that

$$\boldsymbol{U}' \cdot (\boldsymbol{K}^t \cdot \boldsymbol{U}'') = \boldsymbol{U}' \cdot (\boldsymbol{U}'' \cdot (\boldsymbol{K}^t)^T) = \boldsymbol{U}'' \cdot (\boldsymbol{K}^t)^T \cdot \boldsymbol{U}'$$

and, going back to Eq. 2.32, obtain the following:

$$U'' \cdot K^t \cdot U' = U'' \cdot (K^t)^T \cdot U'.$$

Thus,

$$K^t = (K^t)^T. \tag{2.33}$$

This equality means that the tensor K^t is symmetric, that is, $K^t_{ij} = K^t_{ji}$.

The tensor K^c_0 in the expression (2.26) for the torque L is called the conjugate tensor. Its dimensionality is the square of length, and it depends only on the particle's geometry and on the position of the particle's center 0. In general the tensor K^c_0 is neither symmetric nor antisymmetric.

For rotational motion, we can also introduce the velocity tensor V^r_0 and the associative pressure vector P^r_0 satisfying the relations

$$v^r_0 = V^r_0 \cdot \Omega, \quad p^r_0 = \mu_e P^r_0 \cdot \Omega \tag{2.34}$$

and the equations

$$\begin{aligned} &\nabla \cdot V^r_0 = 0, \quad \nabla^2 V^r_0 - \nabla P^r_0 = 0, \\ &V^r_0 = \varepsilon \cdot r_0 \quad \text{on} \quad S_a, \quad V^r_0 \to 0 \quad \text{at} \quad r_0 \to \infty, \end{aligned} \tag{2.35}$$

where ε is the so-called permutation symbol (Levi–Civita symbol) – a tensor with components

$$\varepsilon_{ijk} = \begin{cases} 1, & \text{if} \quad (i, j, k) \quad \text{are} \quad \text{cyclic}, \\ -1, & \text{if} \quad (i, j, k) \quad \text{are} \quad \text{anticyclic}, \\ 0, & \text{otherwise}. \end{cases}$$

These tensors do not depend on μ_e and Ω at any point inside the fluid; they depend only on the choice of the origin 0 and on the surface geometry of the particle. If the origin 0 coincides with the geometrical center of the particle, then

$$V^r_0 = \left(\frac{a}{r}\right)^3 \varepsilon \cdot r, \quad P^r_0 = 0,$$

where r is measured relative to the particle's center.

The stress induced in the fluid by the particle's rotational motion is determined by the tensor

$$T^r_0 = \mu_e \Pi^r_0 \cdot \Omega,$$

where we introduced a tensor of the third rank

$$\Pi^r_0 = -I \cdot P^r_0 + (\nabla V^r_0 + {}^T(\nabla V^r_0)).$$

The force and torque acting on the particle during its rotational motion are

$$F_0^r = \int_{S_a} d\mathbf{S} \cdot T_0^r, \quad L_0^r = \int_{S_a} r_0 \times (d\mathbf{S} \cdot T_0^r).$$

Now, if we define the tensors $K_0^{c^*}$ and K_0^r that are invariable for the given particle by the relations

$$K_0^{c^*} = -\int_{S_a} d\mathbf{S} \cdot \Pi_0^r, \quad K_0^r = -\int_{S_a} r_0 \times (d\mathbf{S} \cdot \Pi_0^r),$$

then expressions for the forces and torques experienced by the particle during its rotational motion can be written as

$$F = -\mu_e K_0^{c^*} \cdot U_0, \quad L = -\mu_e K_0^r \cdot \Omega. \tag{2.36}$$

The tensor K_0^r is called the rotational resistance tensor and $K_0^{c^*}$ – its conjugate tensor. Their dimensionalities are, respectively, L^3 and L^2. The reciprocity theorem implies their symmetry, that is, $K_0^{c^*} = (K_0^c)^T$.

Thus it follows from Eqs. (2.21), (2.26), and (2.36) that an isolated particle undergoing translational and rotational motion of in a fluid that is quiescent at the infinity experiences the force and the torque that are given by

$$F = -\mu_e (K^t \cdot U_0 + (K_0^c)^T \cdot \Omega), \quad L_0 = -\mu_e (K_0^c \cdot U_0 + K_0^r \cdot \Omega). \tag{2.37}$$

They depend on three independent resistance tensors K^t, K_0^r, and K_0^c characterizing the particle's inherent properties: surface shape and size. The tensors $\mathbf{K}^t$ and $\mathbf{K}_0^r$ are symmetric, that is, $K^t = (K^t)^T$ and $K_0^r = (K_0^r)^T$.

If a is the characteristic linear size of the particle, then, introducing dimensionless resistance tensors by dividing each dimensional resistance tensor by its corresponding a^k, we can rewrite the relations (2.37):

$$F = -\mu_e a (\tilde{K}^t \cdot U_0 + a (\tilde{K}_0^c)^T \cdot \Omega), \quad L_0 = -\eta_e a^2 (\tilde{K}_0^c \cdot U_0 + a \tilde{K}_0^r \cdot \Omega). \tag{2.38}$$

The dimensionless tensors $\tilde{K}^t$, $\tilde{K}_0^r$ and $\tilde{K}_0^c$ are called particle shape tensors because, in contrast to K^t, K_0^r, and K_0^c, they do not depend on the particle's size.

The relations (2.37) imply two remarkable properties. The first one is that translational and rotational motions are related in the sense that rotational motion induces translational motion and vice versa. Secondly, the resistance matrices corresponding to tensors K_0^c and $(K_0^c)^T$, which are responsible for the connection between rotational and translational motions, consist of the same elements. The latter property results from the reciprocity theorem, which can be considered as a manifestation of the general mechanical reciprocity principle (another manifestation of this general principle is the well-known Onsager reciprocity principle).

The relations (2.37) can be represented in a compact matrix form. Let us write the vectors of force, torque, translational and rotational velocities as 3×1 matrices, whose elements are the corresponding components in the chosen coordinate system with the vector basis $\boldsymbol{e}_1$, $\boldsymbol{e}_2$, $\boldsymbol{e}_3$:

$$\|\boldsymbol{F}\| = \begin{Vmatrix} F_1 \\ F_2 \\ F_3 \end{Vmatrix}, \quad \|\boldsymbol{L}_0\| = \begin{Vmatrix} L_{10} \\ L_{20} \\ L_{30} \end{Vmatrix}, \quad \|\boldsymbol{U}_0\| = \begin{Vmatrix} U_{10} \\ U_{20} \\ U_{30} \end{Vmatrix}, \quad \|\boldsymbol{\Omega}\| = \begin{Vmatrix} \Omega_1 \\ \Omega_2 \\ \Omega_3 \end{Vmatrix}.$$

From these matrices, we can compose new matrices of generalized force and generalized velocity:

$$\|\mathbf{F}_0\| = \begin{Vmatrix} \|\boldsymbol{F}\| \\ \|\boldsymbol{L}_0\| \end{Vmatrix}, \quad \|\mathrm{u}_0\| = \begin{Vmatrix} \|\boldsymbol{U}_0\| \\ \|\boldsymbol{\Omega}\| \end{Vmatrix}.$$

In a similar manner, we can construct 3×3 resistance matrices

$$\|\boldsymbol{K}_0^{t,r,c}\| = \begin{Vmatrix} K_{11}^{t,r,c} & K_{12}^{t,r,c} & K_{13}^{t,r,c} \\ K_{21}^{t,r,c} & K_{22}^{t,r,c} & K_{23}^{t,r,c} \\ K_{31}^{t,r,c} & K_{32}^{t,r,c} & K_{33}^{t,r,c} \end{Vmatrix}$$

and the grand 6×6 resistance matrix

$$\|\mathbf{K}_0\| = \begin{Vmatrix} \|\boldsymbol{K}^t\| \|\boldsymbol{K}_0^c\|^T \\ \|\boldsymbol{K}_0^c\| \|\boldsymbol{K}_0^r\| \end{Vmatrix}.$$

With these notations, the relation (2.37) takes the form

$$\|\mathbf{F}_0\| = -\mu_e \|\mathbf{K}_0\| \, \|\mathbf{u}_0\|. \tag{2.39}$$

Since the rate of energy dissipation is $\|\mathbf{u}_0\|^T \|\mathbf{K}_0\| \, \|\mathbf{u}_0\|$, the condition that it must be positive implies that the grand resistance matrix is positive definite.

Resistance tensors $\boldsymbol{K}_0^c$ and $\boldsymbol{K}_0^r$ depend on the choice of origin 0. A transition to another coordinate system with the origin P transforms these tensors according to the rule

$$\boldsymbol{K}_p^c = \boldsymbol{K}_0^c - \boldsymbol{r}_{0p} \times \boldsymbol{K}^t, \quad \boldsymbol{K}_p^r = \boldsymbol{K}_0^r - \boldsymbol{r}_{0p} \times \boldsymbol{K}^t \times \boldsymbol{r}_{0p} + \boldsymbol{K}_0^c \times \boldsymbol{r}_{0p} - \boldsymbol{r}_{0p} \times (\boldsymbol{K}^c)^T,$$

where $\boldsymbol{r}_{0p}$ is the vector from 0 to P.

These formulas follow from the requirement that the force of hydrodynamic resistance does not depend on the choice of origin, while the torque and the translational velocity change as

$$\boldsymbol{L}_p = \boldsymbol{L}_0 - \boldsymbol{r}_{0p} \times \boldsymbol{F}, \quad \boldsymbol{U}_p = \boldsymbol{U}_0 + \boldsymbol{\Omega} \times \boldsymbol{r}_{0p}.$$

For particles whose shape possesses central symmetry, resistance tensors have the simplest form when the origin 0 lies on the particle's symmetry axis. It is also possible to simplify the expression for the resistance tensor for a body of an arbitrary shape. It turns out that there exists a point called the center of reaction R, relative to which the conjugate resistance tensor is symmetric, that is, $\boldsymbol{K}_R^c = (\boldsymbol{K}_R^c)^T$. If the tensor $\boldsymbol{K}_0^c$ is known relative to some point 0 inside the particle, the position of the reaction center R can be determined from the relation

$$\boldsymbol{r}_{0R} = [(\boldsymbol{I} : \boldsymbol{K}^t)\boldsymbol{I} - \boldsymbol{K}^t]^{-1} \cdot \boldsymbol{\varepsilon} : \boldsymbol{K}_0^c$$

For some axisymmetric bodies (spheres, ellipsoids), the symmetric tensor is $\boldsymbol{K}_R^c = 0$. In this case the relations (2.37) have an especially simple form:

$$\boldsymbol{F} = -\mu_e \boldsymbol{K}^t \cdot \boldsymbol{U}_R, \quad \boldsymbol{L}_0 = -\mu_e \boldsymbol{K}_R^r \cdot \boldsymbol{\Omega}. \tag{2.40}$$

It follows from the last relation that if the reaction center is taken as the origin, then translational and rotational motions are mutually independent.

For spherical particles, the reaction center coincides with the center of the sphere, and resistance tensors are equal to

$$\boldsymbol{K}^t = 6\pi\mu_e a \boldsymbol{I}, \quad \boldsymbol{K}_R^r = 8\pi\mu_e a^3 \boldsymbol{I}, \quad \boldsymbol{K}_R^c = 0.$$

For a particle with ellipsoidal surface,

$$\frac{x_1^2}{a_1^2} + \frac{x_2^2}{a_2^2} + \frac{x_3^2}{a_3^2} = 1 \tag{2.41}$$

the reaction center coincides with the center of the ellipsoid, and in the coordinate system whose origin is placed at the point R and whose coordinate axes are parallel to the ellipsoid's axes, the resistance tensors are

$$\begin{aligned}
\boldsymbol{K}^t &= 16\pi\left(\frac{1}{\chi + a_1^2\alpha_1}\boldsymbol{e}_1\boldsymbol{e}_1 + \frac{1}{\chi + a_2^2\alpha_2}\boldsymbol{e}_2\boldsymbol{e}_2 + \frac{1}{\chi + a_3^2\alpha_3}\boldsymbol{e}_3\boldsymbol{e}_3\right),\\
\boldsymbol{K}_R^r &= \frac{16\pi}{3}\left(\frac{a_2^2 + a_3^2}{a_2^2\alpha_2 + a_3^2\alpha_3}\boldsymbol{e}_1\boldsymbol{e}_1 + \frac{a_3^2 + a_1^2}{a_2^2\alpha_2 + a_1^2\alpha_1}\boldsymbol{e}_2\boldsymbol{e}_2 + \frac{a_1^2 + a_2^2}{a_1^2\alpha_1 + a_2^2\alpha_2}\boldsymbol{e}_3\boldsymbol{e}_3\right),\\
\boldsymbol{K}_R^c &= 0,
\end{aligned} \tag{2.42}$$

where

$$a_j = \int_0^\infty \frac{dX}{(a_j^2 + X)\Delta(X)} \quad (j = 1, 2, 3), \quad \chi = \int_0^\infty \frac{dX}{\Delta(X)},$$

$$\Delta(X) = [(a_1^2 + X)(a_2^2 + X)(a_3^2 + X)]^{1/2}.$$

If the force and torque acting on the particle are given, then Eq. (2.39) gives the particle's velocity:

$$\|\mathbf{u}_0\| = -\mu_e^{-1}\|\mathbf{K}_0\|^{-1}\|\mathbf{F}_0\|. \tag{2.43}$$

The matrix $\|\mathbf{K}_0\|^{-1} = \|\boldsymbol{B}\|$ is called the mobility matrix and the corresponding tensor $\boldsymbol{B}$ – the mobility tensor.

2.2
Motion of an Isolated Particle in a Moving Fluid

Consider now the translational and rotational motion of a solid particle of an arbitrary shape in a viscous incompressible fluid whose velocity at the infinity is non-zero and equal to $\boldsymbol{v}^\infty$. One encounters such problems when studying motion of fluids with suspended particles inside pipes. The velocity profile of the carrier fluid can be nonuniform over the pipe cross section; this is the case, for example, in Poiseulle and Couette flows. Since the particle size is small and the average distance between particles in low-concentrated mixtures is large as compared to the particle size, we assume a shear velocity profile of the fluid in the vicinity of the particle.

Let us put the origin 0 of the coordinate system inside the particle. The translational velocity of the point 0 will be denoted by $\boldsymbol{U}_0$ and the rotational velocity of the particle about the axis passing through 0 will be denoted by $\boldsymbol{\Omega}$. The fields of velocity $\boldsymbol{v}$ and pressure p inside the fluid are described by Stokes equations (2.1):

$$\nabla \cdot \boldsymbol{v} = 0, \quad \nabla^2 \boldsymbol{v} = \frac{1}{\mu_e} \nabla p \tag{2.44}$$

with inhomogeneous boundary conditions

$$\boldsymbol{v} = \boldsymbol{U}_0 + \boldsymbol{\Omega} \times \boldsymbol{r}_0 \quad \text{on} \quad S_a, \quad \boldsymbol{v} \to \boldsymbol{v}^\infty \quad \text{at} \quad r_0 \to \infty, \tag{2.45}$$

where $\boldsymbol{r}_0$ is the radius vector of a point relative to the origin.

The velocity field at the infinity, $\boldsymbol{v}^\infty(\boldsymbol{r})$, should also satisfy the appropriate hydrodynamic equations (for Poiseulle and Couette flows, these are Stokes equations

(2.44)). We assume that the unperturbed velocity profile at the infinity corresponds to shear flow. Let us introduce the tensor of velocity gradient at the infinity $\boldsymbol{g}^{\infty} = \nabla \boldsymbol{v}^{\infty}$ with components $g_{ij}^{\infty} = \nabla_j v_i^{\infty}$. It is evident that the symmetric part of this tensor $\boldsymbol{E}^{\infty} = 0.5(\boldsymbol{g}^{\infty} + (\boldsymbol{g}^{\infty})^T)$ with components $E_{ij}^{\infty} = 0.5(g_{ij}^{\infty} + g_{ji}^{\infty})$ is the rate-of-strain tensor and the antisymmetric part $\Lambda^{\infty} = 0.5(\boldsymbol{g}^{\infty} - (\boldsymbol{g}^{\infty})^T)$ with components $\Lambda_{ij}^{\infty} = 0.5(g_{ij}^{\infty} - g_{ji}^{\infty})$ is the vorticity tensor. The latter is related to the vorticity vector $\boldsymbol{\Omega}^{\infty}$, which stands for the rotational velocity of the fluid in an unperturbed flow, by the expression $\boldsymbol{\Omega}^{\infty} = -0.5\varepsilon \cdot \Lambda$, or, in the component form, $\boldsymbol{\Omega}_i^{\infty} = 0.5\varepsilon_{ijk}\Lambda_{kj}$. Then the unperturbed velocity, that is, velocity at the infinity, is written as

$$\boldsymbol{v}^{\infty} = \boldsymbol{v}^0 + \boldsymbol{\Omega}^{\infty} \times \boldsymbol{r}_0 + \boldsymbol{E}^{\infty} \cdot \boldsymbol{r}_0 \tag{2.46}$$

or, in the component form,

$$v_i^{\infty} = v_i^0 + \varepsilon_{ijk}\Omega_i^{\infty} x_k + E_{ij}^{\infty} x_j, \tag{2.47}$$

where $\boldsymbol{v}^0$ is the velocity of point 0 as if it were in the fluid, $\boldsymbol{r}_0$ is the radius vector (relative to point 0) of the point under consideration, and x_j are the radius vector components. It should be noted that Ω^{∞} and $\boldsymbol{E}^{\infty}$ are constants and do not depend on the choice of origin 0.

To solve the boundary value problem defined by Eqs. (2.44), (2.45), one must exploit the linearity of the problem and seek the solution as a superposition of particular solutions $(\boldsymbol{v}_0', p_0')$, $(\boldsymbol{v}_0'', p_0'')$, and $(\boldsymbol{v}''', p''')$, the first two of which are solutions of the boundary value problems

$$\nabla \cdot \boldsymbol{v}_0' = 0, \quad \nabla^2 \boldsymbol{v}_0' = \frac{1}{\mu_e} \nabla p_0',$$

$$\boldsymbol{v}_0' = -\boldsymbol{E}^{\infty} \cdot \boldsymbol{r}_0 \quad \text{on} \quad S_a, \quad \boldsymbol{v}_0' \to 0 \quad \text{at} \quad r_0 \to \infty;$$

$$\nabla \cdot \boldsymbol{v}_0'' = 0, \quad \nabla^2 \boldsymbol{v}_0'' = \frac{1}{\mu_e} \nabla p_0'',$$

$$\boldsymbol{v}_0'' = (\boldsymbol{U}_0 - \boldsymbol{v}^{\infty}) + (\boldsymbol{\Omega} - \boldsymbol{\Omega}^{\infty}) \times \boldsymbol{r}_0 \quad \text{on} \quad S_a, \quad \boldsymbol{v}_0'' \to 0 \quad \text{at} \quad r_0 \to \infty,$$

and the third one is equal to

$$\boldsymbol{v}''' = \boldsymbol{v}^0 + \boldsymbol{\Omega}^{\infty} \times \boldsymbol{r}_0 + \boldsymbol{E}^{\infty} \cdot \boldsymbol{r}_0, \quad p''' = 0.$$

in the entire region.

It should be noted that the fields $(\boldsymbol{v}_0', p_0')$ and $(\boldsymbol{v}_0'', p_0'')$ depend on the choice of origin 0, while the field $(\boldsymbol{v}''', p''')$ and the net field $\boldsymbol{v} = \boldsymbol{v}_0' + \boldsymbol{v}_0'' + \boldsymbol{v}'''$, $p = p_0' + p_0'' + p'''$ do not depend on it.

Having determined the velocity and pressure fields, one can find the stresses $\boldsymbol{T}'$, $\boldsymbol{T}''$ and $\boldsymbol{T}'''$. It is evident that $\boldsymbol{T}''' = 2\mu_e \boldsymbol{E}^{\infty}$ in the whole region occupied by the fluid. It follows from Eq. (2.14) that the force and torque experienced by the particle can also be expressed as a sum of corresponding forces and torques:

$$F = F_0' + F_0'' + F''', \quad L_0 = L_0' + L_0'' + L'''.$$

Their components are equal to

$$\begin{aligned} F_0' &= -\mu_e K_0^t : E^\infty, \quad L_0' = -\mu_e K_0^r : E^\infty, \\ F_0'' &= -\mu_e (K^t \cdot (U_0 - v^\infty) + (K_0^c)^T \cdot (\Omega - \Omega^\infty)), \\ L_0'' &= -\mu_e (K_0^c \cdot (U_0 - v^\infty) + K_0^r \cdot (\Omega - \Omega^\infty)), \\ F''' &= 0, \quad L''' = 0. \end{aligned}$$

To summarize, the hydrodynamic force and torque acting on the particle are:

$$\begin{aligned} F &= -\mu_e (K^t \cdot (U_0 - v^\infty) + (K_0^c)^T \cdot (\Omega - \Omega^\infty) + K_0^t : E^\infty), \\ L_0 &= -\mu_e (K_0^c \cdot (U_0 - v^\infty) + K_0^r \cdot (\Omega - \Omega^\infty) + K_0^r : E^\infty), \end{aligned} \tag{2.48}$$

A comparison with the relation (2.37) for the force and torque acting on a particle moving in a quiescent fluid shows that in a shear flow, additional terms appear in the relations for the corresponding force and torque. These terms are proportional to the rate-of-strain tensor of an undisturbed flow E^∞, with third-rank tensors K_0^t and K_0^r being the proportionality factors. These tensors are called the tensors of translational and rotational shear resistance; they characterize the particle's intrinsic properties.

Currently, there exist exact solutions of the problem of motion of spherical and ellipsoidal particles in a shear flow [6]. Presented below are some results for the case when the origin 0 is made to coincide with the reaction center R.

For an ellipsoidal particle (2.41), the tensors K_R^t and K_R^r are

$$\begin{aligned} K_R^t &= 0, \\ K_R^r &= \frac{8\pi}{3}\left(\frac{a_3^2 - a_2^2}{a_2^2\alpha_2 + a_3^2\alpha_3}(e_1e_2e_3 + e_1e_3e_2)\right. \\ &\quad \left. + \frac{a_1^2 - a_3^2}{a_3^2\alpha_3 + a_1^2\alpha_1}(e_2e_3e_1 + e_2e_1e_3) + \frac{a_2^2 - a_1^2}{a_1^2\alpha_1 + a_2^2\alpha_2}(e_3e_1e_2 + e_3e_2e_1)\right). \end{aligned} \tag{2.49}$$

In the particular case of spherical particle, we have $K_R^t = K_R^r = 0$, and the other tensors are given in Section 2.1.

Formulas (2.37) and (2.48) allow us to find the forces and torques acting on particles undergoing translational and rotational motion with given velocities. For a particle freely suspended in the fluid, the force and torque are equal to zero. From these conditions, one can find the translational and rotational velocities of the particle:

$$\boldsymbol{U}_0 - \boldsymbol{v}^\infty = -[(\boldsymbol{K}^t - (\boldsymbol{K}_0^c)^T \cdot (\boldsymbol{K}_0^r)^{-1} \cdot \boldsymbol{K}_0^c]^{-1} \cdot [\boldsymbol{K}_0^t - (\boldsymbol{K}_0^c)^T \cdot (\boldsymbol{K}_0^r)^{-1} \cdot \boldsymbol{K}_0^r)] : \boldsymbol{E}^\infty,$$

$$\boldsymbol{\Omega} - \boldsymbol{\Omega}^\infty = -[(\boldsymbol{K}^t - \boldsymbol{K}_0^c \cdot (\boldsymbol{K}^t)^{-1} \cdot (\boldsymbol{K}_0^c)^T)^{-1} \cdot (\boldsymbol{K}_0^r - \boldsymbol{K}_0^c \cdot (\boldsymbol{K}_0^t)^{-1} \cdot \boldsymbol{K}_0^t)] : \boldsymbol{E}^\infty.$$

The relation (2.48) can be written in a matrix form similar to Eq. (2.39). One complication is that $\boldsymbol{K}_0^t$ and $\boldsymbol{K}_0^r$ are third-rank tensors with elements $\boldsymbol{K}_{ijk}^t$, $\boldsymbol{K}_{ijk}^r$ and they cannot be represented in a matrix form as simply as other terms. Consider the components of vectors $\boldsymbol{F}_0'$ and $\boldsymbol{L}_0'$:

$$F_{0i}' = -\mu_e K_{ijk}^t E_{kj}^\infty, \quad L_{0i}' = -\mu_e K_{ijk}^r E_{kj}^\infty \tag{2.50}$$

in more detail. For convenience's sake, we have dropped the index 0 in the tensors $\boldsymbol{K}_0^t$ and $\boldsymbol{K}_0^r$, and will follow this convention from now on. As far as $(i, j, k) = (1, 2, 3)$, each of the tensors $\boldsymbol{K}^t$ and $\boldsymbol{K}^r$ has 27 elements. Due to the symmetry of the rate-of-strain tensor $\boldsymbol{E}^\infty$, we have $E_{jk}^\infty = E_{kj}^\infty$, $K_{ijk}^t = K_{ikj}^t$, and $K_{ijk}^r = K_{ikj}^r$. Therefore each of the tensors $\boldsymbol{K}^t$ and $\boldsymbol{K}^r$ has only 18 independent components. To represent the last terms in Eq. (2.48) in the matrix form, let us proceed in the following way. First, we shall use the symmetry of tensors $\boldsymbol{K}^t$ and $\boldsymbol{K}^r$ with respect to the last two indices and the symmetry of the tensor $\boldsymbol{E}^\infty$ to replace the pair of indices (jk) with one index l according to the rule below:

tensor indices	(jk)	11	22	33	23, 32	31, 13	12, 21
matrix indices	(l)	1	2	3	4	5	6

Furthermore, since $K_{112}^t = K_{121}^t$, $K_{112}^r = K_{121}^r$, and so on, let us denote $K_{112}^t = K_{121}^t = \frac{1}{2}\Phi_{16}$, $K_{112}^r = K_{121}^r = \frac{1}{2}t_{16}$, etc.

In the same way, we make a transition from the matrix $\boldsymbol{E}^\infty(E_{ij})$ to the matrix $\boldsymbol{S}$. As a result, the tensor $\boldsymbol{K}^t$ with components K_{ijk}^t (enumerated in the table below)

$i=1$			$i=2$			$i=3$		
K_{111}^t	K_{112}^t	K_{111}^t	K_{211}^t	K_{212}^t	K_{213}^t	K_{311}^t	K_{312}^t	K_{313}^t
	K_{122}^t	K_{123}^t		K_{222}^t	K_{223}^t		K_{322}^t	K_{323}^t
		K_{133}^t			K_{233}^t			K_{333}^t

turns into a matrix Φ with components Φ_{ij},

$$
\begin{array}{ccccccccc}
\Phi_{11} & \frac{1}{2}\Phi_{16} & \frac{1}{2}\Phi_{15} & \Phi_{21} & \frac{1}{2}\Phi_{26} & \frac{1}{2}\Phi_{25} & \Phi_{31} & \frac{1}{2}\Phi_{36} & \frac{1}{2}\Phi_{35} \\
 & \Phi_{12} & \frac{1}{2}\Phi_{14} & & \Phi_{22} & \frac{1}{2}\Phi_{24} & & \Phi_{32} & \frac{1}{2}\Phi_{34} \\
 & & \Phi_{13} & & & \Phi_{23} & & & \Phi_{33}
\end{array}
$$

and the tensor $\boldsymbol{E}^\infty$ with components E_{ij},

$$
\begin{array}{ccc}
E_{11} & E_{12} & E_{13} \\
 & E_{22} & E_{23} \\
 & & E_{33}
\end{array}
$$

turns into a matrix $\boldsymbol{S}$ with components

$$
\begin{array}{ccc}
S_1 & S_6 & S_5 \\
 & S_2 & S_4 \\
 & & S_3
\end{array}
$$

The transition from the tensor $\boldsymbol{K}^r$ to a matrix $\tau(\tau_{ij})$ is handled in a similar manner. As a result, the relation (2.50) reduces to

$$
\boldsymbol{F}'_{0i} = -\mu_e K^t_{ijk} E^\infty_{kj} = -\mu_e \Phi_{il} S_l, \quad L'_{0i} = -\eta_e K^r_{ijk} E^\infty_{kj} = -\mu_e \tau_{il} S_l, \tag{2.51}
$$

where $i = 1, 2, 3$ and $l = 1, 2, \ldots, 6$.

Let us now introduce a generalized force vector F'_0 written in the column form, so that $\|F'_0\|^T = \|\boldsymbol{F}'_0, \boldsymbol{L}'_0\|$; a shear vector $\boldsymbol{J}$, also written in the column form, so that $\|\boldsymbol{J}\|^T = \|S_1, S_2, S_3, S_4, S_5, S_6\|$; and a matrix of shear resistance $\|\Psi\|$,

$$
\|\boldsymbol{\Psi}\| = \begin{Vmatrix} \boldsymbol{\Phi} \\ \boldsymbol{\tau} \end{Vmatrix} = \begin{Vmatrix}
\Phi_{11} & \Phi_{12} & \Phi_{13} & \Phi_{14} & \Phi_{15} & \Phi_{16} \\
\Phi_{21} & \Phi_{22} & \Phi_{23} & \Phi_{24} & \Phi_{25} & \Phi_{26} \\
\Phi_{31} & \Phi_{32} & \Phi_{33} & \Phi_{34} & \Phi_{35} & \Phi_{36} \\
\tau_{11} & \tau_{12} & \tau_{13} & \tau_{14} & \tau_{15} & \tau_{16} \\
\tau_{21} & \tau_{22} & \tau_{23} & \tau_{24} & \tau_{25} & \tau_{26} \\
\tau_{31} & \tau_{32} & \tau_{33} & \tau_{34} & \tau_{35} & \tau_{36}
\end{Vmatrix}.
$$

The relation (2.51) may now be written in the matrix form:

$$
\|\mathbf{F}'_0\| = -\mu_e \Psi \|\boldsymbol{J}\| \tag{2.52}
$$

Combining it with (2.39), we get

$$\|\mathbf{F}_0\| = -\mu_e(\|\mathbf{K}_0\|\,\|\mathbf{u}_0\| + \|\boldsymbol{\Psi}\|\,\|J\|). \tag{2.53}$$

For a particle freely suspended in the fluid, we use (2.53) with $F_0 = 0$, obtaining both translational and rotational velocities of the particle:

$$\|u_0\| = \|K_0\|^{-1}\|\boldsymbol{\Psi}\|\,\|J\|. \tag{2.54}$$

Note that the matrices $\|\boldsymbol{K}_0\|$ and $\|\Phi\|$ depend only on the intrinsic properties of the particle, and for a given class of geometrical shapes, they can be determined once and for all.

If the particle's shape possesses some kind of symmetry, then the task of finding the elements of matrices K_0 and Φ is considerably simplified. For example, consider an axisymmetric particle. Introduce a Cartesian coordinate system with the basis $\boldsymbol{e}_1$, $\boldsymbol{e}_2$, $\boldsymbol{e}_3$, where $\boldsymbol{e}_3$ is directed along the symmetry axis. For axisymmetric bodies, the tensors that were introduced in the previous discussion become

$$\boldsymbol{K}^t = a(\boldsymbol{I}-\boldsymbol{e}_3\boldsymbol{e}_3) + b\boldsymbol{e}_3\boldsymbol{e}_3; \quad \boldsymbol{K}_0^c = c(\boldsymbol{\varepsilon}\cdot\boldsymbol{e}_3); \quad \boldsymbol{K}_0^r = d(\boldsymbol{I}-\boldsymbol{e}_3\boldsymbol{e}_3) + e\boldsymbol{e}_3\boldsymbol{e}_3,$$

$$\boldsymbol{K}_0^t = f\boldsymbol{e}_3\boldsymbol{e}_3\boldsymbol{e}_3 + g((\boldsymbol{I}\boldsymbol{e}_3) + (\boldsymbol{I}\boldsymbol{e}_3)^T); \quad \boldsymbol{K}_0^r = h((\boldsymbol{\varepsilon}\cdot\boldsymbol{e}_3\boldsymbol{e}_3) + (\boldsymbol{\varepsilon}\cdot\boldsymbol{e}_3\boldsymbol{e}_3)^T),$$

where the scalar factors $a, b, \ldots$ depend only on the geometrical properties of the particle. Using the definitions of polyadics $\boldsymbol{abc} = a_i b_j c_k$ and $\boldsymbol{Aa} = A_{ij}a_k$, of the scalar product $\boldsymbol{A}\cdot\boldsymbol{a} = A_{ij}a_j$ and of the permuation symbol

$$\boldsymbol{\varepsilon} = \boldsymbol{e}_1\boldsymbol{e}_2\boldsymbol{e}_3 - \boldsymbol{e}_1\boldsymbol{e}_3\boldsymbol{e}_2 + \boldsymbol{e}_2\boldsymbol{e}_3\boldsymbol{e}_1 - \boldsymbol{e}_2\boldsymbol{e}_1\boldsymbol{e}_3 + \boldsymbol{e}_3\boldsymbol{e}_1\boldsymbol{e}_2 - \boldsymbol{e}_3\boldsymbol{e}_2\boldsymbol{e}_1$$

we can rewrite these relations in the matrix form:

$$\boldsymbol{K}^t = \begin{Vmatrix} a & 0 & 0 \\ 0 & a & 0 \\ 0 & 0 & b \end{Vmatrix}, \quad \boldsymbol{K}_0^c = \begin{Vmatrix} 0 & c & 0 \\ -c & 0 & 0 \\ 0 & 0 & 0 \end{Vmatrix},$$

$$\boldsymbol{K}_0^r = \begin{Vmatrix} d & 0 & 0 \\ 0 & d & 0 \\ 0 & 0 & e \end{Vmatrix}, \quad (\boldsymbol{K}_0^c)^T = \begin{Vmatrix} 0 & -c & 0 \\ c & 0 & 0 \\ 0 & 0 & 0 \end{Vmatrix}.$$

The grand resistance matrix $\|\mathbf{K}_0\|$ and the shear resistance matrix $\|\boldsymbol{\Psi}\|$ are

$$\|\mathbf{K}_0\| = \begin{Vmatrix} a & 0 & 0 & 0 & c & 0 \\ 0 & a & 0 & -c & 0 & 0 \\ 0 & 0 & b & 0 & 0 & 0 \\ d & 0 & 0 & 0 & -c & 0 \\ 0 & d & 0 & c & 0 & 0 \\ 0 & 0 & e & 0 & 0 & 0 \end{Vmatrix}, \quad \|\boldsymbol{\Psi}\| = \begin{Vmatrix} \|\boldsymbol{\vartheta}\| & \|\boldsymbol{\Psi}\| \\ \|\mathbf{0}\| & \|\boldsymbol{\chi}\| \end{Vmatrix},$$

where

$$\|\boldsymbol{\vartheta}\| = \begin{Vmatrix} 0 & 0 & 0 \\ 0 & 0 & 0 \\ 0 & 0 & f+2g \end{Vmatrix}, \quad \|\boldsymbol{\Psi}\| = \begin{Vmatrix} 0 & 2g & 0 \\ 2g & 0 & 0 \\ 0 & 0 & 0 \end{Vmatrix}, \quad \|\boldsymbol{\chi}\| = \begin{Vmatrix} 2h & 0 & 0 \\ 0 & -2h & 0 \\ 0 & 0 & 0 \end{Vmatrix},$$

and $\|\mathbf{0}\|$ is a 3×3 zero matrix.

The inverse resistance matrix (mobility matrix) is

$$\|\mathbf{K}_0\|^{-1} = \begin{Vmatrix} m & 0 & 0 & 0 & cna^{-1} & 0 \\ 0 & m & 0 & -cna^{-1} & 0 & 0 \\ 0 & 0 & b^{-1} & 0 & 0 & 0 \\ 0 & -cmd^{-1} & 0 & n & 0 & 0 \\ cmd^{-1} & 0 & 0 & 0 & n & 0 \\ 0 & 0 & 0 & 0 & 0 & e^{-1} \end{Vmatrix},$$

where

$$m = \frac{d}{ad-c^2}, \quad n = \frac{a}{ad-c^2}.$$

Now, using Eq. (2.47), we can find the velocity of a freely suspended axisymmetric particle:

$$U_0 - v^0 = \begin{Vmatrix} kS_2 \\ kS_1 \\ \lambda S_3 \end{Vmatrix}, \quad \boldsymbol{\Omega} - \boldsymbol{\Omega}^\infty = \begin{Vmatrix} \nu S_4 \\ -\nu S_5 \\ 0 \end{Vmatrix}, \tag{2.55}$$

where $k = \dfrac{2(ch-gd)}{ad-c^2}, \quad \lambda = -\dfrac{f+2g}{b}, \quad \nu = \dfrac{2(gc-ah)}{ad-c^2}.$

It is readily seen that an axisymmetric particle freely suspended in a fluid rotates with the same angular velocity as the fluid in a shear flow at the infinity.

Consider now an ellipsoidal particle with semi-axes a_1, a_2 and a_3. Put the origin 0 into the center of the ellipsoid and direct the basis vectors $\{e_i\}$ along the axes of the ellipsoid. Using the corresponding tensors $\boldsymbol{K}^t$, $\boldsymbol{K}^r_R$, $\boldsymbol{K}^c_R$ and the parameters a_1, a_2 and a_3 (see Section 2.1), we get

$$\|\boldsymbol{\Psi}\| = \frac{16\pi}{3}\left\|\begin{matrix} 0 & 0 & 0 & 0 & 0 & 0 \\ 0 & 0 & 0 & 0 & 0 & 0 \\ 0 & 0 & 0 & 0 & 0 & 0 \\ 0 & 0 & 0 & \dfrac{a_3^2-a_2^2}{a_3^2\alpha_3+a_2^2\alpha_2} & 0 & 0 \\ 0 & 0 & 0 & 0 & \dfrac{a_1^2-a_3^2}{a_1^2\alpha_1+a_3^2\alpha_3} & 0 \\ 0 & 0 & 0 & 0 & 0 & \dfrac{a_2^2-a_1^2}{a_2^2\alpha_2+a_1^2\alpha_1} \end{matrix}\right\|. \tag{2.56}$$

Similar relations for a spherical particle could be obtained from Eq. (2.42) and Eq. (2.56) by taking $a_1 = a_2 = a_3 = a$. For circular disk, we would have to take $a_1 = a_2$, $a_3 = 0$.

As an example, consider a circular disk freely suspended in a fluid. The origin of coordinates is placed into center of the disk, the basis $\boldsymbol{i}_1$, $\boldsymbol{i}_2$, $\boldsymbol{i}_3$ is fixed in space, and the unperturbed fluid velocity is given by

$$\boldsymbol{v}^\infty = \frac{1}{2}\dot{\gamma}(x_1\boldsymbol{i}_1 + x_2\boldsymbol{i}_2 - 2x_3\boldsymbol{i}_3).$$

For such flow, we have 1

$$\boldsymbol{v}^0 = 0, \quad \Omega^\infty = 0, \quad \boldsymbol{E}^\infty = -\frac{1}{2}\dot{\gamma}(\boldsymbol{I} - 3\boldsymbol{i}_3\boldsymbol{i}_3).$$

By determining the matrices $\|\mathbf{K}_0\|^{-1}$ and $\|\Phi\|$, one can convince himself that $U_0 = 0$ (the translational velocity vanishes) and that the angular velocity is

$$\boldsymbol{\Omega} = S_4\boldsymbol{e}_1 - S_5\boldsymbol{e}_2.$$

The basis vector $\boldsymbol{e}_3$ of the local coordinate system is normal to the plane of the disk. If ϑ is the angle between the vector $\boldsymbol{i}_3$ and the plane of the disk, and φ is the angle between the projection of $\boldsymbol{i}_3$ on that plane and one of the local axes, then

$$S_4 = E_{23} = \frac{3}{2}\dot{\gamma}\sin\vartheta\cos\vartheta\sin\varphi, \quad S_5 = E_{13} = \frac{3}{2}\dot{\gamma}\sin\vartheta\cos\vartheta\cos\varphi.$$

2.3
Motion of Two Particles in a Fluid

Consider slow motion of two particles a and b in an unbounded viscouss incompressible fluid. The particles' translational and rotational velocities are, respectively, $\boldsymbol{U}_a$, $\mathbf{U}_b$ and $\boldsymbol{\Omega}_a$, $\boldsymbol{\Omega}_b$. The Reynolds numbers of the particles are determined by their parameters and are assumed to be small, so the fluid flow is Stokesian and can be described by Stokes equations

$$\nabla \cdot \boldsymbol{v} = 0, \quad \nabla^2 \boldsymbol{v} = \frac{1}{\mu_e} \nabla p \tag{2.57}$$

with the boundary conditions

$$\begin{aligned} &\boldsymbol{v} = \boldsymbol{U}_a + \boldsymbol{\Omega}_a \times \boldsymbol{r}_a \quad \text{on} \quad S_a, \quad \boldsymbol{v} = \boldsymbol{U}_b + \boldsymbol{\Omega}_b \times \boldsymbol{r}_b \quad \text{on} \quad S_b, \\ &\boldsymbol{v} \to \boldsymbol{v}^{\infty} \quad \text{at} \quad |\boldsymbol{r}_0| \to \infty. \end{aligned} \tag{2.58}$$

Here $\boldsymbol{r}_0$ is the radius vector of the considered point with respect to a local origin 0 that is fixed inside the fluid; $\boldsymbol{r}_\mathrm{a}$ and $\boldsymbol{r}_\mathrm{b}$ are the radius vectors of this point with respect to the centers 0_a and 0_b of particles a and b; $\boldsymbol{v}^{\infty} = \boldsymbol{v}^0 + \Omega^{\infty} \times \boldsymbol{r}_0 + \boldsymbol{E}^{\infty} \cdot \boldsymbol{r}_0$ is the velocity of the fluid far from both the particles (at the infinity).

Exact, approximate, and numerical solutions have been derived for slow translational and rotational motions of two spherical particles in a quiescent fluid and in a shear flow [7–31]. When one particle is much larger then the other, the motion of the smaller particle in close proximity to the larger one can be treated as motion near a plane.

Particle motions in a quiescent fluid ($\boldsymbol{v}^{\infty} = 0$) and in a moving flow ($\boldsymbol{v}^{\infty} \neq 0$) are considered separately.

2.3.1
Fluid is at Rest at the Infinity ($\mathbf{v}^{\infty} = 0$)

The problem of obtaining the fields of velocity $\boldsymbol{v}$ and pressure p in the fluid that are induced by translational and rotational motions of two spherical particles of radii a and b reduces to the following boundary value problem:

$$\begin{aligned} &\nabla \cdot \boldsymbol{v} = 0, \quad \nabla^2 \boldsymbol{v} = \frac{1}{\mu_e} \nabla p \\ &\boldsymbol{v} = \boldsymbol{U}_a + \boldsymbol{\Omega}_a \times \boldsymbol{r}_a \quad \text{on} \quad S_a, \quad \boldsymbol{v} = \boldsymbol{U}_b + \boldsymbol{\Omega}_b \times \boldsymbol{r}_b \quad \text{on} \quad S_b, \quad \text{at} \quad |\boldsymbol{r}_0| \to \infty. \end{aligned} \tag{2.59}$$

Thanks to its linearity, it can be treated as a superposition of simpler problems. If the vectors of translational velocity are represented in terms of two components – along the line of centers and perpendicular to it, the problem is separated into three particular problems: translational motion of particles with different velocities along their line of centers (axisymmetric problem); translational motion of particles with different velocities perpendicular to the line of centers (non-axisymmetric problem); and rotation of particles with different angular velocities. Each of these motions, in its turn, can be divided into two submotions where one of the particles is at rest. It should be noted that by setting the radius of one of the particles to infinity one obtains the solution for the problem of particle motion near a plane wall, which can be considered as an asymptotic solution of the problem of relative motion of two particles with substantially different sizes.

If the particles are drops of fluid whose viscosity differs from that of the surrounding fluid, then particle rotation is a meaningless term, and we should assume $\Omega_a = \Omega_b = 0$. Boundary conditions on the particle surface differ from (2.59), because we must satisfy the condition of equality of velocities and stresses at the interface between the internal and external fluids. As opposed to the surface of a solid particle, drop surface deform under the action of nonuniform stresses, which become especially noticeable when the minimal clearance between drops becomes much smaller than the drop size. All of this greatly complicates the problem. The discussion that follows will apply to solid particles only.

Exact and approximate solutions exist for the three above-mentioned problems, making it possible to determine the forces and torques acting on moving particles for two limiting cases: when the particles are relatively far from each other (the "far asymptotic region"), and when the minimal clearance between particle surfaces is small compared to particle radii (the "near asymptotic region"). An approximate solution suitable for all interparticle distances can be found by using the method of asymptotic expansions.

To derive the exact solution, we introduce the stream function, which allows to make a transition from equations (2.57) to a single fourth order partial differential equation for the stream function. In a curvilinear bi-spherical coordinate system, the region between the two spheres transforms into a region between two parallel plates, and the new boundary value problem is solved by using the method of separation of variables. The variables are given by infinite series whose convergence rate decreases as the gap between particles gets smaller. This is why using an asymptotic solution is the preferred way of finding the hydrodynamic forces and torques acting on a sphere in the vicinity of another sphere (or a plane wall).

In the case of more then two particles, as well as for two particles of non-spherical shape, it is impossible to find the appropriate coordinate system. This is why an exact solution exists only for a system of two spherical particles.

The distinctive feature of the solution for two particles is the singularity of some elements of the resistance matrix (resistance factors) when the minimal interparticle distance tends to zero. This means that most of the influence particles exert on each other is confined to the near asymptotic region. This is why hydrodynamic forces and torques (together with molecular and electrostatic interparticle forces) play a key role in the phenomena of aggregation and coagulation of particles.

When the particles are relatively far from each other, that is, when the interparticle distance is equal to several particle radii, elements of the resistance tensor are nearly equal to the corresponding values for an isolated particle. To derive hydrodynamic forces and torques for the case when the particles are relatively far from each other, we can use the regular method of successive approximations, which allows to solve the boundary value problem (2.59) to any approximation by considering at each step only the boundary conditions on one of the spheres. This method is called the reflection method. It was first applied by Smoluchowski to the problem of hydrodynamic interaction of n particles separated by large distances.

Let us look at translational motion of n identical spherical particles with different velocities $\boldsymbol{U}_k (k = a, b, c, \ldots, n)$. Our goal is to find the forces $\boldsymbol{F}_k$, and torques $\boldsymbol{L}_k$ acting on the particles and ensuring their translational motion with the given velocities $\boldsymbol{U}_k$. The boundary value problem reduces to Eqs. (2.50) with the boundary conditions

$$\boldsymbol{v} = \boldsymbol{U}_k \quad \text{on} \quad S_k, \quad (k = a, b, c, \ldots, n); \quad \boldsymbol{v} \rightarrow 0 \quad \text{at} \quad r \rightarrow \infty.$$

Linearity of the problem allows us to seek the solution in the form of infinite sums:

$$\boldsymbol{v} = \sum_{i=1}^{\infty} \boldsymbol{v}^{(i)}, \quad p = \sum_{i=1}^{\infty} p^{(i)},$$

where each term $(\boldsymbol{v}^{(i)}, p^{(i)})$ obeys Eqs. (2.57) and the zero boundary conditions at the infinity and in its turn is represented as a finite sum of fields $(\boldsymbol{v}_k^{(i)}, p_k^{(i)})$,

$$\boldsymbol{v}^{(i)} = \sum_{k=b}^{n} \boldsymbol{v}_k^{(i)}, \quad p^{(i)} = \sum_{k=b}^{n} p_k^{(i)},$$

also obeying Eqs. (2.57) and the zero boundary conditions at the infinity. Let now choose some particle, for example particle a, as the test particle and obtain the field $(\boldsymbol{v}^{(1)}, p^{(1)})$ that satisfies the boundary condition $\boldsymbol{v}^{(1)} = \boldsymbol{U}_a$ on S_a. The reflection of this field from particle b is derived by using the boundary condition $\boldsymbol{v}^{(2)} = \boldsymbol{U}_b - \boldsymbol{v}^{(1)}$ on S_b. In general, the reflection of $\boldsymbol{v}^{(1)}$ from the other $n-1$ particles is $\boldsymbol{v}_k^{(2)} = \boldsymbol{U}_k - \boldsymbol{v}^{(1)}$, where $k = b, c, \ldots, n$. Thus the total reflection of the velocity field $\boldsymbol{v}^{(1)}$ of particle S_a from the other $n-1$ particles is

$$\boldsymbol{v}^{(2)} = \sum_{k=b}^{n} \boldsymbol{v}_k^{(2)}.$$

It is now possible to estimate (to the first approximation) the influence of $(n-1)$ particles on the test particle via the velocity $\boldsymbol{v}^{(3)}$ resulting from the reflection of the field $\boldsymbol{v}^{(2)}$ from this particle. The field $\boldsymbol{v}^{(3)}$ obeys the boundary condition $\boldsymbol{v}^{(3)} = -\boldsymbol{v}^{(2)}$ on S_a. Therefore terms of the order a/l, where l is the distance between particle centers, appear in the expression for $\boldsymbol{v}^{(3)}$. To summarize, the velocity in the vicinity of particle a is determined up to the terms of order a/l:

$$\boldsymbol{v} \approx \boldsymbol{v}^{(1)} + \boldsymbol{v}^{(3)}.$$

In the same way, one could find velocity to the first approximation for any other particle chosen as the test particle.

The next approximation is derived by continuing the process of reflection, taking the field $\boldsymbol{v}^{(3)}$ instead of $\boldsymbol{v}^{(1)}$. Reflection $\boldsymbol{v}_b^{(4)}$ from particle b, that is, from S_b, is equal to $\boldsymbol{v}_b^{(4)} = -\boldsymbol{v}^{(3)}$, and so on. The net reflection is

$$\boldsymbol{v}^{(4)} = \sum_{k=b}^{n} \boldsymbol{v}_k^{(4)}.$$

The second approximation of the velocity field $\boldsymbol{v}^{(5)}$ in the neighborhood of particle a satisfies the boundary condition $\boldsymbol{v}^{(5)} = -\boldsymbol{v}^{(4)}$ on S_a and ensures accuracy to the order of a^2/l^2. The next approximations give the velocity field as a power series of a/l.

If the particles are far away from each other, one can get a good approximation by assuming the velocity field induced in the fluid by a particle to be the same as the velocity field induced by a point force and (or) by a point-pair force applied at the particle center. Then the resistance force exerted on the given particle by the reflected field can be obtained by considering this field as equivalent to a uniform velocity field with the same magnitude and direction as the velocity that would exist at the particle's location if the particle itself were absent.

The obtained approximations of velocity and pressure fields make it possible to find the force and torque on the particle from Eqs. (2.14) and (2.15):

$$\boldsymbol{F}_a = \boldsymbol{F}_a^{(1)} + \boldsymbol{F}_a^{(3)} + \boldsymbol{F}_a^{(5)} + \ldots; \quad \boldsymbol{L}_a = \boldsymbol{L}_a^{(1)} + \boldsymbol{L}_a^{(3)} + \boldsymbol{L}_a^{(5)} + \ldots.$$

In the case of two particles a and b separated by a large distance, forces and torques have the form of power series in l/a and l/b.

Another limiting solution can be derived in the case when the gap between the surfaces is small as compared to the particles' radii a and b. In this case there appears a small parameter equal to the ratio of the minimum clearance to the particle radius ($\ll 1$ and $\ll k = b/a$, $b \leq a$), which allows to obtain a solution by employing the method of asymptotic expansions in a local cylindrical coordinate system (ar, θ, az) with the origin at the intersection of the line of centers and the spherical surface S_b and the z-axis directed along the line of centers, while the r-axis lies in the plane tangential to S_b. In this coordinate system, the equations of spheres S_a and S_b become

$$z = 1 + \varepsilon - (1 - r^2)^{1/2}, \quad z = -k^{-1} + k^{-1}(1 - k^2 r^2)^{1/2}.$$

If we now introduce new scaled coordinates $Z = r/\varepsilon$ and $R = r/\varepsilon^{1/2}$ instead of z and r, the equations of the spheres assume an asymptotic form:

$$Z = 1 + 0.5R^2 + 0(\varepsilon), \quad Z = 0.5kR^2 + 0(\varepsilon).$$

Stokes equations and boundary conditions at particle surfaces can be transformed in a similar way, after which we can extract the principal terms in asymptotic expansions of the velocity and pressure fields. The concrete form of these expansions will depend on the type of particle motion: translational motion parallel or perpendicular to the line of centers, or rotational motion around these axes.

A distinctive feature of near asymptotic expansions is the singularity of almost all resistance coefficients in the grand resistance matrix, with the exception of coefficients corresponding to the rotation of the particles around their line of centers.

Taking together the far asymptotic and near asymptotic solutions, we can use the method of matching of asymptotic expansions to get approximate solutions for all kinds of particle motions for the entire range of relative distances between the particles. In doing that, we make use of the fact that due to the linearity of Stokes equations, the general equation describing translational motion with velocities $\boldsymbol{U}_a$ and $\boldsymbol{U}_b$ and rotational motion with angular velocities Ω_a and Ω_b can be thought of as a superposition of simple motions in which one of the particles is assumed to be at rest. It was shown in Section 2.1 that forces $\boldsymbol{F}_a$ and torques $\boldsymbol{L}_a$ acting on the particles $\alpha = 1, 2$ (here and later, particles will be indicated by indices 1 and 2) are connected with their velocities through the relation

$$\left\| \begin{matrix} \boldsymbol{F}_1 \\ \boldsymbol{F}_2 \\ \boldsymbol{L}_1 \\ \boldsymbol{L}_2 \end{matrix} \right\| = \left\| \begin{matrix} K_{11}^t & K_{12}^t & (K_{11}^c)^T & (K_{12}^c)^T \\ K_{21}^t & K_{22}^t & (K_{21}^c)^T & (K_{22}^c)^T \\ K_{11}^c & K_{12}^c & K_{11}^r & K_{12}^r \\ K_{21}^c & K_{22}^c & K_{21}^r & K_{21}^r \end{matrix} \right\| \left\| \begin{matrix} \boldsymbol{U}_1 \\ \boldsymbol{U}_2 \\ \Omega_1 \\ \Omega_2 \end{matrix} \right\|. \tag{2.60}$$

The grand resistance matrix contains tensors $\boldsymbol{K}_{ab}^t$, $\boldsymbol{K}_{ab}^c$ and $\boldsymbol{K}_{ab}^r$, whose elements $\boldsymbol{K}_{ij}^{(ab)}$ (where a, $b = 1, 2$ are used to index the particles and i, $j = 1, 2$ indicate the tensor elements) obey the conditions of symmetry, some of which are valid for particles of an arbitrary shape and some follow from the geometrical properties of a system of two spherical particles. The former type of symmetry conditions follows from the reciprocity theorem (see Section 2.1):

$$K_{ij}^{(ab)t} = K_{ji}^{(ab)t}, \quad (K_{ij}^{(ab)c})^T = K_{ji}^{(ba)c}; \quad K_{ij}^{(ab)r} = K_{ji}^{(ba)r}. \tag{2.61}$$

Let us prove the first condition. Consider two problems that involve translational motion of two spherical particles S_1 and S_2 in one and the same fluid. In the first problem particle S_1 moves with the constant velocity $\boldsymbol{U}_1$ and the second particle is at rest: $\boldsymbol{U}_2 = 0$. In the second problem, the opposite is true: the first particle is at rest, $\boldsymbol{U}_1 = 0$, and the second one moves with the constant velocity $\boldsymbol{U}_2$. The velocity and stress fields in the two problems are designated as ($\boldsymbol{v}'$, $\boldsymbol{T}'$) and ($\boldsymbol{v}''$, $\boldsymbol{T}''$). From the reciprocity theorem (2.2), there follows that at $\mu_e' = \mu_e''$,

$$\int_S ds \cdot \boldsymbol{v}' \cdot \boldsymbol{T}'' = \int_S ds \cdot \boldsymbol{v}'' \cdot \boldsymbol{T}'. \tag{2.62}$$

Form the surface S from the surface of a sphere S_∞ with a large radius and the surfaces of two spheres S_1 and S_2 placed inside the sphere S_∞. Since $ds \sim r^{-2}$, while $\boldsymbol{v}' \sim r^{-1}$, $\boldsymbol{v}'' \sim r^{-1}$, and $T \sim r^{-2}$, we have for $r \to \infty$:

$$\int_S = \int_{S_\infty} + \int_{S_1} + \int_{S_2} \sim \int_{S_1} + \int_{S_2}. \tag{2.63}$$

The following boundary conditions are valid at the surfaces S_1 and S_2:

$$\boldsymbol{v}' = \boldsymbol{U}_1, \quad \boldsymbol{v}'' = 0 \quad \text{on} \quad S_1; \quad \boldsymbol{v}' = 0, \quad \boldsymbol{v}'' = \boldsymbol{U}_2 \quad \text{on} \quad S_2,$$

Therefore the relations (2.62) and (2.63) first reduce to

$$\boldsymbol{U}_1 \cdot \int_S d\boldsymbol{s} \cdot \boldsymbol{T}'' = \boldsymbol{U}_2 \cdot \int_S d\boldsymbol{s} \cdot \boldsymbol{T}'$$

and then, due to Eq. (2.14), to

$$\boldsymbol{U}_1 \cdot \boldsymbol{F}_1^{(2)} = \boldsymbol{U}_2 \cdot \boldsymbol{F}_2^{(1)}. \tag{2.64}$$

where $\boldsymbol{F}_1^{(2)}$ is the resistance force acting on particle S_1 and induced by the motion of particle S_2, and $\boldsymbol{F}_2^{(1)}$ – the resistance force acting on particle S_2 and induced by the motion of particle S_1. From Eq. (2.60), it follows that in the case of translational motion ($\Omega_1 = \Omega_2 = 0$) the forces acting on the particles S_1 and S_2 are

$$\boldsymbol{F}_1^{(2)} = \boldsymbol{K}_{12}^t \cdot \boldsymbol{U}_2, \quad \boldsymbol{F}_2^{(1)} = \boldsymbol{K}_{21}^t \cdot \boldsymbol{U}_1. \tag{2.65}$$

A substitution of Eq. (2.65) into Eq. (2.64) yields

$$\boldsymbol{U}_1 \cdot \boldsymbol{K}_{12}^t \cdot \boldsymbol{U}_2 = \boldsymbol{U}_2 \cdot \boldsymbol{K}_{21}^t \cdot \boldsymbol{U}_1. \tag{2.66}$$

Using the properties of the scalar product of a tensor and a vector, one obtains:

$$\boldsymbol{U}_2 \cdot (\boldsymbol{K}_{12}^t \cdot \boldsymbol{U}_1) = \boldsymbol{U}_2 \cdot (\boldsymbol{U}_1 \cdot (\boldsymbol{K}_{21}^t)^T) = \boldsymbol{U}_1 \cdot (\boldsymbol{K}_{21}^t)^T \cdot \boldsymbol{U}_2.$$

and hence,

$$\boldsymbol{U}_1 \cdot \boldsymbol{K}_{12}^t \cdot \boldsymbol{U}_2 = \boldsymbol{U}_1 \cdot (\boldsymbol{K}_{21}^t)^T \cdot \boldsymbol{U}_2$$

and

$$\boldsymbol{K}_{21}^t = (\boldsymbol{K}_{21}^t)^T. \tag{2.67}$$

This means that the tensor $\boldsymbol{K}_{21}^t$ is symmetric, and its elements $\boldsymbol{K}_{ij}^{(ab)t}$ obey the conditions

$$\boldsymbol{K}_{ij}^{(ab)t} = \boldsymbol{K}_{ji}^{(ba)t}.$$

The other relations in (2.61) can be proved in the same way.

Geometry of the system of two spherical particles gives rise to a dependence of tensors $\boldsymbol{K}_{ab}$ on the relative position of the particles, that is, on the vector $\boldsymbol{r}=\boldsymbol{x}_2-\boldsymbol{x}_1$ connecting the particle centers $\boldsymbol{x}_1$ and $\boldsymbol{x}_2$, and on the particle radii a_1 and a_2. It is obvious that physical essence of the problem does not change if we interchange the particle indices 1 and 2 and replace $\boldsymbol{r}$ with $-\boldsymbol{r}$. So, the conditions

$$\boldsymbol{K}_{ab}(\boldsymbol{r}, a_1, a_2) = \boldsymbol{K}_{(3-a,3-b)}(-\boldsymbol{r}, a_2, a_1). \tag{2.68}$$

must hold.

Introduce a new coordinate system with the basis vectors $\boldsymbol{e}=\boldsymbol{r}/r$ (parallel to the line of centers) and $\boldsymbol{i}, \boldsymbol{j}$ (perpendicular to each other and to the line of centers). Then the symmetry conditions (2.68) allow us to express components of the tensor $\boldsymbol{K}_{ab}$ in terms of two scalar functions X_{ab} and Y_{ab}:

$$\begin{aligned}
K_{ij}^{(ab)t} &= X_{ab}^{t} e_i e_j + Y_{ab}^{t}(\delta_{ij} - e_i e_j), \\
K_{ij}^{(ab)c} &= (K_{ji}^{(ab)c})^T = Y_{ab}^{c} \varepsilon_{ijk} e_k, \\
K_{ij}^{(ab)r} &= X_{ab}^{r} e_i e_j + Y_{ab}^{r}(\delta_{ij} - e_i e_j).
\end{aligned} \tag{2.69}$$

where δ_{ij} is the Kronecker symbol and ε_{ijk} – the permutation symbol.

Let us replace dimensional tensors $\boldsymbol{K}_{ab}$ with dimensionless ones:

$$\tilde{K}_{ab}^{t} = \frac{K_{ab}^{t}}{3\pi(a_a + a_b)}, \quad \tilde{K}_{ab}^{c} = \frac{K_{ab}^{c}}{3\pi(a_a + a_b)^2}, \quad \tilde{K}_{ab}^{r} = \frac{K_{ab}^{r}}{3\pi(a_a + a_b)^3},$$

Also, we introduce the dimensionless distance s between the particles and the ratio k of particle radii:

$$s = \frac{2r}{a_1 + a_2}, \quad k = \frac{a_2}{a_1}.$$

The symmetry conditions (2.61) and (2.68) will then rearrange to

$$\begin{aligned}
\tilde{X}_{ab}^{t}(s,k) &= \tilde{X}_{ba}^{t}(s,k) = \tilde{X}_{(3-a,3-b)}^{t}(s,k^{-1}), \\
\tilde{Y}_{ab}^{t}(s,\lambda) &= \tilde{Y}_{ba}^{t}(s,k) = \tilde{Y}_{(3-a,3-b)}^{t}(s,k^{-1}), \\
\tilde{Y}_{ab}^{c}(s,k) &= -\tilde{Y}_{(3-a,3-b)}^{t}(s,k^{-1}), \\
\tilde{X}_{ab}^{r}(s,k) &= \tilde{X}_{ba}^{r}(s,k) = \tilde{X}_{(3-a,3-b)}^{r}(s,k^{-1}), \\
\tilde{Y}_{ab}^{r}(s,k) &= \tilde{Y}_{ba}^{r}(s,k) = \tilde{Y}_{(3-a,3-b)}^{r}(s,k^{-1}).
\end{aligned} \tag{2.70}$$

The conditions (2.70) mean that the forces and torques acting on the particles are specified by 16 scalar functions (4 functions for $\tilde{Y}_{ab}^{c}$ and 3 functions for each of the following: $\tilde{X}^t_{ab}$, $\tilde{Y}^t_{ab}$, $\tilde{X}^r_{ab}$, $\tilde{Y}^r_{ab}$) that depend on $s \in [2,\infty)$ and $k \in [0,1]$. Each set of functions defined by superscripts t, c and r is responsible for a specific type of particle motion. Thus, $\tilde{X}^t_{ab}$ are the resistance coefficients for translational motion of particles along their line of centers; $\tilde{Y}^t_{ab}$ – resistance coefficients for translational motion perpendicular to the line of centers; $\tilde{Y}^c_{ab}$ – resistance coefficients for rotational motion around the axis $\boldsymbol{j}$; $\tilde{X}^r_{ab}$ – for rotational motion around the line of centers (axis $\boldsymbol{e}$); $\tilde{Y}^r_{ab}$ – for rotational motion around the axis $\boldsymbol{i}$.

Thus, scalar functions provide a suitable description of all kinds of relative motions of two solid spherical particles. Therefore, Eq. (2.60) allows to determine the forces and torques that act on particles moving in an arbitrary way as functions of translational and rotational velocities $\boldsymbol{U}_a$ and $\boldsymbol{\Omega}_a$, relative distance s between the particles, and the ratio k of particle radii.

Both exact and asymptotic solutions of all the above-formulated problems about motion of two spherical particles in an unbounded fluid that is quiescent at the infinity are currently well-known. Of utmost interest are the approximate solutions obtained by splicing the far asymptotic solution and the near asymptotic solution because their relatively simple form comes in handy when we try to solve more complicated problems that involve, for example, the hydrodynamic behavior of a particle ensemble. Listed below are the expressions for dimensionless resistance functions that appear in Eq. (2.60), derived for various types of particle motions.

Translational motion along the line of centers (axis $\boldsymbol{e}$):

Far asymptotic region ($s \to \infty$):

$$\tilde{X}^t_{11} = \sum_{k=0}^{\infty} f_{2k}(k) \frac{1}{(1+k)^{2k}} \frac{1}{s^{2k}},$$

$$\tilde{X}^t_{12} = -\frac{1}{(1+k)} \sum_{k=0}^{\infty} f_{2k+1}(k) \frac{1}{(1+k)^{2k+1}} \frac{1}{s^{2k+1}},$$

where

$$f_0 = 1;\ f_1 = 3k;\ f_2 = 9k;\ f_3 = -4k + 27k^2 - 4k^3;$$

$$f_4 = -24k + 81k^2 + 36k^3;\ f_5 = 72k^2 + 243k^3 + 72k^4;$$

$$f_6 = 16k + 108k^2 + 281k^3 + 648k^4 + 144k^5;$$

$$f_7 = 288k^2 + 1620k^3 + 1515k^4 + 1629k^5 + 288k^6;\ \text{etc.}$$

Owing to the first condition (2.70), the following equality is valid:

$$f_{2k+1}(k) = k^{2k+2} f_{2k+1}(k^{-1}).$$

Near asymptotic region ($s \to 2$ or $\xi \to 0$):

It is convenient to introduce the dimensionless distance (clearance) between the particles

$$\xi = \frac{r - a_1 - a_2}{0.5(a_1 + a_2)} = s - 2.$$

At $\xi \ll 1$ and $\xi \ll k$ we have:

$$\tilde{X}^t_{11} = g_1(k)\frac{1}{\xi} + g_2(k)\ln\left(\frac{1}{\xi}\right) + 0(1),$$

$$\tilde{X}^t_{11} = -\frac{2}{1+k}\left[g_1(k)\frac{1}{\xi} + g_2(k)\ln\left(\frac{1}{\xi}\right)\right] + 0(1),$$

where

$$g_1(k) = \frac{2k^2}{(1+k)^3}; \quad g_2(k) = \frac{k(1+7k+k^2)}{5(1+k)^3}.$$

Intermediate region $(2 < s < \infty)$:

$$\tilde{X}^t_{11} = -\frac{g_1}{(1-4s^{-2})} - g_2 \ln(1-4s^{-2}) - g_3(1-4s^{-2})\ln(1-4s^{-2})$$

$$+ f_0(k) - g_1 + \sum_{\substack{m=2 \\ (m-even)}}^{\infty} \left\{\frac{1}{2^m(1+k)^m} f_m(k) - g_1 - \frac{2g_2}{m} + \frac{4g_3}{m\, m_1}\right\}\left(\frac{2}{s}\right)^m;$$

$$-\frac{1}{2}(1+k)\tilde{X}^t_{12} = \frac{2g_1}{s(1-4s^{-2})} - g_2 \ln\left(\frac{s+2}{s-2}\right) + g_3(1-4s^{-2})\ln\left(\frac{s+2}{s-2}\right)$$

$$+4\frac{g_3}{s} + \sum_{\substack{m=2 \\ (m-even)}}^{\infty} \left\{\frac{1}{2^m(1+k)^m} f_m(k) - g_1 - \frac{2g_2}{m} + \frac{4g_3}{m\, m_1}\right\}\left(\frac{2}{s}\right)^m,$$

where

$$m_1 = -2\delta_{m2} + (m-2)(1-\delta_{m2}),$$

$$g_3 = \frac{1}{42}(1 + 18k - 29k^2 + 18k^3 k^4)(1+k)^{-3}.$$

*Translational motion perpendicular to the line of centers (axis **i**):*
Far asymptotic region $(s \to \infty)$:

$$\tilde{Y}^t_{11} = \sum_{k=0}^{\infty} f_{2k}(k)\frac{1}{(1+k)^{2k}}\frac{1}{s^{2k}},$$

$$\tilde{Y}^t_{12} = -\frac{2}{(1+k)}\sum_{k=0}^{\infty} f_{2k+1}(k)\frac{1}{(1+k)^{2k+1}}\frac{1}{s^{2k+1}},$$

where

$$f_0 = 1; \quad f_1 = \frac{3}{2}k; \quad f_2 = \frac{9}{4}k; \quad f_3 = 2k + \frac{27}{8}k^2 + 2k^3;$$

$$f_4 = 6k + \frac{81}{16}k^2 + 18k^3; \quad f_5 = \frac{63}{2}k^2 + \frac{243}{32}k^3 + \frac{63}{2}k^4;$$

$$f_6 = 4k + 54k^2 + \frac{1241}{64}k^3 + 81k^4 + 72k^5;$$

$$f_7 = 144k^2 + \frac{1053}{8}k^3 + \frac{19083}{128}k^4 + \frac{1053}{8}k^5 + 144k^6; \quad \text{etc.}$$

Near asymptotic region ($\xi \ll 1$ *and* $\xi \ll k$):

$$\tilde{Y}^t_{11} = g_2(k)\ln\left(\frac{1}{\xi}\right) + 0(1),$$

$$\tilde{Y}^t_{12} = -\frac{2}{1+k}\left[g_2(k)\ln\left(\frac{1}{\xi}\right)\right] + 0(1),$$

where

$$g_2(k) = \frac{4k(2+k+2k^2)}{15(1+k)^3}.$$

Intermediate region ($2 < s < \infty$):

$$\tilde{Y}^t_{11} = -g_2\ln(1-4s^{-2}) - g_3(1-4s^{-2})\ln(1-4s^{-2}) + f_0(k)$$

$$+\sum_{\substack{m=2\\(m-\text{even})}}^{\infty}\left\{\frac{1}{2^m(1+k)^m}f_m(k) - \frac{2g_2}{m} + \frac{4g_3}{m\,m_1}\right\}\left(\frac{2}{s}\right)^m,$$

$$-\frac{1}{2}(1+k)\tilde{Y}^t_{12} = g_2\ln\left(\frac{s+2}{s-2}\right) + g_3(1-4s^{-2})\ln\left(\frac{s+2}{s-2}\right)$$

$$+4\frac{g_3}{s} + \sum_{\substack{m=2\\(m-\text{odd})}}^{\infty}\left\{\frac{1}{2^m(1+k)^m}f_m(k) - \frac{2g_2}{m} + \frac{4g_3}{m\,m_1}\right\}\left(\frac{2}{s}\right)^m,$$

where

$$g_3(k) = \frac{2}{375}(16 - 45k + 58k^2 - 457k^3 + 16k^4)(1+k)^{-3}.$$

Rotation around the axis perpendicular to the line of centers and to the direction of transverse translational motion (axis $\mathbf{j}$*):*
Far asymptotic region ($s \to \infty$):

$$\tilde{Y}_{11}^{c} = \sum_{k=0}^{\infty} f_{2k+1}(k)\frac{1}{(1+k)^{2k+1}}\frac{1}{s^{2k+1}},$$

$$\tilde{Y}_{12}^{c} = -\frac{4}{(1+k)^{2}}\sum_{k=0}^{\infty} f_{2k}(k)\frac{1}{(1+k)^{2k}}\frac{1}{s^{2k}},$$

where

$$f_0 = f_1 = 0;\ f_2 = -6k;\ f_3 = -9k;$$

$$f_4 = -\frac{27}{2}k^2;\quad f_5 = -12k - \frac{81}{4}k^2 - 36k^3;$$

$$f_6 = -108k^2 - \frac{243}{8}k^3 - 72k^4;$$

$$f_7 = -189k^2 + \frac{8409}{16}k^3 - 243k^4 - 144k^5;\quad \text{etc.}$$

Near asymptotic region ($\xi \ll 1$ *and* $\xi \ll k$):

$$\tilde{Y}_{11}^{c} = g_2(k)\ln\left(\frac{1}{\xi}\right) + 0(1),$$

$$\tilde{Y}_{12}^{c} = -\frac{2}{(1+k)^2}\left[g_2(k)\ln\left(\frac{1}{\xi}\right)\right] + 0(1),$$

where

$$g_2(k) = -\frac{1}{5}\frac{k(4+k)}{(1+k)^2}.$$

Intermediate region ($2 < s < \infty$):

$$\tilde{Y}_{11}^{c} = g_2\ln\left(\frac{s+2}{s-2}\right) + g_3(1-4s^{-2})\ln\left(\frac{s+2}{s-2}\right) + 4\frac{g_3}{s}$$
$$+\sum_{\substack{m=2\\(m-odd)}}^{\infty}\left\{\frac{1}{2^m(1+k)^m}f_m(k) - \frac{2g_2}{m} + \frac{4g_3}{m\,m_1}\right\}\left(\frac{2}{s}\right)^m,$$

$$-\frac{1}{4}(1+k)^2\tilde{Y}_{12}^{c} = -g_2\ln(1-4s^{-2}) - g_3(1-4s^{-2})\ln(1-4s^{-2})$$
$$+\sum_{\substack{m=2\\(m-even)}}^{\infty}\left\{\frac{1}{2^m(1+k)^m}f_m(k) - g_1 - \frac{2g_2}{m} + \frac{4g_3}{m\,m_1}\right\}\left(\frac{2}{s}\right)^m,$$

where

$$g_2(k) = -\frac{1}{250}\frac{(32-33k+83k^2+43k^3)}{(1+k)^2}.$$

Rotation around the line of centers (axis $\boldsymbol{e}$*):*
Far asymptotic region ($s \to \infty$):

$$\tilde{X}^r_{11} = \sum_{k=0}^{\infty} f_{2k}(k)\frac{1}{(1+k)^{2k}}\frac{1}{s^{2k}},$$

$$\tilde{X}^r_{12} = -\frac{8}{(1+k)^3}\sum_{k=0}^{\infty} f_{2k+1}(k)\frac{1}{(1+k)^{2k+1}}\frac{1}{s^{2k+1}},$$

where

$$f_0 = 1;\; f_1 = f_2 = 0;\; f_3 = 8k^3;\; f^4 = f_5 = 0;$$
$$f_6 = 64k^3;\; f_7 = 0;\; f_8 = 768k^5;\; f_9 = 512k^6;\text{ etc.}$$

Near asymptotic region ($\xi \ll 1$ *and* $\xi \ll \mathrm{k}$):

$$\tilde{X}^r_{11} = \frac{k^3}{(1+k)^3}\zeta\left(3,\frac{k}{1+k}\right) - \frac{k^2}{4(1+k)}\xi\ln\left(\frac{1}{\xi}\right) + 0(\xi),$$

$$\tilde{X}^r_{12} = -\frac{8k^3}{(1+k)^6}\zeta\left(3,\frac{k}{1+k}\right) + \frac{2k^2}{(1+k)^4}\xi\ln\left(\frac{1}{\xi}\right) + 0(\xi),$$

where

$$\zeta(z,a) = \sum_{k=0}^{\infty}(k+a)^{-z}.$$

Intermediate region ($2 < s < \infty$):

$$\tilde{X}^r_{11} = \frac{k^2}{2(1+k)}\ln(1-4s^{-2}) + \frac{k^2}{1+k}\frac{1}{s}\ln\left(\frac{s+2}{s-2}\right) + 1$$
$$+\sum_{k=1}^{\infty}\left\{\frac{1}{(1+k)^{2k}}f_{2k}(k) - 2^{2k+1}\frac{1}{k(2k-1)}\frac{k^2}{4(1+k)}\right\}\left(\frac{1}{s}\right)^{2k},$$

$$+\tilde{X}^r_{12} = \frac{4k^2}{(1+k)^4}\ln\left(\frac{s+2}{s-2}\right) + \frac{8k^2}{(1+k)^4}\ln(1-4s^{-2})$$
$$-\frac{8}{(1+k)^3}\sum_{k=1}^{\infty}\left\{\frac{1}{(1+k)^{2k+1}}f_{2k+1}(k) - 2^{2k+2}\frac{1}{k(2k+1)}\frac{k^2}{1+k}\right\}\left(\frac{1}{s}\right)^{2k+1}.$$

Rotation around the axis perpendicular to the line of centers and parallel to the direction of transverse translational motion (axis **i**):

Far asymptotic region ($s \to \infty$):

$$\tilde{Y}^r_{11} = \sum_{k=0}^{\infty} f_{2k}(k) \frac{1}{(1+k)^{2k}} \frac{1}{s^{2k}},$$

$$\tilde{X}^r_{12} = \frac{8}{(1+k)^3} \sum_{k=0}^{\infty} f_{2k+1}(k) \frac{1}{(1+k)^{2k+1}} \frac{1}{s^{2k+1}},$$

where

$$f_0 = 1; \quad f_1 = f_2 = 0; \quad f_3 = 4k^3; \quad f_4 = 12k; \quad f_5 = 18k^4$$
$$f_6 = 27k^2 + 256k^3; \quad f_7 = 72k^4 + \frac{81}{2}k^5 + 72k^6; \quad \text{etc.}$$

Near asymptotic region ($\xi \ll 1$ *and* $\xi \ll k$):

$$\tilde{Y}^r_{11} = g_2(k) \ln\left(\frac{1}{\xi}\right) + 0(1); \quad \tilde{Y}^r_{12} = g_4(k) \ln\left(\frac{1}{\xi}\right) + 0(1),$$

where

$$g_2(k) = \frac{2k}{5(1+k)}; \quad g_4(\lambda) = \frac{4k^2}{5(1+k)^4}.$$

Intermediate region ($2 < s < \infty$):

$$\tilde{Y}^r_{11} = -g_2 \ln(1-4s^{-2}) - g_3(1-4s^{-2}) \ln(1-4s^{-2}) + f_0(k)$$
$$+ \sum_{\substack{m=2 \\ (m-\text{even})}}^{\infty} \left\{ \frac{1}{2^m (1+k)^m} f_m(k) - \frac{2g_2}{m} + \frac{4g_3}{m\, m_1} \right\} \left(\frac{2}{s}\right)^m,$$

$$\frac{1}{8}(1+k)\tilde{Y}^r_{12} = g_4 \ln\left(\frac{s+2}{s-2}\right) + g_5(1-4s^{-2}) \ln\left(\frac{s+2}{s-2}\right)$$
$$+4\frac{g_5}{s} + \sum_{\substack{m=2 \\ (m-\text{odd})}}^{\infty} \left\{ \frac{1}{2^m (1+k)^m} f_m(k) - \frac{2g_4}{m} + \frac{4g_5}{m\, m_1} \right\} \left(\frac{2}{s}\right)^m,$$

where

$$g_3(k) = \frac{(8+6k+33k^2)}{125(1+k)}; \quad g_5(k) = \frac{4k(43-24k+43k^2)}{125(1+k)^4}.$$

When the forces and torques acting on the particles are known, one can find particle velocities via the mobility matrix:

$$\begin{Vmatrix} \mathbf{U}_1 \\ \mathbf{U}_2 \\ \mathbf{\Omega}_1 \\ \mathbf{\Omega}_2 \end{Vmatrix} = \begin{Vmatrix} k_{11}^t & k_{12}^t & (k_{11}^c)^T & (k_{12}^c)^T \\ k_{21}^t & k_{22}^t & (k_{21}^c)^T & (K_{22}^c)^T \\ k_{11}^c & k_{12}^c & k_{11}^r & k_{11}^r \\ k_{21}^c & k_{22}^c & k_{21}^r & k_{21}^r \end{Vmatrix} \begin{Vmatrix} \mathbf{F}_1 \\ \mathbf{F}_2 \\ \mathbf{L}_1 \\ \mathbf{L}_2 \end{Vmatrix}. \tag{2.71}$$

The mobility matrix is the inverse resistance matrix, so, according of the reciprocity theorem, its elements k_{ab}^t, k_{ab}^c, and k_{ab}^r must obey conditions similar to Eq. (2.61), namely,

$$k_{ij}^{(ab)t} = k_{ji}^{(ab)t}; \quad (k_{ij}^{(ab)c})^T = k_{ji}^{(ba)c}; \quad k_{ij}^{(ab)r} = k_{ji}^{(ba)r}. \tag{2.72}$$

They can also be represented in the form similar to (2.69):

$$\begin{aligned} k_{ij}^{(ab)t} &= x_{ab}^t e_i e_j + y_{ab}^t(\delta_{ij} - e_i e_j), \\ k_{ij}^{(ab)c} &= (k_{ji}^{(ab)c})^T = y_{ab}^c \varepsilon_{ijk} e_k, \\ k_{ij}^{(ab)r} &= x_{ab}^r e_i e_j + y_{ab}^r(\delta_{ij} - e_i e_j). \end{aligned} \tag{2.73}$$

If we now introduce the dimensionless mobility tensors

$$\tilde{k}_{ab}^t = 3\pi(a_a + a_b)k_{ab}^t; \quad \tilde{k}_{ab}^c = \pi(a_a + a_b)^2 k_{ab}^c; \quad \tilde{k}_{ab}^r = \pi(a_a + a_b)^3 k_{ab}^r$$

and parameters s and k in the same way as before, the resulting 16 dimensionless scalar functions $\tilde{x}_{ab}^t, \tilde{y}_{ab}^t, \tilde{y}_{ab}^c, \tilde{x}_{ab}^r$,, and $\tilde{y}_{ab}^r$ depending on $s \in [2, \infty)$ and $k \in [0,1]$ will characterize all kinds of translational and rotational motions of two solid spherical particles in a fluid.

Since the grand resistance and mobility matrices are mutually inverse, their components' representations (2.69) and (2.73) in combination with the fact that axisymmetric translational motion (i.e., along the line of centers) is not mutually correlated with either non-axisymmetric motion (i.e., perpendicular to the line of centers) or axisymmetric rotation (i.e., around the line of centers), lead to the following dependence between the elements of both matrices:

$$\begin{Vmatrix} x_{11}^t & \dfrac{2}{1+k} x_{12}^t \\ \dfrac{2}{1+k} x_{12}^t & \dfrac{1}{k} x_{22}^t \end{Vmatrix} = \begin{Vmatrix} X_{11}^t & \dfrac{1+k}{2} X_{12}^t \\ \dfrac{1+k}{2} X_{12}^t & k X_{22}^t \end{Vmatrix}^{-1},$$

$$\begin{Vmatrix} x_{11}^r & \frac{2}{1+k}x_{12}^r \\ \frac{2}{1+k}x_{12}^r & \frac{1}{k}x_{22}^r \end{Vmatrix} = \begin{Vmatrix} X_{11}^r & \frac{(1+k)^3}{8}X_{12}^r \\ \frac{(1+k)^3}{8}X_{12}^r & kX_{22}^r \end{Vmatrix}^{-1},$$

$$\begin{Vmatrix} y_{12}^t & \frac{2}{1+k}y_{12}^t & \frac{3}{2}y_{11}^c & \frac{6}{(1+k)^2}y_{21}^c \\ \frac{2}{1+k}y_{12}^t & \frac{1}{k}y_{22}^t & \frac{6}{(1+k)^2}y_{12}^c & \frac{3}{2k^2}y_{22}^c \\ \frac{3}{2}y_{11}^c & \frac{6}{(1+k)^2}y_{12}^c & \frac{3}{4}y_{11}^r & \frac{6}{(1+k)^3}y_{12}^r \\ \frac{6}{(1+k)^2}y_{21}^c & \frac{3}{2k^2}y_{22}^c & \frac{6}{(1+k)^2}y_{12}^r & \frac{3}{4k^3}y_{22}^r \end{Vmatrix} =$$

$$= \begin{Vmatrix} Y_{12}^t & \frac{1+k}{2}Y_{12}^t & \frac{2}{3}Y_{11}^c & \frac{(1+k)^2}{6}Y_{21}^c \\ 2(1+k)Y_{12}^t & kY_{22}^t & \frac{(1+k)^3}{6}Y_{12}^c & \frac{2}{3}k^2Y_{22}^c \\ \frac{3}{2}Y_{11}^c & \frac{(1+k)^2}{6}Y_{12}^c & \frac{4}{3}Y_{11}^r & \frac{(1+k)^3}{6}Y_{12}^r \\ \frac{(1+k)^2}{6}Y_{21}^c & \frac{2}{3}k^2Y_{22}^c & \frac{(1+k)^3}{6}Y_{12}^r & \frac{4}{3}k^3Y_{22}^r \end{Vmatrix}^{-1}.$$

2.3.2
Fluid is Moving at the Infinity ($\boldsymbol{v}^\infty \neq 0$)

Let us now consider the motion of two spherical particles in an unbounded fluid whose velocity at the infinity is the one that is typical for a shear flow [32–38]:

$$\boldsymbol{v}^\infty = \boldsymbol{v}^0 + \boldsymbol{\Omega}^\infty \times \boldsymbol{r}_0 + \boldsymbol{E}^\infty \cdot \boldsymbol{r}_0. \tag{2.74}$$

The velocity and pressure fields are described by the boundary value problem (2.57), (2.58):

$$\nabla \cdot \boldsymbol{v} = 0, \quad \nabla^2 \boldsymbol{v} = \frac{1}{\mu_e}\nabla p, \tag{2.75}$$

$$\begin{aligned} &\boldsymbol{v} = \boldsymbol{U}_1 + \boldsymbol{\Omega}_1 \times \boldsymbol{r}_1 \quad \text{on} \quad S_1, \quad \boldsymbol{v} = \boldsymbol{U}_2 + \boldsymbol{\Omega}_2 \times \boldsymbol{r}_2 \quad \text{on} \quad S_2; \\ &\boldsymbol{v} \to \boldsymbol{v}^\infty \quad \text{at} \quad |\boldsymbol{r}_0| \to \infty. \end{aligned} \tag{2.76}$$

where $\boldsymbol{v}^\infty$ is given by Eq. (2.74).

The boundary value problem is handled in the same manner as in the case of single-particle motion (see Section 2.2). Think of the fluid velocity as a sum $\boldsymbol{v}=\boldsymbol{v}^{\infty}+\boldsymbol{v}^{*}$, where $\boldsymbol{v}^{\infty}$, $\boldsymbol{v}^{*}$ satisfy Eqs. (2.75) and $\boldsymbol{v}^{*}$ also obeys the boundary conditions

$$\boldsymbol{v}^{*}=(\boldsymbol{U}_1-\boldsymbol{v}_1^{\infty})+(\boldsymbol{\Omega}_1-\boldsymbol{\Omega}^{\infty})\times\boldsymbol{r}_1-\boldsymbol{E}^{\infty}\cdot\mathbf{r}_1 \quad \text{on} \quad S_1,$$
$$\boldsymbol{v}^{*}=(\boldsymbol{U}_2-\boldsymbol{v}_2^{\infty})+(\boldsymbol{\Omega}_2-\boldsymbol{\Omega}^{\infty})\times\boldsymbol{r}_2-\boldsymbol{E}^{\infty}\cdot\boldsymbol{r}_2 \quad \text{on} \quad S_2,$$

$$\boldsymbol{v}^{*}\to 0 \quad \text{at} \quad |\boldsymbol{r}_0|\to\infty.$$

Here $\boldsymbol{v}_1^{\infty}$ and $\boldsymbol{v}_2^{\infty}$ are the unperturbed velocities at the particle surfaces.
The fluid velocity disturbance is, in its turn, represented as a sum,

$$\boldsymbol{v}^{*}=\boldsymbol{v}_1^{t}+\boldsymbol{v}_1^{r}+\boldsymbol{v}_1^{s}+\boldsymbol{v}_2^{t}+\boldsymbol{v}_2^{r}+\boldsymbol{v}_2^{s},$$

whose components with subscript 1 vanish on the sphere S_1 and components with subscript 2 vanish on S_2. The superscripts correspond to the categorization of the flows that has been done in Section 2, so t signifies translational, r – rotational,and s – shear flows and these flows obey the boundary conditions at particle surfaces S_i $(i=1,\ 2)$:

$$\boldsymbol{v}_i^{t}=\boldsymbol{U}_i-\boldsymbol{v}_i^{\infty},\quad \boldsymbol{v}_i^{r}=(\boldsymbol{\Omega}_i-\boldsymbol{\Omega}^{\infty})\times\mathbf{r}_i,\quad \boldsymbol{v}_i^{s}=-\boldsymbol{E}^{\infty}\cdot\boldsymbol{r}_i \quad \text{on} \quad S_i.$$

Because of the linearity of the boundary value problem, the force and the torque acting on particles can be represented as sums

$$\boldsymbol{F}_i=\sum_j\sum_k\boldsymbol{F}_{ij}^{k}+\boldsymbol{K}_i^{t}:\boldsymbol{E}^{\infty},\quad \boldsymbol{L}_i=\sum_j\sum_k\boldsymbol{L}_{ij}^{k}+\boldsymbol{K}_i^{r}:\boldsymbol{E}^{\infty},\tag{2.77}$$

where $i,\ j=1,\ 2$; $k=t,\ r$, and the summands are given by

$$\boldsymbol{F}_{ij}^{t}=-\mu_e(\boldsymbol{K}_{ij}^{t}\cdot(\boldsymbol{U}_j-\boldsymbol{v}_j^{\infty})),\quad \boldsymbol{F}_{ij}^{r}=-\mu_e((\boldsymbol{K}_{ji}^{c})^{T}\cdot(\boldsymbol{\Omega}_j-\boldsymbol{\Omega}^{\infty})),$$
$$\boldsymbol{L}_{ij}^{t}=-\mu_e(\boldsymbol{K}_{ij}^{c}\cdot(\boldsymbol{U}_j-\boldsymbol{v}_j^{\infty})),\quad \boldsymbol{L}_{ij}^{r}=-\mu_e(\boldsymbol{K}_{ij}^{r}\cdot(\boldsymbol{\Omega}_j-\boldsymbol{\Omega}^{\infty})).$$

For particles separated by large distances, the tensors $\boldsymbol{K}_{ij}^{t}$, $\boldsymbol{K}_{ij}^{r}$, $\boldsymbol{K}_{ij}^{c}$ and $(\boldsymbol{K}_{ji}^{c})^{T}$ are determined by the following relations:

$$\boldsymbol{K}_{11}^{t}=6\pi a_1\begin{Vmatrix}1+\dfrac{9}{64}\dfrac{a_1a_2}{h^2} & 0 & 0\\ 0 & 1+\dfrac{9}{64}\dfrac{a_1a_2}{h^2} & 0\\ 0 & 0 & 1+\dfrac{9}{64}\dfrac{a_1a_2}{h^2}\end{Vmatrix},$$

$$K_{11}^{r} = 8\pi a_1^3 \begin{Vmatrix} 1+\dfrac{3}{64}\dfrac{a_2 a_1^3}{h^4} & 0 & 0 \\ 0 & 1+\dfrac{3}{64}\dfrac{a_2 a_1^3}{h^4} & 0 \\ 0 & 0 & 1 \end{Vmatrix},$$

$$K_{21}^{t} = -\frac{9}{4}\pi a_2 \begin{Vmatrix} \dfrac{a_1}{h} & 0 & 0 \\ 0 & \dfrac{a_1}{h} & 0 \\ 0 & 0 & \dfrac{a_1}{h} \end{Vmatrix}, \quad K_{21}^{r} = \frac{1}{2}\pi \frac{a_2^3 a_1^3}{h^3} \begin{Vmatrix} 1 & 0 & 0 \\ 0 & 1 & 0 \\ 0 & 0 & -2 \end{Vmatrix},$$

$$K_{11}^{c} = -\frac{9}{16}\pi \frac{a_1^4 a_2}{h^3} \begin{Vmatrix} 0 & 0 & 0 \\ -1 & 0 & 0 \\ 0 & 0 & 0 \end{Vmatrix}, \quad K_{21}^{c} = -\frac{3}{2}\pi \frac{a_1 a_2^3}{h^2} \begin{Vmatrix} 0 & 1 & 0 \\ -1 & 0 & 0 \\ 0 & 0 & 0 \end{Vmatrix},$$

where $2h$ is the distance between the particle centers and a_1, a_2 are the radii of the particles. The other matrices are obtained from the ones above by permutation of indices 1 and 2 and (or) transposition of a matrix.

Finally, one can obtain the grand resistance matrix

$$\|\mathbf{K}\| = \begin{Vmatrix} \|K_{ij}^{t}\| & \|(K_{ji}^{c})^{T}\| \\ \|K_{ij}^{c}\| & \|K_{ij}^{r}\| \end{Vmatrix}.$$

Let us introduce the shear resistance matrix, just like we did in Section 2.2:

$$\|\boldsymbol{\Psi}\| = \begin{Vmatrix} \|\boldsymbol{\vartheta}_1\| & \|\boldsymbol{\Psi}_1\| \\ \|\boldsymbol{\vartheta}_2\| & \|\boldsymbol{\Psi}_2\| \\ \|\mathbf{0}\| & \|\chi_1\| \\ \|\mathbf{0}\| & \|\chi_2\| \end{Vmatrix}.$$

For the particles spaced far away from each other, the elements of this matrix are equal to

$$\|\boldsymbol{\vartheta}_i\| = \begin{Vmatrix} 0 & 0 & 0 \\ 0 & 1 & 0 \\ 0 & 0 & F_i' \end{Vmatrix}, \quad \|\boldsymbol{\Psi}_i\| = \begin{Vmatrix} 0 & 2G_i & 0 \\ 2G_i & 1 & 0 \\ 0 & 0 & F_i' \end{Vmatrix},$$

$$\|\boldsymbol{\chi}_i\| = \begin{Vmatrix} 2H_i & 0 & 0 \\ 0 & -2H_i & 0 \\ 0 & 0 & 0 \end{Vmatrix},$$

where

$$F_1' = -\pi a_1 a_2 \left\{ \frac{15}{4}\left(\frac{a_2}{h}\right)^2 + \frac{45}{16}\left(\frac{a_1}{h}\right)^3 + \frac{3}{8}\left(\frac{a_2}{h}\right)^4 + \frac{9}{64}\left(\frac{a_1}{h}\right)^5 \right\},$$

$$G_1 = -\pi a_1 a_2 \left\{ \frac{3}{16}\left(\frac{a_2}{h}\right)^4 + \frac{9}{128}\left(\frac{a_1}{h}\right)^5 \right\},$$

$$H_1 = \pi a_2 \left\{ \frac{5}{16} a_2^2 \left(\frac{a_2}{h}\right)^4 + \frac{9}{128} a_1^2 \left(\frac{a_1}{h}\right)^5 \right\}.$$

The expressions for F_2' and G_2 are derived from F_1' and G_i by interchanging the indices 1 and 2 and changing the sign, and the expression for H_2 – by simply interchanging the indices 1 and 2.

Introducing the generalized force vector F, the particle velocity u, and the shear rate J, as we did in Section 2.1, we can rewrite Eq. (2.77) in the matrix form:

$$\|\mathbf{F}\| = -\mu_e(\|\mathbf{K}\|\|\mathbf{u}\| + \|\boldsymbol{\Psi}\|\|J\|). \tag{2.78}$$

2.4 Multi-Particle Motion

Let us now generalize the problems examined in Sections 2.1–2.3, for the case of motion of more than two particles in an unbounded fluid [39].

Let there be n solid particles with arbitrary surface shapes and spaced far apart. We will be indicating i-th particle ($i = 1, 2, \mathbf{K}, n$) with the symbol S_i, where S_i will also imply the particle's surface. The symbol 0_i will stand for the center (local origin) of i-th particle; $\boldsymbol{U}_i$ will denote the translational velocity of the point 0_i and Ω_i – the particle's angular velocity. The velocity of the fluid at the infinity is

$$\boldsymbol{v}^\infty = \boldsymbol{v}^0 + \boldsymbol{\Omega}^\infty \times \boldsymbol{r}_0 + \boldsymbol{E}^\infty \cdot \boldsymbol{r}_0, \tag{2.79}$$

where 0 is an origin of coordinates placed into an arbitrary point inside the fluid, $\boldsymbol{r}_0$ is the radius vector from the origin 0 to the given point, and $\boldsymbol{v}^0$ is the velocity of an undisturbed flow at the point 0.

The velocity and pressure fields are represented as

$$\boldsymbol{v} = \boldsymbol{v}^{\infty} + \sum_{i=1}^{n} \boldsymbol{v}_i, \quad p = \sum_{i=1}^{n} p_i, \tag{2.80}$$

where $\boldsymbol{v}^{\infty}$, $\boldsymbol{v}_i$, p_i obey Stokes equations and $\boldsymbol{v}_i$ also satisfies the boundary conditions

$$\boldsymbol{v}_i = \boldsymbol{U}_i + \boldsymbol{\Omega}_i \times \boldsymbol{r}_i - \boldsymbol{v}^0 - \boldsymbol{\Omega}^{\infty} \times \boldsymbol{r}_0 - \boldsymbol{E}^{\infty} \cdot \boldsymbol{r}_0 \quad \text{at} \quad S_i, \tag{2.81}$$

$$\boldsymbol{v}_i = 0 \quad \text{at} \quad S_j (j \neq i), \quad \boldsymbol{v}_i \to 0 \quad \text{at} \quad |\boldsymbol{r}_i| \to \infty. \tag{2.82}$$

Here $\boldsymbol{r}_i$ is the radius vector connecting 0_i with the given point.

If $\boldsymbol{v}_i^{\infty}$ is the value of the undisturbed velocity at the point 0_i (i.e., the value the velocity would have if the particle were absent), then the condition (2.81) can be rewritten as

$$\boldsymbol{v}_i = (\boldsymbol{U}_i - \boldsymbol{v}_i^{\infty}) + (\boldsymbol{\Omega}_i - \boldsymbol{\Omega}^{\infty}) \times \boldsymbol{r}_i - \boldsymbol{E}^{\infty} \cdot \boldsymbol{r}_i \quad \text{at} \quad S_i. \tag{2.83}$$

In the next step we decompose $(\boldsymbol{v}_i, p_i)$ into summands corresponding to shear $(\boldsymbol{v}_i', p_i')$, translational, $(\boldsymbol{v}_i'', p_i'')$ and rotational $(\boldsymbol{v}_i''', p_i''')$ motions:

$$\boldsymbol{v}_i = \boldsymbol{v}_i' + \boldsymbol{v}_i'' + \boldsymbol{v}_i''', \quad p_i = p_i' + p'' + p_i'''$$

where each component must obey Stokes equation and satisfy the boundary conditions

$$\boldsymbol{v}_i' = -\boldsymbol{E}^{\infty} \cdot \boldsymbol{r}_i; \quad \boldsymbol{v}_i'' = \boldsymbol{U}_i - \boldsymbol{v}_i^{\infty}; \quad \boldsymbol{v}_i''' = (\boldsymbol{\Omega}_i - \boldsymbol{\Omega}^{\infty}) \times \boldsymbol{r}_i \quad \text{on} \quad S_i. \tag{2.84}$$

If the solutions of the corresponding boundary value problems are known, one can find the hydrodynamic force and the torque relative to the point 0 acting on i-th particle:

$$\begin{aligned} \boldsymbol{F}_i &= -\mu_e \left\{ \sum_{j=1}^{n} [\boldsymbol{K}_{ij}^t \cdot (\boldsymbol{U}_i - \boldsymbol{v}_i^{\infty}) + (\boldsymbol{K}_{ji}^c)^T \cdot (\boldsymbol{\Omega}_i - \boldsymbol{\Omega}^{\infty})] + \boldsymbol{\Phi}_i : \boldsymbol{E}^{\infty} \right\}, \\ \boldsymbol{L}_i &= -\mu_e \left\{ \sum_{j=1}^{n} [\boldsymbol{K}_{ij}^c \cdot (\boldsymbol{U}_i - \boldsymbol{v}_i^{\infty}) + \boldsymbol{K}_{ij}^r \cdot (\boldsymbol{\Omega}_i - \boldsymbol{\Omega}^{\infty})] + \boldsymbol{\tau}_i : \boldsymbol{E}^{\infty} \right\}. \end{aligned} \tag{2.85}$$

The tensors $\boldsymbol{K}_{ij}^t$, $\boldsymbol{K}_{ij}^r$, $\boldsymbol{K}_{ij}^c$, and $(\boldsymbol{K}_{ji}^c)^T$ are determined from the expression for the net stress at the particle surface S_i induced by translational motion $\boldsymbol{T}_i''$, rotational motion $\boldsymbol{T}_i'''$, and the shear flow $\boldsymbol{T}_i'$:

$$K^t_{ij} = -\int_{S_i} \mathbf{\Pi}''_j \cdot ds; \quad K^r_{ij} = -\int_{S_i} \boldsymbol{r}_i \times (\mathbf{\Pi}'''_j \cdot ds); \quad K^c_{ij} = -\int_{S_i} \boldsymbol{r}_i \times (\mathbf{\Pi}''_j \cdot ds)$$

$$\Phi_i = -\sum_{j=1}^{n} \int_{S_i} ds \cdot \mathbf{\Pi}'_j; \quad \boldsymbol{\tau}_i = -\sum_{j=1}^{n} \int_{S_i} \boldsymbol{r}_i \times (ds \cdot \mathbf{\Pi}'_j).$$

Third-rank tensors $\mathbf{\Pi}''_j, \mathbf{\Pi}'''_j$ (triadics) and the fourth-rank tensor (tetradic) $\mathbf{\Pi}'_j$ characterize the properties of fields $(\boldsymbol{v}''_i, p''_i), (\boldsymbol{v}'''_i, p'''_i)$, and $(\boldsymbol{v}'_i, p'_i)$. The first two were determined in Section 2.1, and the third one is defined by the expression $\boldsymbol{T}'_i = \mu_e \mathbf{\Pi}'_j : \boldsymbol{E}^\infty$. The dyadics $\boldsymbol{K}^t_{ij}, \boldsymbol{K}^r_{ij}, \boldsymbol{K}^c_{ij}$ and the triadics Φ_i, $\boldsymbol{\tau}_i$ characterize geometrical properties of the momentary configuration of a particle system: sizes, shapes, and positions of the centers 0_i.

To write the relations (2.85) in a matrix form, let us introduce $3n \times 1$ matrices (column vectors)

$||\boldsymbol{F}||$, $||\boldsymbol{L}||$, $||\boldsymbol{U} - \boldsymbol{v}^\infty||$, and $||\Omega - \Omega^\infty||$, and such that

$$||\boldsymbol{F}|| = |||\boldsymbol{F}_1||||\boldsymbol{F}_2||\ldots||\boldsymbol{F}_2|||^T,$$

which includes in consecutive order the parameters of n particles, and $3n \times 3n$ matrices $||\boldsymbol{K}^t||$, $||\boldsymbol{K}^r||$, $||\boldsymbol{K}^c||$:

$$||\boldsymbol{K}|| = \begin{Vmatrix} ||\boldsymbol{K}_{11}|| & ||\boldsymbol{K}_{12}|| & \ldots & ||\boldsymbol{K}_{1n}|| \\ \ldots & \ldots & & \ldots \\ ||\boldsymbol{K}_{n1}|| & ||\boldsymbol{K}_{n2}|| & \cdots & ||\boldsymbol{K}_{nn}|| \end{Vmatrix}.$$

Each of the elementary matrices $||\boldsymbol{K}_{ij}||$ is a 3×3 matrix, and the matrices $\left\|\boldsymbol{K}^t_{ij}\right\|$ and $\left\|\boldsymbol{K}^r_{ij}\right\|$ are symmetric. The matrices $||\Phi_i||$ and $||\tau_i||$ are 3×6 matrices, from which we can form a $6n \times 6$ combined matrix (see Section 2.2):

$$||\boldsymbol{\Psi}|| = \begin{Vmatrix} ||\Phi_1|| \\ \vdots \\ ||\Phi_n|| \\ ||\boldsymbol{\tau}_1|| \\ \vdots \\ ||\boldsymbol{\tau}_n|| \end{Vmatrix}.$$

The above-mentioned matrices allow us to construct the $6n \times 6n$ grand resistance matrix:

$$||\mathbf{K}|| = \begin{Vmatrix} ||K^t|| & ||(K^c)^T|| \\ ||K^c|| & ||K^r|| \end{Vmatrix},$$

as well as generalized matrices of forces ($6n \times 1$), velocities ($6n \times 1$), and shear (6×1):

$$||\mathbf{F}|| = \begin{Vmatrix} ||F|| \\ ||L|| \end{Vmatrix}; \quad ||\mathbf{u}|| = \begin{Vmatrix} ||(U - v^\infty)|| \\ ||(\boldsymbol{\Omega} - \boldsymbol{\Omega}^\infty)|| \end{Vmatrix}; \quad ||J|| = \begin{Vmatrix} S_1 \\ \vdots \\ S_6 \end{Vmatrix}.$$

Now the equations (2.85) can be written in a compact matrix form:

$$||\mathbf{F}|| = -\mu_e(||\mathbf{K}||\ ||\mathbf{u}|| + ||\boldsymbol{\Psi}||\ ||J||). \tag{2.86}$$

If the particles are freely suspended in the fluid, then setting $||F|| = ||\mathbf{0}||$ in Eq. (2.86), we obtain the velocities of particles:

$$||\mathbf{u}|| = -||\mathbf{K}||\ ||\boldsymbol{\Psi}||\ ||J||. \tag{2.87}$$

The grand resistance matrix $||\mathbf{K}||$ is symmetric and non-singular, so there exists an inverse matrix and consequently, it is possible to obtain the velocities of suspended particles. The elements of matrices $||\mathbf{K}||$ and $||\boldsymbol{\Psi}||$ do not depend on dynamic parameters (velocities); they depend only on the geometric properties of the particles and on the configuration of a particle system. At the present time, they can be determined only by solving problems on hydrodynamic interaction of two spherical particles as a function of the relative distance between particles and the ratio of their radii. Some examples of pair interactions were given in Section 2.3.

2.5
Flow of a Fluid Through a Random Bed of Particles

Consider a slow Stokesian flow of viscous incompressible fluid through a layer of buoyant (cloud) or fixed in space (porous medium) rigid spherical particles with a given distribution of particles over radii [40–43]. Our goal is to determine the force of particle resistance and the effective viscosity of the disperse medium taking into account hydrodynamic interaction between particles, that is, the hindered motion of particles [44].

The velocity and pressure in the fluid are described by Stokes equations

$$\nabla \cdot \boldsymbol{u} = 0, \tag{2.88}$$

$$\nabla \cdot \boldsymbol{T} = \mu_e \nabla^2 \boldsymbol{u} - \nabla(p - \rho \mathbf{g} \cdot \boldsymbol{r}) = 0, \tag{2.89}$$

where $\boldsymbol{T}$ is the stress tensor with components

$$\tau_{ij} = -(p - \rho \mathbf{g} \cdot \boldsymbol{r})\delta_{ij} + \mu_e \left(\frac{\partial u_i}{\partial X_j} + \frac{\partial u_j}{\partial X_i} \right) = 0. \tag{2.90}$$

There are also the boundary conditions $\boldsymbol{u} = \boldsymbol{u}_s$ that must hold on particle surfaces; $\boldsymbol{u}_s$ is the particles' velocity. The fluid's velocity at the infinity is given. If the particles are at rest or their positions in space are fixed, then $\boldsymbol{u}_s = 0$. For buoyant particles, their velocity $\boldsymbol{u}_s$ is a sum of translational and rotational velocities and is found from the relations for the force and torque exerted on the particle by the surrounding fluid (see Section 2.1). Hydrodynamic interactions between particles can be taken into account by averaging over the particle ensemble.

Following [42,43], we assume that $\boldsymbol{u}$, p and ρ are defined in the whole region under consideration, including the particles. We further assume that that $\boldsymbol{u}$ is a continuous function, whereas p, ρ, and derivatives of $\boldsymbol{u}$ are discontinuous at the particle surface S. Introducing the Saffman function

$$H(X) = \begin{cases} 1, & \text{if } X \text{ is in fluid,} \\ 0, & \text{if } X \text{ is in particle.} \end{cases} \tag{2.91}$$

and averaging it over the particle ensemble, we get

$$\langle H(X) \rangle = 1 - \varphi = \eta, \tag{2.92}$$

where φ is the volume concentration (volume fraction) of particles and η is the porosity (volume fraction occupied by fluid) of the disperse medium.

From now on, averaging over the particle ensemble will be designated by angle brackets, and the mean-flow-rate values of hydrodynamic parameters – by a horizontal bar on the top. The Saffman function gives the relation between these averages; in particular, the mean-flow-rate velocity and pressure in the fluid are equal to

$$\overline{\boldsymbol{u}(\boldsymbol{X})} = \frac{\langle H(\boldsymbol{X})\boldsymbol{u}(\boldsymbol{X}) \rangle}{\langle H(\boldsymbol{X}) \rangle} = \frac{\langle \boldsymbol{u}(\boldsymbol{X}) \rangle}{\eta}, \quad \overline{p(\boldsymbol{X})} = \frac{\langle H(\boldsymbol{X})\, p(\boldsymbol{X}) \rangle}{\langle H(\boldsymbol{X}) \rangle} = \frac{\langle p(\boldsymbol{X}) \rangle}{\eta}. \tag{2.93}$$

Since velocity is a continuous function in the entire region, we can introduce the average cross-sectional velocity $\bar{\boldsymbol{u}}$ in addition to the ensemble average $\langle \boldsymbol{u} \rangle$. For a porous medium, this velocity is called the speed of filtration. For $\varphi \ll 1$ (a low-concentrated suspension or a highly permeable medium), $\bar{\boldsymbol{u}} \approx \langle \boldsymbol{u} \rangle$.

Before averaging Stokes equations, we must first average the stress tensor $\boldsymbol{T}$. We have:

$$\langle H\tau_{ij} \rangle = -\langle H(p - \rho \mathbf{g} \cdot \boldsymbol{r}) \rangle > \delta_{ij} + \mu_e \left\langle \frac{\partial u_i}{\partial X_j} + \frac{\partial u_j}{\partial X_i} \right\rangle. \tag{2.94}$$

Using the obvious equalities

$$\left\langle H\frac{\partial u_i}{\partial X_j}\right\rangle = \frac{\partial}{\partial X_j}\langle Hu_i\rangle - \left\langle u_i\frac{\partial H}{\partial X_j}\right\rangle, \tag{2.95}$$

$$\left\langle (1-H)\frac{\partial u_i}{\partial X_j}\right\rangle = \frac{\partial}{\partial X_j}\langle (1-H)u_i\rangle + \left\langle u_i\frac{\partial H}{\partial X_j}\right\rangle \tag{2.96}$$

and the property

$$\partial H/\partial X_j = \delta(\boldsymbol{X}-\boldsymbol{X}_s)$$

of the Saffman function (where $\boldsymbol{X}_s$ is a point on the particle's surface), and adding the relations (2.95) and (2.96), we can write

$$\left\langle H\frac{\partial u_i}{\partial X_j}\right\rangle + \left\langle (1-H)\frac{\partial u_i}{\partial X_j}\right\rangle = \frac{\partial \bar{u}_i}{\partial X_j}.$$

Interchanging the indices,

$$\left\langle H\frac{\partial u_j}{\partial X_i}\right\rangle + \left\langle (1-H)\frac{\partial u_j}{\partial X_i}\right\rangle = \frac{\partial \langle u_j\rangle}{\partial X_i},$$

and adding both relations, we obtain the following:

$$\left\langle H\left(\frac{\partial u_i}{\partial X_j}+\frac{\partial u_j}{\partial X_i}\right)\right\rangle + \left\langle (1-H)\left(\frac{\partial u_i}{\partial X_j}+\frac{\partial u_j}{\partial X_i}\right)\right\rangle = \frac{\partial \langle u_i\rangle}{\partial X_j}+\frac{\partial \langle u_j\rangle}{\partial X_i}. \tag{2.97}$$

Inside rigid particles, the velocity remains constant, and in the fluid, $H=1$. Therefore the second term on the left-hand side vanishes and Eq. 2.97 becomes

$$\left\langle H\left(\frac{\partial u_i}{\partial X_j}+\frac{\partial u_j}{\partial X_i}\right)\right\rangle = \frac{\partial \langle u_i\rangle}{\partial X_j}+\frac{\partial \langle u_j\rangle}{\partial X_i}. \tag{2.98}$$

At $i=j$ this equality reduces to $\langle H\nabla\cdot\boldsymbol{u}\rangle = \nabla\cdot\langle\boldsymbol{u}\rangle$, and we get the averaged continuity equation,

$$\nabla\cdot\langle\boldsymbol{u}\rangle = 0. \tag{2.99}$$

Hence, Eq. (2.94) reduces to

$$\langle H\tau_{ij}\rangle = -(1-\varphi)(\bar{p}-\rho\boldsymbol{g}\cdot\boldsymbol{r})\delta_{ij} + \mu_e\left(\frac{\partial \langle u_i\rangle}{\partial X_j}+\frac{\partial \langle u_j\rangle}{\partial X_i}\right). \tag{2.100}$$

The ensemble-averaged momentum equation (2.89) has the form

$$\langle H\nabla \cdot \boldsymbol{T}\rangle = 0. \tag{2.101}$$

Using the well-known tensor equality

$$\nabla \cdot (H\boldsymbol{T}) = H\nabla \cdot \boldsymbol{T} + \boldsymbol{T} \cdot \nabla H,$$

we rewrite Eq. (2.101) as

$$\nabla \cdot \langle (H\boldsymbol{T})\rangle - \langle \boldsymbol{T} \cdot \nabla H\rangle = 0. \tag{2.102}$$

Let us substitute Eq. (2.100) into Eq. (2.102) and then use the averaged continuity equation (2.99) to get

$$\mu_e \nabla^2 \langle \boldsymbol{u}\rangle - \langle \boldsymbol{T} \cdot \nabla H\rangle - \nabla(1-\varphi)(\bar{p} - \rho \boldsymbol{g} \cdot \boldsymbol{r}) = 0. \tag{2.103}$$

According to Eq. (2.91), the function H undergoes a jump at the particle surface, so ∇H is a delta function. In order to understand the meaning of the second term in Eq. (2.103), let us take an integral of this term over a finite volume V, transform it into a surface integral according to Gauss's theorem, and use the property of the delta function. After some manipulations, we get

$$-\int \langle \boldsymbol{T} \cdot \nabla H\rangle d\boldsymbol{r} = -\int \nabla \cdot \langle H\boldsymbol{T}\rangle d\boldsymbol{r} = -\int \langle H\boldsymbol{T}\rangle \cdot \boldsymbol{n}\, dS = -\left\langle \int \boldsymbol{T} \cdot \boldsymbol{n}\, dS \right\rangle, \tag{2.104}$$

where $\boldsymbol{n}$ is the unit normal vector.

The quantity $-\langle H\boldsymbol{T}\rangle \cdot \boldsymbol{n} ds$ is equal to the force exerted by the fluid on a particle surface element, and Eq. (2.104) is the average force exerted on the fluid by the particles in the given volume. Hence, the second term in Eq. (2.103) has the meaning of volume force exerted on the fluid by the particles.

Now, consider a disperse phase consisting of N identical spherical particles. The positions of their centers will be defined by radius vectors $\boldsymbol{r}_1, \boldsymbol{r}_2, \ldots, \boldsymbol{r}_N$. Let p_N $(\boldsymbol{r}_1, \boldsymbol{r}_2, \ldots, \boldsymbol{r}_N)$ be the N-particle PDF. If we assume the velocity $\boldsymbol{u}$ and function H to depend only on $\boldsymbol{r}_1, \boldsymbol{r}_2, \ldots, \boldsymbol{r}_N$, then the ensemble average of a function G $(\boldsymbol{r}_1, \boldsymbol{r}_2, \ldots, \boldsymbol{r}_N)$, where the averaging is carried out over the ensemble of particle configurations, is given by the formula

$$\langle G\rangle = \int G(\boldsymbol{r}_1, \boldsymbol{r}_2, \ldots, \boldsymbol{r}_N)\, p_N(\boldsymbol{r}_1, \boldsymbol{r}_2, \ldots, \boldsymbol{r}_N) d\boldsymbol{r}_1 d\boldsymbol{r}_2 \ldots d\boldsymbol{r}_N. \tag{2.105}$$

Of utmost interest are the one-particle PDF p_1 $(\boldsymbol{r}_1)$ and the two-particle PDF p_2 $(\boldsymbol{r}_1, \boldsymbol{r}_2)$. These are marginal PDFs that can be obtained from a given N-particle PDF p_N $(\boldsymbol{r}_1, \boldsymbol{r}_2, \ldots, \boldsymbol{r}_N)$ using the formulas

$$p_1(\boldsymbol{r}_1) = \int p_N(\boldsymbol{r}_1, \boldsymbol{r}_2, \ldots, \boldsymbol{r}_N) d\boldsymbol{r}_2 \ldots d\boldsymbol{r}_N, \tag{2.106}$$

$$p_2(\boldsymbol{r}_1, \boldsymbol{r}_2) = \int p_N(\boldsymbol{r}_1, \boldsymbol{r}_2, \ldots, \boldsymbol{r}_N) d\boldsymbol{r}_3 \ldots d\boldsymbol{r}_N. \tag{2.107}$$

In the case of random independent single-particle distributions, the N-particle distribution is the product of single-particle distributions:

$$p_N(\boldsymbol{r}_1, \boldsymbol{r}_2, \ldots, \boldsymbol{r}_N) = p_1(\boldsymbol{r}_1)\, p_1(\boldsymbol{r}_2) \ldots p_1(\boldsymbol{r}_N).$$

For identical particles, one can introduced the distribution of the number of particles,

$$n(\boldsymbol{r}_1) = N\, p_1(\boldsymbol{r}), \tag{2.108}$$

such that $nd\,\boldsymbol{r}$ is equal to the probable (i.e., ensemble-averaged) number of particles whose centers are located in the volume element $d\,\boldsymbol{r}$. The Saffman function H for this case is

$$H(\boldsymbol{r}; \boldsymbol{r}_1, \boldsymbol{r}_2, \ldots, \boldsymbol{r}_N) = 1 - \sum_{j=1}^{N} \mathrm{H}(a - |\boldsymbol{r} - \boldsymbol{r}_j|), \tag{2.109}$$

where H (X) is the Heaviside function. Then

$$\begin{aligned} \langle H \rangle = 1 - \varphi &= \int H\, p_N(\boldsymbol{r}_1, \boldsymbol{r}_2, \ldots, \boldsymbol{r}_N) d\boldsymbol{r}_1 d\boldsymbol{r}_2 \ldots d\boldsymbol{r}_N \\ &= 1 - N \int \mathrm{H}(a - |\boldsymbol{r} - \boldsymbol{r}_1|)\, p_1(\boldsymbol{r}_1) d\boldsymbol{r}_1 \\ &= 1 - N \int\limits_{|\boldsymbol{r} - \boldsymbol{r}_1| < a} p_1(\boldsymbol{r}_1) d\boldsymbol{r}_1 = 1 - \int\limits_{|\boldsymbol{r} - \boldsymbol{r}_1| < a} n(\boldsymbol{r}_1) d\boldsymbol{r}_1, \end{aligned} \tag{2.110}$$

Hence, the volume concentration of particles is

$$\varphi = \int\limits_{|\boldsymbol{r} - \boldsymbol{r}_1| < a} n(\boldsymbol{r}_1) d\boldsymbol{r}_1, \tag{2.111}$$

and for identical particles with a uniform distribution $n = const$,

$$\varphi = 4\pi a^3 n/3. \tag{2.112}$$

From the ensemble of probable particle configurations, we can choose a subensemble containing those configurations in which particle 1 occupies one and the same position in space. We can then introduce the PDF of all other particles. This

PDF is called the conditional PDF and is equal to

$$p_N(\boldsymbol{r}_1, \boldsymbol{r}_2, \ldots, \boldsymbol{r}_N)/\, p_1(\boldsymbol{r}_1). \tag{2.113}$$

Averaging of functions that relies on the use of such conditional PDFs will be designated as

$$\langle G \rangle_1 = \int G(\boldsymbol{r}_1, \boldsymbol{r}_2, \ldots, \boldsymbol{r}_N) \frac{p_N(\boldsymbol{r}_1, \boldsymbol{r}_2, \ldots, \boldsymbol{r}_N)}{p_1(\boldsymbol{r}_1)} d\boldsymbol{r}_2 \ldots d\boldsymbol{r}_N. \tag{2.114}$$

The conditional two-particle PDF and the corresponding distribution of the number of particles of type 2 are, respectively, $p_2\,(\boldsymbol{r}_1, \boldsymbol{r}_2)/p_1\,(\boldsymbol{r}_1)$ and $n\,(\boldsymbol{r}_2) = N\,p_2\,(\boldsymbol{r}_1, \boldsymbol{r}_2)/p_1\,(\boldsymbol{r}_1)$. For a random and independent distribution of identical particles, we can take $p_2\,(\boldsymbol{r}_1, \boldsymbol{r}_2) = 0$ at $|\mathbf{r}_1 - \mathbf{r}_2| < 2a$; $p_2\,(\mathbf{r}_1, \mathbf{r}_2) = p_1\,(\mathbf{r}_1)\,p_1\,(\mathbf{r}_2)$ at $|\mathbf{r}_1 - \mathbf{r}_2| > 2a$; and $n\,(\mathbf{r}_2) = 0$ at $|\mathbf{r}_1 - \mathbf{r}_2| < 2a$; $n\,(\mathbf{r}_2) = const$ at $|\mathbf{r}_1 - \mathbf{r}_2| > 2a$. Such an approximation is possible for non-interacting particles only. The PDF for interacting particles is determined from Liouville or Fokker–Planck equations (for more details, see Sections 3.7–3.9).

Going back to Eq. (2.103), let us look at the second term. We have

$$\nabla H = \nabla\left(1 - \sum_{j=1}^{N} H(a - |\boldsymbol{r} - \boldsymbol{r}_j|)\right) = \sum_{j=1}^{N} \frac{\boldsymbol{r} - \boldsymbol{r}_j}{|\boldsymbol{r} - \boldsymbol{r}_j|} \delta(|\boldsymbol{r} - \boldsymbol{r}_j| - a)$$

and

$$\begin{aligned}
\langle \boldsymbol{T} \cdot \nabla H \rangle &= \sum_{j=1}^{N} \int p_N(\boldsymbol{r}_1, \boldsymbol{r}_2, \ldots, \boldsymbol{r}_N)\delta(|\boldsymbol{r} - \boldsymbol{r}_j| - a)\boldsymbol{T} \cdot \frac{\boldsymbol{r} - \boldsymbol{r}_j}{|\boldsymbol{r} - \boldsymbol{r}_j|} d\boldsymbol{r}_1 \ldots d\boldsymbol{r}_N \\
&= N \int p_N \delta(|\boldsymbol{r} - \boldsymbol{r}_1| - a)\boldsymbol{T} \cdot \frac{\boldsymbol{r} - \boldsymbol{r}_1}{|\boldsymbol{r} - \boldsymbol{r}_1|} d\boldsymbol{r}_1 \ldots d\boldsymbol{r}_N \\
&= \int n(\boldsymbol{r}_1)\delta(|\boldsymbol{r} - \boldsymbol{r}_1| - a)\langle \boldsymbol{T} \rangle_1 \cdot \frac{\boldsymbol{r} - \boldsymbol{r}_1}{|\boldsymbol{r} - \boldsymbol{r}_1|} d\boldsymbol{r}_1 \\
&= \int_{\substack{r_1 = r - an \\ r\ \mathit{fixed}}} n(\boldsymbol{r} - a\boldsymbol{n})\langle \boldsymbol{T} \rangle_1 \cdot \boldsymbol{n} a^2 d\boldsymbol{n}.
\end{aligned} \tag{2.115}$$

The derivation relies on property 2 of the PDF (see Section 1.4) and also uses the fact that a unit normal vector to the particle surface is equal to $\boldsymbol{n} = (\boldsymbol{r} - \boldsymbol{r}_1)/|\boldsymbol{r} - \boldsymbol{r}_1|$ and that $d\boldsymbol{n}$ is an element of a solid angle on a unit sphere. For $n = const$, we have

$$\langle \boldsymbol{T}_1 \cdot \nabla H \rangle = n a^2 \int_{\substack{r_1 = r - an \\ r\ \mathit{fixed}}} \langle \boldsymbol{T} \rangle_1 \cdot \boldsymbol{n} d\boldsymbol{n}. \tag{2.116}$$

and now Eq. (2.103) takes the form

$$\mu_e \nabla^2 \langle \boldsymbol{u} \rangle - \nabla(1-\varphi)(\bar{p} - \rho \boldsymbol{g} \cdot \boldsymbol{r}) - na^2 \int\limits_{\substack{r_1 = r - an \\ r\ fixed}} \langle \boldsymbol{T} \rangle_1 \cdot \boldsymbol{n} d\boldsymbol{n} = 0. \tag{2.117}$$

$\langle \boldsymbol{T} \rangle_1$ is obtained from the relation (2.100), which, when reformulated for the case of conditional averaging, will be written as

$$\langle \tau_{ij} \rangle = -(1-\varphi)(\bar{p}_1 - \rho \boldsymbol{g} \cdot \boldsymbol{r})\delta_{ij} + \mu_e \left(\frac{\partial \langle u_i \rangle_1}{\partial X_j} + \frac{\partial \langle u_j \rangle_1}{\partial X_i} \right). \tag{2.118}$$

When deriving the last relation, we remembered that $\varphi_1 = 0$ at the surface of the sphere. The subscript 1 implies that we are considering the situation when particle 1 is fixed.

The equation for $\langle \boldsymbol{u} \rangle_1$ is similar to (2.103):

$$\mu_e \nabla^2 \langle \boldsymbol{u} \rangle_1 - \nabla(1-\varphi_1)(\bar{p}_1 - \rho \boldsymbol{g} \cdot \boldsymbol{r}) - \langle \boldsymbol{T} \cdot \nabla H \rangle_1. \tag{2.119}$$

The derivation of $\langle \boldsymbol{T} \cdot \nabla H \rangle_1$ is similar to the derivation of Eq. (2.115):

$$\langle \boldsymbol{T} \cdot \nabla H \rangle_1 = \int\limits_{\substack{r_2 = r - an \\ r_1\ and\ r\ fixed}} n_1(\boldsymbol{r} - a\boldsymbol{n}) \langle \boldsymbol{T} \rangle_{1,2} \cdot \boldsymbol{n} a^2 d\boldsymbol{n}, \tag{2.120}$$

with the only difference that in the right-hand side, n is replaced by n_1 and $\langle \boldsymbol{T} \rangle_1$ is replaced by $\langle \boldsymbol{T} \rangle_{1,2}$. Here $\langle \boldsymbol{T} \rangle_{1,2}$ is the outcome of conditional averaging of stress with particles 1 and 2 fixed, and the dependence of $\langle \boldsymbol{T} \cdot \nabla H \rangle_{1,2}$ on $\langle \boldsymbol{u} \rangle_{1,2}$ has the same form as the dependence of $\langle \boldsymbol{T} \cdot \nabla H \rangle_1$ on $\langle \boldsymbol{u} \rangle_1$. We can continue this process by fixing (in consecutive order) the positions of particles 1, 2, 3, etc. The result is an infinite system of equations, where each finite subsystem is unclosed.

Considering the first approximation only, we shall assume

$$na^2 = \int\limits_{\substack{r_1 = r - an \\ r\ fixed}} \langle \boldsymbol{T} \rangle_1 \cdot \boldsymbol{n} d\boldsymbol{n} = F(\langle \boldsymbol{u} \rangle) \tag{2.121}$$

to close the equations (2.99) and (2.117).

The functional dependence (2.121) generalizes the assumption made in [40]. A medium which obeys the condition (2.121) is called the Brinkman medium.

The form of the function $F(\langle \boldsymbol{u} \rangle)$ is different for the cases when the particles are fixed in space (as in a porous medium) and when they are suspended in the fluid (as in a suspension). In the first case the function is assumed to have the form $F(\langle \boldsymbol{u} \rangle) = A\langle \boldsymbol{u} \rangle + B\nabla^2 \langle \boldsymbol{u} \rangle$ and in the second case – $F(\langle \boldsymbol{u} \rangle) = B\nabla^2 \langle \boldsymbol{u} \rangle + C\boldsymbol{g}$.

Let us fix particle 1 and consider the fluid flowing around it. The presence of other particles is accounted for by introducing the effective fluid viscosity μ_{eff} and the upstream velocity $\bar{\boldsymbol{u}}$ at the infinity. Both of these quantities have to be determined.

The velocity and pressure fields are described by the equations

$$\nabla \cdot \langle \boldsymbol{u} \rangle_1 = 0,$$
$$-\nabla(\bar{p}_1 - \rho \boldsymbol{g} \cdot \boldsymbol{r}) + \mu_{eff} \nabla^2 \langle \boldsymbol{u} \rangle_1 - \mu_{eff} \alpha^2 \langle \boldsymbol{u} \rangle_1 = 0, \tag{2.122}$$

where $\mu_{eff}\,\alpha^2 = A/(1-\varphi)$, $\mu_{eff} = (\mu_e - B)/(1-\varphi)$, and the parameter α has to be determined.

These equations should be solved under the following boundary conditions:

$$\langle \boldsymbol{u} \rangle_1 = 0 \quad \text{at} \quad \boldsymbol{r} = \boldsymbol{r}_1 + a\boldsymbol{n}; \quad \langle \boldsymbol{u} \rangle_1 \to \bar{\boldsymbol{u}} \quad \text{at} \quad \boldsymbol{r} \to \infty, \tag{2.123}$$

The unperturbed velocity $\bar{\boldsymbol{u}}$ must obey the equations similar to Eqs. (2.122):

$$\nabla \cdot \bar{\boldsymbol{u}} = 0, \quad -\nabla \bar{p}_0 + \mu_{eff} \nabla^2 \bar{\boldsymbol{u}} - \mu_{eff} \alpha^2 \bar{\boldsymbol{u}} = 0, \tag{2.124}$$

where $\bar{p}_0 = \bar{p} - \rho \boldsymbol{g} \cdot \boldsymbol{r}$.

Solving equations (2.124) as discussed in [43], we get

$$\alpha a = \frac{9\varphi/4 + 3(8\varphi - 3\varphi^2)^{1/2}/4}{1 - 3\varphi/2} \tag{2.125}$$

and $\mu_{eff}(\varphi)$. The derivation of the latter dependence is lengthy and is not presented here.

One can see from the second equation (2.122) that if the characteristic linear scale L of the problem exceeds $1/\alpha$, then the term $\mu_{eff} \nabla^2 \bar{\boldsymbol{u}}$ is negligibly small as compared to $\mu_{eff} \alpha^2 \bar{\boldsymbol{u}}$. The equation then reduces to the Darcy equation for a low-permeable porous medium:

$$-\nabla \bar{p} - \frac{\mu_{eff}}{k} \bar{\boldsymbol{u}} = 0, \tag{2.126}$$

where $k = 1/\alpha^2$ is the permeability of the porous medium.

In the second case, the problem reduces to the system of equations

$$\nabla \cdot \langle \boldsymbol{u} \rangle = 0,$$
$$-\nabla(1-\varphi)(\bar{p} - \rho \boldsymbol{g} \cdot \boldsymbol{r}) + \mu_e \nabla^2 \langle \boldsymbol{u} \rangle - F(\langle \boldsymbol{u} \rangle) = 0 \tag{2.127}$$

$$\nabla \cdot \langle \boldsymbol{u} \rangle_1 = 0,$$
$$-\nabla(1-\varphi)(\bar{p}_1 - \rho \boldsymbol{g} \cdot \widehat{\boldsymbol{r}}) + \mu_e \nabla^2 \langle \boldsymbol{u} \rangle_1 - F(\langle \boldsymbol{u} \rangle_1) = 0 \tag{2.128}$$

with the boundary conditions

$$\langle \boldsymbol{u} \rangle_1 = \boldsymbol{V} + a\,\widehat{\boldsymbol{r}} \times \boldsymbol{\Omega} \quad \text{at} \quad \boldsymbol{r} = \boldsymbol{r}_1 + a\,\widehat{\boldsymbol{r}}; \quad \langle \boldsymbol{u} \rangle_1 \to \langle \boldsymbol{u} \rangle \quad \text{at} \quad \boldsymbol{r} \to \infty, \tag{2.129}$$

where $\boldsymbol{r}_1$ is the particle's radius vector, and $\boldsymbol{V}$ and $\boldsymbol{\Omega}$ are the translational and rotational velocities the test particle 1 freely suspended in the fluid. They are determined by the condition that the force and torque acting on the particle are both equal to zero:

$$a^2 \int\limits_{\substack{r = r_1 + a\,\widehat{r} \\ r_1 \text{ fixed}}} \langle \boldsymbol{T} \rangle_1 \cdot \widehat{\boldsymbol{r}}\, d\,\widehat{\boldsymbol{r}} + \frac{4}{3}\pi a^3 \rho_p \boldsymbol{g} = 0, \quad \int\limits_{\substack{r = r_1 + a\,\widehat{r} \\ r_1 \text{ fixed}}} \widehat{\boldsymbol{r}} \times \langle \boldsymbol{T} \rangle_1 \cdot \widehat{\boldsymbol{r}}\, d\,\widehat{\boldsymbol{r}} = 0, \tag{2.130}$$

The functional $F(\langle \boldsymbol{u} \rangle)$ is given by

$$F(\langle \boldsymbol{u} \rangle) = B\nabla^2 \langle \boldsymbol{u} \rangle + C\boldsymbol{g}.$$

The corresponding equations look simpler than equations (2.122) and (2.124). In particular, the second equation (2.122) assumes the form of a standard Stokes equation

$$-\nabla\left(\bar{p}_1 - \rho \boldsymbol{g} \cdot \boldsymbol{r} + \frac{C\boldsymbol{g} \cdot \boldsymbol{r}}{1 - \varphi} \right) + \mu_{eff} \nabla^2 \langle \boldsymbol{u} \rangle_1 = 0, \tag{2.131}$$

where $\mu_{eff} = (\mu_e - B)/(1 - \varphi)$ is the effective viscosity.

Skipping the solution of this equation, we give only the formula for the coefficient of effective viscosity:

$$\frac{\mu_{eff}}{\mu_e} = \frac{1}{1 - 5\varphi/2}. \tag{2.132}$$

At $\varphi \ll 1$, the equation (2.132) gives the well-known Einstein's formula:

$$\frac{\mu_{eff}}{\mu_e} \approx 1 + \frac{5}{2}\varphi.$$

References

1 Lamb, H. (1945) *Hydrodynamics*, Dover, New York.

2 Levich, V.G. (1962) *Physicochemical Hydrodynamics*, Prentice–Hall,Englewood Cliffs, NJ.

3 Happel, J. and Brenner, H. (1983) *Low Reynolds Number Hydrodynamics*, Martinus Nijhof, The Hague.

4 Kim, S. and Karrila, S.J. (1991) *Microhydrodynamics*, Butterworth–Heinemann, Boston.

5 Probstein, R.F. (1995) *Physicochemical Hydrodynamics*, Wiley, New York.

6 Brenner, H. (1966) Hydrodynamic Resistance of Particles at small Reynolds Numbers. *Adv. Chem. Eng.*, **6**, 287–438.

7 Stimson, M. and Jefferey, G.B. (1926) The Motion of two Spheres in a viscous Fluid. *Proc. Roy. Soc. A*, **111**, 110.

8 Brenner, H. (1961) The slow Motion of a Sphere through a viscous Fluid toward a plane Wall. *Chem. Eng. Sci.*, **16**, 242–251.

9 Goldman, A.J., Cox, R.G. and Brenner, H. (1966) The slow Motion of two identical arbitrary oriented Spheres through a viscous Fluid. *Chem. Eng. Sci.*, **21**, 1151–1170.

10 Goldman, A.J., Cox, R.G. and Brenner, H. (1967) Slow viscous Motion of a Sphere parallel to a plane Wall. Part I. Motion through a quiescent Fluid. Part. II. Couette Flow. *Chem. Eng. Sci.*, **22**, 637, 653.

11 O'Neil, M.E. and Stewartson, K. (1967) On the slow Motion of a Sphere parallel a nearby plane Wall. *J. Fluid Mech.*, **27**, 705–724.

12 Cooley, M.D. and O'Neil, M.E. (1968) On the slow Rotation of a Sphere about a Diameter parallel to a nearby plane Wall. *J. Inst. Math. Applics.*, **4**, 163–173.

13 O'Neil, M.E. (1969) On asymmetric slow viscous Flows caused by the Motion of two Spheres almost in Contact. *Proc. Camb. Phil. Soc. A*, **65**, 543–556.

14 Cooley, M.D. and O'Neil, M.E. (1969) On the slow Motion generated in a viscous Fluid by the approach of a Sphere to a plane Wall or stationary Sphere. *Mathematika*, **16** (1), 37–49.

15 Cooley, M.D. and O'Neil, M.E. (1969) On the slow Motion of two Spheres in Contact along their Line of Centers through a viscous Fluid. *Proc. Camb. Phil. Soc.*, **66**, 407–415.

16 Davis, M.H. (1969) The slow Translation and Rotation of two unequal Spheres in a viscous Fluid. *Chem. Eng. Sci.*, **24**, 1769–1776.

17 O'Neil, M.E. (1970) Exact Solutions of the Equations of slow viscous Flow generated by the asymmetrical Motion of two equal Spheres. *Appl. Sci. Res.*, **21**, 452–466.

18 O'Neil, M.E. and Majumdar, S.R. (1970) Asymptotic Flow viscous Motions caused by the Translation or Rotation of two Spheres. Part I. The Determination of Exact Solutions for any Values of the Ratio Radii and Separation Parameters. Part II. Asymptotic Forms of the Solutions when the Minimum Clearance between the Spheres approaches Zero. *ZAMP*, **21**, 164, 180.

19 Goren, S.L. (1970) The normal Force exerted by creeping Flow on a small Sphere touching a Plane. *J. Fluid Mech.*, **41**, 619–625.

20 Goren, S.L. and O'Neil, M.E. (1971) On the hydrodynamic Resistance to a Particle of a dilute Suspension when in the Neighborhood of a large Obstacle. *Chem. Eng. Sci.*, **26**, 325–338.

21 Wakiya, S. (1971) Slow Motion in Shear Flow of a Doublet of two Spheres in Contact. *J. Phys. Soc. Japan*, **31**, 158–1587.

22 Wacholder, E. and Weihs, D. (1972) Slow Motion of a Fluid Sphere in the Vicinity of another Sphere or a Plane Wall. *Chem. Eng. Sci.*, **27**, 1817–1828.

23 Haber, S., Hetsroni, G. and Solan, A. (1973) On the low Reynolds Number Motion of two Droplets. *Int. J. Multiphase Flow*, **1**, 57–71.

24 Zinchenko, A.Z. (1978) To Calculation of hydrodynamic Interactions of Droplets by Low Reynolds Number. *Appl. Math. Mech.*, (5), 955–959.

25 Zinchenko, A.Z. (1980) Slow asymmetric motion of two Droplets in viscous Medium. *Appl. Math. Mech.*, (1), 49–59.

26 Schmitz, R. and Felderhof, R. (1982) Mobility Matrix for two spherical Particles with hydrodynamic Interaction. *Physica A*, **116**, 163–177.

27 Jeffrey, D.J. and Onishi, Y. (1984) The Forces and Couples acting on two nearly touching Spheres in Low-Reynolds-Number Flow. *ZAMP*, **35**, 634–641.

28 Jeffrey, D.J. and Onishi, Y. (1984) Calculation of the Resistance and Mobility Functions for two unequal Spheres in Low-Reynolds-Number Flow. *J. Fluid Mech.*, **139**, 261–290.

29 Kim, S. and Mifflin, R.T. (1985) The Resistance and Mobility Functions of two equal Spheres in Low Reynolds Number Flow. *Phys. Fluids*, **28**, 2033–2044.

30 Fuentes, Y.O., Kim, S. and Jeffrey, D.J. (1988) Mobility Functions for two unequal viscous Drops in Stokes Flow. Part 1. Axisymmetric Motions. *Phys. Fluids*, **31**, 2445–2455.

31 Fuentes, Y.O., Kim, S. and Jeffrey, D.J. (1989) Mobility Functions for two unequal viscous Drops in Stokes Flow. Part 2. Nonaxisymmetric Motions. *Phys. Fluids A*, **1**, 61–76.

32 O'Neil, M.E. (1968) A Sphere in Contact with a Plane Wall in a slow linear Shear Flow. *Chem. Eng. Sci.*, **23**, 1293–1298.

33 Lin, G.L., Lee, K.J. and Sather, N.F. (1970) Slow Motion of two Spheres in a Shear Field. *J. Fluid Mech.*, **43**, 35–47.

34 Nir, A. and Acrivos, A. (1973) On the creeping Motion of two arbitrary-sized touching Spheres in a linear Shear Field. *J. Fluid Mech.*, **59**, 209–223.

35 Hetsroni, G. and Haber, S. (1978) Low Reynolds Number Motion of two Drops submerged in an unbounded arbitrary Velocity Field. *Int. J. Multiphase Flow*, **4**, 1–17.

36 Zinchenko, A.Z. (1983) Hydrodynamic Interactions of two identical Liquid Spheres in linear Flow Field. *Appl. Math. Mech.*, (1), 56–63.

37 Martinov, S.I. (1998) Hydrodynamic Interactions of Particles. *Fluid Dyn.*, (2), 112–119, (in Russian).

38 Martinov, S.I. (2000) Interactions of Particles in Flow with parabolic Velocity Profile. *Fluid Dyn.*, (1), 84–91, (in Russian).

39 Brenner, H. and O'Neil, M.E. (1972) On the Stokes Resistance of multiparticle Systems in a linear shear Field. *Chem. Eng. Sci.*, **27**, 1421–1439.

40 Brinkman, H.C. (1947) A Calculation of the viscous Force exerted by a flowing Fluid on a dense Swarm of Particles. *Appl. Sci. Res.*, **A1**, 27–34.

41 Tamm, C.K.W. (1969) The Drag on a Cloud of spherical Particles in low Reynolds Number Flow. *J. Fluid Mech.*, **38**, 537–546.

42 Saffman, P.G. (1971) On the Boundary Condition at the Surface of a Porous Medium. *Studies in Appl. Math*, **30** (2), 93–101.

43 Lundgren, T.S. (1972) Slow Flow through stationary Random Beds and Suspensions of Spheres. *J. Fluid Mech.*, **51**, 273–299.

44 Brenner, H. (1974) Rheology of dilute Suspension of axisymmetric Brownian Particles. *Int. J. Multiphase Flow*, **1**, 195–341.

3
Brownian Motion of Particles

Chaotic motion of microparticles (whose size can range from 1 nm to 10 μm) suspended in a fluid was first discovered by Brown in 1827 and is now called Brownian motion. A systematic study of Brownian motion was initiated only in the early 20th century by Einstein, Smoluchowski, Perrin, Langevin, and Lorentz. Their works showed that Brownian motion is caused by very frequent collisions of particles with molecules of the surrounding fluid undergoing incessant random motion (thermal motion). The motion of molecules is so irregular that Brownian motion can be described only probabilistically, under the assumption of frequent and statistically independent impacts of molecules against a particle. More recently discovered phenomena similar to Brownian motion, for instance, fluctuations of current within conductors, are also caused by thermal motion of molecules and electrons.

Brownian motion of a particle in a fluid is taking place under the action of two forces: a random force caused by collisions with molecules, and a systematic force caused by viscous friction. The statistical average of kinetic energy of thermal motion of a particle is equal to $3k\vartheta/2$, where k is the Boltzmann constant and ϑ – the absolute temperature. This follows from the law of uniform distribution of kinetic energy over degrees of freedom (equipartition theorem). Thus, when a particle is moving along a straight line, this energy equals $k\vartheta/2$.

The first works on Brownian motion were devoted to the motion of an isolated particle. The results obtained in these works are applicable for suspensions with very low volume concentration of particles (infinitely dilute suspensions), when the average distance between particles is large compared to particle sizes, so that inter-particle interactions (hydrodynamic interactions, molecular interactions, collisions) can be neglected. Later works on Brownian motion took into account pair interactions between particles, which made their results applicable for higher (but not too high!) volume concentrations φ. The more recent works attempt to take into account multiparticle interactions as well.

We begin our discussion of Brownian motion with the analysis of random motion of an individual particle in a quiescent fluid, neglecting external forces and

Statistical Microhydrodynamics. Emmanuil G. Sinaiski and Leonid Zaichik

ISBN: 978-3-527-40656-2

interactions with other particles and with the surrounding fluid. Such a motion is called random walk. Further on, we shall discuss particle motion under the action of systematic and random forces. Finally, we go to the case of particle motion affected by the flow of the surrounding fluid and by interparticle interactions. Works [1–12] will provide the reader with extra information about Brownian motion and the common mathematical techniques employed in research.

3.1
Random Walk of an Isolated Particle

The case of Brownian motion of an isolated particle boils down to the random walk problem, which can generally be formulated in the following way. A particle driven by some external influence undergoes successive displacements $\mathbf{r}_1, \mathbf{r}_2, \ldots, \mathbf{r}_i, \ldots$, each of which is independent from all previous ones (in terms of both absolute magnitude and direction). The probability for a displacement to lie in the interval $(\mathbf{r}_i, \mathbf{r}_i + d\mathbf{r}_i)$ is specified by some probability density distribution $\tau_i(\mathbf{r}_i)$. (Possible types of distributions will be considered later on.) Our task is to find the probability that after N steps the particle will be located in the interval $(\mathbf{R}, \mathbf{R} + d\mathbf{R})$.

Consider first the simplest variant of random walk – a series of steps of equal length along a straight line, where each step can be directed forward or backward with the same probability of 0.5. The step length is assumed to be 1, so after N steps the particle can wind up at any one of the points with coordinates $-N, -N+1, \ldots, -1, 0, 1, \ldots, N-1, N$. The probability $P(m, N)$ that after N steps the particle will be found at the point m is given by the Bernoulli distribution

$$P(m, N) = C^m_{(m+N)/2}\left(\frac{1}{2}\right)^N, \tag{3.1}$$

where $C^m_{(m+N)/2} = \frac{N!}{\left(\frac{1}{2}(N+m)\right)!\left(\frac{1}{2}(N-m)\right)!}$ is the binomial coefficient, and the numbers m and N have the same parity.

The average displacement and the root-mean-square deviation of the particle are, respectively,

$$\langle m \rangle = 0, \ \sqrt{\langle m^2 \rangle} = \sqrt{N}.$$

In the case when $N \gg 1$ and $m \ll N$, which is of primary interest to us, Eq. (3.1) gives the following asymptotic distribution:

$$P(m, N) = \left(\frac{2}{\pi N}\right)^{1/2} \exp\left\{-\frac{m^2}{2N}\right\}. \tag{3.2}$$

Now, consider the same problem but with steps of length l. Let us introduce the particle displacement $X = ml$ and consider the intervals $\Delta X \gg l$ along the straight line. Then the probability to find the particle in the coordinate interval (X, $X+\Delta X$) after N steps is

$$P(X,N)\Delta X = P(m,N)\frac{\Delta X}{2l} = \frac{1}{(2\pi N l^2)^{1/2}}\exp\left\{-\frac{X^2}{2Nl^2}\right\}\Delta X.$$

If the particle accomplishes ν displacements in a unit time, the probability to find it in the interval (X, $X+\Delta X$) at the moment t is

$$P(X,t)\Delta X = \frac{1}{2(\pi D t)^{1/2}}\exp\left\{-\frac{X^2}{4Dt}\right\}\Delta X. \tag{3.3}$$

where $D=\nu l^2 2$.

It was assumed up to this point that the straight line along which the particle is moving is unbounded. Consider now the case of a bounded straight line. Suppose a boundary is placed at the point ($X_1 = m_1 l > 0$) which either reflects or absorbs the particle. If the boundary is reflective, the probability for the particle to be at the point $m < m_1$ after N steps is

$$P(m,N;m_1) = P(m,N) + P(2m_1 - m, N).$$

The second term on the right-hand side is the probability to find the particle at the mirror-reflected point $2m_1 - m$ after N steps; keep in mind that the probability $P(m,n)$ is still determined by the formula (3.1). In the limiting case of $N \gg 1$ and $m \ll N$ we arrive at the formula

$$P(m,N) = \left(\frac{2}{\pi N}\right)^{1/2}\left(\exp\left\{-\frac{m^2}{2N}\right\} + \exp\left\{-\frac{(2m_1-m)^2}{2N}\right\}\right).$$

Introducing the same notation as the one implied by Eq. (3.3), we obtain the following:

$$P(X,t;X_1) = \frac{1}{2(\pi D t)^{1/2}}\left(\exp\left\{-\frac{X^2}{4Dt}\right\} + \exp\left\{-\frac{(2X_1-X)^2}{4Dt}\right\}\right). \tag{3.4}$$

Note that the distribution (3.4) obeys the well-known boundary condition $(\partial P/\partial X)_{X=X_1}$ that is responsible for the vanishing of the flux at the plane wall.

For an absorbing boundary, the probability that after N steps the particle will be found at a point $m < m_1$ is equal to

$$P(m,N;m_1) = P(m,N) - P(2m_1 - m, N).$$

The minus sign before the second term expresses the fact that the absorbing boundary prohibits the particle to get into a mirror-reflected point. The limiting case

$N \gg 1$ and $m \ll N$ results in the formula

$$P(X,t;X_1) = \frac{1}{2(\pi Dt)^{1/2}}\left(\exp\left\{-\frac{X^2}{4Dt}\right\} - \exp\left\{-\frac{(2X_1 - X)^2}{4Dt}\right\}\right). \tag{3.5}$$

The distribution (3.5) satisfies the condition $P(X_1, t; X_1) = 0$, which corresponds to a well-known boundary condition at the completely absorbing wall.

The derived expressions make it possible to determine the probabilistic flux of particles deposited on the wall. In order to do so, it is necessary to find the probability that after N steps the particle will arrive at the point m_1 and that while performing these N steps, it will never cross or touch the boundary, that is, the point $m = m_1$. This probability is $A(m_1, N) = m_1/NP(m_1, N)$. In the limiting case of large N it is equal to

$$A(X_1,t) = \frac{X_1}{Nt}\frac{1}{(\pi Dt)^{1/2}}\exp\left\{-\frac{X^2}{4Dt}\right\}\Delta X, \tag{3.6}$$

where $X_1 = m_1 l$; $N = nt$; $D = nl^2/2$; l is the step length; n is the number of displacements in a unit time.

If the initial number of particles ($t = 0$, $X = 0$) is known, we can find the rate of deposition particles on the absorbing wall. At the moment t, the ratio of the number of deposited particles per unit time to the initial number of particles will be equal to

$$q(X_1,t) = \frac{X_1}{t}\frac{1}{2(\pi Dt)^{1/2}}\exp\left\{-\frac{X_1^2}{4Dt}\right\}. \tag{3.7}$$

The distribution (3.7) obeys the condition

$$q(X_1,t) = -D\left(\frac{\partial P}{\partial X}\right)_{X=X_1}, \tag{3.8}$$

where P is determined by the relation (3.5).

If P stands for the concentration of a matter, then Eq. (3.8) can be interpreted as a probability flux or, from the physical viewpoint, as a diffusion flux of matter.

Consider now the general random walk problem. The particle accomplishes N displacements, whose positions are determined by radius vectors $\boldsymbol{r}_1, \boldsymbol{r}_2, \ldots, \boldsymbol{r}_N$, so after N displacements the particle will be found at the point $\boldsymbol{R} = \sum_{i=1}^{N} \boldsymbol{r}_i$. Each displacement is assigned its own probability density $\tau_i(\boldsymbol{r}_i)$. Thus $\tau_i(\boldsymbol{r}_i)d\boldsymbol{r}_i$ is the probability for i-th displacement vector to belong to the range $(\boldsymbol{r}_i, \boldsymbol{r}_i + d\boldsymbol{r}_i)$. In the coordinate form, this probability is written as

$$\tau_i(\boldsymbol{r}_i)d\boldsymbol{r}_i = \tau_i(X_i, Y_i, Z_i)dX_i\,dY_i\,dZ_i, \quad (i = 1, 2, \ldots, N).$$

The problem is to find the probability $p_N(\boldsymbol{R})d\boldsymbol{R}$ that after N displacements the particle will be found in the interval $(\boldsymbol{R}_i, \boldsymbol{R}_i + d\boldsymbol{R}_i)$. The solution of this problem obtained by

Markov's method has the following form:

$$p_N(\boldsymbol{R}) = \frac{1}{8\pi^3}\int_{-\infty}^{\infty} A_N(\boldsymbol{\rho})\exp(-i\rho\cdot\boldsymbol{R})d\boldsymbol{\rho},$$
$$A_N(\boldsymbol{\rho}) = \prod_{j=1}^{N}\int_{-\infty}^{\infty}\tau_j(\boldsymbol{r}_j)\exp(i\boldsymbol{\rho}\cdot\boldsymbol{r}_j)d\boldsymbol{r}_j, \tag{3.9}$$

where $\boldsymbol{\rho}$ is an N-dimensional vector and $A_N(\boldsymbol{\rho})$ is the Fourier transform of the function $\prod_{j=1}^{N}\tau_j(\boldsymbol{r}_j)$.

Hence, the solution of the general random walk problem depends on the form of probability density $\tau_i(\boldsymbol{r}_i)$ for each displacement. Let us take a look at some distributions commonly encountered in applications.

3.1.1 Isotropic Distribution

A distribution is isotropic if it does not depend on the direction of the displacement vector $\boldsymbol{r}_i$, and depends only on its length (or, which is the same, on the square of length $|\boldsymbol{r}_i^2| = r^2$ if all displacements have the same length). Clearly, such a distribution possesses a spherical symmetry. That is why such distributions are sometimes called spherical distributions of displacement directions. Also, distributions of displacements must to be identical for all displacements, so that we can drop the index after τ, writing

$$\tau_j(\boldsymbol{r}_j) = \tau(\boldsymbol{r}^2). \tag{3.10}$$

Then the second relation (3.9) yields

$$A_N(\boldsymbol{\rho}) = \left\{\int_{-\infty}^{\infty}\tau(r^2)\exp(i\boldsymbol{\rho}\cdot\boldsymbol{r})d\boldsymbol{r}\right\}^N.$$

Switching to a spherical coordinate system and integrating, we get:

$$A_N(\boldsymbol{\rho}) = \left\{\int_{-\infty}^{\infty}\int_{-1}^{1}\int_{0}^{2\pi} r^2\tau(r^2)\exp(i|\boldsymbol{\rho}|\cdot rt)d\omega\,dt\,dr\right\}^N = \left\{4\pi\int_0^{\infty}\frac{\sin(|\boldsymbol{\rho}|r)}{|\boldsymbol{\rho}|r}r^2\tau(r^2)dr\right\}^N$$
$$= \left\{4\pi\int_0^{\infty}\left(1-\frac{1}{6}|\boldsymbol{\rho}|^2r^2+\ldots\right)r^2\tau(r^2)dr\right\}^N.$$

Tending N to ∞, we obtain the following expression:

$$A_N(\boldsymbol{\rho}) = \exp\left\{-\frac{N|\boldsymbol{\rho}^2|\langle r^2\rangle}{6}\right\},$$

where $\langle r^2\rangle$ is the root-mean-square displacement.

A substitution of this expression into the first relation (3.9) and a subsequent integration results in

$$p(\boldsymbol{R}) = p_{N\to\infty}(\boldsymbol{R}) = \left(\frac{3}{2\pi N\langle r^2\rangle}\right)^{3/2} \exp\left\{-\frac{3|\boldsymbol{R}|^2}{2N\langle r^2\rangle}\right\}. \tag{3.11}$$

Suppose the particle accomplishes ν displacements per second. If particle makes N displacements during the time t, then $N = \nu t$. Designating $D = \nu\langle r^2\rangle/6$, we can rewrite Eq. (3.11) as

$$p(\boldsymbol{R}) = \frac{1}{(4\pi Dt)^{3/2}} \exp\left\{-\frac{|\boldsymbol{R}|^2}{4Dt}\right\}. \tag{3.12}$$

So, after time t the particle will be found in spherical shell volume element between $\boldsymbol{R}$ and $\boldsymbol{R} + d\boldsymbol{R}$ with probability $p(\boldsymbol{R})d\boldsymbol{R}$, where $p(\boldsymbol{R})$ is given by Eq. (3.12).

3.1.2
Gaussian Distribution

Let the displacements have a probability density described by a Gaussian distribution:

$$\tau_j(\boldsymbol{r}_j) = \left(\frac{3}{2\pi\langle r_j^2\rangle}\right)^{3/2} \exp\left\{-\frac{3|\boldsymbol{r}_j|^2}{2\langle r_j^2\rangle}\right\}, \tag{3.13}$$

where $\langle r_j^2\rangle$ is the mean square of j-th displacement. Since the distribution (3.13) depends on $|\boldsymbol{r}_j|$, the direction of each displacement vector can be arbitrary. A substitution of Eq. (3.13) into the second relation (3.9) leads to

$$\begin{aligned} A_N(\rho) &= \prod_{j=1}^{N} \left(\frac{3}{2\pi\langle r_j^2\rangle}\right)^{3/2} \int_{-\infty}^{\infty}\int_{-\infty}^{\infty}\int_{-\infty}^{\infty} \exp(i(\rho_1 X_j + \rho_2 Y_j + \rho_3 Z_j) \\ &\quad -3(X_j^2 + Y_j^2 + Z_j^2)/2\langle r_j^2\rangle) dX_j + dY_j + dZ_j \\ &= \prod_{j=1}^{N} \exp\left(-(\rho_1^2 + \rho_2^2 + \rho_3^2)\langle r_j^2\rangle/6\right) = \exp\left(-|\boldsymbol{\rho}|^2 N\langle r_j^2\rangle/6\right), \end{aligned}$$

where $\langle r^2\rangle = \frac{1}{N}\sum_{j=1}^{N}\langle r_j^2\rangle$.

Now it follows from the first relation (3.9) that

$$p_N(\boldsymbol{R}) = \left(\frac{3}{2\pi N\langle r^2\rangle}\right)^{3/2} \exp\left\{-\frac{3|\boldsymbol{R}|^2}{2N\langle r^2\rangle}\right\}. \tag{3.14}$$

Hence, the assumption of a Gaussian distribution of probability density of displacements allows to obtain an exact solution of the general random walk problem for $N \gg 1$.

3.1.3
An Arbitrary Distribution $\tau(r)$ in the Limiting Case $N \gg 1$

We assume that all $\tau_j(\boldsymbol{r}_j)$ are identical and equal to $\tau(\boldsymbol{r})$. In this case one can get an exact expression for the probability density $p_N(\boldsymbol{R})$ at $N \to \infty$. Eq. (3.9) gives us

$$A_N(\boldsymbol{\rho}) = \left\{ \int_{-\infty}^{\infty}\int_{-\infty}^{\infty}\int_{-\infty}^{\infty} \tau(X, Y, Z)\exp(i(\rho_1 X + \rho_2 Y + \rho_3 Z)dX\,dY\,dZ \right\}^N.$$

Expanding the exponent as a Taylor series and recalling the expressions for moments of the distribution $\tau(\boldsymbol{r})$, we get:

$$\begin{aligned} A_N(\boldsymbol{\rho}) &= \left\{ \int_{-\infty}^{\infty}\int_{-\infty}^{\infty}\int_{-\infty}^{\infty} (1 + i(\rho_1 X + \rho_2 Y + \rho_3 Z) - \frac{1}{2}(\rho_1^2 X^2 + \rho_2^2 Y^2 + \rho_3^2 Z^2 \right. \\ &\quad \left. + 2\rho_1\rho_2 XY + 2\rho_2\rho_3 YZ + 2\rho_3\rho_1 ZX) + \ldots)\tau(X, Y, Z)dX\,dY\,dZ \right\}^N \\ &= 1 + i(\rho_1\langle X\rangle + \rho_2\langle Y\rangle + \rho_3\langle Z\rangle) - \frac{1}{2}(\rho_1^2\langle X^2\rangle + \rho_2^2\langle Y^2\rangle + \rho_3^2\langle Z^2\rangle \\ &\quad + 2\rho_1\rho_2\langle XY\rangle + 2\rho_2\rho_3\langle YZ\rangle + 2\rho_3\rho_1\langle ZX\rangle) + \ldots)^N. \end{aligned}$$

At $N \to \infty$, the derived expression tends to an exponential expression, so that

$$A_N(\boldsymbol{\rho}) = \exp\left\{ iN(\rho_1\langle X\rangle + \rho_2\langle Y\rangle + \rho_3\langle Z\rangle) - \frac{1}{2}NQ(\rho) \right\},$$

where $Q(\rho)$ denotes the following quadratic form:

$$Q(\boldsymbol{\rho}) = \rho_1^2\langle X^2\rangle + \rho_2^2\langle Y^2\rangle + \rho_3^2\langle Z^2\rangle + 2\rho_1\rho_2\langle XY\rangle + 2\rho_2\rho_3\langle YZ\rangle + 2\rho_3\rho_1\langle ZX\rangle.$$

Let us choose a coordinate system whose axes coincide with the principal directions of the quadratic form (ξ, η, ζ). Then $Q(\rho)$ assumes the canonical form

$$Q(\rho) = \rho_\chi^2\langle\chi^2\rangle + \rho_\eta^2\langle\eta^2\rangle + \rho_\xi^2\langle\zeta^2\rangle,$$

where $\langle\xi^2\rangle$, $\langle\eta^2\rangle$, and $\langle\zeta^2\rangle$ are the eigenvalues of the symmetric matrix of the quadratic form. In the new coordinate system, the vector $\boldsymbol{R}$ has components (R_ξ, R_η, R_ζ). Then Eq. (3.9) gives the following expression for the case $N \to \infty$:

$$\begin{aligned} p(\boldsymbol{R}) &= \frac{1}{8\pi^3}\int_{-\infty}^{\infty}\int_{-\infty}^{\infty}\int_{-\infty}^{\infty} \exp\left\{ -\frac{1}{2}N(\rho_\xi^2\langle\xi^2\rangle + \rho_\eta^2\langle\eta^2\rangle + \rho_\xi^2\langle\zeta^2\rangle) \right. \\ &\quad \left. - i(\rho_\xi(R_\xi - N\langle\xi\rangle) + \rho_\eta(R_\eta - N\langle\eta\rangle) + \rho_\zeta(R_\zeta - N\langle\zeta\rangle)) \right\} d\rho_\xi\, d\rho_\eta\, d\rho_\zeta \\ &= \frac{1}{(8\pi^3 N^3\langle\xi^2\rangle\langle\eta^2\rangle\langle\zeta^2\rangle)^{1/2}}\exp\left\{ -\frac{R_\xi - N\langle\xi\rangle}{2N\langle\xi^2\rangle} - \frac{R_\eta - N\langle\eta\rangle}{2N\langle\eta^2\rangle} - \frac{R_\zeta - N\langle\zeta\rangle}{2N\langle\zeta^2\rangle} \right\}. \end{aligned}$$

The obtained distribution $p(\boldsymbol{R})$ shows that the particle experiences a systematical drift $N\langle\xi\rangle$, $N\langle\eta\rangle$, $N\langle\zeta\rangle$ along the principal axes, onto which fluctuations $N\langle\xi^2\rangle$, $N\langle\eta^2\rangle$, $N\langle\zeta^2\rangle$ get superimposed.

3.2
Random Walk of an Ensemble of Particles

The previous section dealt with the motion of an isolated particle. Since interactions of the particle with the surrounding fluid and with the other particles were not taken into account, we may as well consider an ensemble of particles that were all initially located at one and the same point and then started a random walk (we again disregard the influence of the surrounding fluid and interparticle interactions). The results of the previous section can be applied directly to this case, but their interpretation should be changed. For instance, the expression (3.9) and all subsequent relations have the meaning of fraction of the total number of particles that will be found at time t in the spatial interval $(\boldsymbol{R}, \boldsymbol{R}+d\boldsymbol{R})$ given that all particles were initially located at the point $\boldsymbol{R}=0$. In this formulation of the problem the total number of particles is conserved and is equal to the initial value.

We are mostly interested in the limiting case of an extremely large number of particles $N \to \infty$. Then particle positions can be regarded as continuous random variables, and a continuous probability distribution can be introduced. Besides, since the random walk is a Markowian process, the results of Sections 1.11 and 1.12 can be used, in particular the differential Fokker–Planck equation for the PDF of a random variable in a Markowian process. For an isotropic distribution $\tau(r^2)$ considered in the previous section, the Fokker–Planck equation assumes the form

$$\frac{\partial p}{\partial t} = D\left\{\frac{\partial^2 p}{\partial X^2} + \frac{\partial^2 p}{\partial Y^2} + \frac{\partial^2 p}{\partial Z^2}\right\} = D\Delta p, \tag{3.15}$$

because, due to the symmetry of the distribution $\tau(\boldsymbol{r}^2)$, the drift $\boldsymbol{A}$ is zero and the diffusion tensor is reduced to $\boldsymbol{D} = D\boldsymbol{I}$, where $\boldsymbol{I}$ is a unit tensor with components δ_{ij}.

In the more general case 3 (see Section 3.1), the distribution $\tau(\boldsymbol{r})$ that controls individual particle displacements does not possess any special symmetry properties. Therefore moments of the first order that determine the drift vector are nonzero, and moments of the second order that determine the symmetric diffusion tensor cannot be thought of as components of a diagonal matrix. But when the coordinate system is chosen in such a way that its axes coincide with principal axes of the diffusion tensor, the diffusion tensor reduces to $\boldsymbol{D} = ||D_{ij}\delta_{ij}||$ and the Fokker–Planck equation to

$$\frac{\partial p}{\partial t} = A_1\frac{\partial p}{\partial X} + A_2\frac{\partial p}{\partial Y} + A_3\frac{\partial p}{\partial Z} + D_{11}\frac{\partial^2 p}{\partial X^2} + D_{22}\frac{\partial^2 p}{\partial Y^2} + D_{33}\frac{\partial^2 p}{\partial Z^2}, \tag{3.16}$$

where $A_1 = -\nu\langle X\rangle$, $A_2 = -\nu\langle Y\rangle$, $A_3 = -\nu\langle Z\rangle$, $D_{11} = \nu\langle X^2\rangle/2$, $D_{11} = \nu\langle Y^2\rangle/2$, $D_{33} = \nu\langle Z^2\rangle/2$, and ν is the number of displacements per unit time.

Eq. (3.16) can be rewritten as a continuity equation:

$$\frac{\partial p}{\partial t} + \nabla\cdot\boldsymbol{j} = 0,$$

where j is probability flux with components

$$j_X = \left\{-A_1 P - D_{11}\frac{\partial p}{\partial X}\right\}, \quad j_Y = \left\{-A_2 P - D_{22}\frac{\partial p}{\partial Y}\right\}, \quad j_Z = \left\{-A_3 P - D_{33}\frac{\partial p}{\partial Z}\right\}.$$

These fluxes have the meaning of probable numbers of particles crossing the unit elements of surfaces perpendicular to axes X, Y and Z in a unit time.

If that all particles are initially concentrated at the point (X_0, Y_0, Z_0), the solution of Eq. (3.17) is

$$p(X, Y, Z) = \frac{1}{8(\pi t)^{3/2}(D_{11}D_{22}D_{33})^{1/2}} \times \exp\left\{-\frac{(X-X_0+A_1 t)^2}{4D_{11}t} - \frac{(Y-Y_0+A_2 t)^2}{4D_{22}t} - \frac{(Z-Z_0+A_3 t)^2}{4D_{33}t}\right\}$$

3.3
Brownian Motion of a Free Particle in a Quiescent Fluid

Brownian motion of a free particle is defined as the particle's motion in the fluid under the action of two forces: the random force caused by collisions between the particle and molecules of the fluid, and the systematic force of hydrodynamic resistance.

Assume that the fluid is at rest, the external force field is absent, and interparticle interactions are negligibly small. The last assumption means that the suspension is infinitely dilute.

The modern theory of Brownian motion is based on the Langevin stochastic equation (see Section 1.13):

$$m\dot{\boldsymbol{u}} = -h\boldsymbol{u} + \boldsymbol{F}^{fl}(t), \quad \boldsymbol{u} = \dot{\boldsymbol{X}}(t). \tag{3.17}$$

Here $\boldsymbol{X}$ is the particle's position in some fixed coordinate system; $\boldsymbol{u}$ is the particle's velocity; m – the particle's mass; $h = 6\pi\mu_e$ – the coefficient of hydrodynamic resistance; $\boldsymbol{F}^{fl}$ – the random force; μ_e – the coefficient of dynamic viscosity; a – the radius of the particle (the particle is assumed to be spherical).

As to random force, it is assumed that: $\boldsymbol{F}^{fl}$ varies much faster than $\boldsymbol{u}$, therefore it is independent of $\boldsymbol{u}$; the average value is $\langle \boldsymbol{F}^{fl} \rangle = 0$; the correlation is $\langle \boldsymbol{F}^{fl}(t)\, \boldsymbol{F}^{fl}(t+\tau)\rangle = \boldsymbol{F}\delta(\tau)$. The last two conditions follow from Eqs. (1.14) and (1.15).

First, we should note that the characteristic time of velocity change that happens due to the action of a systematic force on the particle has the order of m/h. Since $m \sim a^3$ and $h \sim a$, we have $m/h \sim a^2$. The particle size is small, so m/h is small as well. At the same time it is much longer than the characteristic time of change of the random force. Hence, for $t \gg m/h$, the inertia term in the Langevin equation can be dropped, and Eq. (3.11) reduces to

$$0 = -h\boldsymbol{u} + \boldsymbol{F}^{fl}(t).$$

The parameter $t_\nu = m/h$ is called the characteristic time of dynamic (viscous) relaxation of the particle. If the process is taking place on a time scale much longer then the characteristic time of random force fluctuation and yet shorter than t_ν, the particle's velocity varies with time, and the corresponding PDF should be a function of $\boldsymbol{u}$. For $t \gg t_\nu$, the PDF is a function of the particle's position $\boldsymbol{r}$ and the initial velocity $\boldsymbol{u}_0$, while being independent of $\boldsymbol{u}$. Further on, our main attention will be focused on these two limiting cases.

Consider the motion of a particle during the time interval $\Delta t = t - t_0$ long enough compared to the period of random force fluctuation (and thereby to the period of random acceleration of the particle), yet short enough compared to the time of change of the particle's velocity. Since the Langevin equation is the equation for velocity $\boldsymbol{u}$ in a Markowian process, one has to specify the velocity $\boldsymbol{u}_0$ at the previous instant of time and then, using the properties of $\boldsymbol{F}^{fl}$, determine statistical properties of the distribution p. In this case one should consider the dependence of p not on particle position $\mathbf{X}$ but on particles velocity $\mathbf{u}$, in other words, $p = p(\boldsymbol{u}, t|\boldsymbol{u}_0, t_0)$. Then $p(\boldsymbol{u}, t|\boldsymbol{u}_0, t_0)d\boldsymbol{u}$ gives the probability that the particle's velocity at time t will belong to the interval $(\boldsymbol{u}, \boldsymbol{u}+d\boldsymbol{u})$ given that at the previous (initial) moment $t_0 = t - \Delta t$ it was equal to $\boldsymbol{u}_0$, $\boldsymbol{u} - \Delta\boldsymbol{u}$. The PDF defined in this way should satisfy the initial condition

$$p(\boldsymbol{u}, t_0|\boldsymbol{u}_0, t_0) = \delta(\boldsymbol{u}-\boldsymbol{u}_0) = \delta(u_X - u_{X0})\delta(u_Y - u_{Y0})\delta(u_Z - u_{Z0}).$$

Consider now the condition that the distribution $p(\boldsymbol{u}, t|\boldsymbol{u}_0, t_0)$ must satisfy at $t \to \infty$. Look at the Fokker–Planck equation. We already wrote it for the coordinate distribution $p(\mathbf{Z}, t|\mathbf{Y}, t')$ in Section 1.12. A transition to the velocity distribution leads to a similar equation, but with different drift and diffusion coefficients. This is self-evident, because the term "coordinates" usually implies generalized coordinates, which includes velocities.

The one-dimensional Fokker–Planck equation is

$$\frac{\partial p}{\partial t} = -\frac{\partial (A_u p)}{\partial u} + \frac{\partial^2 (D_u p)}{\partial u^2}. \tag{3.18}$$

The coefficient A_u stands for the particle's acceleration, therefore, according to Eq. (3.17), it is equal to

$$A_u = -fu, \quad f = h/m = 6\pi\mu_e a/m.$$

Now Eq. (3.18) for $D_u = const$ takes the form

$$\frac{\partial p}{\partial t} = -\frac{\partial (fup)}{\partial u} + D_u \frac{\partial^2 p}{\partial u^2}, \tag{3.19}$$

This form allows us to find statistical moments $\langle u \rangle$ and $\langle u^2 \rangle$ without actually solving the equation. To this end, we multiply both sides successively by u and

u^2 and integrate the result over u from $-\infty$ up to $+\infty$, assuming that at $u \to \pm\infty$, $p \to 0$ faster than the function u^{-n}. The results are

$$\langle u \rangle = u_0 e^{-ft}, \quad \langle (u - \langle u \rangle)^2 \rangle = D_u (1 - e^{-ft})/f.$$

For $t \gg t_v = 1/f = m/6\pi\mu_e a$, we have

$$\langle u \rangle = 0, \quad \langle u^2 \rangle = D_u/f.$$

The characteristic time of viscous relaxation for the particle is $t_v = 2a^2\rho_p/9\rho_e\nu_e$, where ρ_p is the particle's density and $\nu_e = \mu_e/\rho_e$ is the coefficient of kinematic viscosity. Smallness of the particle's size results in a short viscous relaxation time and, consequently, very short transient period for the particle's velocity. The energy of thermal motion of the particle is $m\langle u^2 \rangle/2 = \mathrm{k}\vartheta/2$, so $\langle u^2 \rangle = \mathrm{k}\vartheta/m = D_u/f$ and $D_u = f\,\mathrm{k}\vartheta/m$.

Introduce new variables $\nu = ue^{ft}$, $\Phi = p^{-3ft}$, and $d\tau = D_u e^{2ft} dt$. In these variables, Eq. (3.19) transforms to a standard diffusion equation

$$\frac{\partial \Phi}{\partial \tau} = \frac{\partial^2 \Phi}{\partial \nu^2},$$

whose solution in the original variables is

$$p(\boldsymbol{u}, t | \boldsymbol{u}_0, t_0) = \frac{\exp\left\{-\frac{f(u - u_0 e^{-ft})^2}{2D_u(1 - e^{-2ft})}\right\}}{\left(\frac{2\pi D_u}{f}(1 - e^{-2ft})\right)^{1/2}}.$$

In the limiting case of $t \gg 1/f$, we have

$$p(\boldsymbol{u}, t | \boldsymbol{u}_0, t_0) = \left(\frac{f}{2\pi D_u}\right)^{1/2} \exp\left\{-\frac{fu^2}{2D_u}\right\} = \left(\frac{m}{2\pi \mathrm{k}\vartheta}\right)^{1/2} \exp\left\{-\frac{mu^2}{2\mathrm{k}\vartheta}\right\}. \tag{3.20}$$

So, after a sufficiently long time a Maxwellian velocity distribution will be established.

Consider now the PDF of displacements $p(\boldsymbol{r}, t | \boldsymbol{r}_0, \boldsymbol{u}_0, t_0)$. Recall that $p(\boldsymbol{r}, t | \boldsymbol{r}_0, \boldsymbol{u}_0, t_0)d\boldsymbol{r}$ is the probability for the particle's displacement at time t to be in the interval $(\boldsymbol{r}, \boldsymbol{r} + d\boldsymbol{r})$ given that $\boldsymbol{r}_0$ was the initial displacement and $\boldsymbol{u}_0$ – the initial velocity. In the one-dimensional case, this problem boils down to the Fokker–Planck equation

$$\frac{\partial p}{\partial t} = -\frac{\partial (Ap)}{\partial X} + \frac{\partial^2 (Dp)}{\partial X^2}, \tag{3.21}$$

where $A = u$. Solving this equation is tricky because p now depends on X. Therefore, in order to get $p(\boldsymbol{r}, t|\boldsymbol{r}_0, \boldsymbol{u}_0, t_0)$, we proceed as follows. Since $\dot{\boldsymbol{r}} = \boldsymbol{u}$, we can write

$$\boldsymbol{r}-\boldsymbol{r}_0 = \int_0^t \boldsymbol{u}(t)dt. \tag{3.22}$$

It follows from the Langevin equation (3.17) that

$$\boldsymbol{u}-\boldsymbol{u}_0 e^{-ft} = e^{-ft}\int_0^t e^{f\xi}F^{fl}(\xi)d\xi. \tag{3.23}$$

At $t \gg 1/f$, the left-hand side of Eq. (3.23) is equal to $\boldsymbol{u}$. Since statistical properties should be the same for both sides of Eq. (3.23), then, as we showed earlier, at $t \gg 1/f$ (or, which is the same, at $t \to \infty$) the distribution

$$e^{-ft}\int_0^t e^{f\xi}F^{fl}(\xi)d\xi.$$

should be Maxwellian, in other words,

$$\lim_{t\to\infty}\left\{e^{-ft}\int_0^t e^{f\xi}F^{fl}(\xi)d\xi\right\} = \left(\frac{m}{2\pi k\vartheta}\right)^{3/2}\exp\left\{-\frac{m|u|^2}{2k\vartheta}\right\}. \tag{3.24}$$

The difference between the distributions (3.24) and (3.20), namely, the different exponents in the first term of the product, is explained by different dimensionalities of these distributions.

From Eq. (3.22) and Eq. (3.23), we find that

$$\boldsymbol{r}-\boldsymbol{r}_0 = \frac{1}{f}\boldsymbol{u}_0(1-e^{-ft}) + \int_0^t e^{-f\eta}d\eta + \int_0^\eta e^{f\xi}\boldsymbol{F}^{fl}(\xi)d\xi.$$

Integration by parts gives

$$\boldsymbol{r}-\boldsymbol{r}_0-\frac{1}{f}\boldsymbol{u}_0(1-e^{-ft}) = \int_0^t \psi(\xi)\boldsymbol{F}^{fl}(\xi)d\xi, \tag{3.25}$$

where

$$\psi(\xi) = \frac{1}{f}(1-e^{-f(\xi-t)}). \tag{3.26}$$

Now, consider the random variable

$$\boldsymbol{R} = \int_0^t \psi(\xi)\boldsymbol{F}^{fl}(\xi)d\xi.$$

The random function $\boldsymbol{F}^{fl}(\xi)$ varies extremely rapidly as compared to any other quantity. The period of fluctuation of $\boldsymbol{F}^{fl}(\xi)$ is of the same order as the characteristic time of interaction between the particle and molecules of the surrounding fluid. Therefore a Maxwellian distribution of the random variable $\boldsymbol{R}$

$$p(\boldsymbol{R}) = \left(\frac{1}{4\pi D_u \int_0^t \psi^2(\xi)d\xi}\right)^{3/2} \exp\left\{-\frac{|\boldsymbol{R}|^2}{4D_u \int_0^t \psi^2(\xi)d\xi}\right\}, \tag{3.27}$$

is established very quickly in the course of Brownian motion.

Substituting Eq. (3.26) into Eq. (3.27), we get

$$\begin{aligned} p(\boldsymbol{r}, t|\boldsymbol{r}_0, \boldsymbol{u}_0, t_0) &= \left(\frac{m f^2}{2\pi k\vartheta[2ft - 3 + 4e^{-ft} - e^{-2ft}]}\right)^{3/2} \\ &\times \exp\left\{-\frac{m f^2 |\boldsymbol{r}-\boldsymbol{r}_0-\boldsymbol{u}_0[1-e^{-ft}/f]|^2}{2\pi k\vartheta[2ft-3+4e^{-ft}-e^{-2ft}]}\right\}. \end{aligned} \tag{3.28}$$

It is readily seen that at $t \gg 1/f$, this equation can be simplified:

$$p(\boldsymbol{r}, t|\boldsymbol{r}_0, \boldsymbol{u}_0, t_0) = \frac{1}{(4\pi Dt)^{3/2}} \exp\left\{-\frac{|\boldsymbol{r}-\boldsymbol{r}_0|^2}{4Dt}\right\}, \tag{3.29}$$

where $D = k\vartheta/mf = k\vartheta/6\pi a\mu_e$.

Eq. (3.29) gives us the particle's root-mean-square displacement in the X-direction:

$$\langle (X-X_0)^2 \rangle = \frac{1}{3}\langle (|\boldsymbol{r}-\boldsymbol{r}_0|)^2 \rangle = \int_0^\infty |\boldsymbol{r}-\boldsymbol{r}_0|^2 p(\boldsymbol{r}, t|\boldsymbol{r}_0, \boldsymbol{u}_0, t_0)d\boldsymbol{r} = 2D_{br}t, \tag{3.30}$$

where D_{br} is called the coefficient of Brownian diffusion.

The relation (3.30) was first obtained by Einstein. It is valid only for $t \gg 1/f$. The expression for root-mean-square displacement for an arbitrary time t can be derived from (3.28):

$$|\boldsymbol{r}-\boldsymbol{r}_0|^2 = \frac{|\boldsymbol{u}_0|^2}{f^2}(1-e^{-ft})^2 + \frac{3k\vartheta}{m f^2}(2ft - 3 + 4e^{-ft} - e^{-2ft}).$$

Averaging this relation over $\boldsymbol{u}_0$ and taking into account the fact that $\langle|\boldsymbol{u}_0|^2\rangle = 3k\vartheta/m$, we get

$$\langle|\boldsymbol{r}-\boldsymbol{r}_0|^2\rangle = \frac{6k\vartheta}{mf^2}(ft-1+e^{-ft}). \tag{3.31}$$

The relation (3.30) follows from Eq. (3.31) at $t\to\infty$. An equation describing the initial stage of Brownian motion can be obtained from (3.31) by setting $t\to 0$:

$$\langle|\boldsymbol{r}-\boldsymbol{r}_0|^2\rangle = \frac{3k\vartheta}{m}t^2 = \langle|u_0|^2\rangle t^2. \tag{3.32}$$

3.4
Brownian Motion of a Particle in an External Force Field

Consider Brownian motion of a particle under the action of two systematic forces: the drag force $-h\boldsymbol{u}$ and the external force $\boldsymbol{F}_e$ (gravity force, electric force), and one random force $\boldsymbol{F}^{fl}(t)$. We assume that the characteristic time of random force fluctuations is much shorter then the characteristic times of variations of systematic forces.

Let us examine the particle's motion in the time interval $\Delta t = t - t_0$ that is long enough as compared to the period of random force fluctuations (i.e., to the duration of the accelerating force), but short enough in comparison with the time it takes for physical parameters, such as the particle's velocity, to change by a noticeable amount. The particle's motion is described by the Langevin equation:

$$m\dot{\boldsymbol{u}} = -h\boldsymbol{u} + \boldsymbol{F}_e + \boldsymbol{F}^{fl}(t), \quad \boldsymbol{u} = \dot{\boldsymbol{X}}(t). \tag{3.33}$$

The Fokker–Planck equation for the PDF $p(\boldsymbol{r}, \boldsymbol{u}, t|\boldsymbol{r}_0, \boldsymbol{u}_0, t_0)$ in this case reduces to

$$\frac{\partial p}{\partial t} + \boldsymbol{u}\cdot\nabla_r p + \nabla_u\cdot(-f\boldsymbol{u} + \boldsymbol{F}_e) = D\nabla_r^2 p + D_u\nabla_u^2 p. \tag{3.34}$$

The subscripts r and u in the operator ∇ mean that the derivatives are taken with respect to r_i and u_i, respectively.

Since during the time Δt the particle's position changes by $\Delta\boldsymbol{r} = \boldsymbol{u}\Delta t$, the mean-square displacement is $\langle(\Delta\boldsymbol{r})^2\rangle \sim |\boldsymbol{u}|^2(\Delta t)^2$ and (see Eq. 1.123)

$$D = \frac{1}{2}\lim_{\Delta t\to 0}\frac{\langle(\Delta\boldsymbol{r})^2\rangle}{\Delta t} = 0.$$

The obtained estimate allows to ignore the first summand on the right-hand side of equation (3.34):

$$\frac{\partial p}{\partial t} + \boldsymbol{u}\cdot\nabla_r p + \nabla_u\cdot p\left(-f\boldsymbol{u} + \frac{\boldsymbol{F}_e}{m}\right) = D_u\nabla_u^2 p. \tag{3.35}$$

It is convenient to rewrite this equation as

$$\frac{\partial p}{\partial t}+\boldsymbol{u}\cdot\nabla_r p+\frac{\boldsymbol{F}_e}{m}\cdot\nabla_u p=-f\nabla_u\cdot(p\boldsymbol{u})+D_u\nabla_u^2 p, \tag{3.36}$$

so that the left-hand side represents the total (substantial) derivative:

$$\frac{Dp(\boldsymbol{r},\boldsymbol{u},t|\boldsymbol{r}_0,\boldsymbol{u}_0,t_0)}{Dt}=\frac{\partial p}{\partial t}+\boldsymbol{u}\cdot\nabla_r p+\boldsymbol{F}_e\cdot\nabla_u p.$$

In the one-dimensional case, Eq. (3.35) has the form

$$\frac{\partial p}{\partial t}+u\frac{\partial p}{\partial X}+\frac{\partial}{\partial u}\left\{p\left(-fu+\frac{F_e}{m}\right)\right\}=D_u\frac{\partial^2 p}{\partial u^2}. \tag{3.37}$$

An attempt to solve Eq. (3.37) for the general case run into difficulties. Therefore we consider simple case when the external force is absent: $\boldsymbol{F}_e=0$. Then Eq. (3.37) reduces to

$$\frac{\partial p}{\partial t}+\boldsymbol{u}\cdot\nabla_r p-f\boldsymbol{u}\cdot\nabla_u p=3fp+D_u\nabla_u^2 p, \tag{3.38}$$

which can be solved by the method of characteristics. We write the two characteristic vector equations for Eq. (3.38):

$$\frac{\partial \boldsymbol{u}}{\partial t}=-f\boldsymbol{u},\quad \frac{\partial r}{\partial t}=\boldsymbol{u}.$$

The two families of solutions are:

$$\boldsymbol{u}e^{ft}=\boldsymbol{I}_1,\quad \boldsymbol{r}+\frac{\boldsymbol{u}}{f}=\boldsymbol{I}_2$$

where different $\boldsymbol{I}_1$ and $\boldsymbol{I}_2$ are different vectors. The general solution of Eq. (3.38) can be written as $p=p(\boldsymbol{I}_1,\boldsymbol{I}_2)$. In new variables $\nu=\boldsymbol{u}e^{ft}$, $\boldsymbol{R}=\boldsymbol{r}+\boldsymbol{u}/f$, and $\Phi=pe^{-3ft}$, Eq. (3.38) assumes the form

$$\frac{\partial\Phi}{\partial t}=D_u\left(e^{2ft}\nabla_\nu^2\Phi+\frac{2}{f}e^{ft}(\nabla_\nu\cdot\nabla_R)\Phi+\frac{1}{f^2}\nabla_R^2\Phi\right)$$

or, in the coordinate form

$$\frac{\partial\Phi}{\partial t}=D_u\left\{e^{2ft}\Delta_\xi\Phi+\frac{2}{f}e^{ft}\sum_{i=1}^{3}\frac{\partial^2\Phi}{\partial\xi_i\partial X_i}+\frac{1}{f^2}\Delta_X\Phi\right\}, \tag{3.39}$$

where the vector $\boldsymbol{v}$ has coordinates (ξ_1, ξ_2, ξ_3) and the vector $\boldsymbol{R}$ has coordinates (X_1, X_2, X_3). We look for the solution in the form

$$\Phi = \prod_{i=1}^{3} \Phi_i(\xi_i, X_i).$$

By substituting it into Eq. (3.39), it is easy to convince ourselves that Φ_i obeys the equation

$$\frac{\partial \Phi_i}{\partial t} = \varphi^2(t)\frac{\partial^2 \Phi_i}{\partial \xi_i^2} + 2\varphi(t)\psi(t)\frac{\partial^2 \Phi_i}{\partial \xi_i \partial X_i} + \psi^2(t)\frac{\partial^2 \Phi_i}{\partial X_i^2}, \tag{3.40}$$

where $\varphi(t) = D_u^{1/2} e^{ft}, \quad \psi(t) = D_u^{1/2}/f.$

The solution of this equation with the source-type initial condition, at the point $\xi = X = 0$, is

$$\Phi_i(\xi_i, X_i) = \frac{1}{2\pi\Delta^{1/2}} \exp\left\{-\frac{\alpha\xi_i^2 + 2\beta\xi_i X_i + \gamma X_i^2}{2\Delta}\right\},$$

where

$$\Delta = \alpha\gamma - \beta^2, \quad \alpha = 2\int_0^t \psi^2(t)dt, \quad \beta = -2\int_0^t \varphi(t)\psi(t)dt, \quad \gamma = 2\int_0^t \varphi^2(t)dt.$$

After some math, one finally obtains:

$$p(\boldsymbol{r}, \boldsymbol{u}, t|\boldsymbol{r}_0, \boldsymbol{u}_0, t_0) = \frac{e^{3ft}}{8\pi^3\Delta^{3/2}} \exp\left\{-\frac{[\alpha|\boldsymbol{v}-\boldsymbol{v}_0|^2 + 2\beta(\boldsymbol{v}-\boldsymbol{v}_0)\cdot(\boldsymbol{R}-\boldsymbol{R}_0) + \gamma|\boldsymbol{R}-\boldsymbol{R}_0|^2]}{2\Delta}\right\},$$

where $\boldsymbol{v} - \mathrm{v}_0 = e^{ft}\boldsymbol{u} - \boldsymbol{u}_0$; $\boldsymbol{R} - \boldsymbol{R}_0 = \boldsymbol{r} + \boldsymbol{u}/f - \boldsymbol{r}_0 - \boldsymbol{u}_0 - \boldsymbol{u}_0/f$; $\alpha = 2D_\mathrm{u}t/f^2$, $\beta = -2D_u(e^{ft} - 1)/f^2$, $\gamma = D_u(e^{2ft} - 1)/f$.

3.5 The Smoluchowski Equation

Consider Brownian motion of a free particle when $t \gg 1/f$. As was shown above, this is the time it takes for the particle velocity to reach its equilibrium value (another term for t is "dynamic relaxation time"). The particle's motion is considered inertialess, and the probability density of particle transition from $\boldsymbol{r}$ to $\boldsymbol{r} + d\boldsymbol{r}$ in time $\Delta t \gg 1/f$ is given by Eq. (3.29), namely,

$$\tau(\Delta\boldsymbol{r}) = \frac{1}{(4\pi D\Delta t)^{3/2}} \exp\left\{-\frac{|\Delta\boldsymbol{r}|^2}{4D\Delta t}\right\}, \tag{3.41}$$

where $D = \mathrm{k}\vartheta/mf$.

Then the problem of Brownian motion of a free particle reduces to the random walk problem (see Section 3.1). In this case the particle's motion can be characterized by a PDF of displacements $p(\mathbf{r}, t/\mathbf{r}_0, t_0)$ and the corresponding Fokker–Planck equation is considered in the displacement space $\mathbf{r}$ rather than in the phase space $(\mathbf{r}, \mathbf{u})$. If the external force is absent, the Fokker–Planck equation reduces to the ordinary diffusion equation

$$\frac{\partial p}{\partial t} = D\nabla_r^2 p. \tag{3.42}$$

The presence of external force field means that the particle's velocity will be influenced by an external systematic force $\mathbf{F}_e$ in accordance with the Langevin equation (3.33). If the force $\mathbf{F}_e$ does not change noticeably on distances of the order $(D/f)^{1/2}$, then any initial velocity distribution has enough time to turn into a Maxwellian distribution, and there is no significant change of r: since $|\mathbf{Dr}| = |\mathbf{Du}|/f \sim (D/f)^{1/2}$. Then the PDF $p(\mathbf{r}, \mathbf{u}, t/\mathbf{r}_0, \mathbf{u}_0, t_0)$ can be written as

$$p(\mathbf{r}, \mathbf{u}, t|\mathbf{r}_0, \mathbf{u}_0, t_0) = \left\{-\frac{m}{2\pi k\vartheta}\right\}^{3/2} \exp\left\{-\frac{m|\mathbf{u}|^2}{2k\vartheta}\right\} p(\mathbf{r}, t|\mathbf{r}_0, t_0).$$

As a result, the particle acquires the average drift velocity along the direction of the external force $\mathbf{F}_e$. Since particle inertia is neglected, this drift velocity is equal to $\mathbf{F}_e/h$. Then on the left-hand side of Eq. (3.42) there appears an additional convective term $\nabla\cdot(\mathbf{F}_e/h)$, and the equation for the PDF $p(\mathbf{r}, t/\mathbf{r}_0, t_0)$ becomes

$$\frac{\partial p}{\partial t} = \nabla_r\cdot\left(D\nabla_r p - \frac{\mathbf{F}_e}{mf} p\right). \tag{3.43}$$

Equations (3.42) and (3.43) are called the Smoluchowski equations. By analogy with Eq. (1.153), Eq. (3.43) can be rewritten in the divergent form:

$$\frac{\partial p}{\partial t} + \nabla_r \mathbf{j} = 0, \quad \mathbf{j} = -D\nabla_r p + \frac{\mathbf{F}_e}{mf} p, \tag{3.44}$$

where $\mathbf{j}$ has the meaning of diffusion flux.

Let L be the characteristic linear size of the region where the Brownian motion is considered. Then the time it takes for a stationary distribution p to establish is estimated by $t_s \sim L^2/D = 6\pi a\mu_e L^2/kT$. For $t \gg t_s$, Eq. (3.44) reduces to

$$\nabla_r \mathbf{j} = \nabla_r\cdot\left(-D\nabla_r p + \frac{\mathbf{F}_e}{mf} p\right) = 0, \tag{3.45}$$

from which there follows

$$\mathbf{j} = -D\nabla_r p + \frac{\mathbf{F}_e}{mf} p = \mathbf{const}. \tag{3.46}$$

Many external fields, such as gravitational and electrical fields, are potential fields. Then $\boldsymbol{F}_e/m = -\nabla_r \phi$, and the relation (3.46) assumes the form

$$\boldsymbol{j} = -D\nabla_r p - \frac{1}{f} p\nabla_r \Phi = -D\exp\left(-\frac{\phi}{Df}\right)\nabla_r\left\{p\exp\left(\frac{\phi}{Df}\right)\right\}.$$

Integrating this relation between two points A and B, we get:

$$\boldsymbol{j} = Df\left\{p\exp\left(\frac{\phi}{Df}\right)\right\}_B^A \left\{\int_A^B f\exp\left(\frac{\phi}{Df}\right)d\boldsymbol{X}\right\}^{-1}. \tag{3.47}$$

The expression (3.47) allows to determine the flux of probability density for the given values of p at the points A and B.

3.6
Brownian Motion of a Particle in a Moving Fluid

The analysis of Brownian motion in the previous sections was limited to the simple case of a quiescent fluid with of interactions between particles. Starting from this section, we gradually begin to take both factors (fluid motion and interparticle interactions) into account. Let us first consider the motion of one particle in a moving fluid while neglecting interparticle interactions, which is permissible when the volume concentration of particles in the suspension is negligibly small, in other words, when the suspension is infinite dilute [13].

Of primary practical interest is the motion of suspensions in tubes, capillaries, and porous media, where the flow velocity is characterized by considerable inhomogeneity over the cross section. As a rule, the characteristic linear scale of such inhomogeneity by far exceeds the particle size and the average distance between particles. Therefore, when considering a particle's motion, one can assume the velocity field in the neighborhood of the particle to be linear (which is typical for a shear flow).

One example of a purely shear flow is the Couette flow. This is a flow between two parallel plane walls undergoing translational motion in their respective planes, each plane moving with its own constant velocity. Let us direct our Y-axis perpendicular to the walls, while placing the origin of coordinates at the lower wall. The velocity of the lower wall is assumed to be zero. Then the velocity profile is

$$\boldsymbol{V}(Y) = \dot{\gamma}Y, \tag{3.48}$$

where $\dot{\gamma}$ is called the rate of shear.

One example of a non-shear flow is a stationary flow in a pipe or in a channel; it is known as the Poiseuille flow with a parabolic velocity profile. But when one is interested in the flow in the vicinity of a particle, whose size is much smaller then the diameter of the tube, the flow can be modeled as a shear flow with the linear velocity profile of the type (3.48). By the same token, we can approximate a flow with an arbitrary velocity profile by a shear flow with a linear profile. Since we are interested in the effect of fluid motion on the Brownian motion of the particle, let us assume that the particle is buoyant in a shear flow with the velocity profile (3.48) without specifying the particulars of the flow.

The particle's motion should be described by the Langevin equation (3.33), but since in the present case the particle's velocity $\boldsymbol{u}(t)$ differs from the flow velocity $\boldsymbol{V}$ of the fluid, the equation takes the form

$$\frac{d\boldsymbol{u}}{dt} = -f(\boldsymbol{u}-\boldsymbol{V}) + \boldsymbol{F}^{fl}(t). \tag{3.49}$$

We use the same designations as before: $f = h/m$; h is the coefficient of viscous friction; m is the particle mass. The fluctuating force $F^{fl}(t)$ should satisfy the conditions (1.156) of random Gaussian white noise, namely,

$$\langle F_i^{fl}(t)\rangle = 0, \quad \langle F_i^{fl}(t_1)F_j^{fl}(t_2)\rangle = 2\mathrm{k}Tf\,\delta_{ij}\delta(t_1-t_2), \tag{3.50}$$

where F_i^{fl} are components of the fluctuating force and δ_{ij} is the Kronecker symbol.

For simplicity's sake, we consider a two-dimensional fluid flow, so that $i, j = X, Y$, and the flow velocity $\boldsymbol{V}$ at the point $\boldsymbol{X}(t)$, where the particle is located at the time t, is

$$\boldsymbol{V}[\boldsymbol{X}(t)] = (\dot{\gamma}Y, 0). \tag{3.51}$$

Now the conditions (3.50) reduce to

$$\begin{aligned}
&\langle F_X^{fl}(t)\rangle = \langle F_Y^{fl}(t)\rangle = 0\\
&\langle F_X^{fl}(t_1)F_X^{fl}(t_2)\rangle = 2\mathrm{k}Tf\,\delta(t_1-t_2),\\
&\langle F_X^{fl}(t_1)F_Y^{fl}(t_2)\rangle = 0.
\end{aligned} \tag{3.52}$$

By integrating Eq. (3.49), we get the velocity components u and v of the particle:

$$\begin{aligned}
u(t) &= u(0)e^{-ft} + \int_0^t e^{-f(t-\tau)}(\dot{\gamma}hY(\tau) + F_X^{fl}(\tau))d\tau,\\
v(t) &= v(0)e^{-ft} + \int_0^t e^{-f(t-\tau)}F_Y^{fl}(\tau)d\tau.
\end{aligned} \tag{3.53}$$

Assume for simplicity that $X(0) = Y(0) = 0$ (i.e., the particle is initially located at the origin). Since $v(\mathrm{t}) = dY/dt$, we obtain from Eq. (3.53):

$$Y(t) = \int_0^t v(\tau)d\tau = \frac{1}{f}v(0)(1-e^{-ft}) + \frac{1}{f}\int_0^t (1-e^{-f(t-\tau)})F_Y^{fl}(\tau)d\tau. \tag{3.54}$$

A substitution of this relation into Eq. (3.53) gives

$$u(t) = u(0)e^{-ft} + \frac{\dot{\gamma}}{f}v(0)(1-e^{-ft}) - \dot{\gamma}v(0)te^{-ft}$$
$$+\frac{\dot{\gamma}}{f}\int_0^t(1-e^{-f(t-\tau)})F_Y^{fl}(\tau)d\tau - \dot{\gamma}\int_0^t d\tau\int_0^t e^{-f(t-\sigma)}F_Y^{fl}(\sigma)d\sigma + \int_0^t e^{-f(t-\tau)}F_X^{fl}(\tau)d\tau$$

and

$$X(t) = \int_0^t u(\tau)d\tau = \frac{1}{f}u(0)(1-e^{-ft}) + \frac{\dot{\gamma}}{f}v(0)t(1+e^{-ft})$$
$$-\frac{2\dot{\gamma}}{f^2}v(0)(1-e^{-ft}) + \frac{\dot{\gamma}}{f}\int_0^t d\tau\int_0^\tau(1+e^{-f(t-\sigma)})F_X^{fl}(\sigma)d\sigma$$
$$-\frac{2\dot{\gamma}}{f^2}\int_0^t(1-e^{-f(t-\tau)})F_Y^{fl}(\tau)d\tau + \frac{1}{f}\int_0^t(1-e^{-f(t-\tau)})F_X^{fl}(\tau)d\tau. \quad (3.55)$$

The expressions (3.54) and (3.55) for random variables – particle coordinates $X(t)$ and $Y(t)$ – allow to determine the particle's statistical characteristics – the moments. We are mainly interested in the first and second moments. Using the first property (3.52) of a fluctuating force, one gets the mean values of particle coordinates:

$$\langle X(t)\rangle = \frac{1}{f}\langle u(0)\rangle(1-e^{-ft}) + \frac{\dot{\gamma}}{f}\langle v(0)\rangle t(1+e^{-ft}) - \frac{2\dot{\gamma}}{f^2}\langle v(0)\rangle(1-e^{-ft}),$$
$$\langle Y(t)\rangle = \frac{1}{f}\langle v(0)\rangle(1-e^{-ft}). \quad (3.56)$$

In order to derive the second moments, that is, root-mean-squares of particle displacements along the coordinate axes, it is necessary to use all statistical properties (3.52) of the fluctuating force. We have:

$$\langle X^2(t)\rangle = \frac{1}{f^2}\langle u^2(0)\rangle(1-e^{-ft})^2 + \frac{\dot{\gamma}^2}{f^2}\langle v^2(0)\rangle t^2(1+e^{-ft})^2$$
$$+\frac{4\dot{\gamma}^2}{f^4}\langle v^2(0)\rangle(1-e^{-ft})^2 + \frac{2\dot{\gamma}}{f^2}\langle u(0)v(0)\rangle t(1-e^{-2ft})$$
$$-\frac{4\dot{\gamma}}{f^3}\langle u(0)v(0)\rangle(1-e^{-ft})^2 - \frac{4\dot{\gamma}^2}{f^3}\langle v^2(0)\rangle t(1-e^{-2ft})$$
$$+2\dot{\gamma}^2 D\left\{\frac{1}{3}t^3 + \frac{4}{f^3}(1-e^{-ft}) + \frac{1}{4f^3}(1-e^{-2ft})\right.$$
$$\left.-\frac{2}{f}t^2e^{-ft}\left(1+\frac{1}{4}e^{-ft}\right) - \frac{4}{f^2}te^{-ft}\left(1+\frac{1}{8}e^{-ft}\right)\right\}$$
$$+2D\left(1+\frac{4\dot{\gamma}^2}{f^2}\right)\left\{t - \frac{2}{f}(1-e^{-ft}) + \frac{1}{2f}(1-e^{-2ft})\right\}$$
$$-\frac{8\dot{\gamma}^2 D}{f}\left\{\frac{1}{2}t^2 - \frac{1}{4f^2}(1-e^{-2ft}) + \frac{1}{2f}te^{-2ft}\right\} \quad (3.57)$$

and

$$\langle Y^2(t)\rangle = \frac{1}{f^2}\langle v^2(0)\rangle(1-e^{-ft}) + 2Dt - \frac{4}{f}D(1-e^{-ft}) + \frac{1}{f}D(1-e^{-2ft}). \tag{3.58}$$

where $D = D_{br} = k\vartheta/mf$ is Einstein's coefficient of Brownian diffusion in a quiescent fluid.

We mentioned in Section 3.3 that the velocity PDF becomes Maxwellian very quickly, regardless of the initial distribution. Hence, it is safe to assume that the initial velocity distribution is Maxwellian (see Eq. 3.20). Then

$$\langle u(0)\rangle = \langle v(0)\rangle = \langle u(0)v(0)\rangle = 0, \quad \langle u^2(0)\rangle = \langle v^2(0)\rangle = \frac{k\vartheta}{m}. \tag{3.59}$$

From Eq. (3.56), it follows that

$$\langle X(t)\rangle = \langle Y(t)\rangle = 0. \tag{3.60}$$

Substituting Eq. (3.59) into Eq. (3.57) and Eq. (3.58), we find root-mean-square displacements of the particle. At $t \gg t_v = 1/f$, they have the following simple form:

$$\begin{aligned}\langle X^2(t)\rangle &= \frac{\dot{\gamma}^2 D}{f}\left(1-\frac{1}{f}\right)^2 + \frac{2}{3}\dot{\gamma}^2 Dt\left(t-\frac{3}{f}\right)^2 + 2D\left(t-\frac{1}{f}\right) + \frac{3\dot{\gamma}^2 D}{2f^3}\\ &= 2Dt\left\{1 + \frac{2\dot{\gamma}^2}{f} - \frac{3\dot{\gamma}^2}{2f}t + \frac{1}{3}\dot{\gamma}^2 t^2\right\} \approx 2Dt + \frac{2}{3}D\dot{\gamma}^2 t^3.\end{aligned} \tag{3.61}$$

$$\langle Y^2(t)\rangle \approx 2D\left(t-\frac{1}{f}\right) \approx 2Dt. \tag{3.62}$$

In a quiescent fluid, $\dot{\gamma} = 0$ and Eq. (3.61) and Eq. (3.62) lead to

$$\langle X^2(t)\rangle = \langle Y^2(t)\rangle = 2Dt. \tag{3.63}$$

We shall follow Einstein, defining the coefficient of Brownian diffusion for one-dimensional motion (along the X-axis) in a quiescent fluid as

$$D_{br} = \frac{\langle X^2(t)\rangle}{2t}. \tag{3.64}$$

A comparison with Eq. (3.61) and Eq. (3.62) shows that in a shear flow, root-mean-square displacements along the X- and Y-axis obey different laws. Thus, the root-mean-square displacement perpendicular to the direction of the flow is described by Einstein's formula (3.64):

$$\langle Y^2(t)\rangle = 2D_{br}t, \tag{3.65}$$

and the gradient of chemical potential equals

$$\nabla\mu = \left\{\frac{\partial\mu}{\partial n}\right\}_{p,\vartheta} \nabla n + \left\{\frac{\partial\mu}{\partial P}\right\}_{n,\vartheta} \nabla p.$$

If there is no external force and the particle motion occurs under the action of diffusion only, then p and ϑ are distributed uniformly in the suspension and Eq. (3.72) gives the thermodynamic force acting on a particle:

$$\boldsymbol{F}_{th} = -\left\{\frac{\partial\mu}{\partial n}\right\}_{p,\vartheta} \nabla n.$$

In a similar way, we can introduce the chemical potential per molecule, $\mu_0(n, p, \vartheta)$, and the corresponding thermodynamic force acting on a molecule:

$$\boldsymbol{F}^c_{th} = -\nabla\mu_0(n, p, \vartheta).$$

If the external force acts selectively on particles only, then $\nabla\mu_0 = 0$. For the case of diffusion with no external forces, we get the following expression for $\boldsymbol{F}^c_{th}$:

$$\boldsymbol{F}^c_{th} = -\left\{\frac{\partial\mu_0}{\partial n}\right\}_{p,\vartheta} \nabla n = -\frac{n}{n_0}\boldsymbol{F}_{th}.$$

The last relation implies $n_0\boldsymbol{F}^c_{th} + n\boldsymbol{F}_{th} = 0$, in other words, the net thermodynamic force acting on particles and molecules in a unit volume is zero. Hence, the diffusion driving force gives rise to a particle flux as though each particle were driven by the effective force

$$\boldsymbol{F}_{eff} = \boldsymbol{F}_{th}\left(1 + \frac{n\mathrm{V}}{n_0\mathrm{V}_0}\right) = \frac{\boldsymbol{F}_{th}}{1-\varphi}, \quad \varphi = n\mathrm{V},$$

and the molecules of fluid were exempt from the effect of this force. For an infinitely dilute suspension ($\varphi \to 0$) one can assume $\boldsymbol{F}_{eff} = \boldsymbol{F}_{th}$, in other words, the effect of the diffusion driving force on fluid molecules can be neglected.

It is easy to generalize the above results (which assumed the particles to be identical) for the case when the suspension contains particles of different types. Let μ_i, n_i $(n_1, n_2, \ldots)$, V_i, φ_i denote the chemical potential, number concentration, volume, and volume concentration particles of i-th type, and let μ_0 be the chemical potential of a fluid molecule. Suppose a force given by the potential φ_i acts on each particle of i-th type. Then the chemical potential per particle of i-th type is $\mu_i + \varphi_i$ and the force acting on each particle is

$$\boldsymbol{F}_{th\,i} = -\sum_{j=1}^{s}\left\{\frac{\partial\mu_i}{\partial n_j}\right\}_{p,\vartheta} \nabla n_j, \quad (i = 1, 2, \ldots, s),$$

whereas the force acting on a fluid molecule is

$$\boldsymbol{F}^{c}_{th} = -\sum_{j=1}^{s}\left\{\frac{\partial \mu_0}{\partial n_j}\right\}_{p,\vartheta}\nabla n_j = -\frac{1}{n_0}\sum_{i=1}^{s} n_i \boldsymbol{F}_{th\,i}.$$

The effective force acting on a particle of *i*-th type (and causing the diffusion of particles relative to the fluid) can be represented as

$$\boldsymbol{F}_{eff\,i} = \boldsymbol{F}_{th\,i} - \frac{\mathrm{v}_i}{\mathrm{v}_0}\boldsymbol{F}^{c}_{th} = \boldsymbol{F}_{th\,i} + \frac{\mathrm{v}_i}{1-\varphi}\boldsymbol{F}^{c}_{th}\sum_{j=1}^{s} n_j \boldsymbol{F}_{th\,j},$$

where $\varphi = 1 - n_0\mathrm{v}_0$.

For a very dilute suspension ($N \ll N_0$), the chemical potential of the solute (i.e., of the particles) can be expressed in terms of the PDF of particle positions $p(\boldsymbol{X}) = p(\boldsymbol{X}_1, \boldsymbol{X}_2, \ldots, \boldsymbol{X}_N)$. In the absence of external forces, it is equal to

$$\mu = \mu^{(0)}(p, \vartheta) + \mathrm{k}\vartheta \ln(\boldsymbol{X}). \tag{3.73}$$

The thermodynamic force acting on a particle located at $\boldsymbol{X}_k$ is (see Eq. (3.72))

$$\boldsymbol{F}_k = -\mathrm{k}\vartheta \nabla_k p = -\mathrm{k}\vartheta \frac{\partial p(\boldsymbol{X}_1, \boldsymbol{X}_2, \ldots, \boldsymbol{X}_N)}{\partial \boldsymbol{X}_k}; \quad (k = 1, 2, \ldots, N). \tag{3.74}$$

As we noted earlier, for a dilute suspension, we only need to consider the pair interactions between particles. So we are only interested in pair PDFs $p(\boldsymbol{X}_1, \boldsymbol{X}_2) = p_{12}$. If the suspension is statistically homogeneous, then $p(\boldsymbol{X}_1, \boldsymbol{X}_2) = p_{12}(\boldsymbol{r})$, where $\boldsymbol{r} = \boldsymbol{X}_1 - \boldsymbol{X}_2$.

Consider the relative motion of two particles under the action of forces $\boldsymbol{F}_1$ and $\boldsymbol{F}_2$, respectively. The particles are assumed to be spherical, with radii a_1 and a_2, and their motions are assumed to be slow, and happening at low Reynolds numbers, so the velocities of particle centers, $\boldsymbol{u}_1$ and $\boldsymbol{u}_2$, are both linear superpositions of forces (see Section 2.3)

$$\boldsymbol{u}_1 = \boldsymbol{b}_{11}\cdot\boldsymbol{F}_1 + \boldsymbol{b}_{12}\cdot\boldsymbol{F}_2, \quad \boldsymbol{u}_2 = \boldsymbol{b}_{21}\cdot\boldsymbol{F}_1 + \boldsymbol{b}_{22}\cdot\boldsymbol{F}_2. \tag{3.75}$$

Here $\boldsymbol{b}_{ij}$ are the mobility tensors, whose components depend on the viscosities of both the particles and the surrounding fluid, on the geometrical properties of both particles (it means the radii in the case of spherical particles), and on the vector $\boldsymbol{r}$ between the particle centers. For rigid particles, the mobility tensors are

$$\boldsymbol{b}_{ij} = \frac{1}{3\pi\mu_e(a_i + a_j)}\left\{A_{ij}(r)\frac{\boldsymbol{rr}}{r^2} + B_{ij}(r)\left(\boldsymbol{I} - \frac{\boldsymbol{rr}}{r^2}\right)\right\}, \tag{3.76}$$

where μ_e is the coefficient of dynamic viscosity of the surrounding fluid, $r = |\boldsymbol{r}|$, $\boldsymbol{rr} = \mathrm{r}_i\mathrm{r}_j$, $\boldsymbol{I}$ is the unit tensor, and A_{ij} and B_{ij} are dimensionless mobility coefficients that depend on two dimensionless parameters:

$$s = \frac{2r}{a_1 + a_2}, \qquad k = \frac{a_2}{a_1}.$$

The functions A_{ij} (s, k) and B_{ij} (s, k) are derived by solving the hydrodynamic problem on slow (Stokesian) motion of two particles in an unbounded quiescent fluid under the action of forces $\boldsymbol{F}_1$ and $\boldsymbol{F}_2$ (see Section 2.3). The linearity of Stokes equations allows to obtain the general solution as a superposition of solutions of particular problems about particle motion along and perpendicular to the line of centers. According to the reciprocity theorem, mobility coefficients have the properties

$$\begin{aligned} A_{11}(s,k) &= A_{22}(s,k^{-1}), & B_{11}(s,k) &= B_{22}(s,k^{-1}),\\ A_{12}(s,k) &= A_{21}(s,k), & A_{12}(s,k^{-1}) &= A_{21}(s,k^{-1}),\\ B_{12}(s,k) &= B_{21}(s,k), & B_{12}(s,k^{-1}) &= B_{21}(s,k^{-1}). \end{aligned} \tag{3.77}$$

It follows from Eq. (3.77) that in order to obtain mobility tensor components, one only has to determine A_{11} (s, k), B_{11} (s, k) on the intervals $0 \le k < \infty$, $2 \le s < \infty$ and A_{12} (s, k), B_{12} (s, k) on the intervals $0 \le k < \infty$, $2 \le s < \infty$.

At the current moment, almost all coefficients A_{ij} (s, k) and B_{ij} (s, k) are known for the case of solid or liquid spherical particles. As a rule, they have the form of infinite series whose convergence decreases as we diminish the gap between particles. There also exist approximate asymptotic solutions that are valid for relatively large and relatively small distances between particle surfaces. Since we are concerned with low-concentrated suspensions, we can use an asymptotic solution in the far asymptotic region ($s \gg 1$, particles are separated by large distances):

$$\begin{aligned} A_{11} &\approx 1 - \frac{60k^3}{s^4(1+k)^4} + \frac{32k^3(15-4k^2)}{s^6(1+k)^6} - \frac{192k^3(5-22k^2+3k^4)}{s^8(1+k)^8} + 0(s^{-10}),\\ A_{12} &\approx \frac{3}{2s} - \frac{2(1+k^2)}{s^3(1+k)^2} + \frac{1200k^3}{s^7(1+k)^6} + 0(s^{-10}),\\ B_{11} &\approx 1 - \frac{68k^5}{s^6(1+k)^6} + 0(s^{-10}), \quad B_{12} \approx \frac{3}{4s} + \frac{1+k^2}{s^3(1+k)^2} + 0(s^{-10}). \end{aligned}$$

In the other limiting case of $s \to 2$ (near asymptotic region, particles almost touch) the following approximations can be used:

$$\begin{aligned} A_{11}(2,k) &= \frac{2}{1+k} A_{12}(2,k) = \frac{1}{k} A_{22}(2,k),\\ A_{11} - \frac{4}{1+k} A_{12} + \frac{1}{k} A_{22} &\sim \frac{(1+k)^3}{2k^2}(s-2). \end{aligned}$$

At $s \to 2$, the coefficients B_{11}, B_{12}, and B_{22} behave as

$$B_{11}, B_{12}, B_{22} \sim const + 0\left(\frac{1}{\ln(s-2)^{-1}}\right).$$

We can find the coefficient of relative Brownian diffusion of two particles from the relations above, following the same procedure that was used by Einstein. In a homogeneous suspension at thermodynamic equilibrium, the flux of particles 2 relative to particles 1 is equal (with the opposite sign) to the relative convective flux caused by the thermodynamic force acting on the particles. Therefore the diffusion flux is equal to $-\boldsymbol{D}(\boldsymbol{r}) \cdot \nabla p(\boldsymbol{r})$. Using the relations (3.74) and (3.75), we obtain the following expression for the convective flux:

$$(\boldsymbol{u}_2 - \boldsymbol{u}_1)\, p(\boldsymbol{r}) = -\mathrm{k}\vartheta \nabla\, p(\boldsymbol{r}) \cdot (\boldsymbol{b}_{11} - \boldsymbol{b}_{21} - \boldsymbol{b}_{12} + \boldsymbol{b}_{22}).$$

Equating the diffusion flux to the convective one, we get the diffusion coefficient. It is a tensor and is sometimes referred to as the tensor of mutual diffusion, or interdiffusion tensor:

$$\boldsymbol{D}(\boldsymbol{r}) = \mathrm{k}\vartheta \cdot (\boldsymbol{b}_{11} - \boldsymbol{b}_{21} - \boldsymbol{b}_{12} + \boldsymbol{b}_{22}). \tag{3.78}$$

Substituting into Eq. (3.78) the expressions (3.76) for mobility tensors $\boldsymbol{b}_{ij}$, we can write

$$\boldsymbol{D}(\boldsymbol{r}) = D^{(0)}\left\{G(s,k)\frac{\boldsymbol{rr}}{r^2} + H(s,k)\left(\boldsymbol{I} - \frac{\boldsymbol{rr}}{r^2}\right)\right\}, \tag{3.79}$$

where

$$D^{(0)} = \frac{\mathrm{k}\vartheta}{6\pi\mu_e}\left(\frac{1}{a_1} + \frac{1}{a_2}\right),$$

$$G(s,k) = \frac{kA_{11} + A_{22}}{1+k} - \frac{4kA_{12}}{(1+k)^2}, \; H(s,k) = \frac{kB_{11} + B_{22}}{1+k} - \frac{4kB_{12}}{(1+k)^2}. \tag{3.79a}$$

The coefficient $D^{(0)}$ can be interpreted as the coefficient of relative diffusion of particles 2 and 1 taking place in the absence of mutual hydrodynamic interactions, that is, in the course of unhindered particle motion. So, the hindered motion of particles is taken into account by the factor inside the brackets in the formula (3.79).

By analogy with the properties (3.77), we can write similar properties for $G(s, k)$ and $H(s, k)$:

$$G(s,k) = G(s,k^{-1}), \quad H(s,k) = H(s,k^{-1}).$$

If the particles are spaced far apart, then, taking $s \to \infty$, we get $G \to 1$, $H \to 1$, $\mathbf{D} \to D_0\mathbf{I}$. For small interparticle distances ($s \to 2$), we have

$$G \sim \frac{(1+k)^2}{2k}(s-2); \quad H(s)-H(2) \sim \frac{const}{(\ln(s-2)^{-1})}.$$

The coefficient of Brownian diffusion also depends on the volume concentration φ of particles. For a dilute suspension ($\varphi \ll 1$) containing identical solid spherical particles of radius a, this dependence reduces to

$$\mathbf{D} = (1+1.45\varphi)\frac{\mathrm{k}\vartheta}{6\pi\mu_e a}\mathbf{I}. \tag{3.80}$$

3.8 Brownian Diffusion with Hydrodynamic Interactions and External Forces

In this section, we consider Brownian motion of particles, taking into account both hydrodynamic and molecular interactions of particles and the gravity force [16]. Settling of particles under the action of gravity is called sedimentation. As in Section 3.7, we assume the particles to be rigid and spherical, and the suspension – low-concentrated, so we need to consider only pair interactions between particles. The suspension contains m types of particles of different sizes and densities. The symbols $a_i, \rho_i, n_i, \varphi_i, \mathbf{U}_i^{(0)}$ ($i = 1, 2, \ldots, m$) will denote, respectively, the radius, density, number concentration, volume concentration, and the velocity of gravitational settling of particles of i-th type. The symbol ρ_e will denote the density of the surrounding fluid. The velocity of settling obeys Stokes law:

$$\mathbf{U}_i^{(0)} = \frac{2a_i^2(\rho_i-\rho_e)}{9\mu_e}\mathbf{g}, \quad \mathbf{U}_j^{(0)} = \gamma k^2 \mathbf{U}_i^{(0)} \tag{3.81}$$

where $\mathbf{g}$ is the gravitational acceleration and the dimensionless parameters γ and k are defined as

$$\gamma = \frac{\rho_j-\rho_e}{\rho_i-\rho_e}g, \quad k = \frac{a_j}{a_i}.$$

Since only pair interactions are taken into account, we can introduce a pair PDF $p(\mathbf{X}_i, \mathbf{X}_j)$ such that $p(\mathbf{X}_i, \mathbf{X}_j)d\mathbf{X}_i, d\mathbf{X}_j$ is the probability to find i-th particle in the volume element $d\mathbf{X}_i$ centered at $\mathbf{X}_i$ and j-th particle – in the volume element $d\mathbf{X}_j$ centered at $\mathbf{X}_j$. Assuming the suspension to be statistically homogeneous, we get $p(\mathbf{X}_i, \mathbf{X}_j) = p_{ij}(\mathbf{r})$, where $\mathbf{r} = \mathbf{X}_j - \mathbf{X}_i$. The distribution $p_{ij}(\mathbf{r})$ obeys the condition $p_{ij}(\mathbf{r}) \to 1$ at $r \to \infty$, meaning that large interparticle distances are the most likely.

The molecular Van der Waals force is a central force acting along the line of centers of i-th and j-th particles and is equal to

$$\boldsymbol{F}_{ij}^{m} = \nabla_r \varphi_{ij}(r) = \frac{\boldsymbol{r}}{r}\frac{d\phi_{ij}}{dr}, \tag{3.82}$$

where φ_{ij} is the potential of molecular interaction.

The gravitational force acting on particles causes their sedimentation with the average velocity given by Eq. (3.81). Hence

$$\boldsymbol{F}_i^g = 6\pi\mu_e a_i \boldsymbol{U}_i^{(0)}, \quad \boldsymbol{F}_j^g = 6\pi\mu_e a_j \boldsymbol{U}_j^{(0)}. \tag{3.83}$$

The forces of hydrodynamic interaction have been discussed in the previous section. According to the formulas (3.75), the relative velocity of two sedimenting particles is

$$\mathbf{V}_{ij}(\boldsymbol{r}) = \boldsymbol{u}_j - \boldsymbol{u}_i = 6\pi\mu_e a_i \boldsymbol{U}_i^{(0)}(\boldsymbol{b}_{ji} - \boldsymbol{b}_{ii}) + 6\pi\mu_e a_j \boldsymbol{U}_j^{(0)}(\boldsymbol{b}_{jj} - \boldsymbol{b}_{ij}). \tag{3.84}$$

where $\boldsymbol{b}_{ij}$ are mobility tensors.

For particles placed far apart, when the hydrodynamic interaction is negligible, the relative sedimentation velocity is equal to

$$\mathbf{V}_{ij}^{(0)}(\boldsymbol{r}) = \boldsymbol{U}_j^{(0)} - \boldsymbol{U}_i^{(0)} = (k^2\gamma - 1)\boldsymbol{U}_i^{(0)}. \tag{3.85}$$

Then, substituting the relations (3.76) into Eq. (3.84) and using Eq. (3.85), we obtain:

$$\mathbf{V}_{ij}(\boldsymbol{r}) = \mathbf{V}_{ij}^{(0)}(\boldsymbol{r})\left\{L(s,k)\frac{\boldsymbol{rr}}{r^2} + M(s,k)\left(\boldsymbol{I} - \frac{\boldsymbol{rr}}{r^2}\right)\right\}, \tag{3.86}$$

where

$$\begin{aligned} L(s,k) &= \frac{k^2\gamma A_{jj} - A_{ii}}{k^2\gamma - 1} + \frac{2(1-k^3\gamma)A_{ij}}{k^2\gamma - 1}, \\ M(s,k) &= \frac{k^2\gamma B_{jj} - B_{ii}}{k^2\gamma - 1} + \frac{2(1-k^3\gamma)B_{ij}}{(1+k)(k^2\gamma - 1)}. \end{aligned} \tag{3.87}$$

The interdiffusion tensor is given by the relations (3.78) and (3.79):

$$\boldsymbol{D}_{ij}(\boldsymbol{r}) = kT(\boldsymbol{b}_{ii} + \boldsymbol{b}_{jj} - \boldsymbol{b}_{ij} - \boldsymbol{b}_{ji}). \tag{3.88}$$

The pair PDF $p_{ij}(\boldsymbol{r})$ is described by the Fokker–Planck equation,

$$\begin{aligned} \frac{\partial p_{ij}}{\partial t} &= -\nabla\cdot(\boldsymbol{V}_{ij} p_{ij}) + \nabla\cdot\left\{-\boldsymbol{D}_{ij}\cdot(p_{ij}\cdot \boldsymbol{F}_{ij}^{m} + \nabla p_{ij})\right\} \\ &= -\nabla\cdot(\boldsymbol{V}_{ij} p_{ij}) + \nabla\cdot\left\{p_{ij}\boldsymbol{D}_{ij}\cdot\nabla\left(\frac{\phi_{ij}}{kT}\right)\right\} + \nabla\cdot(\boldsymbol{D}_{ij}\cdot\nabla p_{ij}). \end{aligned} \tag{3.89}$$

This equation can be rewritten in the form

$$\frac{\partial p_{ij}}{\partial t} + \nabla \cdot \boldsymbol{J} = 0, \tag{3.90}$$

if we introduce the probability density flux $\boldsymbol{j}(\boldsymbol{r})$:

$$\boldsymbol{J}(\boldsymbol{r}) = \left\{ \boldsymbol{V}_{ij} - \boldsymbol{D}_{ij} \cdot \nabla \left(\frac{\phi_{ij}}{k\vartheta} \right) \right\} p_{ij} - \boldsymbol{D}_{ij} \cdot \nabla p_{ij}. \tag{3.91}$$

The boundary conditions at the infinity is

$$p_{ij}(\boldsymbol{r}) \to 1 \text{ at } r \to \infty \tag{3.92}$$

The second boundary condition is that the component of flux along the line of centers must be zero at the surface of a sphere of radius $a_i + a_j$ (i.e., the radius of the particles' contact surface):

$$\boldsymbol{r} \cdot \boldsymbol{J}(\boldsymbol{r}) = 0 \text{ at } r = a_i + a_j. \tag{3.93}$$

At the sphere surface ($r = a_i + a_j$), the conditions $\boldsymbol{r} \cdot \boldsymbol{V}_{12} = 0$ and $\boldsymbol{r} \cdot \boldsymbol{D}_{ij} = 0$ should hold. The former results from the fact that the particles are rigid, and the latter follows from the estimation $|\boldsymbol{r} \cdot \boldsymbol{D}_{ij}| \sim r - (a_i + a_j)$ at $r \to (a_i + a_j)$. Therefore the condition (3.93) transforms to

$$\boldsymbol{r} \cdot \boldsymbol{D}_{ij} \cdot \nabla p_{ij} = 0. \tag{3.94}$$

The first term on the right-hand side of Eq. (3.89) has the meaning of convective transport of particles due to gravity, and the remaining two terms stand for the Brownian diffusion due to the forces of hydrodynamic and molecular interaction between particles.

In the general case, solution of Eq. (3.89) presents considerable difficulties. Let us look at the possible ways in which it can be simplified. The ratio between the first (convective) and the third (diffusional) terms is equal by the order of magnitude to the Peclet number

$$Pe_{ij} = \frac{(a_i + a_j) V_{ij}^{(0)}}{2 D_{ij}^{(0)}}, \tag{3.95}$$

where

$$D_{ij}^{(0)} = \frac{k\vartheta}{6\pi\mu_e} \left(\frac{1}{a_i} + \frac{1}{a_j} \right) \tag{3.96}$$

is the coefficient of unhindered (i.e. occurring in the absence of interparticle interactions) diffusion.

The Peclet number is strongly dependent on particle size: $Pe_{ij} \sim (a_i + a_j)^4$. For Brownian particles, $a_i \sim 0.01 - 10\,\mu m$, so the Peclet number can vary in a wide range – from very low to very high values – depending the properties of both the fluid and the particles. For example, for a water suspension at normal temperature, where $a_i = 0.5\,\mu m$, $a_j = 1\,\mu m$, $\rho_i = \rho_j = 2\rho_e$, the Peclet number equals $Pe_{ij} = 1.5$.

When $Pe_{ij} \gg 1$, the diffusional terms in Eq. (3.89) become negligibly small as compared to the convective term, while at $Pe_{ij} \ll 1$, diffusion prevails over convection. Consider these two limiting cases in more detail.

3.8.1 High Peclet Numbers: $Pe_{ij} \gg 1$

From estimation given above, it is evident that this case applies to relatively large particles. The condition $Pe_{ij} \gg 1$ means that the term describing Brownian diffusion in the Fokker–Planck equation (3.89) can be dropped. For simplicity's sake, we also neglect molecular interactions by setting $\boldsymbol{f}_{ij} = 0$. Then Eq. (3.89) reduces to a simple transport equation

$$\frac{\partial p_{ij}}{\partial t} + \nabla\cdot(\boldsymbol{V}_{ij}\, p_{ij}) = 0 \tag{3.97}$$

and we find from Eq. (3.86) that

$$\nabla\cdot\boldsymbol{V}_{ij} = \frac{2\boldsymbol{V}_{ij}^{(0)}\cdot\boldsymbol{r}}{(a_i + a_j)r} W(s), \tag{3.98}$$

where $s = 2r/(a_i + a_j)$ and

$$\begin{aligned} W(s) &= \frac{2(L-M)}{s} + \frac{dL}{ds} \\ &= \frac{120k^3(\gamma-1)}{(1+k)^4(\gamma k^2-1)s^5} + \frac{24k^3}{(1+k)^6 s^7}\left\{\frac{27(\gamma-k^2)}{(\gamma k^2-1)s^5} - 80\right\} + \frac{12000k^3(\gamma k^3-1)}{(1+k)^7(\gamma k^2-1)s^8} \\ &\quad + \frac{64k^3}{(1+k)^8 s^9}\left\{\frac{100(\gamma k^4-1) + 63(\gamma-k^4) - 405(\gamma-1)k^2}{(\gamma k^2-1)}\right\} + 0(s^{-10}). \end{aligned} \tag{3.99}$$

Introduce a function $q(r)$ by specifying the following relation:

$$\frac{d(\ln q)}{dr} = \frac{2W(r)}{(a_i + a_j)L(r)}. \tag{3.100, a}$$

Then Eq. (3.98) transforms to

$$\nabla\cdot\boldsymbol{V}_{ij} = -\frac{\boldsymbol{r}\cdot\boldsymbol{V}_{ij}}{rq}\frac{dq}{dr}$$

and Eq. (3.97) to

$$\frac{\partial p_{ij}}{\partial t}+\nabla\cdot(\mathbf{V}_{ij}\,p_{ij})=\frac{\partial p_{ij}}{\partial t}+(\nabla\cdot\mathbf{V}_{ij})\,p_{ij}+\mathbf{V}_{ij}\cdot\nabla p_{ij}$$
$$=\left(\frac{\partial}{\partial t}+\mathbf{V}_{ij}\cdot\nabla\right)\left\{\frac{p_{ij}(\mathbf{r},t)}{q(r)}\right\}=\frac{D}{Dt}\left\{\frac{p_{ij}(\mathbf{r},t)}{q(r)}\right\}=0, \qquad (3.100,\text{b})$$

where D/Dt is the substantial derivative. We can see from Eq. (3.100) that $p_{ij}(\mathbf{r},t)/q(r)$ remains constant for a point mass moving along a trajectory in the $\mathbf{r}$-space. In particular, for trajectories coming from the infinity, where the condition $p_{ij}=1$ should holds, we can write

$$p_{ij}(\mathbf{r},t)=\frac{q(r)}{q(\infty)}, \qquad (3.101)$$

and p_{ij} then follows from the relations (3.100, a) and (3.102):

$$p_{ij}(s)=\exp\left\{\int_s^\infty\frac{W(s)}{L(s)}ds\right\}=\exp\left\{\int_s^\infty\left(\frac{2(L-M)}{sL}+\frac{1}{L}\frac{dL}{ds}\right)ds\right\}, \qquad (3.102)$$

where $W(s)$ and $L(s)$ are given by Eq. (3.87) and Eq. (3.99).

At $s\gg 1$, the expression (3.102) for p_{ij} can be written as a series

$$p_{ij}(s)=1+\frac{30k^3(\gamma-1)}{(1+k)^4(\gamma k^2-1)}s^{-4}+\frac{72k^3(\gamma k^3-1)(\gamma-1)}{(1+k)^5(\gamma k^2-1)^2}s^{-5}$$
$$+\frac{4k^3}{(1+k)}\left\{\frac{45(\gamma k^3-1)^2(\gamma-1)}{(\gamma k^2-1)^3}+\frac{27(\gamma-k^2)}{(\gamma k^2-1)}-80\right\}s^{-6}+0(s^{-7}), \qquad (3.103)$$

which represents the far asymptotic expansion of p_{ij} and is valid for particles spaced far apart. The near asymptotic expansion ($s\to 2$) can not be derived in the current case, because the approximate equation (3.103) is valid in the region outside the spherical diffusion boundary layer of thickness $\delta_D\sim(a_i+a_j)/\sqrt{Pe_{ij}}$ adjacent to the sphere of radius $r=a_i+a_j$, or $s=2$.

3.8.2
Small Peclet Numbers, $Pe_{ij}\ll 1$

The condition $Pe_{ij}\ll 1$ means that one can neglect the convective term in Eq. (3.89), leaving only the diffusional ones, so the Fokker–Planck equation reduces to

$$\frac{\partial p_{ij}}{\partial t}=\nabla\cdot\left\{p_{ij}\mathbf{D}_{ij}\cdot\nabla\left(\frac{\phi_{ij}}{k\vartheta}\right)\right\}+\nabla\cdot(\mathbf{D}_{ij}\cdot\nabla p_{ij}). \qquad (3.104)$$

We should note that the convective flux of particles is also absent when all particles are identical, that is, when they all have the same radius ($a_i=a_j$, or $k=1$) and density ($\rho_i=\rho_j$, or $\gamma=1$).

We shall limit ourselves to stationary processes only. Such an assumption is justified for suspensions that are stable, that is, when there is no particle coagulation. Then the equation (3.104) takes the form

$$\nabla\cdot\left\{ p_{ij}\boldsymbol{D}_{ij}\cdot\nabla\left(\frac{\phi_{ij}}{\mathrm{k}\vartheta}\right)\right\} + \nabla\cdot(\boldsymbol{D}_{ij}\cdot\nabla p_{ij}) = 0. \tag{3.105}$$

Since the molecular force potential ϕ_{ij} depends only on r, the distribution p_{ij} also depends on r only, and the equation reduces to

$$\frac{d}{dr}\left\{\frac{\boldsymbol{r}\cdot\boldsymbol{D}_{ij}\cdot\boldsymbol{r}}{r^2}\left[p_{ij}\frac{d}{dr}\left(\frac{\phi_{ij}}{\mathrm{k}\vartheta}\right)+\frac{d p_{ij}}{dr}\right]\right\} = 0. \tag{3.106}$$

The solution of Eq. (3.106) with the boundary conditions (3.92) and (3.94) is the Boltzmann distribution

$$p_{ij}(\boldsymbol{r}) = \exp\left(-\frac{\phi_{ij}(\boldsymbol{r})}{\mathrm{k}\vartheta}\right). \tag{3.107}$$

The expression (3.107) is a pair PDF for the limiting case $Pe_{ij}=0$. A correction to the solution for the case of nonzero but small values of Pe_{ij} can be derived by using the method of small perturbations, assuming

$$p_{ij}(\boldsymbol{r}) = \exp\left(-\frac{\phi_{ij}(\boldsymbol{r})}{\mathrm{k}\vartheta}\right)\left\{1 + Pe_{ij}\,p_{ij}^{(1)}(\boldsymbol{r}) + 0(Pe_{ij})\right\}. \tag{3.108}$$

Substituting Eq. (3.108) into the stationary variant of Eq. (3.89), one gets the equation for $p_{ij}^{(1)}(\boldsymbol{r})$:

$$\nabla_r(e^{-\phi_{ij}/\mathrm{k}\vartheta}\boldsymbol{D}_{ij}\cdot\nabla p_{ij}^{(1)}) = \nabla_r(\boldsymbol{V}_{ij}e^{-\phi_{ij}/\mathrm{k}\vartheta}), \tag{3.109}$$

whose solution will be sought in the form

$$p_{ij}^{(1)}(\boldsymbol{r}) = \frac{\boldsymbol{r}\cdot\boldsymbol{V}_{ij}^{(0)}}{r\boldsymbol{V}_{ij}^{(0)}}Q(s).$$

Then the relation (3.108), together with the expressions (3.86), (3.88), and (3.98) for $\boldsymbol{V}_{ij}$, $\boldsymbol{D}_{ij}$, and $\nabla\cdot\boldsymbol{V}_{ij}$ gives the following equation for $Q(s)$:

$$\frac{d}{ds}\left(s^2G\frac{dQ}{ds}\right)-\frac{d}{ds}\left(\frac{\phi_{ij}(\boldsymbol{r})}{\mathrm{k}\vartheta}\right)s^2G\frac{dQ}{ds}-2HQ = s^2W-\frac{d}{ds}\left(\frac{\phi_{ij}(\boldsymbol{r})}{\mathrm{k}\vartheta}\right)s^2L. \tag{3.110}$$

The second integral is taken over the volume $|\boldsymbol{X}+\boldsymbol{r}_k-\boldsymbol{X}|>a$, and since particles are spaced far apart, the point $\boldsymbol{X}$ is located within the fluid, and the equation

$$\frac{\partial \sigma_{ij}}{\partial X_j}=\mu_e \nabla^2 u_i$$

is valid for all values of k. Then Eq. (3.133) assumes the form

$$\begin{aligned}&\sum_{k=1}^{N}\frac{n}{N!}\int f(\boldsymbol{X}_0+\boldsymbol{r}_k,\Theta_{N-1})P(\Theta_{N-1}|\boldsymbol{X}_0+\boldsymbol{r}_k)d\Theta_{N-1}\\&\quad+\frac{1}{N!}\int\limits_{|\boldsymbol{X}_0+\boldsymbol{r}_k-\boldsymbol{X}|\le a}\mu_e\nabla^2 u(\boldsymbol{X},\Theta_N)P(\Theta_N)d\Theta_N=0.\end{aligned}\tag{3.134}$$

In a statistically homogeneous suspension, all summands in the first term in the expression (3.134) are the same, and, since the final expression should not depend on $\boldsymbol{r}_k$, we can take $\boldsymbol{r}_k=0$. In the second integral, we take $\boldsymbol{X}=\boldsymbol{X}_0$. Now, comparing Eq. (3.127) to Eq. (3.134) and remembering that $P(\Theta_N, \boldsymbol{X}_0)=0$ for all $\boldsymbol{r}_k<2a$, we obtain:

$$\begin{aligned}\langle \boldsymbol{V}_2\rangle&=\frac{1}{N!}\int\limits_{r_k>a}\frac{1}{6}a^2(\nabla^2\boldsymbol{u}(\boldsymbol{X},\Theta_N))_{\boldsymbol{X}=\boldsymbol{X}_0}\{P(\Theta_N|\boldsymbol{X}_0)-P(\Theta_N)\}d\Theta_N\\&\quad-\frac{1}{(N-1)!}\int\frac{na^2}{6\mu_e}f(\boldsymbol{X}_0,\Theta_{N-1})P(\Theta_{N-1}|\boldsymbol{X}_0)d\Theta_{N-1}.\end{aligned}\tag{3.135}$$

Transformation of the first term on the right-hand side of Eq. (3.135) can be performed using the same procedure that took us from Eq. (3.130) to Eq. (3.131). To transform the second term, we observe that the viscous friction force acting on an isolated sphere in an unbounded fluid is equal to 2/3 of the drag force $6\pi\mu_e a\boldsymbol{U}_0$. And since the volume content of the disperse phase is $\varphi=4\pi a^3 n/3$, the second summand is estimated by $-0.5\varphi\boldsymbol{U}_0$, which is accurate to the order of φ. The resulting expression for $\langle \boldsymbol{V}_2\rangle$ (which has the same accuracy) is

$$\langle \boldsymbol{V}_2\rangle=\int\limits_{r>a}\frac{1}{6}a^2(\nabla^2\boldsymbol{u}(\boldsymbol{X},\boldsymbol{X}_0+\boldsymbol{r}))_{\boldsymbol{X}=\boldsymbol{X}_0}\{P(\boldsymbol{X}_0+\boldsymbol{r}|\boldsymbol{X}_0)-P(\boldsymbol{X}_0+\boldsymbol{r})\}d\boldsymbol{r}+\frac{1}{2}\varphi\boldsymbol{U}_0.\tag{3.136}$$

So, the relations (3.131), (3.136), (3.128) give approximate expressions for $\langle \boldsymbol{V}_1\rangle$, $\langle \boldsymbol{V}_2\rangle$, and $\langle \boldsymbol{W}\rangle$, and thereby for the particle's average sedimentation velocity (given by Eq. (3.125)) in a low-concentrated suspension. To calculate the integrals that appear in these relations, one has to know the probability distribution for two spherical particles as well as the fluid velocity field induced by two settling particles.

In real suspensions, particles can come very close together, which results in collisions and aggregation. But in the limiting case of an infinitely dilute suspension ($\varphi\to 0$), particle collisions are extremely unlikely, and therefore it is quite natural to assume that

$$P(\boldsymbol{X}+\boldsymbol{r}|\boldsymbol{X}) = \begin{cases} n \text{ at } r \geq 2a, \\ 0 \text{ at } r < 2a. \end{cases} \tag{3.137}$$

A more correct method of finding $P(\boldsymbol{X}+\boldsymbol{r}|\boldsymbol{X})$ is to solve the Liouville equation (1.142) for the pair probability distribution:

$$\frac{\partial P(\boldsymbol{X}+\boldsymbol{r}|\boldsymbol{X})}{\partial t} = -(\boldsymbol{U}_1-\boldsymbol{U}_2)\cdot\nabla P - P\nabla\cdot(\boldsymbol{U}_1-\boldsymbol{U}_2), \tag{3.138}$$

where $\boldsymbol{U}_1$ and $\boldsymbol{U}_2$ are the sedimentation velocities or two particles with centers at the points $\boldsymbol{X}$ and $\boldsymbol{X}+\boldsymbol{r}$. If the particles have nearly equal sizes, then $\boldsymbol{U}_1 \approx \boldsymbol{U}_2$, and it follows from Eq. (3.138) that $P(\boldsymbol{X}+\boldsymbol{r}|\boldsymbol{X}) \approx const = n$.

By the same token, the unconditional probability distribution can be regarded as constant, that is, $P(\boldsymbol{X}+\boldsymbol{r}) = n$. Then

$$P(\boldsymbol{X}+\boldsymbol{r}|\boldsymbol{X}) - P(\boldsymbol{X}+\boldsymbol{r}) = \begin{cases} 0 \text{ at } r \geq 2a, \\ -n \text{ at } r < 2a. \end{cases} \tag{3.139}$$

The relations (3.131), (3.136) and (3.128) lead us to write

$$\langle \boldsymbol{V}_1 \rangle = -n \int\limits_{r<2a} \boldsymbol{u}(\boldsymbol{X}_0, \boldsymbol{X}_0+\boldsymbol{r})d\boldsymbol{r},$$

$$\langle \boldsymbol{V}_2 \rangle = -n \int\limits_{a<r<2a} \frac{1}{6}a^2(\nabla^2 \boldsymbol{u}(\boldsymbol{X}, \boldsymbol{X}_0+\boldsymbol{r}))_{\boldsymbol{X}=\boldsymbol{X}_0} d\boldsymbol{r} + \frac{1}{2}\varphi \boldsymbol{U}_0,$$

$$\langle \boldsymbol{W} \rangle = n \int \boldsymbol{W}(\boldsymbol{X}_0, \boldsymbol{X}_0+\boldsymbol{r})d\boldsymbol{r},$$

where $\boldsymbol{u}(\boldsymbol{X}_0, \boldsymbol{X}_0+\boldsymbol{r})$, $(\nabla^2 \boldsymbol{u}(\boldsymbol{X}, \boldsymbol{X}_0+\boldsymbol{r}))_{\boldsymbol{X}=\boldsymbol{X}0}$ and $\boldsymbol{W}(\boldsymbol{X}_0, \boldsymbol{X}_0+\boldsymbol{r})$ are determined at the point $(\boldsymbol{X}_0)$ under the condition that the point $(\boldsymbol{X}_0+\boldsymbol{r})$ is the center of another particle.

It is convenient to change designations, by taking $\boldsymbol{X}$ as the center of the test particle and $\boldsymbol{X}+\boldsymbol{r}$ as the point where function has to be computed. Then $\langle \boldsymbol{V}_1 \rangle$, $\langle \boldsymbol{V}_2 \rangle$, and $\langle \boldsymbol{W} \rangle$ will be rewritten as

$$\langle \boldsymbol{V}_1 \rangle = -n \int\limits_{r<2a} \boldsymbol{u}(\boldsymbol{X}+\boldsymbol{r}, \boldsymbol{X})d\boldsymbol{r},$$

$$\langle \boldsymbol{V}_2 \rangle = -n \int\limits_{a<r<2a} \frac{1}{6}a^2(\nabla_r^2 \boldsymbol{u}(\boldsymbol{X}+\boldsymbol{r}, \boldsymbol{X}))_{\boldsymbol{X}=\boldsymbol{X}_0} d\boldsymbol{r} + \frac{1}{2}\varphi \boldsymbol{U}_0,$$

$$\langle \boldsymbol{W} \rangle = n \int \boldsymbol{W}(\boldsymbol{X}+\boldsymbol{r}, \boldsymbol{X})d\boldsymbol{r}, \tag{3.140}$$

where

$$W(X + r, X) = U(X + r, X) - U_0 - u(X + r, X) - a^2 \nabla_r^2 u(X + r, X)/6. \tag{3.141}$$

A solution of the Stokesian problem about the motion of an isolated sphere in an unbounded fluid gives us the following relations:

$$u(X + r, X) = \begin{cases} U_0 & \text{at } r \le a, \\ U_0\left(\dfrac{3a}{4r} + \dfrac{a^3}{4r^3}\right) + r\dfrac{r \cdot U_0}{r^2}\left(\dfrac{3a}{4r} - \dfrac{3a^3}{4r^3}\right) & \text{at } r \ge a, \end{cases}$$

$$\nabla_r^2 u = U_0 \frac{4a}{2r^3} - r\frac{r \cdot U_0}{r^2}\frac{9a}{2r^3}. \tag{3.142}$$

Then

$$\int_{r \le a} u(X + r, X) dr = \frac{4}{3}\pi a^3 U_0, \qquad \int_{a < r \le 2a} u(X + r, X) dr = 6\pi a^3 U_0,$$

$$\int_{a < r < 2a} \frac{1}{6} a^2 (\nabla_r^2 u(X + r, X))_{X = X_0} dr = 0.$$

Let us plug the derived values into the first two relations (3.140):

$$\langle V_1 \rangle = -\frac{22}{3}\pi a^3 n U_0 = -\frac{11}{2}\varphi U_0, \quad \langle V_2 \rangle = \frac{1}{2}\varphi U_0.$$

From Eq. (3.141), one gets $W(X + r, X)$:

$$W(X + r, X) = U(X + r, X) - U_0 - U_0\left(\frac{3a}{4r} + \frac{a^3}{2r^3}\right) - r\frac{r \cdot U_0}{r^2}\left(\frac{3a}{4r} - \frac{3a^3}{2r^3}\right). \tag{3.143}$$

Here $U(X + r, X)$ is the sedimentation velocity of each of the two spheres whose centers are separated by the (vector) distance r, and U_0 is the sedimentation velocity of an isolated sphere.

Two special cases are possible when we consider the motion of two spheres: the line of centers can be parallel or perpendicular to the direction of gravity. In other words, the spheres settle with equal velocities either along or perpendicular to the line of centers. In the general case the solution is a superposition of the two particular solutions:

$$U(X + r, X) = b_1 r \frac{r \cdot U_0}{r^2} + b_2\left(U_0 - r\frac{r \cdot U_0}{r^2}\right), \tag{3.144}$$

where b_1 and b_2 are the mobility coefficients of the spheres. They depend on the relative distance r/a between sphere centers. Eq. (3.141) gives us the expression for $\langle W \rangle$:

$$\langle \mathbf{W} \rangle = \varphi U_0 \int_2^{\infty} \left\{ b_1 + 2b_2 - 3\left(1 + \frac{a}{r}\right) \right\} \frac{r^2}{a^2} d\left(\frac{r}{a}\right).$$

Using the expressions for mobility coefficients (see Section 3.7), we get

$$\langle \mathbf{W} \rangle = -1.55\varphi U_0. \tag{3.145}$$

Finally, substituting the derived values of $\langle \mathbf{V}_1 \rangle$, $\langle \mathbf{V}_2 \rangle$, and $\langle \mathbf{W} \rangle$ into Eq. (3.125), we get

$$\langle U \rangle = U_0(1 - 6.55\varphi). \tag{3.146}$$

where U_0 is given by Eq. (3.112).

3.10 Particle Sedimentation in a Polydisperse Dilute Suspension, with Hydrodynamic and Molecular Interactions and Brownian Motion of Particles

Consider the sedimentation of particles in a suspension containing particles of different sizes [18,19]. Let a_i, ρ_I, n_i, φ_i and U_i^0 denote the radius, density, number and volume concentrations, and unhindered sedimentation velocity of i-th particle ($i = 1, 2, \ldots, m$). The velocities U_i^0 are given by Eq. (3.112). We assume that the volume concentration $\varphi = \sum_{i=1}^{m} \varphi_i$ is small and particle interact with each other, in other words, the molecular forces described by the potential $\varphi_{ij}(r, a_i, a_j)$ and the hydrodynamic forces are taken into account. Smallness of φ allows us to consider pair interactions only, with the accuracy to the order of φ.

For a suspension of identical particles, the average sedimentation velocity is given by the approximate formula (it takes into account only hydrodynamic interactions) derived in the previous section:

$$\langle U \rangle = U_0(1 + S\varphi + 0(\varphi^2)). \tag{3.147}$$

where the sedimentation coefficient is equal to $S = -6.55$.

For suspension of particles of different sizes, the average sedimentation velocity of i-th particle can be represented in a similar form:

$$\langle U_i \rangle = U_i^0 \left(1 + \sum_{j=1}^{m} S_{ij} \varphi_j + 0(\varphi^2) \right). \tag{3.148}$$

Sedimentations coefficients S_{ij} that describe interactions between particles i and j depend on the ratio $k = a_j/a_i$, on the parameter $\gamma = (\rho_j - \rho_e)/(\rho_j - \rho_e)$ that characterizes the relation between the densities ρ_k of particles and ρ_e of the surrounding fluid, on the Peclet number Pe_{ij} (see Eq. (3.95), and on the parameter of molecular interaction. In the general case S_{ij} consists of three terms that describe the effects of the external gravity force, molecular interactions, and Brownian motion, respectively. To determine these coefficients, one has to use the results obtained in Sections 3.7–3.9.

Let $\boldsymbol{F}_i$ and $\boldsymbol{F}_j$ be the forces acting on the particles i and j. These forces are the net forces that include both gravitational and molecular forces. The molecular force depends on the distance r between the particle centers and tends to zero at $r \to \infty$. The force of gravity $\boldsymbol{F}_i^0$ is a downward force acting on the particle i. Then the quantity

$$\boldsymbol{U}_i^a = \boldsymbol{b}_{ii} \cdot \boldsymbol{F}_i + \boldsymbol{b}_{ij} \cdot \boldsymbol{F}_j - \frac{\boldsymbol{F}_i^0}{6\pi\mu_e a_i}. \tag{3.149}$$

gives the velocity increment acquired by i-th particle due to the presence of the particle j. In order to determine the mean value of this increment, it is necessary to average Eq. (3.149) over all possible positions of i-th particle, whose probability distribution is given by the pair distribution $n_j p_{ij}(\boldsymbol{r})$. But it was shown above that this procedure results in divergent integrals, because components A_{ij} and B_{ij} of mobility tensors $\boldsymbol{b}_{ii}$ and $\boldsymbol{b}_{ij}$ behave as r^{-1} and r^{-3} at $r \to \infty$. To avoid the difficulty of handling divergent integrals, we adopt the same approach that was used in Section 3.9.

Let $\langle \boldsymbol{U}_i \rangle$ be the velocity of i-th particle centered at the point $\boldsymbol{X}$ averaged over all possible configurations of the other particles. Also, let $\langle \boldsymbol{u} \rangle$ denote the velocity of the fluid at this point, also averaged over all possible particle configurations. The point $\boldsymbol{X}$ can be located within the fluid or inside a spherical particle (in this case the volume occupied by the particle should be treated as if it were filled with fluid). For a single particle ($\varphi \to 0$) of radius a, the velocity $\langle \boldsymbol{u} \rangle$ has the meaning of constant velocity of the surrounding fluid, and the force $\boldsymbol{F}^*$ will cause the particle to move relative to the fluid with the velocity

$$\langle \boldsymbol{U} \rangle - \langle \boldsymbol{u} \rangle = \frac{\boldsymbol{F}^*}{6\pi\mu_e a}.$$

If we also take into account hydrodynamic interaction of the test particle with other particles (in the pair interaction approximation), then we should get the following formula, which is accurate to the order of φ (see Eq. (3.141):

$$\begin{aligned}\langle \boldsymbol{U}_i^a \rangle = n_j \int\limits_{r \geq a_j} &\{(\boldsymbol{b}_{ii} \cdot \boldsymbol{F}_i + \boldsymbol{b}_{ij} \cdot \boldsymbol{F}_j - \boldsymbol{U}_i^0)\, p_{ij}(\boldsymbol{r}) - \boldsymbol{u}(\boldsymbol{X}|\boldsymbol{X} + \boldsymbol{r}, a_j) \\ &- \frac{1}{6} a_i^2 \nabla_r^2 \boldsymbol{u}(\boldsymbol{X}|\boldsymbol{X} + \boldsymbol{r}, a_j)\} d\boldsymbol{r} - \left(1 - \frac{a_i^2}{2a_j^2}\right) \varphi_j \boldsymbol{U}_j^0,\end{aligned} \tag{3.150}$$

where $u(X|X+r, a_j)$ is the fluid velocity at the point X under the condition that the point $X+r$ is the center of a particle of radius a_j; U_i^0 and U_j^0 are the sedimentation velocities of individual particles of radii a_i and a_j settling under the action of forces F_i^0 and F_j^0; and finally, $U_i^0 = F_i^0/6\pi\mu_e a_i$ and $U_j^0 = F_j^0/6\pi\mu_e a_j$. The last term describes the change of the velocity of i-th particle due to the viscous friction force acting on j-th particle.

Using the relation (3.142) for the fluid's velocity at the point X induced by the motion of the particle centered at $X+r$, one obtains:

$$u(X|X+r, a_j) + \frac{1}{6}a_i^2\nabla_r^2 u(X|X+r, a_j) = U_j^0\left\{\frac{3a_j}{4r}\left(I+\frac{rr}{r^2}\right)+\frac{a_j(a_i+a_j)}{4r^3}\left(I-\frac{3rr}{r^2}\right)\right\},$$

where I is the unit tensor and rr is the dyadic product.

If we now use the expressions (3.176) for mobility tensors while taking into account the condition $p_{ij}=0$ at $r<a_i+a_j$, the relation (3.150) will provide the general expression for the additional velocity of i-th particle resulting from pair interactions with all the other particles:

$$\langle \bar{U}_i^a\rangle = \sum_{j=1}^{m}\varphi_j\left\{\left(\frac{1+k}{2k}\right)^3 (U_i^0 \cdot J' + U_j^0 \cdot J'' + K_{ij}) - \left(1+\frac{3}{k}+\frac{1}{k^2}\right)U_j^0\right\}, \tag{3.151}$$

where

$$J' = \frac{3}{4\pi}\int_{s\geq 2}\left\{A_{ii}\frac{ss}{s^2}+B_{ii}\left(I-\frac{ss}{s^2}\right)-I\right\}p_{ij}(s)ds,$$

$$J'' = \frac{3}{4\pi}\frac{2k}{1+k}\int_{s\geq 2}\left[\left\{A_{ij}\frac{ss}{s^2}+B_{ij}\left(I-\frac{ss}{s^2}\right)\right\}p_{ij}(s) - \left\{\frac{3}{4s}\left(I+\frac{ss}{s^2}\right)+\frac{1+k^2}{(1+k)^2 s^3}\left(I-\frac{3ss}{s^2}\right)\right\}\right]ds,$$

$$K_{ij} = \frac{3}{4\pi}\int_{s\geq 2}\left[\left\{A_{ii}\frac{ss}{s^2}+B_{ii}\left(I-\frac{ss}{s^2}\right)\right\}\cdot\frac{(F_i-F_i^0)}{6\pi\mu_e a_i} + \frac{2k}{1+k}\left\{A_{ij}\frac{ss}{s^2}+B_{ij}\left(I-\frac{ss}{s^2}\right)\right\}\cdot\frac{(F_i-F_i^0)}{6\pi\mu_e a_i}\right]p_{ij}(s)ds,$$

$k=a_j/a_i$, $s=2r/(a_i+a_j)$, and $\mathbf{S}$ is the vector connecting the centers of particles i, j.

Since the left part of Eq. (3.151) depends linearly on the forces F_i and F_j, we can identify constituents of the additional velocity $\langle U_i^a\rangle$ that account for the effects of gravity, molecular interactions between particles, and Brownian motion of particles, denoting them by $\langle U_i^{Ga}\rangle$, $\langle U_i^{Ia}\rangle$, and $\langle U_i^{Ba}\rangle$. Then

$$\langle U_i^a\rangle = \langle U_i^{Ga}\rangle + \langle U_i^{Ia}\rangle + \langle U_i^{Ba}\rangle. \tag{3.152}$$

Taking our hint from the representation (3.148) of the average sedimentation velocity and the expression (3.152) for the additional particle velocity, we can express the sedimentation coefficients as sums of terms S_{ij}^G, S_{ij}^I,S_{ij}^B that describe the influence of gravity, molecular interactions between particles, and Brownian motion of particles:

$$S_{ij} = S_{ij}^G + S_{ij}^I + S_{ij}^B. \tag{3.153}$$

If the forces $\boldsymbol{F}_i$ and $\boldsymbol{F}_j$ are gravitational, they do not depend on $\mathbf{s}$, and

$$\boldsymbol{F}_i = \boldsymbol{F}_i^0 = 6\pi\mu_e a_i \boldsymbol{U}_i^0, \quad \boldsymbol{F}_j = \boldsymbol{F}_j^0 = 6\pi\mu_e a_j \boldsymbol{U}_j^0, \quad \boldsymbol{U}_j^0 = \gamma k^2 \boldsymbol{U}_j^0, \quad \boldsymbol{K}_{ij} = 0.$$

Then

$$\langle U_i^{Ga}\rangle = U_i^0 \sum_{j=1}^{m} \varphi_j \left\{ \left(\frac{1+k}{2k}\right)^3 (J' + \gamma k^2 J'') - \gamma(k^2 + 3k + 1)\boldsymbol{I} \right\}. \tag{3.154}$$

If $\boldsymbol{F}_i$ and $\boldsymbol{F}_j$ are forces of molecular interactions, then

$$\boldsymbol{F}_i = -\boldsymbol{F}_j = \boldsymbol{F}_{ij}(r), \quad \boldsymbol{F}_i^0 = -\boldsymbol{F}_j^0 = 0,$$

$$\begin{aligned} \boldsymbol{K}_{ij} = \frac{1}{8\pi^2\mu_e a_j} \int_{s\geq 2} \Bigg\{ \left(A_{ii} - \frac{2}{1+k} A_{ij}\right) \frac{\mathbf{ss}}{s^2} \\ + \left(B_{ii} - B_{ij}\frac{2}{1+k}\right)\left(\boldsymbol{I} - \frac{\mathbf{ss}}{s^2}\right) \cdot \boldsymbol{F}_{ij}\, p_{ij}(\mathbf{s}) \Bigg\} ds. \end{aligned} \tag{3.155}$$

Forces of molecular interaction (Van der Waals forces) are central forces having the property $\boldsymbol{F}_i(\boldsymbol{r}) = -\boldsymbol{F}_j(-\boldsymbol{r}) = \boldsymbol{F}_{ij}(\boldsymbol{r})$ and characterized by the potential $\phi_{ij}(r)$, so

$$\boldsymbol{F}_{ij}(\boldsymbol{r}) = -\nabla_{X_i}\phi_{ij}(r) = \frac{2}{a_i + a_j} \frac{\mathbf{s}}{s} \frac{d\phi_{ij}}{ds}.$$

Then

$$\langle U_i^I\rangle = \frac{1}{8\pi^2\mu_e a_i^2} \sum_{j=1}^{m} \varphi_j \frac{(1+k)^2}{4k^3} \int_{s\geq 2} \left(A_{ii} - \frac{2}{1+k} A_{ij}\right) \frac{\mathbf{s}}{s} \frac{d\phi_{ij}}{ds} p_{ij}(\mathbf{s}) ds. \tag{3.156}$$

If we take into account hydrodynamic interactions, molecular interactions, and Brownian motion, the following expression will determine the relative velocity of two particles i and j in the gravitational field:

$$\begin{aligned}\boldsymbol{V}_{ij}(\boldsymbol{r}) = \boldsymbol{V}_{ij}^{(0)}(\boldsymbol{r})\left\{L(s,k)\frac{\boldsymbol{rr}}{r^2}+M(s,k)\left(\boldsymbol{I}-\frac{\boldsymbol{rr}}{r^2}\right)\right\} \\ -\frac{D_{ij}^0}{k\vartheta}\left\{G(s,k)\frac{\boldsymbol{rr}}{r^2}+H(s,k)\left(\boldsymbol{I}-\frac{\boldsymbol{rr}}{r^2}\right)\right\}\cdot\nabla\phi_{ij} \\ -D_{ij}^0\times\left\{G(s,k)\frac{\boldsymbol{rr}}{r^2}+H(s,k)\left(\boldsymbol{I}-\frac{\boldsymbol{rr}}{r^2}\right)\right\}\cdot\nabla\ln(p_{ij}(\boldsymbol{r})),\end{aligned} \tag{3.157}$$

where $\boldsymbol{V}_{ij}^{(0)}$ is the relative velocity of free (unhindered) particles i and j in the absence of interactions (see Eq. (3.85); D_{ij}^0 is the coefficient of unhindered (i.e. without interparticle interactions; see also Eq. (3.96) relative Brownian diffusion of particles i and j; the functions L, M, G, and H are given by Eq. (3.79), a) and Eq. (3.87); P_{ij} is a pair PDF described by the Fokker–Planck equation (3.89), which in the current case reduces to

$$\frac{\partial p_{ij}}{\partial t} + \boldsymbol{V}_{ij}\cdot\nabla p_{ij} = \nabla\cdot(\boldsymbol{D}p_{ij}). \tag{3.158}$$

It should be noted that Brownian motion exerts the same effect on particle sedimentation as the external force $\boldsymbol{F}_{ij}^B$ introduced earlier in Section 3.7, which is known as the thermodynamic force. This force causes i-th particle to acquire the velocity

$$\boldsymbol{b}_{ii}\cdot\boldsymbol{F}_{ij}^B-\boldsymbol{b}_{ij}\cdot\boldsymbol{F}_{ij}^B = k\vartheta(\boldsymbol{b}_{ii}-\boldsymbol{b}_{ij})\cdot\nabla_r\log(p_{ij}(r)),$$

which, after averaging over the ensemble of j-th particles having the distribution $n_j p_{ij}(r)$, becomes

$$\langle\boldsymbol{U}_i^B\rangle = k\vartheta\sum_{j=1}^{m}\frac{1}{4}(a_i+a_j)^2 n_j\int_{s\geq 2}(\boldsymbol{b}_{ii}-\boldsymbol{b}_{ij})\cdot\nabla_s p_{ij}\boldsymbol{ds}.$$

Using the properties of the mobility tensor, one can transform this relation to

$$\begin{aligned}\langle\bar{\boldsymbol{U}}_i^B\rangle = \frac{3}{4\pi a_i}\sum_{j=1}^{m}\varphi_j\frac{D_{ij}^0(1+k)}{2k^2}\int_{s\geq 2}\left(\frac{A_{ii}-B_{ii}}{s}+\frac{1}{2}\frac{dA_{ii}}{ds}\right. \\ \left.-\frac{2(A_{ij}-B_{ij})}{(1+k)s}-\frac{1}{1+k}\frac{dA_{ij}}{ds}\right)\frac{\boldsymbol{s}\cdot\boldsymbol{V}_{ij}^0}{sV_{ij}^0}(1-p_{ij})\boldsymbol{ds}.\end{aligned} \tag{3.159}$$

The derived components of the average sedimentation velocity of i-th particle allow us to find all sedimentation coefficients in their general form:

$$S_{ij}^{G} = \left(\frac{1+k}{2k}\right)^{3}\left\{\frac{\mathbf{g}\cdot\mathbf{J}'\cdot\mathbf{g}}{g^{2}} + \gamma k^{2}\frac{\mathbf{g}\cdot\mathbf{J}''\cdot\mathbf{g}}{g^{2}}\right\} - \gamma(k^{2}+3k+1),$$

$$S_{ij}^{I} = \frac{3}{8\pi}\frac{\gamma k^{2}-1}{Pe_{ij}}\frac{(1+k)^{2}}{4k^{2}}\int\limits_{s\geq 2}\left(A_{ii} - \frac{2}{1+k}A_{ij}\right)\frac{\mathbf{s}}{s}\frac{d(\phi_{ij}/kT)}{ds}\,p_{ij}(\mathbf{s})d\mathbf{s},$$

$$S_{ij}^{B} = \frac{3}{8\pi}\frac{\gamma k^{2}-1}{Pe_{ij}}\frac{(1+k)^{2}}{4k^{2}}\int\limits_{s\geq 2}\left(\frac{A_{ii}-B_{ii}}{s} + \frac{1}{2}\frac{dA_{ii}}{ds}\right.$$
$$\left. - \frac{2(A_{ij}-B_{ij})}{(1+k)s} - \frac{1}{1+k}\frac{dA_{ij}}{ds}\right)\frac{\mathbf{s}\cdot\mathbf{V}_{ij}^{0}}{sV_{ij}^{0}}(1-p_{ij})d\mathbf{s}.$$

In order to obtain the sedimentation coefficients S_{ij}, one should first solve the Fokker–Planck equation (3.158) to find p_{ij}, and then proceed to calculate the corresponding integrals, which in the general case can be done by using numerical methods. In the limiting cases $Pe_{ij} \ll 1$ and $Pe_{ij} \gg 1$, the problem can be considerably simplified (see Section 3.8). For example, at $Pe_{ij} \gg 1$ and assuming the absence of molecular interactions ($\phi_{ij} = 0$), the sedimentation coefficients are equal to

$$S_{ij}(k,\gamma) = \int\limits_{2}^{\infty}\left\{\left(\frac{1+k}{2k}\right)^{3}\left(A_{ii} + 2B_{ii} - 3\right)p_{ij}\right.$$
$$\left. + \frac{1}{4}\gamma(1+k)^{2}\left((A_{ij}+2B_{ij})\,p_{ij} - \frac{3}{8}\right)\right\}s^{2}ds - \gamma(k^{2}+3k+1),$$

(3.160)

where p_{ij} is given by (3.102)

$$p_{ij}(s) = \exp\left\{\int\limits_{s}^{\infty}\frac{W(s)}{L(s)}ds\right\} = \exp\left\{\int\limits_{s}^{\infty}\left(\frac{2(L-M)}{sL} + \frac{1}{L}\frac{dL}{ds}\right)ds\right\}.$$

Of special interest is the particular case of identical particles ($k = 1, \gamma = 1$). In this case, $p_{ij}(\mathbf{r}) = p_{ij}(-\mathbf{r})$, and the contribution of molecular interactions to the average sedimentation velocity is zero. In addition, the average relative velocity of particles is also equal to zero, and therefore the convective term in the Fokker–Planck equation vanishes. Then the distribution $p_{ij}(\mathbf{r})$ becomes spherically symmetrical and turns into the Boltzmann distribution (3.107), and the sedimentation coefficient becomes

$$S = -6.55 + \int\limits_{2}^{\infty}(A_{11} + 2B_{11} - 3 + A_{12} + 2B_{12})_{k=1}\left\{\exp\left(-\frac{\phi(\mathbf{r})}{kT}\right) - 1\right\}s^{2}ds.$$

(3.161)

Tab. 3.1 Coefficients S_{ij} for $Pe_{ij} \gg 1$, $\Phi_{ij}=0$, for various values of k and γ.

	γ							$\lim_{\gamma\to\infty} \frac{S_{ij}}{\gamma}$
k	−2	−1	−0.5	0	0.6	1	1.5	
0.25	−1.96	−2.00	−2.20	−2.56	−3.31	−3.83	−4.73	1.97
0.5	−2.51	−2.27	−2.28	−2.53	−3.41	−4.29	−6.77	1.03
1.0				$S_{ij}=-2.52-0.13\gamma$ $(\gamma\neq 0)$				
2.0	3.18	−0.34	−1.89	−2.44	−9.85	−9.81	−11.16	−3.79
4.0	26.63	10.05	2.03	−2.66	−19.55	−24.32	−32.71	−16.78

(the indices i and j are no longer needed).

For particles of substantially different sizes ($k \ll 1$ or $k \gg 1$), the expression for the average sedimentation velocity is dominated by the gravitational component, and the following approximate relations can be used for arbitrary Peclet numbers:

$$S_{ij} = -\frac{5}{2} - \gamma + 0(k) \ \text{at}\, k << 1, \tag{3.162}$$

$$S_{ij} = -\gamma(k^2 + 3k + 1) + 0(k^{-1}) \ \text{at}\, k \gg 1. \tag{3.163}$$

At $\gamma \gg 1$ and $Pe_{ij} \gg 1$, we have

$$S_{ij} \sim \gamma \left[\frac{1}{4}(1+k)^2 \int_2^{\infty} \left\{ (A_{ij} + 2B_{ij})(p_{ij})_{\gamma\to\infty} - \frac{3}{8} \right\} s^2 ds - (k^2 + 3k + 1) \right]. \tag{3.164}$$

Some typical sedimentation coefficient values are given in Table 3.1.

3.11
Transport Coefficients in Disperse Media

The problem of finding the viscosity of a suspension is closely related to the general problem of finding transport coefficients characterizing the ability of the medium as a whole to transfer some substance: momentum, heat, mass, electric charge, and so on in response to the emergence of a spatial gradient $\boldsymbol{G}$ of some physical parameter, such as velocity, temperature, concentration of neutral or charged particles (see references [20–23]). In the equilibrium state of the medium, the fluxes of these parameters are absent. Any deviation from equilibrium caused by the gradient $\boldsymbol{G}$ gives rise to a directed flux $\boldsymbol{Q}$, which is related to $\boldsymbol{G}$ by a linear dependence:

$$\boldsymbol{Q} = \boldsymbol{K}\cdot\boldsymbol{G}. \tag{3.165}$$

The proportionality factor $\boldsymbol{K}$ is called the transport coefficient. If $\boldsymbol{G}$ is the gradient of a scalar quantity such as concentration, then $\boldsymbol{K}$ is a second-rank tensor (for example, diffusion tensor). When $\boldsymbol{G}$ is the gradient of a vector quantity such as momentum, $\boldsymbol{K}$ is a tensor of the forth rank. The factor $\boldsymbol{K}$ depends on the structure of the given medium. Transport coefficients have the simplest form when the medium is isotropic.

If the medium is microscopically heterogeneous, that is, if it consists of the continuous phase (the fluid) and the disperse phase (solid, liquid, or gaseous microparticles suspended in the fluid), and, in addition, the disperse phase itself is statistically homogeneous, then the linear phenomenological relation (3.165) is valid for the average values of the gradient $\langle \boldsymbol{G} \rangle$ and the flux $\langle \boldsymbol{Q} \rangle$:

$$\langle \boldsymbol{Q} \rangle = \boldsymbol{K}_{eff} \cdot \langle \boldsymbol{G} \rangle. \tag{3.166}$$

The proportionality factor $\boldsymbol{K}_{eff}$ is a macroscopic parameter of the disperse medium and, in contrast to $\boldsymbol{K}$, it is called the effective transport coefficient. Eq. (3.166) adequately describes the transfer of various substances. In a disperse medium, $\boldsymbol{K}_{eff}$ depends on the transport coefficients of each phase and also on statistical properties of the interface, most notably, on the volume concentration φ of the disperse phase. Since the disperse phase consists of particles randomly arranged in space and having different geometrical shapes, spatial orientations, sizes, and densities, statistical properties of the medium are characterized the distribution of particles over their sizes, orientations, and other parameters.

The simplest (and the most conducive to quantitative analysis) case is that of a monodisperse medium, that is, a medium where the disperse phase consists of identical spherical particles of radius a. The medium is also assumed to be statistically homogeneous. In this case statistical properties can be completely determined by the PDF $p(\boldsymbol{X}_1, \boldsymbol{X}_2, \ldots, \boldsymbol{X}_n)$.

The quantity $p(\boldsymbol{X}_1, \boldsymbol{X}_2, \ldots, \boldsymbol{X}_n)\, d\boldsymbol{X}_1, d\boldsymbol{X}_2, \ldots, d\boldsymbol{X}_n)$ is the joint probability for the center of the first particle $\boldsymbol{X}_1$ to be in the volume element $d\boldsymbol{X}_1$, the center of the second particle $\boldsymbol{X}_2$ – in $d\boldsymbol{X}_2, \ldots,$ and the center of n-th particle $\boldsymbol{X}_n$ – in $d\boldsymbol{X}_n$.

Most publications on disperse systems deal with the case of an infinitely small volume fraction (concentration) φ, because in this case the large average distance between particles allows to consider only pair interactions between particles. For larger φ, it becomes necessary to consider multiparticle interactions as well.

There are several properties of disperse media that will hold for all values of φ. One of these states that for a statistically homogeneous system, the average flux is maximum when the local transport properties are uniform. We shall demonstrate this property for the transport of heat, even though the same approach would be applicable to the transport of any other substance.

Denote the local scalar coefficient of heat transport (aka thermal conductivity) by K. The local temperature field is the sum $\langle \boldsymbol{G} \rangle\ \boldsymbol{X} + \vartheta$ of the average value and the fluctuation. The quantities $\boldsymbol{K}$ and ϑ are stationary random functions of position $\boldsymbol{X}$. Since flux obeys the equation $\nabla \cdot \boldsymbol{Q} = 0$, we get

$$\langle \boldsymbol{Q} \cdot \nabla \vartheta \rangle = \langle \nabla \cdot (\vartheta \boldsymbol{Q}) \rangle = \nabla \cdot \langle (\vartheta \boldsymbol{Q}) \rangle = 0.$$

Then the component of the mean flux in the direction of the mean temperature gradient is equal to

$$\langle \boldsymbol{G}\rangle\cdot\langle \boldsymbol{Q}\rangle = \langle \boldsymbol{G}\rangle\cdot(\langle K\rangle\langle \boldsymbol{G}\rangle + \langle K\nabla\vartheta\rangle) = \langle K\rangle\langle \boldsymbol{G}\rangle\cdot\langle \boldsymbol{G}\rangle + \langle(\boldsymbol{Q}-K\nabla\vartheta)\cdot\nabla\vartheta\rangle$$
$$= \langle K\rangle\langle \boldsymbol{G}\rangle\cdot\langle \boldsymbol{G}\rangle - \langle K(\nabla\vartheta)^2\rangle.$$

Since $K \geq 0$ and $(\nabla\vartheta)^2 \geq 0$, the product $\langle \boldsymbol{G}\rangle\cdot\langle \boldsymbol{Q}\rangle$ reaches its maximum value for given $\boldsymbol{K}$ and $\langle \mathbf{G}\rangle$ when the fluctuation of the temperature gradient $\nabla\vartheta$ about the mean value is zero. Now, using Eq. (3.166), we have

$$\langle \boldsymbol{G}\rangle\cdot\langle \boldsymbol{Q}\rangle = \langle \boldsymbol{G}\rangle\cdot \boldsymbol{K}_{eff}\cdot\langle \boldsymbol{G}\rangle \leq \langle K\rangle\langle \boldsymbol{G}\rangle\cdot\langle \boldsymbol{G}\rangle,$$

from which it follows that for a statistically isotropic medium,

$$K_{eff} \leq \langle K\rangle. \tag{3.167}$$

The expression (3.167) turns into equality when $\nabla\vartheta = 0$ in the entire region, which is possible only when the local conductivity is uniform along all straight lines parallel to $\langle \mathbf{G}\rangle$ (or in all directions in the case of an isotropic medium).

For transport processes, the local flux vector $\boldsymbol{Q}$ has definite values at all points of the medium, and is a linear function of the local gradient $\mathbf{G}$. However, the transport coefficients are different for each phase. The functions $\boldsymbol{Q}$ and $\boldsymbol{G}$ are stationary random functions of spatial position. Let us assume the medium to be homogeneous and consider a volume V of the medium sufficiently large to hold many particles of the disperse phase. Then one can employ the ergodic hypothesis, which states that ensemble averaging produces the same result as averaging over the volume V:

$$\langle \boldsymbol{Q}\rangle = \frac{1}{V}\int_V \boldsymbol{Q}dV, \quad \langle \boldsymbol{G}\rangle = \frac{1}{V}\int_V \boldsymbol{G}dV. \tag{3.168}$$

The volume V consists of the volume occupied by the continuous phase and the volume (many small volumes, to be precise) occupied by the disperse phase. Suppose the medium to is isotropic and let K_0 and αK_0 denote scalar transport coefficients (coefficients of heat conductivity, for example) of the continuous and disperse phases. Then we have for these phases: $\boldsymbol{Q}_0 = K_0\boldsymbol{G}$ and $\boldsymbol{Q} = \alpha K_0\boldsymbol{G}$. Substituting these relations into Eq. (3.168), we obtain

$$\langle \boldsymbol{Q}\rangle = \langle \boldsymbol{Q}_0\rangle + \frac{1}{V}\sum \boldsymbol{S}, \tag{3.169}$$

where $\langle \boldsymbol{Q}_0\rangle$ is the average flux in the continuous phase due to the average gradient (we sum over all particles of the disperse medium). The quantity $\boldsymbol{S}$ is a parameter characterizing the particle and is equal to

$$\boldsymbol{S} = \int_{V_0}(\boldsymbol{Q}-\boldsymbol{Q}_0)dV = \int_{A_0}\boldsymbol{X}(\boldsymbol{Q}-\boldsymbol{Q}_0)\cdot\boldsymbol{n}dA, \tag{3.170}$$

where V_0 and A_0 are the particle's volume and surface area, Q is the local flux at a point where the gradient is equal to G, and Q_0 is the value of flux in the volume occupied by the particle as if it were filled with the continuous phase. Physically, the parameter S corresponds to the additional effect that would be produced by the volume V_0 if it were originally filled with the fluid that was suddenly replaced by the material of the particle.

If the particles (either solid or liquid) are suspended in an ambient incompressible viscous fluid with the viscosity coefficient μ_e, and $\langle Q \rangle$ is the tensor of viscous Newtonian stresses with components $\sigma_{ij,}$then S is a tensor whose components are given by

$$S_{ij} = \int_A \left\{ \sigma_{ik} X_j n_k - \frac{1}{3} \sigma_{lk} X_l n_k \delta_{ij} - \mu_e (u_i n_j + u_j n_i) \right\} dA. \tag{3.171}$$

Here X_i are the coordinates of a point on the particle surface; n_i are components of the vector $\boldsymbol{n}$, which is an outward normal vector to the particle surface; δ_{ij} are components of the unit tensor; u_i are components of the velocity vector at the particle surface. The quantities S_{ij} have the meaning of intensities of force dipoles distributed over the particle surface.

In the case of solid particles, the zero relative velocity condition $u_i = 0$ must be satisfied at the particle surface. Then the third integrand term in Eq. (3.171) vanishes. If the particles are liquid, then both normal and tangential stresses of the internal and external liquids should be equal at their surfaces; the particle surfaces (interfaces) are mobile; and $u_i \neq 0$. Our analysis will be limited to solid particles.

If the particles are identical, then Eq. (3.169) can be rewritten as

$$\langle Q \rangle = \langle Q_0 \rangle + n \langle S \rangle, \tag{3.172}$$

where n is the number concentration of particles.

Thus the average flux of substance induced by the applied gradient $\langle G \rangle$ consists of two terms corresponding to the contributions of the two phases.

The expression (3.172) is valid for various two-phase systems, for example: a disperse medium composed of small particles buoyant or sedimenting in the fluid; a porous medium – a fixed array of small solid particles, through which the fluid is flowing. The transferred substance could be mass, heat, charge, and so forth. The meaning of the parameter S is different in each case. Thus, in a sedimentation process, $Q_0 = 0$ and S is the translational velocity of the particle, whereas in the case of fixed particles (porous medium) $Q_0 = 0$ is still true, but S is now the force exerted on the particle by the surrounding fluid. In both cases mentioned above, S can be represented as a surface integral, but of course its form is different from Eq. (3.170).

The value of S for each particle depends on the size, shape, and spatial orientation of that particle as well as the other particles of the disperse system. Therefore in order to determine G and Q, one has to know the distribution of particles over size, shape, and spatial orientation. These distributions characterize statistical properties and microstructure of the disperse medium. The properties of a disperse medium consisting of spherical particles are the easiest to understand. If the particle's shape is not spherical, this leads to the orientational effect, that is, the ability of the particle to assume a certain orientation in the space under the action of an applied gradient of some substance (temperature, electric potential, velocity, and so on). It is this effect that is responsible for the majority of instances of non-Newtonian behavior of the disperse medium. That being said, non-Newtonian properties can manifest themselves even in a medium consisting of spherical particles, especially in a highly concentrated disperse medium which is capable of forming non-spherical structures.

Suppose the disperse medium contains particles of the same composition and shape, and particle orientation is characterized by some vector $\boldsymbol{l}$ (orientation vector). For example, if the particle shape is that of a thin cylinder, then $\boldsymbol{l}$ is a unit vector directed along the axis of the cylinder. Let $p(\boldsymbol{l}, \Theta_N|\boldsymbol{X}_0)$ denote the distribution of joint probability density of orientation $\boldsymbol{l}$ and configuration Θ_N for N particles in the volume V given that the center of the test particle is located at the point $\boldsymbol{X}_0$. Then the average value $\langle \boldsymbol{S} \rangle$ equals

$$\langle \boldsymbol{S} \rangle = \int S(\boldsymbol{l}, \Theta_N)\, p(\boldsymbol{l}, \Theta_N|\boldsymbol{X}_0) d\boldsymbol{l} d\Theta_N. \tag{3.173}$$

Consider some special cases when $\langle \boldsymbol{S} \rangle$ can be determined explicitly.

3.11.1 Infinitely Dilute Suspension with Non-interacting Particles

The volume concentration of particles is assumed to be low ($\varphi \ll 1$), so the average distance between particles is large, though still small compared to the linear scale of the system. Then, in the first approximation, the distribution of macroscopic parameters (velocity, temperature, electric potential, etc.) in the vicinity of a particle can be assumed identical to the distribution of the same parameters in the single-particle case. Then the value of $\boldsymbol{S}$ does not depend on the particle configuration Θ_N, and the relation (3.173) becomes

$$\langle \boldsymbol{S} \rangle = \int S(\boldsymbol{l})\, p(\boldsymbol{l}) d\boldsymbol{l}. \tag{3.174}$$

In order to determine $\boldsymbol{S}(\boldsymbol{l})$ for a given particle configuration relative to a coordinate system fixed in space, one has to solve the corresponding boundary value problem and to determine the distribution of velocity, temperature, electric potential, and other parameters inside and outside the particle in the ambient fluid under the condition that the particle is alone and the gradient $\boldsymbol{G}$ of the substance under consideration on a large distance from the particle is given. Then $\langle \boldsymbol{S} \rangle$ can be found from Eq. (3.174) for a given distribution $p(\boldsymbol{l})$.

If the particles are spherical, there is no need to do the averaging. The case of solid spherical particles has been studied extensively for infinitely dilute suspensions. In particular, the temperature distribution in an isolated solid spherical particle of radius a suspended in a homogeneous unbounded fluid with a given temperature gradient at the infinity is given by

$$\vartheta = \frac{3}{(\alpha + 2)} \boldsymbol{G} \cdot \boldsymbol{X}.$$

From Eq. (3.170), there follows

$$\langle \boldsymbol{S} \rangle = \frac{4}{3} \pi a^3 \frac{3(\alpha - 1)}{(\alpha + 2)} K_0 \boldsymbol{G}.$$

Now Eq. (3.172) yields the average flux:

$$\langle \boldsymbol{Q} \rangle = K_0 \left\{ 1 + \frac{3(\alpha - 1)}{(\alpha + 2)} \varphi \right\} \boldsymbol{G}.$$

In the absence of particles, the flux is equal to $\boldsymbol{K_0 G}$. So, the last expression shows that the presence of particles in the fluid increases the effective heat conductivity:

$$\kappa_{eff} = \kappa_0 \left\{ 1 + \frac{3(\alpha - 1)}{(\alpha + 2)} \varphi \right\}.$$

This formula was first derived by Maxwell for effective electric conductivity of infinite dilute suspensions. A similar expression for the effective viscosity coefficient was obtained by Einstein for spherical solid microparticles:

$$\mu_{eff} = \mu_e^0 (1 + 2.5\varphi),$$

and by Taylor – for spherical liquid microparticles with the internal and external liquid viscosities μ_i and μ_e:

$$\mu_{eff} = \mu_e \left\{ 1 + \frac{2.5\mu_i + \mu_e}{\mu_i + \mu_e} \varphi \right\}.$$

Determination of the heat conduction coefficient for an infinitely dilute suspension containing ellipsoidal particles presents a more difficult problem, because to solve the corresponding internal and external boundary value problems one has to use ellipsoidal harmonics. In this problem, the effective heat conduction coefficient is not a scalar but a tensor. In a coordinate system whose axes coincide with principal axes of the ellipsoid, the tensor matrix has diagonal form. For an infinitely dilute suspension containing ellipsoidal particles with semi-axes a_1, a_2, a_3 and with the same spatial orientation, the diagonal elements of the effective heat conductivity tensor are

$$\kappa_{i\,eff} = \kappa_0\left\{1+\frac{(\alpha-1)}{0.5(\alpha-1)H_i}\varphi\right\}, \quad (i=1,2,3), \tag{3.175}$$

where

$$H_i = \int_0^\infty \frac{a_1, a_2, a_3 dX}{(a_i^2+X)((a_1^2+X)(a_2^2+X)(a_3^2+X))^{1/2}}.$$

For elongated particles ($a_1 \gg a_2,\ a_3$), Eq. (3.175) gives the following asymptotic expressions:

$$H_1 = \frac{2a_2a_3}{a_1^2}\ln\left\{\frac{4a_1}{a_2+a_3}\right\}, \quad H_2 \sim \frac{2a_3}{a_2+a_3}, \quad H_3 \sim \frac{2a_2}{a_2+a_3}.$$

The contribution of other particles to the average flux can be found if we know $p(\boldsymbol{l})$ – the probability density distribution of particles over their orientations. If bulk flow is the only factor that may change the orientation of a particle, and Brownian motion is not taken into account, the sought-for function will follow as the solution of the Liouville equation:

$$\frac{\partial p}{\partial t} = -\nabla\cdot(\dot{\boldsymbol{l}}p). \tag{3.176}$$

We can account for Brownian motion by adding the diffusion term to the right-hand side:

$$\frac{\partial p}{\partial t} = -\nabla\cdot(\dot{\boldsymbol{l}}p) + D_{br}\nabla^2 p. \tag{3.177}$$

where $\dot{\boldsymbol{l}} = dl/dt$ and D_{br} is the scalar coefficient of Brownian diffusion.

If the shape of particles is non-spherical, particles will rotate in the nonuniform velocity field of the surrounding fluid. Then the orientation vector is determined in terms of parameters characterizing fluid motion and the shape of the particle:

$$\dot{\boldsymbol{l}} = \Omega\times\boldsymbol{l} + \gamma_f(\boldsymbol{l}\cdot\boldsymbol{E} - \boldsymbol{l}(\boldsymbol{l}\cdot\boldsymbol{E}\cdot\boldsymbol{l})),$$

where Ω is the angular velocity of local rotation of the fluid flow as a whole, $\boldsymbol{E}$ is the rate of strain associated with the fluid flow, γ_f is the particle's shape factor, which tends to unity (from below) as the particle's length-to-width ratio tends to infinity.

If the angular velocity Ω of the bulk flow is zero (pure straining motion), the particle orientates itself along the principal axis of the rate-of-strain tensor that corresponds to the maximum principal value $\boldsymbol{E}_1$. In the idealized case all particles will acquire the same orientation ($\langle \boldsymbol{l} \rangle = \boldsymbol{l}$), and the probability density distribution $p(\boldsymbol{l})$ will become the delta function $p(\boldsymbol{l}) = \delta(\langle \boldsymbol{l} \rangle = \boldsymbol{l})$. But in general, the tendency of particles to orient themselves along the "maximum" principal axis of the rate-of-strain tensor is moderated by random particle rotations that arise because of rotational Brownian diffusion; this diffusion is characterized by rotational Brownian diffusivity D_{br}^{r}. As a result, particle orientations are spread about the average value $\langle \boldsymbol{l} \rangle$; the standard deviation is $\left(D_{br}^{r}/E_1\right)^{1/2}$.

From the practical point of view, we are often interested in the behavior of particles in a shear bulk flow. As was noted earlier, any flow with a nonuniform velocity profile can be thought of as a shear flow in the neighborhood of a microparticle. Besides, flows that take place in rheological measuring devices (viscometers) are, as a rule, shear flows. Under the influence of shear, nonspherical particles get involved in rotational motion with a nonuniform angular velocity relative to the average orientation vector $\langle \boldsymbol{l} \rangle$; the latter is directed along the tangent to streamlines of the bulk flow. The rotational velocity increases with the shape factor.

Among other factors influencing the orientation of particles, we should mention external electric and magnetic fields, which act selectively on the particles possessing electric or magnetic properties (in other words, they act on the particles that are polarizable or magnetizable). When the external field is applied, particles align themselves along the strength vector of the field.

The propensity of particles to change their spatial orientation under the action of various forces (hydrodynamical, electrical, magnetic) may cause a change of rheological properties of the suspension. In particular, transport coefficients should be interpreted as tensor, rather than scalar, quantities, and the disperse medium by itself can manifest non-Newtonian properties. The latter effect becomes noticeable for highly concentrated suspensions, where the particles cannot be considered as independent from each other and one must account properly for interparticle interactions.

3.11.2
The Influence of Particle Interactions on Transport Coefficients

As was noted in Section 3.10, particle interactions are important even for dilute suspensions. Therefore we begin our analysis with infinitely dilute suspensions ($\varphi \ll 1$) and, for simplicity's sake, consider only the case of identical spherical particles of radius a, because it is for this case that the most important results have been obtained.

Interaction between particles is described by the relation (3.173). For identical spherical particles, $\boldsymbol{S}$ depends only on configuration of the surrounding particles, in other words, $\boldsymbol{S} = \boldsymbol{S}(\Theta_N)$. Then Eq. (3.173) reduces to

$$\langle \boldsymbol{S} \rangle = \int \boldsymbol{S}(\Theta_N)\, p(\Theta_N|\boldsymbol{X}_0)\, d\Theta_N. \tag{3.178}$$

Let us recall that the configuration Θ_N is a set of position vectors $\boldsymbol{r}_1, \boldsymbol{r}_2, \ldots, \boldsymbol{r}_N$ for N particles in the considered volume. Then $p(\Theta_N|\boldsymbol{X}_0)d\Theta_N$ is the probability to find the particles in their respective intervals $(\boldsymbol{r}_1, \boldsymbol{r}_1 + d\boldsymbol{r}_1), (\boldsymbol{r}_2, \boldsymbol{r}_2 + d\boldsymbol{r}_2), \ldots, (\boldsymbol{r}_N, \boldsymbol{r}_N + d\boldsymbol{r}_N)$ given that the test particle is located at $\boldsymbol{X}_0$. If $\boldsymbol{S}_0$ is the value of $\boldsymbol{S}$ in the absence of particle interactions, then the relation (3.172) for the average flux $\langle \boldsymbol{Q} \rangle$ takes the form

$$\langle \boldsymbol{Q} \rangle = \boldsymbol{Q}_0 + n\boldsymbol{S}_0 + n\int \{\boldsymbol{S}(\Theta_N) - \boldsymbol{S}_0\}\, p(\Theta_N|\boldsymbol{X}_0)\, d\Theta_N. \tag{3.179}$$

If we use Eq. (3.179) instead of Eq. (3.172), terms of order φ will appear in the expressions for transport coefficients (for example, in the formulas for κ_{eff} and μ_{eff}). We should, of course, consider only pair interactions between particles and replace the functions $\boldsymbol{S}(\Theta_N)$ and $p(\Theta_N|\boldsymbol{X}_0)$ by $\boldsymbol{S}(\boldsymbol{r})$ and $p(\boldsymbol{r}|\boldsymbol{X}_0)$, where $\boldsymbol{r}$ is the vector from the center of the test particle to the center of the neighboring particle. The pair distribution $p(\boldsymbol{r}|\boldsymbol{X}_0)$ obeys the condition $p(\boldsymbol{r}|\boldsymbol{X}_0) \rightarrow n$ at $r \rightarrow \infty$ and satisfies the Liouville equation

$$\frac{\partial p}{\partial t} = -\nabla \cdot (\boldsymbol{V} p). \tag{3.180}$$

where $\boldsymbol{V}(\boldsymbol{r},t)$ is the velocity of relative motion of the centers of our two particles. This velocity is found by solving the hydrodynamic boundary value problem of relative Stokesian motion of two spherical particles subject to external forces and forces of molecular interaction. If Brownian motion is also important, as is the case for small Peclet numbers, then p is a solution of the Fokker–Planck equation

$$\frac{\partial p}{\partial t} = -\nabla \cdot (\boldsymbol{V} P) + \nabla \cdot (\boldsymbol{D}_{br} \cdot \nabla p), \tag{3.181}$$

where $\boldsymbol{D}_{br}$ is the tensor of Brownian diffusion.

When calculating the integral in Eq. (3.179), we face the same convergence problem that we encountered earlier, in Sections 3.9 and 3.10, because the expressions for $\boldsymbol{S}$ (these could be temperature field or velocity field dipoles modeling the particles that are spaced far apart) behave as r^{-3} and r^{-1}, respectively, as r goes to infinity ($r \rightarrow \infty$).

To overcome this difficulty, we resort to the method described in Section 3.9 Application of this method to the problem of heat transfer in a disperse medium results in the following expression for the third term on the right-hand side of Eq. (3.179):

$$\int \left\{ (\boldsymbol{S}(\boldsymbol{r}) - \boldsymbol{S}_0)\, p(\boldsymbol{r}|\boldsymbol{X}_0) - \frac{4}{3}\pi a^3 n \left(\frac{3(\alpha - 1)}{(\alpha + 2)} \right) K_0 (\boldsymbol{G}(\boldsymbol{r}) - \langle \boldsymbol{G} \rangle) \right\} d\boldsymbol{r},$$

where $\boldsymbol{G}(\boldsymbol{r})$ is the temperature gradient induced at the point $\boldsymbol{X}_0$ by the presence of a particle at the point $\boldsymbol{r}$, given a certain constant temperature gradient $\langle \boldsymbol{G} \rangle$ at the infinity. Integration is carried out over the whole range of $\boldsymbol{r}$. In the absence of more specific information, the unnormalized conditional PDF $P(\boldsymbol{r}|\boldsymbol{X}_0)$ is approximated by the step function (see Eq. (3.137):

$$P(\mathbf{r}|\mathbf{X}_0) = \begin{cases} n \text{ at } r \geq 2a, \\ 0 \text{ at } r < 2a, \end{cases} \tag{3.182}$$

meaning that all physically possible arrangements of particles relative to the test particle have the same likelihood.

In order to find the real PDF $P(\mathbf{r}|\mathbf{X}_0)$, it is necessary to solve two closely related problems at the same time: the problem of relative motion of two spheres, and the problem of determination of $P(\mathbf{r}|\mathbf{X}_0)$ from the conservation equation (3.180).

In the approximation (3.182), the effective heat conduction coefficient is given by

$$\kappa_{i\,eff} = \kappa_0 \left\{ 1 + \frac{3(\alpha-1)}{\alpha+2}\varphi + \frac{3(\alpha-1)^2}{(\alpha+2)^2}\varphi^2 \right\}. \tag{3.183}$$

Similarly, we can write the expression for the effective viscosity coefficient of a suspension containing identical solid spherical particles:

$$\mu_{eff} = \mu_e(1 + 2.5\varphi + 7.6\varphi^2). \tag{3.184}$$

If we properly account for the Brownian motion of microparticles, μ_{eff} will be somewhat smaller:

$$\mu_{eff} = \mu_e(1 + 2.5\varphi + 6.2\varphi^2). \tag{3.185}$$

The presence of particles in the fluid creates additional stresses in the suspension. If Brownian motion is neglected, the components of the average stress tensor $\langle \boldsymbol{T} \rangle$ are given by

$$\langle T_{ij} \rangle = \delta_{ij}\langle T_{kk} \rangle + 2\mu_e\langle E_{ij} \rangle + \langle T_{ij}^p \rangle. \tag{3.186}$$

The first term on the right represents the components of the spherical part $-\langle p \rangle \boldsymbol{I}$ of the stress tensor, where p is the pressure. The sum of the second and the third term represents the stress deviator. This deviator, in its turn, consists of the terms $2\mu_e\langle E_{ij} \rangle$ and $\langle T_{ij}^p \rangle$ – respective contributions from the ambient fluid and from the suspended particles; the latter is equal to

$$\langle T_{ij}^p \rangle = \frac{1}{V}\sum S_{ij} = \frac{1}{V}\sum \int_{A_0} \left\{ \sigma_{ik}X_j n_k - \frac{1}{3}\sigma_{ik}X_i n_k \delta_{ij} - \mu_e(u_i n_j + u_j n_i) \right\} dA,$$

where σ_{ik} are components of the viscous stress tensor, u_i are components of the fluid's velocity, n_i are components of the outer normal $\boldsymbol{n}$ to the particle surface A_0. Summation is performed over all particles in the given volume V.

For identical spherical particles undergoing Brownian motion, the tensor $\langle T_{ij}^{p} \rangle$ is

$$\langle \boldsymbol{T}^{p} \rangle = -nk\vartheta \boldsymbol{I} + n(\langle \boldsymbol{S}_{T}^{H} \rangle + \langle \boldsymbol{S}_{T}^{B} \rangle + \langle S_{T}^{E} \rangle), \tag{3.187}$$

where $-nk\vartheta \boldsymbol{I}$ is the spherical part that arises because of the thermal motion of particles; n is the average number concentration of particles; the three summands in parenthesis are the respective contributions from hydrodynamic interactions $\langle \boldsymbol{S}^{H} \rangle$, from Brownian motion $\langle \boldsymbol{S}^{B} \rangle$, and from the motion invoked by non-hydrodynamical forces $\langle \boldsymbol{S}^{E} \rangle$.

3.12 Concentrated Disperse Media

Up till now we have been discussing infinitely dilute (low-concentrated) suspensions, in other words, the disperse phase (particles) had a very low volume concentration ($\varphi \ll 1$). This approximation made it possible to consider only individual particle motions and pair interactions. Let us go on to consider concentrated disperse systems, where multiparticle (i.e. involving more then two particles) interactions become important because the probability of interactions between the test particle and n neighboring particles is on the order of φ^{n}.

The primary goal of both theoretical and applied research in the mechanics of disperse media is to determine macroscopic properties of the medium: the average sedimentation velocity, the rate of particle coagulation, transport coefficients, and rheology. These parameters depend on the microstructure of the medium, which in its turn is determined by the distribution of particles in space and time and depends on the interaction forces between the particles and the ambient fluid. The forces acting on particles can be Brownian (thermodynamic), hydrodynamic, molecular, electrostatic, and external. If the distribution of particles, the pattern of motion of boundaries (including interfaces), and the physicochemical properties of the continuous and the disperse phase are all given, one can determine the behavior of the disperse phase by solving the corresponding boundary value problem. By averaging the derived solutions over all possible particle configurations or over the volume enclosing a large number of particles, one can find the macroscopic parameters of the medium.

One way of solving such problems is the so-called Stokesian dynamics method, which represents a development of the molecular dynamics method commonly used in the theory of rarefied gases. Before we proceed further, let us recall the history of research in this field.

The first works by Einstein, Brenner, Batchelor, and others were restricted to the case of infinitely dilute suspensions. They assumed a single-particle motion approximation at first, and later on, went to consider particle-pair interactions at low Reynolds numbers. Their research produced transport coefficients accurate to the order of φ and φ^{2} for suspensions containing identical solid and liquid spherical particles. Despite their limited range of application, the obtained results allowed to establish some fundamental properties of disperse systems and to outline the direction of further research.

Great progress was achieved in solving the problem on hydrodynamic interaction of two spherical (solid or liquid) particles at low Reynolds numbers. Exact solutions were obtained, as well as asymptotic solutions for particles separated by large and small distances. Recent enhancements of the method of boundary integral equations have made it possible to obtain solutions (usually numerical ones) for problems involving hydrodynamic interaction of two liquid particles with deformable interfaces. There are also some papers that deal with hydrodynamic interaction of more then two particles [24–29].

Another important problem is to find out how the microstructure of a disperse medium changes in space and in time. Solution of this problem presents great difficulties, because microstructure depends on the relative motion of particles in the suspension, which is affected by interparticle interactions, by interactions of particles with the surrounding fluid (whose bulk flow may be nonuniform), by external forces and by Brownian motion. The problem of determination of microstructure has special importance for the case of concentrated suspensions. One of the main challenges is the necessity to account for multiparticle interactions. Although the exact solution of the multiparticle problem is nowhere in sight, a substantial progress has been achieved in this field as well.

The methods of modeling the dynamics of microscopic structure can be classified into four categories on the basis of the external and internal linear scale of the system in question [30,31].

1. *Modeling on the molecular level* relies on the method of nonequilibrium dynamics, which boils down to examining the dynamics of a finite but sufficiently large number of particles (molecules) based on the equations of motion of particles in a vacuum or in an ambient medium,

$$m_k \frac{d^2 \boldsymbol{r}_k}{dt^2} = -\frac{\partial}{\partial \boldsymbol{r}_k} \sum_{i \neq k} V_{ik}^m (|\boldsymbol{r}_k - \boldsymbol{r}_i|), \tag{3.188}$$

where we take into account only the forces of interaction between particles (particles can be of the same type or different types) characterized by the Lenard–Jones potential

$$V_{ij}^m = 4\varepsilon \left\{ c_{ij} \left(\frac{r}{\sigma}\right)^{-12} - d_{ij} \left(\frac{r}{\sigma}\right)^{-6} \right\}, \tag{3.189}$$

where the parameters ε and σ have the dimensionality of energy and length, respectively, and c_{ij} and d_{ij} are the parameters describing the interaction. The first term on the right-hand side corresponds to the attractive and the second – to the repulsive force.

In order to determine macroscopic parameters, it is necessary to perform averaging over all particles. In particular, the average velocity of a medium containing N particles in the volume V enclosing the point $\boldsymbol{X}$ is

$$\langle \boldsymbol{u}(\boldsymbol{X})\rangle = \frac{1}{N}\sum_{X_k\ v}\frac{d\boldsymbol{X}_k}{dt}. \tag{3.190}$$

The mean value of the stress tensor is derived by averaging the corresponding forces and torques acting on particles:

$$\langle \boldsymbol{T}(\boldsymbol{X})\rangle = \frac{1}{V}\sum_k m_k\left[\frac{d\boldsymbol{X}_k}{dt}-\langle \boldsymbol{u}(\boldsymbol{X})\rangle\right]\left[\frac{d\boldsymbol{X}_k}{dt}-\langle \boldsymbol{u}(\boldsymbol{X})\rangle\right]+\sum_{k<i}\boldsymbol{r}_{ki}\boldsymbol{F}_{ki}, \tag{3.191}$$

where $\boldsymbol{r}_{ki}$ is the vector between the centers of particles k and i, and $\boldsymbol{F}_{ki}$ are the interparticle (intermolecular) forces. The effectiveness of the method is limited by the computational ability of computers employed to solve a large number of ordinary second-order differential equations (3.188).

2. *Modeling of a medium containing big compound molecules (macromolecules)*, for example, proteins or polymer chains. The dynamics of such a system is modeled in the same manner as before. The difference is that we must also take into consideration the presence of chemical forces and the non-spherical geometry of particles, as both factors can affect the formation of macromolecular structure.
3. *Modeling of colloidal systems*, including media containing macromolecules and colloidal particles. The formation of microstructure of these systems is affected by molecular and electrostatic interparticle forces in accordance with the Derjaguin–Landau–Verwey–Overbeek (DLVO) theory; it also depends on the hydrodynamic interactions of particles with other particles and with the surrounding fluid. As a rule, the size of colloidal particles does not exceed several microns; this is why Brownian motion has a strong effect on the particles. In literature, such particles are sometimes referred to as Brownian particles. Since the particle motion occurs at low Reynolds numbers, hydrodynamic equations are taken in the Stokesian approximation, hence the term "Stokesian dynamics method" for the method of modeling such systems. The essence of the method is the same as the two previous cases; the principal difference is the need to take into account hydrodynamic and Brownian (thermodynamic) forces. Since Brownian forces are random, the equations of particle motion are the Langevin stochastic equations rather than Newtonian equations. The first papers on this subject considered Brownian motion, taking into account only the simplest hydrodynamic forces. The applicability of these works was limited to infinitely dilute

suspensions; this is the domain of the so-called Brownian dynamics.

4. *Modeling of granular systems.* Granular systems are defined as coarsely dispersed systems, where the disperse phase may consist of sand particles, grains, coal particles, pebbles, and so forth. Flows of granular media (granular flows) are characterized by relatively large Reynolds numbers, which means that one has to take into account the inertia of particles. The distinguishing feature of such flows is that the structure of the medium is strongly affected by gravity, particle collisions, and the possibility of deposition of particles on obstacles. In order to determine the hydrodynamic forces acting on particles, it is necessary to solve the complete Navier–Stokes equations and then to use the molecular dynamics method to model the dynamics of the disperse phase. In addition to hydrodynamic forces, it is necessary to take into account interparticle forces, which are responsible for the elastic and frictional properties of particles.

Since we are mostly interested in problems belonging to category 3, let us focus on the method of Stokesian dynamics [30,32,33].

Consider a volume V of the disperse medium, containing N particles suspended in the fluid that has viscosity μ_e and density ρ_e. As usual, we assume that at $V \to \infty$ and $N \to \infty$, the number concentration of particles N/V remains constant. Under the action of hydrodynamic forces and torques $\boldsymbol{F}^H$ (these torques arise when the flow velocity of the suspension is not uniform, as is the case for Couette and Poiseuille flows, and (or) particles are not spherical) caused by the particle's motion relative to the fluid, deterministic non-hydrodynamic forces and torques $\boldsymbol{F}^E$ (molecular or external) and stochastic (Brownian) forces and torques $\boldsymbol{F}^B$, particles acquire the velocity $\boldsymbol{U}$. The vectors $\boldsymbol{F}$ and $\boldsymbol{U}$ are generalized 6N-dimensional vectors. Both are written in the column form. The vector $\boldsymbol{F}$ includes forces and torques (in the successive order); $\boldsymbol{U}$ includes (also in the successive order) translational velocities of the particles' centers of mass and the particles' angular velocities. Denote by $\boldsymbol{m}$ a generalized $N \times N$ matrix whose rows are composed of particle masses and moments of inertia. Then the Langevin equations for a system of N particles can be written in a compact form

$$\boldsymbol{m} \cdot \frac{d\boldsymbol{U}}{dt} = \boldsymbol{F}^H + \boldsymbol{F}^E + \boldsymbol{F}^B, \quad \boldsymbol{U} = \frac{d\boldsymbol{X}}{dt}. \tag{3.192}$$

Of primary importance in many applications is the behavior of suspensions in non-uniform velocity fields (examples include flows in pipes and viscometers). On a microscopic scale, any flow of the carrier fluid can be considered as a shear flow. Let $\boldsymbol{u}^\infty$ and $\mathbf{g}^\infty = \nabla \cdot \boldsymbol{u}^\infty = \partial u_i^\infty / \partial \boldsymbol{X}_j$ denote, respectively, the velocity and the tensor of unperturbed velocity gradient in the carrier fluid. Then the symmetric tensor $\boldsymbol{S}^\infty = 0.5(\mathbf{g}^\infty + (\mathbf{g}^\infty)^T)$ and the antisymmetric tensor $\Lambda^\infty = 0.5(\mathbf{g}^\infty - (\mathbf{g}^\infty)^T)$ are connected with the rate-of-strain tensor E^{∞} and components of angular velocity of the fluid Ω_i^{∞} by the relations $E^{\infty} = S^{\infty}$, $\Omega_i^{\infty} = 0.5\varepsilon_{ijk}\Lambda_{kj}^{\infty}$, were ε_{ijk} is the permutation symbol.

We shall characterize the configuration of our N-particle ensemble by the generalized vector X, which includes spatial coordinates X_{α}^{S} and orientation angles X_{α}^{0} (the angles refer to the chosen coordinate system). We also introduce the vector of generalized velocity U^{∞} consisting of translational velocity components $U_{\alpha}^{t\infty} = E^{\infty}\cdot X_{\alpha}$ of particles indexed by a and located at X_{α} and components of the angular velocity $U_{\alpha}^{r\infty} = \Omega_{\alpha}^{\infty}$ of the same particles. Then the hydrodynamic force acting on a particle that moves with the generalized velocity U consists of two terms. The first term is due to the motion of the particle relative to the bulk flow and is proportional to the velocity perturbation of a-th particle at the point X. The second term is the force that would be exerted by the carrier fluid if the particle were moving with the velocity of the bulk flow (i.e., with $U = U^{\infty}$), namely,

$$F^{H} = -R_{FU}\cdot(U - U^{\infty}) + R_{FE} : E^{\infty}. \tag{3.193}$$

Here $R(X)$ is the resistance tensor, whose subscripts indicate the source of resistance: R_{FU} is caused by the action of the hydrodynamic force and torque on the particle as it moves relative to the fluid; R_{FE} is the effect of shear of bulk flow on the particle suspended in the carrier fluid; $R{:}E$ is the double-dot scalar product (i.e., complete scalar product of tensors). R is defined as the generalized resistance that is due to both translational and rotational velocities of the particle. When we consider hydrodynamic forces, it is sometimes convenient to use the mobility tensor instead of the resistance tensor. The two tensors are related by $B = R^{-1}$.

The deterministic force F^{E} can be an external force (gravitational, electric, and so on), a molecular attractive (Van der Waals) force, or an electrostatic repulsive force produced by electric double layers on particle surfaces. The stochastic (Brownian) force F^{B} is produced by thermal fluctuations in the fluid and obeys the following statistical conditions (see Section 1.13):

$$\langle F^{B}\rangle = 0, \quad \langle F^{B}(0)F^{B}(t)\rangle = 2k\vartheta R\delta(t) \tag{3.194}$$

Eqs. (3.192)–(3.194) describe the Brownian motion of an N-particle system when the time scale exceeds the characteristic time of viscous relaxation $t_v = m/6\pi\mu_e a$ (where m is the particle mass), but is much smaller than the characteristic time t_{con} required for the alteration of particle configuration. Since $m \sim a^3$, we conclude that $t_v \sim a^2$ and is negligibly small ($\sim 10^{-13}s$), so the first assumption is always valid.

Integration of the Langevin equation over the time interval $t_v \ll \Delta t_v \ll t_{con}$ gives the particle's displacement (a change of spatial coordinates and orientation angles) with the accuracy up to $O(\mathbf{D}t^2)$:

$$\Delta X = Pe\Big\{U^{\infty} + R_{FU}^{-1}\cdot(R_{FE} : E^{\infty} + (\dot{\gamma}^{*})^{-1}F^{E})\Big\}\Delta t + \nabla\cdot R_{FU}^{-1}\Delta t + X(\Delta t).$$

$$\langle X\rangle = 0, \quad \langle X(\Delta t)X(\Delta t)\rangle = 2R_{FU}^{-1}\Delta t = \frac{2D_{br}\Delta t}{k\vartheta}, \tag{3.195}$$

where $\boldsymbol{X}(\Delta t)$ is a random displacement of the particle in time Δt that is caused by Brownian motion, $\boldsymbol{D}_{br} = \mathrm{k}\vartheta \boldsymbol{R}_{FU}^{-1} = \mathrm{k}\vartheta \boldsymbol{B}$ is the tensor of Brownian diffusion, and $\boldsymbol{B}$ is the mobility tensor. Hence, the average displacement is zero and the root-mean-square displacement is proportional to the mobility $\boldsymbol{B}_{FU} = \boldsymbol{R}_{FU}^{-1}$. We have introduced dimensionless variables in Eq. (3.195). The linear size of a particle was chosen as the characteristic length, and the characteristic time of Brownian diffusion a^2/D_{br}^0 (where $D_{br}^0 = \mathrm{k}\vartheta/6\pi\mu_e a$ is the coefficient of unhindered Brownian diffusion) – as the characteristic time. The quantity $6\pi\mu_e a^2\dot{\gamma}$ was chosen as the characteristic hydrodynamic force, where $\dot{\gamma} = |\boldsymbol{E}^\infty|$ is the rate of shear in the bulk flow. Let $|\boldsymbol{F}^E|$ denote the characteristic value of molecular interaction. Then two dimensionless parameters will appear in the right-hand side of the first equation (3.195): the Peclet number $Pe = \dot{\gamma}a^2/D_{br}^0 = 6\pi\mu_e a^3\dot{\gamma}/\mathrm{k}\vartheta$ and the hydrodynamic interaction parameter $\dot{\gamma}^* = 6\pi\mu_e a^2\dot{\gamma}/|\mathbf{F}^\mathrm{E}|$. The first parameter is ratio between the hydrodynamic and Brownian forces, and the second parameter – the ratio between the hydrodynamic and external (molecular) forces. Yet another dimensionless parameter is the volume concentration φ of the disperse phase.

It follows from the relation (3.195) that particle displacement consists of three terms, each term characterizing one factor that causes particle motion. The first component, $(\boldsymbol{U}^\infty + \boldsymbol{R}_{FU}^{-1}\cdot\boldsymbol{R}_{FE} : \boldsymbol{E}^\infty)\Delta t$, is the contribution from the hydrodynamic force induced by the shear flow. The second contribution, $(\boldsymbol{R}_{FU}^{-1}\cdot\boldsymbol{F}^E)\Delta t$, comes from the external (molecular) force. The third term accounts for Brownian motion and includes $\boldsymbol{X}(\Delta t)$ – the random displacement in time Δt and $(\nabla\cdot\boldsymbol{R}_{FU}^{-1})\Delta t = (\nabla\cdot\boldsymbol{D}_{br})\Delta t/\mathrm{k}\vartheta$ – displacement caused by the variation of the diffusion tensor $\boldsymbol{D}_{br}$ that changes together with spatial configuration of particles.

Hence, the behavior of suspension containing identical spherical particles depends on three dimensionless parameters: $Pe, \dot{\gamma}^*$, and φ. The presence of non-spherical particles of different sizes leads to the appearance of additional dimensionless parameters that characterize particle shapes and volume fractions φ_i of particles of i-th type. If there are several external forces of different nature (molecular, electrostatic, gravitational, etc.), we will need additional parameters to characterize the relative contribution of each force.

For all problems that belong to the domain of Stokesian dynamics, we use the low Reynolds numbers approximation, in other words, we assume that $Re = \rho_e a^2\dot{\gamma}/\mu_e \ll 1$.

The Peclet number expresses the relative contribution of Brownian diffusion to particle displacement. For $Pe \gg 1$, Brownian diffusion has no significant effect on particle motion. The last two terms in Eq. (3.195) can be neglected, and instead of a^2/D_{br}^0, we should take $\dot{\gamma}^{-1}$ (for a shear flow) or a/U_0 (for sedimentation in a quiescent flow) as the characteristic time. In the other limiting case, $Pe \ll 1$, Brownian motion dominates, and the last two terms play the primary role in particle dynamics. Problems of this type can be studied within the framework of Brownian dynamics and do not need a detailed analysis of hydrodynamic forces.

Another method involves the use of a continuous multiparticle PDF $p(\boldsymbol{X})$, where $\boldsymbol{X} = (X_1, X_2, \ldots, X_n)$. Due to the influence of the bulk shear flow a selected particle from the N-particle ensemble will acquire the velocity

$$V = U^{\infty} + R_{FU}^{-1} \cdot (R_{FE} : E^{\infty} + F^{E} - \mathrm{k}\vartheta \nabla \ln p). \tag{3.196}$$

The terms in parenthesis give the contributions from the external flow, the external force, and the Brownian motion, respectively. PDF p obeys the Liouville equation:

$$\frac{\partial p}{\partial t} + \nabla \cdot (V p) = 0. \tag{3.197}$$

Note that if we integrate this Eq. (3.197) over a short time interval Δt under the condition that $t_v \ll \Delta t \ll t_c$, this equation will reduces to the Langevin equation (3.195), which proves the equivalence of both methods.

We see that Eq. (3.195) describes the evolution of an ensemble of N particles suspended in a volume V of the fluid and subject to hydrodynamic, external, and Brownian forces for a given initial that particle configuration $\boldsymbol{X}(0)$.

In order to determine macroscopic parameters of the suspension, one must average the corresponding quantities over the particle ensemble. The problem of particle sedimentation is mostly concerned with obtaining the sedimentation velocity averaged over all particles and all particle configurations. Such an averaging is equivalent to averaging the velocity over the time t during which the dynamic simulation is performed:

$$\langle U \rangle = \left\langle \frac{1}{N} \sum_{k=1}^{N} U_k \right\rangle. \tag{3.198}$$

Assuming $U^{\infty} = 0$ and taking the force of gravity F^{g} as our F^{E}, we get for a suspension of identical spherical particles:

$$\langle U \rangle = \langle R_{FU}^{-1} \cdot F^{g} \rangle = \langle B \rangle \cdot F^{g}, \tag{3.199}$$

and the problem reduces to that of averaging the mobility tensor.

Another problem that can be solved by the same method is the one of fluid filtering at a constant average velocity $U^{\infty} = \langle U^{\infty} \rangle$ through a porous medium that has a given particle configuration. Solving this problem, we obtain the forces $\boldsymbol{F}$ that must be applied in order to preserve the current particle configuration, and the resistance matrix of the porous medium, equal to the inverse permeability matrix K_{pr}^{-1},

$$K_{pr}^{-1} = \frac{N}{V} \langle R_{FU} \rangle. \tag{3.200}$$

The problem of determining rheological properties of the suspension presents the greatest practical interest. As we mentioned in the previous section, the presence of particles in the fluid causes additional stresses, so the stress tensor assumes the form

$$\langle T_{ij}\rangle = \delta_{ij}\langle T_{kk}\rangle + 2\mu_e\langle E_{ij}\rangle + \langle T_{ij}^p\rangle. \tag{3.201}$$

where the last term is the additional contribution to the stress tensor due to the presence of particles. In its turn, it consists of three components, which characterize, respectively, the mechanical (contact) stress arising from the interaction of particles with the shear flow; the "elastic" stress induced by the action of molecular, electrostatic, and external forces on the particles; and the Brownian stress arising, as its name suggests, from the Brownian motion of particles:

$$\boldsymbol{T}^p = \frac{N}{V}(\langle \boldsymbol{S}^H\rangle + \langle \boldsymbol{S}^E\rangle + \langle \boldsymbol{S}^B\rangle), \tag{3.202}$$

where

$$\langle \boldsymbol{S}^H\rangle = -\langle \boldsymbol{R}_{SU}\cdot(\boldsymbol{U}-\boldsymbol{U}^\infty) - \boldsymbol{R}_{SE} : \boldsymbol{E}^\infty\rangle; \quad \langle \boldsymbol{S}^E\rangle = \langle \boldsymbol{X}\boldsymbol{F}^E\rangle;$$
$$\langle \boldsymbol{S}^B\rangle = -k\vartheta\nabla\cdot(\boldsymbol{R}_{SU}\cdot\boldsymbol{R}_{FU}^{-1}).$$

The resistance tensors $\boldsymbol{R}_{SU}$ and $\boldsymbol{R}_{SE}$ have the same meaning as the above-introduced tensors $\boldsymbol{R}_{FU}$ and $\boldsymbol{R}_{FE}$. The difference is that $\boldsymbol{R}_{FU}$ and $\boldsymbol{R}_{FE}$ are proportionality factors between the generalized force $\boldsymbol{F}$ and generalized relative velocity $\boldsymbol{U} - \boldsymbol{U}^\infty$ and the rate-of-strain tensor of the unperturbed flow $\boldsymbol{E}^\infty$:

$$\boldsymbol{F} = \boldsymbol{R}_{FU}\cdot(\boldsymbol{U}-\boldsymbol{U}^\infty) + \boldsymbol{R}_{FE}\cdot(-\boldsymbol{E}^\infty), \tag{3.203}$$

In contrast, $\boldsymbol{R}_{SU}$ and $\boldsymbol{R}_{SE}$ are proportionality factors between the force dipole density tensor $\boldsymbol{S}$ at the particle surface and the quantities $\boldsymbol{U} - \boldsymbol{U}^\infty$ and $\boldsymbol{E}^\infty$:

$$\boldsymbol{S} = \boldsymbol{R}_{SU}\cdot(\boldsymbol{U}-\boldsymbol{U}^\infty) + \boldsymbol{R}_{SE}\cdot(-\boldsymbol{E}^\infty). \tag{3.204}$$

The relations (3.203) and (3.204) can be written in the matrix form:

$$\begin{pmatrix} \boldsymbol{F} \\ \boldsymbol{S} \end{pmatrix} = -\mathbf{R}\cdot\begin{pmatrix} \boldsymbol{U}-\boldsymbol{U}^\infty \\ -\boldsymbol{E}^\infty \end{pmatrix} \tag{3.205}$$

where $\mathbf{R}$ is the grand resistance matrix

$$\mathbf{R} = \begin{pmatrix} \boldsymbol{R}_{FU} & \boldsymbol{R}_{FE} \\ \boldsymbol{R}_{SU} & \boldsymbol{R}_{SE} \end{pmatrix}.$$

Eq. (3.205) can be resolved with respect to $\boldsymbol{U} - \boldsymbol{U}^\infty$ and $\boldsymbol{E}^\infty$:

$$\begin{pmatrix} \boldsymbol{U}-\boldsymbol{U}^\infty \\ -\boldsymbol{E}^\infty \end{pmatrix} = \boldsymbol{B}\begin{pmatrix} \boldsymbol{F} \\ \boldsymbol{S} \end{pmatrix}, \tag{3.206}$$

where the matrix $\boldsymbol{B} = \boldsymbol{R}^{-1}$ is called the grand mobility matrix and is equal to

$$\boldsymbol{B} = \begin{pmatrix} \boldsymbol{B}_{FU} & \boldsymbol{B}_{SU} \\ \boldsymbol{B}_{FE} & \boldsymbol{B}_{SE} \end{pmatrix}.$$

with $\boldsymbol{B}$'s in parenthesis denoting the corresponding mobilities.

Finally, we must broach the subject of finding the elements of the grand resistance and mobility matrices. With this goal in mind, consider the general problem of motion of an N-particle ensemble suspended in the fluid.

Consider one solid particle of an arbitrary shape involved in translational and rotational motion in an incompressible fluid of viscosity μ_e whose shear velocity $\boldsymbol{V}^{\infty}$ at the infinity is given. Let us choose some point 0 inside the particle as the origin of a local coordinate system. Let $\boldsymbol{U}$ and Ω denote the translational velocity of the point 0 and the angular velocity of the particle, and let $\boldsymbol{E}^{\infty}$ and Λ^{∞} be the symmetric and anti-symmetric parts of the undisturbed velocity gradient tensor $\mathbf{g}^{\infty} = \nabla \cdot \boldsymbol{V}^{\infty} = \nabla_j V_i^{\infty}$:

$$\boldsymbol{E}^{\infty} = 0.5(\nabla \cdot \boldsymbol{V}^{\infty} + (\nabla \cdot \boldsymbol{V}^{\infty})^T), \quad \Lambda^{\infty} = 0.5(\nabla \cdot \boldsymbol{V}^{\infty} - (\nabla \cdot \boldsymbol{V}^{\infty})^T),$$

where $(\nabla \cdot \boldsymbol{V}^{\infty})^T = \nabla_i V_j^{\infty}$; $\boldsymbol{E}^{\infty}$ is the rate-of-strain tensor; and Λ^{∞} determines the angular velocity of an unperturbed flow: $\Omega_i^{\infty} = 0.5\varepsilon_{ijk}\Lambda_{kj}^{\infty}$. Then the fluid's velocity far away from the particle is

$$V_i^{\infty} = V_i^0 + \varepsilon_{ijk}\Omega_j^{\infty} X_k + E_{ij}^{\infty} X_j, \tag{3.207}$$

where $\boldsymbol{V}^0$ is the unperturbed velocity at the point 0 as though the particle were absent, and $\boldsymbol{X}_j$ are coordinates of the point 0. If the particle's motion is characterized by low Reynolds numbers (as determined from the particle's parameters), the hydrodynamic problem is formulated by writing the Stokes equations

$$\nabla \cdot \boldsymbol{u} = 0, \quad \nabla p = \mu_e \nabla^2 \boldsymbol{u}. \tag{3.208}$$

with the boundary conditions at the particle surface,

$$\boldsymbol{u} = \boldsymbol{U} + \Omega \times \boldsymbol{X} \tag{3.209}$$

and far away from the particle,

$$\boldsymbol{u} \to \boldsymbol{V}^{\infty} \text{ at } |\boldsymbol{X}| \to \infty. \tag{3.210}$$

Since the equations (3.208)–(3.210) are linear, the general solution is a superposition of solutions of simpler problems – the problem of translational/rotational motion of a particle in a quiescent fluid; and the problem of a motionless particle in a shear flow. As result, we obtain the force $\boldsymbol{F}$ and torque $\boldsymbol{L}$ acting on the particle and the distribution of stresses $\boldsymbol{S}$ on the particle surface. Linearity of the problems implies that these quantities should be linear functions of relative translational and angular velocities and the rate-of-strain tensor of the unperturbed flow (see Section 2.2):

$$\begin{aligned}
\boldsymbol{F} &= \boldsymbol{R}_{FU}\cdot(\boldsymbol{U}-\boldsymbol{V}^{\infty}) + \boldsymbol{R}_{F\Omega}\cdot(\Omega-\Omega^{\infty}) + \boldsymbol{R}_{FE}\cdot(-\boldsymbol{E}^{\infty}),\\
\boldsymbol{L} &= \boldsymbol{R}_{LU}\cdot(\boldsymbol{U}-\boldsymbol{V}^{\infty}) + \boldsymbol{R}_{L\Omega}\cdot(\Omega-\Omega^{\infty}) + \boldsymbol{R}_{LE}\cdot(-\boldsymbol{E}^{\infty}),\\
\boldsymbol{S} &= \boldsymbol{R}_{SU}\cdot(\boldsymbol{U}-\boldsymbol{V}^{\infty}) + \boldsymbol{R}_{S\Omega}\cdot(\Omega-\Omega^{\infty}) + \boldsymbol{R}_{SE}\cdot(-\boldsymbol{E}^{\infty}).
\end{aligned} \tag{3.211}$$

Resistance tensors $\boldsymbol{R}$ can be regarded as intrinsic geometrical properties of the particle. In particular, for particles possessing an axial symmetry, all components of the resistance tensor are expressed through five independent constants.

If we combine the vectors $\boldsymbol{F}$ and $\boldsymbol{L}$ into one generalized vector $\boldsymbol{F}$ and the vectors $\boldsymbol{U}-\boldsymbol{V}^{\infty}$ and $\Omega-\Omega^{\infty}$ – into one generalized vector $\boldsymbol{U}$, the relation (3.211) will take the form (3.205). The particle's motion perturbs the velocity field of the ambient fluid, which in its turn affects the particle's motion. In order to properly account for the mutual influence of the particle and the surrounding fluid and for the presence of other particles, we should write Stokes equations in the integral representation:

$$u_i(\boldsymbol{X}) = u_i^{\infty}(\boldsymbol{X}) - \frac{1}{8\pi\mu_e}\sum_{\alpha=1}^{N}\int_{S_\alpha} J_{ij}(\boldsymbol{X}-\boldsymbol{Y})\,f_j(\boldsymbol{Y})\,dS_Y, \tag{3.212}$$

where $u_i^{\infty}(\boldsymbol{X})$ is the velocity of the fluid at $\boldsymbol{X}$ in the absence of the particle; $\boldsymbol{Y}$ is a point on the particle surface; $J_{ij}(\boldsymbol{r}) = \delta_{ij}/r + \boldsymbol{rr}/r^3$ is the Oseen tensor also known as the stokeslet; $f_j(y) = \sigma_{jk}(\boldsymbol{Y})n_k(\boldsymbol{Y})$ is j-th component of the force acting on a unit surface area at the point $\boldsymbol{Y}$; n_k is the component of outer normal to the particle surface. Summation in Eq. (3.212) is performed over all particles a. The force and torque exerted by the fluid on the particle a are given by

$$F_i^{\alpha} = -\int_{S_\alpha} f_i(\boldsymbol{Y})\,dS_y, \quad L_i^{\alpha} = -\int_{S_\alpha} \varepsilon_{ijk}(Y_j - X_j^{\alpha})\,f_k(\boldsymbol{Y})\,dS_y, \tag{3.213}$$

where X_j^{α} are coordinates of the particle center, relative to which the torques have been defined.

Integral equations (3.212) form the basis of the boundary integral method, which is a standard way to solve problems on motion and interaction of nondeformable and deformable particles. For deformable particles, this may be the only method available, while for nondeformable (solid) particles, the approximate method of moments if the more preferable technique. Expanding the stokeslet in the integrand of Eq. (3.212) in a Taylor series in the vicinity of the point $\boldsymbol{X}^{\alpha}$, we get:

$$\begin{aligned}
u_i(\boldsymbol{X}) - u_i^{\infty}(\boldsymbol{X}) &= -\frac{1}{8\pi\mu_e}\sum_{\alpha=1}^{N}\Bigg\{\int_{S_\alpha} J_{ij}^{\alpha}(\boldsymbol{X}-\boldsymbol{X}^{\alpha})\,f_j(\boldsymbol{Y})\,dS_Y\\
&+ \int_{S_\alpha} \frac{\partial}{\partial Y_k} J_{ij}^{\alpha}(\boldsymbol{X}-\boldsymbol{X}^{\alpha})\Big|_{\boldsymbol{Y}=\boldsymbol{X}^{\alpha}} (Y_k - X_k^{\alpha})\,f_j(\boldsymbol{Y})\,dS_Y\\
&+ \int_{S_\alpha} \frac{\partial^2}{\partial Y_k \partial Y_l} J_{ij}^{\alpha}(\boldsymbol{X}-\boldsymbol{X}^{\alpha})\Big|_{\boldsymbol{Y}=\boldsymbol{X}^{\alpha}} (Y_k - X_k^{\alpha})(Y_l - X_l^{\alpha})\,f_j(\boldsymbol{Y})\,dS_Y + \ldots\Bigg\}.
\end{aligned} \tag{3.214}$$

Designate the n-th order moment (multipole) of surface force density as

$$Q^{\alpha}_{ij,n} = -\int_{S_\alpha} (Y_i - X^{\alpha}_i)\, f_j(\mathbf{Y})\, dS_y, \quad n = 0, 1, 2, \ldots$$

Because of Eq. (3.213), the zero order moment (monopole) is the force acting on the particle. The first order moment (dipole) has a symmetric and an antisymmetric part. The antisymmetric part is the torque acting on the particle, and the symmetric part is equal to the surface force dipole S^{α}_{ij} (see Eq. (3.171)), also known as the stresslet:

$$S^{\alpha}_{ij} = -\frac{1}{2}\int_{S_\alpha} \left\{ (Y_i - X^{\alpha}_i)\, f_j(\mathbf{Y}) + (Y_j - X^{\alpha}_j)\, f_i(\mathbf{Y}) - \frac{2}{3}\delta_{ij}(Y_k - X^{\alpha}_k)\, f_k(\mathbf{Y}) \right\} dS_y. \tag{3.215}$$

The expression (3.214) represents a multipole expansion of the fluid velocity perturbation $\boldsymbol{u}'(\boldsymbol{X})$ at the point $\boldsymbol{X}$ caused by the presence of particles in the fluid. The number of terms in the series (3.215) is limited by the required accuracy. The dipole approximation is adequate enough for all practical purposes. In the case of translational motion of an individual spherical particle of radius a in an infinite fluid, velocity perturbation can be represented as the perturbation created by the point force $J^{\alpha}_{ij}F^{\alpha}_j$ applied at the particle center, and by the quadrupole of intensity $a^2\nabla^2 J^{\alpha}_{ij}/6$ characterizing the finiteness of the particle's size. The velocity perturbation for a spherical particle is

$$u'_i(\boldsymbol{X}) = u_i(\boldsymbol{X}) - u^{\infty}_i(\boldsymbol{X}) = \frac{1}{8\pi\mu_e}\left(1 + \frac{1}{6}a^2\nabla^2\right) J^{\alpha}_{ij}(\boldsymbol{X} - \boldsymbol{X}^{\alpha})\, F^{\alpha}_j. \tag{3.216}$$

In the general case of translational and rotational motion of an N-particle system, the expression for fluid velocity at the point $\boldsymbol{X}$ can be written as

$$u_i(\boldsymbol{X}) = u^{\infty}_i(\boldsymbol{X}) + \frac{1}{8\pi\mu_e}\sum_{\alpha=1}^{N}\left\{ \left(1 + \frac{1}{6}a^2\nabla^2\right) J_{ij}F^{\alpha}_j \right.$$

$$\left. + R_{ij}L^{\alpha}_j + \left(1 + \frac{1}{10}a^2\nabla^2\right) K_{ijk}S_{jk} + \ldots \right\}, \tag{3.217}$$

where $R_{ij} = 0.25\varepsilon_{lkj}(\nabla_k J_{il} - \nabla_l J_{ik})$ is the couplet (rotlet), and $K_{ijk} = (\nabla_k J_{il} - \nabla_l J_{ik})/2$.

Hence, the formula (3.217) represents the fluid's velocity at $\boldsymbol{X}$ as a sum of the unperturbed velocity and the velocity increment induced by the multipoles that model the influence of particles on the fluid. On the other hand, perturbations of the fluid velocity field give rise to perturbations of the motion of particles themselves. In order to determine the velocities of particles, one should use the relations that follow from the Faxen law:

$$\begin{aligned} U_i^\alpha - u_i^\infty(\boldsymbol{X}^\alpha) &= \frac{F_i^\alpha}{6\pi\mu_e a} + \left(1 + \frac{1}{6}a^2\nabla^2\right) u_i'(\boldsymbol{X}^\alpha), \\ \Omega_i^\alpha - \Omega_i^\infty &= \frac{L_i^\alpha}{8\pi\mu_e a^3} + \frac{1}{2}\varepsilon_{ijk}\nabla_j u_k'(\boldsymbol{X}^\alpha), \\ -E_{ij}^\infty &= \frac{S_{ij}^\alpha}{(20/3)\pi\mu_e a^3} + \left(1 + \frac{1}{10}a^2\nabla^2\right) E_{ij}'(\boldsymbol{X}^\alpha). \end{aligned} \tag{3.218}$$

Here u_i' is the perturbation of the fluid's velocity field exerted by i-th particle immersed in the fluid; $E_{ij}' = 0.5(\nabla_j u_i' + \nabla_i u_j')$ are components of the rate-of-strain tensor of the perturbed flow; E_{ij}^∞ are components of the rate-of-strain tensor of the unperturbed flow.

The relations (3.218) can be written in a matrix form similar to Eq. (3.206):

$$\begin{pmatrix} \boldsymbol{U}_\alpha - \boldsymbol{u}^\infty(\boldsymbol{x}) \\ \boldsymbol{U}_b - \boldsymbol{u}^\infty(\boldsymbol{x}) \\ \vdots \\ \Omega_\alpha - \Omega^\infty \\ \Omega_b - \Omega^\infty \\ \vdots \\ -\boldsymbol{E}^\infty \\ -\boldsymbol{E}^\infty \\ \vdots \end{pmatrix} = \begin{pmatrix} \boldsymbol{a}_{aa} & \boldsymbol{a}_{ab} & \cdots & \boldsymbol{b}_{aa}^T & \boldsymbol{b}_{ab}^T & \cdots & \boldsymbol{g}_{aa}^T & \boldsymbol{g}_{ab}^T & \cdots \\ \boldsymbol{a}_{ba} & \boldsymbol{a}_{bb} & \cdots & \boldsymbol{b}_{ba}^T & \boldsymbol{b}_{bb}^T & \cdots & \boldsymbol{g}_{ba} & \boldsymbol{g}_{bb} & \cdots \\ \vdots & \vdots & & \vdots & \vdots & & \vdots & \vdots & \\ \boldsymbol{b}_{aa} & \boldsymbol{b}_{ab} & \cdots & \boldsymbol{c}_{aa} & \boldsymbol{c}_{ab} & \cdots & \boldsymbol{h}_{aa}^T & \boldsymbol{h}_{ab}^T & \cdots \\ \boldsymbol{b}_{ba} & \boldsymbol{b}_{bb} & \cdots & \boldsymbol{c}_{ba} & \boldsymbol{c}_{bb} & \cdots & \boldsymbol{h}_{ba}^T & \boldsymbol{h}_{bb}^T & \cdots \\ \vdots & \vdots & & \vdots & \vdots & & \vdots & \vdots & \\ \boldsymbol{g}_{aa} & \boldsymbol{g}_{ab} & \cdots & \boldsymbol{h}_{aa} & \boldsymbol{h}_{ab} & \cdots & \boldsymbol{m}_{aa} & \boldsymbol{m}_{ab} & \cdots \\ \boldsymbol{g}_{ba} & \boldsymbol{g}_{bb} & \cdots & \boldsymbol{h}_{ba} & \boldsymbol{h}_{bb} & \cdots & \boldsymbol{m}_{ba} & \boldsymbol{m}_{bb} & \cdots \\ \vdots & \vdots & & \vdots & \vdots & & \vdots & \vdots & \end{pmatrix} \cdot \begin{pmatrix} \boldsymbol{F}_a \\ \boldsymbol{F}_b \\ \vdots \\ \boldsymbol{L}_a \\ \boldsymbol{L}_b \\ \vdots \\ \boldsymbol{S}_a \\ \boldsymbol{S}_b \\ \vdots \end{pmatrix}. \tag{3.219}$$

Let us confine ourselves to dipoles in the multipole expansion (3.217). Then we have 11 unknowns (degrees of freedom) for each particle: three components of both translational and rotational velocities, and five independent parameters of the symmetric stresslet with a zero trace. The corresponding mobility matrix shall be denoted by $\boldsymbol{B}^\infty$. Matrix elements of $\boldsymbol{B}^\infty$ are determined from the relations (3.217)–(3.218) and are written in a dimensionless form: all linear quantities are divided by the particle radius a, and elements of the matrices $\boldsymbol{a}$, $\boldsymbol{b}$, $\boldsymbol{c}$, $\boldsymbol{g}$, $\boldsymbol{h}$ and $\boldsymbol{m}$ are divided by $6\pi\mu_e a^n$, where we take $n = 1$ for the matrix $\boldsymbol{a}$, $n = 2$ for the matrix $\boldsymbol{b}$, and $n = 3$ for the other matrices. Also, r is the distance between the centers of particles a and b, and $e_i = r_i/r$ are components of the unit vector along the line of centers of these two particles. Mobility matrix elements are equal to (see Eq. 2.69)

$$\begin{aligned} a_{ij}^{ab} &= x_{ab}^a e_i e_j + y_{ab}^a(\delta_{ij} - e_i e_j), \\ b_{ij}^{ab} &= y_{ab}^b \varepsilon_{ijk} e_k, \\ c_{ij}^{ab} &= x_{ab}^c e_i e_j + y_{ab}^c(\delta_{ij} - e_i e_j), \\ g_{ijk}^{ab} &= x_{ab}^g\left(e_i e_j - \frac{1}{3}\delta_{ij}\right) e_k + y_{ab}^g(e_i\delta_{jk} + e_j\delta_{ik} - 2e_i e_j e_k), \end{aligned}$$

$$h^{ab}_{ijk} = y^h_{ab}(e_i\varepsilon_{jkl}e_l + e_j\varepsilon_{ikl}e_l),$$

$$m^{ab}_{ijkl} = \frac{2}{3}x^m_{ab}\left(e_ie_j - \frac{1}{3}\delta_{ij}\right)\left(e_ke_l - \frac{1}{3}\delta_{kl}\right) + \frac{1}{2}y^m_{ab}(e_i\delta_{jl}e_k + e_j\delta_{il}e_k + e_i\delta_{jk}e_l$$
$$+ e_j\delta_{ik}e_l - 4e_ie_je_ke_l) + \frac{1}{2}z^m_{ab}(\delta_{ik}\delta_{jl} + \delta_{jk}\delta_{il} - \delta_{ij}\delta_{kl} + e_ie_j\delta_{kl} + \delta_{ij}e_ke_l$$
$$+ e_ie_je_ke_l - e_i\delta_{jl}e_k - e_j\delta_{il}e_k + e_i\delta_{jk}e_l + e_j\delta_{ik}e_l).$$

In the approximation defined by (3.218) and (3.219), the scalar mobility functions x_{ab}, y_{ab}, and z_{ab} are equal to

$$\begin{aligned}
x^a_{11} &= x^a_{22} = 1; \quad x^a_{12} = x^a_{21} = \frac{3}{2}r^{-1} - r^{-3},\\
y^a_{11} &= y^a_{22} = 1; \quad y^a_{12} = y^a_{21} = \frac{3}{4}r^{-1} + \frac{1}{2}r^{-3},\\
y^b_{11} &= -y^b_{22} = 0; \quad y^b_{12} = -y^b_{21} = -\frac{3}{4}r^{-2},\\
x^c_{11} &= x^c_{22} = \frac{3}{4}; \quad x^c_{12} = x^c_{21} = \frac{3}{4}r^{-3},\\
y^c_{11} &= y^c_{22} = \frac{3}{4}; \quad y^c_{12} = y^c_{21} = -\frac{3}{8}r^{-3},\\
x^g_{11} &= -x^g_{22} = 0; \quad x^g_{12} = -x^g_{21} = -\frac{9}{4}r^{-2} - \frac{18}{5}r^{-4},\\
y^g_{11} &= -y^g_{22} = 0; \quad y^g_{12} = -y^g_{21} = \frac{6}{5}r^{-4},\\
y^h_{11} &= y^h_{22} = 0; \quad y^h_{12} = y^h_{21} = -\frac{9}{8}r^{-3},\\
x^m_{11} &= x^m_{22} = \frac{9}{10}; \quad x^m_{12} = x^m_{21} = -\frac{9}{2}r^{-3} + \frac{54}{5}r^{-5},\\
y^m_{11} &= y^m_{22} = \frac{9}{10}; \quad y^m_{12} = y^m_{21} = \frac{9}{4}r^{-3} - \frac{36}{5}r^{-5},\\
z^m_{11} &= z^m_{22} = \frac{9}{10}; \quad z^m_{12} = z^m_{21} = -\frac{9}{5}r^{-5},
\end{aligned} \tag{3.220}$$

Elements of the matrix $\boldsymbol{B}^{\infty}$ have the form of a power series expansion of the relative distance r and are determined up to the order of r^{-5}. The Faxen law used in the derivation is valid for low-concentrated suspensions, that is, for $r \gg 1$.

When constructing elements of the resistance matrix $\boldsymbol{R}$, one should bear in mind that the case of two particles spaced far apart should be distinguished from that of particles placed close together, because some elements of the matrix $\boldsymbol{R}_{UF}$ (and of the corresponding mobility matrix $\boldsymbol{B}_{UF}$) show different behaviors for these two cases. For example, the resistance matrix element responsible for the relative motion of two particles along their line of centers is finite at $r \to \infty$ and singular (i.e., goes to infinity) when the gap between particles disappears ($\Delta \to 0$). This circumstance should be taken into account when constructing elements of the matrix $\boldsymbol{R}$.

References

1 Chandrasekhar, S. (1943) *Stochastic Problems in Physics and Astronomy. Rev. Mod. Phys.*, **15**, 1–89.

2 Einstein, A. and Smoluchowski, M. (1936) *Brownian Motion*, ONTI, Moscow (in Russian).

3 Landau, L.D. and Lifshiz, E.M. (1964) *Statistical Physics*, Nauka, Moscow (in Russian).

4 Tunizki, N.N., Kaminski, V.A. and Timashov, S.F. (1972) *Methods of Physicochemical Kinetics*, Chimia, Moscow (in Russian).

5 Klyatskin, V.I. (1975) *Statistical Description of Dynamic Systems with Fluctuating Parameters*, Nauka, Moscow (in Russian).

6 Happel, J. and Brenner, H. (1983) *Low Reynolds Number Hydrodynamics*, Martinus Nijhof, The Hague.

7 Klyatskin, V.I. (1980) *Stochastic Equations and Waves in Randomly Inhomogeneous Media*, Nauka, Moscow (in Russian).

8 Leontovitch, M.A. (1983) *Introduction to Thermodynamics. Statistical Physics*, Nauka, Moscow (in Russian).

9 Gardiner, C.W. (1985) *Handbook of Stochastic Methods*, 2nd ed., Spriner–Verlag.

10 Landau, L.D. and Lifshitz, E.M. (1999) *Hydrodynamics*, Nauka, Moscow (in Russian).

11 Van Kampen, N.G. (1992) *Stochastic Processes in Physics and Chemistry*, North Holland, Amsterdam.

12 Russel, W.B. (1981) *Brownian Motion of Small Particles Suspended in Liquids. Ann. Rev. Fluid Mech.*, **13**, 425–455.

13 Katayama, Y. and Terauti, R. (1966) *Brownian Motion of Single Particle under Shear Flow. Eur. J. Phys.*, **17**, 136–140.

14 Batchelor, G.K. (1976) *Brownian Diffusion of Particles with Hydrodynamic Interaction. J. Fluid Mech.*, **74**, 1–29.

15 Batchelor, G.K. (1977) *The Effect of Brownian Motion on the Bulk Stress in a Suspension of Spherical Particles. J. Fluid Mech.*, **83**, 97–117.

16 Batchelor, G.K. (1983) *Diffusion in a Dilute Polydisperse System of Interacting Spheres. J. Fluid Mech.*, **131**, 155–175.

17 Batchelor, G.K. (1972) *Sedimentation in a Dilute Dispersion of Spheres. J. Fluid Mech.*, **52**, 245–268.

18 Batchelor, G.K. (1982) *Sedimentation in a Dilute polydisperse Dispersion of Interacting Spheres. Part 1. General theory. J. Fluid Mech.*, **119**, 379–408.

19 Batchelor, G.K. and Wen, C.-S. (1982) *Sedimentation in a Dilute Polydisperse Dispersion of Interacting Spheres. Part 2. Numerical results. J. Fluid Mech.*, **124**, 495–528.

20 Batchelor, G.K. (1970) *The Stress System in a Suspension of Force-free Particles. J. Fluid Mech.*, **41**, 545–570.

21 Jeffrey, D.J. (1973) *Conduction through a Random Suspension of Spheres. Proc. Roy. Soc. A.*, **335**, 355.

22 Batchelor, G.K. (1974) *Transport Properties of Two-phase Materials with Random Structure. Ann. Rev. Fluid Mech.*, **6**, 227–255.

23 O'Brien, R.W. (1979) *A Method for the Calculation of the Effective Transport Properties of Suspensions of Interacting Particles. J. Fluid Mech.*, **91**, 17–39.

24 Kynch, G.L. (1959) *The Slow Motion of two or more Spheres through a Viscous Fluid. J. Fluid Mech.*, **5**, 193–208.

25 Mazur, P. and van Saarloos, W. (1982) *Many-Sphere Hydrodynamic Interaction and Mobilities in a Suspension. Physica A.*, **115**, 21–57.

26 Martynov, S.I. (1998) *Hydrodynamic Interactions of Particles. Fluid Dyn.* **33** (2), 245–251.

27 Martynov, S.I. (2000) *Particle Interactions in a Flow with a Parabolic Velocity Profile. Fluid Dynamics* **35** (1), 68–74.

28 Martynov, S.I. (2002) *Viscous Flow Past a Periodic Array of Spheres. Fluid Dyn.* **37** (6), 889–895.

29 Baranov, V.E. and Martynov, S.I. (2004) *Effect of Hydrodynamic Interaction of a Large Number of Particles on their Sedimentation Rate in a Viscous Fluid. Fluid Dyn.* **39** (1), 136–147.

30 Brady, J.F. and Bossis, G. (1988) *Stokesian Dynamics. Ann. Rev. Fluid Mech.*, **20**, 111–157.

31 Koplik, J. (1995) *Continuum Deductions from Molecular Hydrodynamics. Ann. Rev. Fluid Mech.*, **27**, 257–292.

32 Durlofsky, L. and Brady, J.F. (1987) *Dynamic Simulation of Hydrodynamically Interacting Particles. J. Fluid Mech.*, **180**, 21–49.

33 Phung, T.H., Brady, J.F. and Bossis, G. (1996) *Stokesian Dynamics Simulation of Brownian Suspensions. J. Fluid Mech.*, **313**, 181–207.

4
Turbulent Flow of Fluids

4.1
General Information on Laminar and Turbulent Flows

All fluid flows are divided into two distinct categories: smooth layered flows known as laminar flows, and chaotic flows, also called turbulent. The latter term means that the flow in any chosen direction can be described by average values only. Fluid velocity and other parameters – pressure, temperature, concentrations of dissolved substances (if the fluid is a multicomponent mixture) – fluctuate randomly about their average values, displaying a very irregular spatial and temporal behavior. The fluctuations are characterized by different periods and different amplitudes of both hydrodynamic and non-hydrodynamic parameters. This indicates a complex internal structure of the turbulent flow, which distinguishes it from the laminar flow and explains the difference in properties of these two types of flows. As compared to the laminar flow, the turbulent flow has better ability to transport momentum and heat, to spread impurities and products of chemical reactions throughout the volume, to transport particles, to promote interparticle interactions (collisions, coagulation), and so on.

The transport of each substance (momentum, heat, matter, etc.) is characterized by its own coefficient. Transport coefficients for a turbulent flow are called effective (or turbulent) transport coefficients to distinguish them from their ordinary counterparts that describe a laminar flow. Thus, momentum transport is characterized by the coefficient of turbulent kinematic viscosity ν_t, which is defined by a constitutive (phenomenological) proportionality relation between the stress tensor and the rate-of-strain tensor. Similarly, we can define the effective coefficients of heat conductivity, diffusion, and other coefficients.

The kinetic molecular theory, which focuses on the internal molecular structure of substances, is the proper tool to derive ordinary transport coefficients of liquids and gases. In contrast, turbulent transport coefficients follow from the internal structure of the flow. In a sense, there is similarity between these two derivations, as both rely on the same statistical methods.

Nevertheless, the kinetic molecular theory (i.e., statistical theory of molecular ensembles) is vastly different from statistical hydrodynamic models of a viscous fluid. First, under the standard assumptions the kinetic molecular theory makes about molecular interactions, total kinetic energy of a set of moving molecules is

Statistical Microhydrodynamics. Emmanuil G. Sinaiski and Leonid Zaichik

ISBN: 978-3-527-40656-2

constant in time, whereas in hydromechanics, viscous dissipation causes kinetic energy of a real fluid to change as the fluid moves. The second difference is that molecular ensembles are, by their nature, discrete, and their evolution is described by ordinary differential equations (see the molecular dynamics method in Section 3.11), while hydrodynamics describes flows of continuous media by partial differential equations. Turbulent flows must always obey the basic conservation laws, which are implicitly present in the equations of hydro- and thermodynamics.

The reader will find the basics of turbulent flow theory and the relevant mathematical techniques explained in [1–11].

4.2
The Momentum Equation for Viscous Incompressible Fluids

An isothermal flow of a viscous incompressible fluid is completely determined by specifying four quantities in each point $\boldsymbol{X}$ at the moment t: the three components u_i of velocity $\boldsymbol{u}(\boldsymbol{X}, t)$ and the pressure $p(\boldsymbol{X}, t)$. These quantities are completely described by the continuity equation

$$\nabla \cdot \boldsymbol{u} = 0, \tag{4.1}$$

(which translates to

$$\frac{\partial U_i}{\partial X_i} = 0 \tag{4.2}$$

in the component form in a Cartesian coordinate system), and the momentum equation

$$\rho \frac{D\boldsymbol{u}}{Dt} = -\nabla \cdot \boldsymbol{T} + \boldsymbol{F}, \tag{4.3}$$

where $\boldsymbol{T}$ is the stress tensor

$$\boldsymbol{T} = \left(-p + \frac{2\mu}{3}\Delta \cdot \boldsymbol{u}\right) I + 2\mu \boldsymbol{E}, \tag{4.4}$$

$\boldsymbol{I}$ – the unit tensor, $\boldsymbol{E}$ – the rate-of-strain tensor with components

$$E_{ij} = \frac{1}{2}\left(\frac{\partial U_i}{\partial X_j} + \frac{\partial U_j}{\partial X_i}\right). \tag{4.5}$$

Eq. (4.3) can be rearranged as follows:

$$r_e \frac{D\boldsymbol{u}}{Dt} = -\nabla p + m_e \mathrm{D}\boldsymbol{u} + \boldsymbol{F}, \tag{4.6}$$

or, in the component form in a Cartesian coordinate system,

$$\rho_e \frac{Du_i}{Dt} = -\frac{\partial p}{\partial X_i} + \mu_e \sum_{k=1}^{3} \frac{\partial^2 u_i}{\partial X_k^2} + F_i, \quad (i = 1, 2, 3). \tag{4.7}$$

Here ρ_e is the fluid density, $\boldsymbol{F}$ – the density of external forces acting on the unit volume of the fluid, μ_e – the dynamic viscosity coefficient of the fluid. D/Dt is the substantial derivative

$$\frac{D}{Dt} = \frac{\partial}{\partial t} + u_k \frac{\partial}{\partial X_k}. \tag{4.8}$$

Here and further on, a repeated index implies summation over all values of the index.

Now Eq. (4.7) can be rewritten as

$$\frac{\partial u_i}{\partial t} + u_k \frac{\partial u_i}{\partial X_k} = -\frac{1}{\rho_e}\frac{\partial p}{\partial X_i} + \nu_e \sum_{k=1}^{3} \frac{\partial^2 u_i}{\partial X_k^2} + f_i, \quad (i = 1, 2, 3), \tag{4.9}$$

where we have introduced $\nu_e = \mu/\rho_e$ – the coefficient of kinematic viscosity of the fluid; $f_i = F_i/\rho_e$ – the density of external forces acting on a unit mass.

Eqs. (4.1) and (4.9) are called the Navier–Stokes equations. Hydrodynamics of viscous incompressible fluids is based on these equations.

The momentum equation can be written in another form. Introduce the vorticity vector

$$\boldsymbol{\omega} = \nabla \times \boldsymbol{u}, \tag{4.10}$$

with components

$$\omega_k = \varepsilon_{kji} \frac{\partial u_i}{\partial X_j}, \quad (i, j, k = 1, 2, 3). \tag{4.11}$$

Here ε_{kji} are permutation symbols (components of a complete antisymmetric tensor).

If the external force is absent, then taking the curl ($\nabla\times$) of both sides of Eq. (4.6) and using the component expression for the curl,

$$(\nabla\times)_k = \varepsilon_{kji} \frac{\partial}{\partial X_j},$$

we eliminate the pressure from the equations and after some algebra obtain

$$\frac{\partial \omega_k}{\partial t} + u_i \frac{\partial \omega_k}{\partial X_i} - \omega_i \frac{\partial u_k}{\partial X_i} = \nu_e \Delta \omega_k, \quad (k = 1, 2, 3). \tag{4.12}$$

Eqs. (4.12) and (4.11) help to determine the components of velocity $\boldsymbol{u}$ and vorticity ω. Applying the operator $\nabla\cdot$ to Eq. (4.6) and using Eq. (4.1) with $f=0$, one gets the Poisson equation for pressure,

$$\Delta p = -\rho_e \frac{\partial^2 (u_i u_j)}{\partial X_i \partial X_j} \tag{4.13}$$

whose general solution is

$$p(\boldsymbol{X}) = \frac{\rho_e}{4\pi} \int \frac{\partial^2 [u_i(\boldsymbol{X}') u_j(\boldsymbol{X}')]}{\partial X_i' \partial X_j'} \frac{d\boldsymbol{X}}{|\boldsymbol{X}-\boldsymbol{X}'|} + F(\boldsymbol{X}), \tag{4.14}$$

where $F(\boldsymbol{X})$ is an arbitrary harmonic function and integration is performed over the whole volume of the fluid. If the flow is taking place in an unbounded fluid, then from the requirement of finiteness of both pressure and velocity at the infinity there follows $F = const$, and since pressure is always determined up to a hydrostatic constant, it is natural to take $F = 0$.

Suppose the flow is stationary and the external forces are absent. Then ratio between the inertia term and the viscous friction term in Eq. (4.9) is equal by the order of magnitude to the Reynolds number

$$\mathrm{Re} = \frac{UL}{\nu_e}, \tag{4.15}$$

where U is the characteristic velocity and L – the characteristic linear scale.

The Reynolds number is an important flow parameter that determines the relative role of inertia and friction forces. At $\mathrm{Re} \ll 1$ the inertia force in the momentum equation can be neglected and the Navier–Stokes equations reduce to linear Stokes equations, which considerably simplifies their solution.

At $\mathrm{Re} \gg 1$, the inertial force exceeds the friction force everywhere with the exception of thin layers of thickness $\delta \sim L/\mathrm{Re}^{1/2}$ (they are called viscous boundary layers) adjacent to the rigid boundaries of the flow volume. Other peculiarity of flows with high Reynolds numbers is that the effect of inertia forces leads to energy transfer from large-scale components of the flow to small-scale components, causing formation of sharp local inhomogeneities in the flow.

For stationary flows of viscous incompressible fluid in the absence of external forces, the Reynolds number is the only similarity criterion. For non-stationary flows characterized by time scale T different from L/U and for flows that are considerably affected by external forces such as gravity, there appear additional similarity criteria.

In order to solve the Navier–Stokes equations, it is necessary to specify boundary conditions. The zero relative velocity condition holds at rigid boundary surfaces (i.e., the flow velocity $\boldsymbol{u}$ must be equal to the velocity of the boundary's motion). For a boundary between two fluid phases we require the equality of velocities and stresses on both sides of the boundary. The effect exerted by the boundary conditions will be different at low and high Reynolds numbers. At low Reynolds numbers, small changes (perturbations) of boundary conditions (for example, velocity values or shape of the boundary) lead to small changes of flow parameters, whereas at high Reynolds numbers, even small variations of boundary conditions may completely change the character of the flow.

Remember that the term "high Reynolds number" is intrinsically ambiguous. For example, while the transition from a laminar flow to a turbulent one is known to occur at high Reynolds numbers Re_{cr} (they are referred to as critical values), it cannot be characterized by one universal value of Re_{cr}, because the values of Re_{cr} can differ considerably for various flow types.

4.3 The Equations of Heat Inflow, Heat Conduction and Diffusion

In problems involving non-isothermal flows, one has to add to the system of equations (4.1)–(4.6) an equation describing the temperature change – the heat inflow equation, which represents the physical law of energy conservation:

$$\rho_e \frac{D}{Dt}\left(\frac{u^2}{2}+e\right) = \nabla\cdot(\boldsymbol{T}\cdot\boldsymbol{u}) - \nabla\cdot\boldsymbol{q} + \rho\boldsymbol{F}\cdot\boldsymbol{u} + \rho q_\varepsilon, \qquad (4.16)$$

where e is the internal energy density, $\boldsymbol{q}$ – the heat flux vector, q_ε – the density of internal heat sources (Joule heat, heat due to chemical reactions, and so on).

If the heat flux $\boldsymbol{q}$ is caused by heat conduction, then using Fourier's law $\boldsymbol{q} = -\kappa\nabla\vartheta$ and employing the Navier–Stokes equations, we write Eq. (4.16) as

$$\rho_e \frac{D}{Dt}\left(\frac{u^2}{2}+e\right) = \boldsymbol{T}:\boldsymbol{E} + \nabla\cdot(\kappa\nabla\vartheta) + \rho q_\varepsilon, \qquad (4.17)$$

where κ is thermal conductivity (aka heat conduction coefficient), $\boldsymbol{T}:\boldsymbol{E}$ is a complete (double-dot) scalar product of stress and rate-of-strain tensors,

$$\boldsymbol{T}:\boldsymbol{E} = T_{ij}E_{ij} = -p\frac{\partial u_i}{\partial x_i} + \rho_e\varepsilon$$

and ε is the so-called dissipative function characterizing the rate of dissipation of a part of kinetic energy into heat:

$$\varepsilon = \frac{1}{2}\nu_e \sum_{i,j}\left(\frac{\partial u_i}{\partial x_j} + \frac{\partial u_j}{\partial x_i}\right)^2. \qquad (4.18)$$

In other words, ε is the amount of heat released in a unit volume per unit time due to viscous effects. For incompressible flows, $\boldsymbol{T}:\boldsymbol{E} = \rho_e/\varepsilon$. The internal energy density of a fluid is equal to $e = c_v\vartheta + e_0$, where c_v is the thermal capacity (specific heat at constant volume) and e_0 is an additive constant. If thermal expansion of the fluid is neglected and in the absence of internal sources of heat ($q_\varepsilon = 0$), the equation of heat inflow becomes

$$\frac{\partial \vartheta}{\partial t} + \boldsymbol{u} \cdot \nabla \vartheta = \chi \Delta \vartheta + \frac{\varepsilon}{c_p}, \tag{4.19}$$

where $\chi = \kappa/\rho_e c_p$ is thermal diffusivity. The derivation of Eq. (4.19) hinges upon the thermodynamic relation $p/\rho_e = (c_p - c_v)\vartheta$.

The second term on the right-hand side of Eq. (4.19) has the meaning of the amount of heat released in a unit volume per unit time due to internal friction forces. In the majority of hydrodynamic problems this term is negligibly small and can be dropped. As a result, Eq. (4.19) reduces to a well-known convective heat conduction equation

$$\frac{\partial \vartheta}{\partial t} + \boldsymbol{u} \cdot \nabla \vartheta = \chi \Delta T. \tag{4.20}$$

Keep in mind that while it is safe to neglect the heat produced as a result of viscous dissipation when calculating temperature distribution, viscous dissipation by itself (or, to be more precise, specific dissipation ε) has proved to be a very important physical parameter of the turbulent flow. It will be shown later that this parameter characterizes the local structure of turbulence. A more casual term "energy dissipation" is often used for this parameter.

The heat conduction equation (4.20) will have the same look as the diffusion equation if temperature $\vartheta(\boldsymbol{X}, t)$ is replaced by impurity concentration $C(\boldsymbol{X}, t)$ and thermal diffusivity χ – by the coefficient of molecular diffusion D_m:

$$\frac{\partial C}{\partial t} + \boldsymbol{u} \cdot \nabla C = D_m \Delta C. \tag{4.21}$$

If the impurity present in the fluid (it can be either a soluble or an insoluble substance) does not exert any influence on the dynamics of the flow and on the rheological properties of the fluid, it is called passive impurity. So, the assumption that the impurity is passive means that velocity $\boldsymbol{u}$ in the diffusion equation (4.21) can be determined independently from the hydrodynamic equations. A similar assertion is true for the heat conduction equation when the viscosity coefficient does not depend on temperature.

The form of the heat conduction and diffusion equations suggests the introduction of an additional similarity parameter known as the Peclet number. For first equation, it is called the thermal Peclet number, and for the second one – the diffusional Peclet number:

$$\mathrm{Pe}_T = \frac{UL}{\chi}, \quad \mathrm{Pe}_D = \frac{UL}{D_m}. \tag{4.22}$$

These parameters characterize the ratio of the convective term to the thermal diffusivity term or to the diffusion term. Thus the Peclet number has the same physical meaning as the Reynolds number. Heat conduction or diffusion term dominates at $Pe \ll 1$, while convective transfer prevails at $Pe \gg 1$. As in the case of high Reynolds numbers, a thin boundary layer called thermal or diffusion boundary layer forms at the surface of a body placed into the fluid if $Pe \gg 1$. Two other dimensionless parameters can be introduced in addition to the Peclet number: the Prandtl number $Pr = \nu_e/\chi$ and the Schmidt number $Sc = \nu_e/D$. These parameter are related to the Reynolds and Peclet numbers through the expressions

$$Pe_T = Re\ Pr, \quad Pe_D = Re\ Sc. \tag{4.23}$$

4.4 The Conditions for the Beginning of Turbulence

The equations of motion (4.1)–(6.6) and heat inflow (4.19) are employed in calculations of various flows, in particular, flows in pipes and flows bypassing solid objects. But the obtained solutions show poor agreement with actual behavior of flows. For example, the flow in a circular pipe, known as the Poiseuille flow, shows a good match between the theory and the experiment only at sufficiently low Reynolds numbers. The theoretical solution of the problem about the boundary layer on a plate (i.e., on a flat surface), known as the Blasius solution, agrees with experimental data only at sufficiently low values of UX/ν_e, where U is the velocity of the bulk flow and X is the distance from the leading edge of the plate. The situation is similar for many other flows.

As a rule, theoretical solutions of hydrodynamic equations can adequately describe the flows observed in experiments only under some special conditions. When these conditions are not met, the flow can change its character. A smooth variation of hydrodynamic parameters, which is suggested by the theoretical solution, can turn into disorderly fluctuations (both spatial and temporal) of these parameters as the flow changes from laminar to turbulent.

The general condition for the beginning of turbulence has been established by Reynolds. Namely, the flow remains laminar until the Reynolds number $Re = UL/\nu_e$ stays below some critical value Re_{cr}, whereas at $Re > Re_{cr}$, the flow becomes turbulent.

It is easy to see why the Reynolds number is responsible for the transition from the laminar to the turbulent flow if we recall that this is equal by the order of magnitude to the ratio between the inertial and the viscous term in the momentum equation (4.6) and that at $Re \gg 1$, the inertial term becomes dominant. Inertial forces may draw together fluid volumes initially located far apart, which promotes formation of sharp inhomogeneities in the flow. Viscous forces, on the other hand, tend to smooth small inhomogeneities. Therefore at $Re \ll 1$, when viscous forces prevail over the inertial ones, there are no sharp inhomogeneities and the flow is laminar. As the Reynolds number increases, inertial forces prevail, the smoothing effect of viscosity becomes less pronounced, there appear small-scaled regions with sharp flow inhomogeneities, and disordered fluctuations are observed in the flow. At first, these fluctuations decay shortly after their birth, but starting from some Reynolds number Re_{cr} called the critical number, these fluctuations increase and the flow becomes turbulent.

Experimental research shows that Re_{cr} is not a universal constant that would apply to all kinds of flows. Even repeated measurements for one and the same flow under identical conditions produce different values of Re_{cr}. Such a spread in the values of Re_{cr} is explained by the fact that the critical Reynolds number depends not only on the type of the flow but also on the amount of perturbation ("initial turbulence") of the laminar flow at the entrance to the pipe or at the leading edge of the plate. These perturbations decay at $\mathrm{Re} < \mathrm{Re}_{cr}$ but increase at $\mathrm{Re} > \mathrm{Re}_{cr}$. The value of Re_{cr} for a pipe flow is approximately equal to 2800. Perturbations can also be caused by small protuberances and other irregularities on the wall surface.

An increase of Re_{cr} helps postpone the emergence of turbulence. This can be achieved by special technical procedures, for example, by grinding the wall surface, by optimization of hydrodynamic conditions at the entrance to the pipe, and so on. Multiple studies of flows in boundary layers around rigid bodies have produced similar conclusions about the conditions for the beginning of turbulence.

4.5
Hydrodynamic Instability

Any theoretical analysis of the conditions for the emergence of turbulence should begin with the observation that the velocity and pressure fields in laminar and turbulent flows are solutions of hydrodynamic equations with the corresponding initial and boundary conditions. The laminar flow is described by stationary solutions of these equations, while the turbulent flow should be described by non-stationary equations. Hydrodynamic equations have a stationary solution for any Reynolds number. However, not every solution corresponds to an actually existing flow. This is due to the fact that real flows should also be stable in addition to obeying the hydrodynamic equations. This means that small perturbations appearing in the flow must decay with time without changing the general character of the flow. If perturbations increase with time, then such a flow cannot exist for a long time. At some point it will evolve into a different flow.

This is precisely what happens when a laminar flow becomes turbulent. Therefore we should expect the critical Reynolds number to be a criterion of stability, in other words, at $\mathrm{Re} < \mathrm{Re}_{cr}$ the laminar flow will be stable, while at $\mathrm{Re} > \mathrm{Re}_{cr}$ it will become unstable, and small perturbations that always exist in a flow will eventually make it turbulent. So our task is to formulate the mathematical problem of stability of hydrodynamic equations (4.1) and (4.6) describing the laminar fluid flow in order to obtain the theoretical value of Re_{cr}.

One valuable theoretical method of examining stability of the flow is the method of small perturbations, whose essence consists in the following. Let $U_i(\mathbf{x}, t)$ and

$P(\boldsymbol{x}, t)$ be particular solutions of the Navier–Stokes equations, and let u_i' and p' be small perturbations of these fields appearing in the flow at the initial moment, such that $u_i' \ll U_i$ and $p' \ll P$. The resulting fields of velocity $u_i = U_i + u_i'$ and pressure $p = P + p'$ also obey the equations (4.1) and (4.6). Consider a flow in the absence of external forces. Substituting $u_i = U_i + u_i'$ and $p = P + p'$ into the equations and neglecting second-order terms, one gets the linear equations for perturbations u' and p_i':

$$\frac{\partial u_i'}{\partial X_i} = 0, \tag{4.24}$$

$$\frac{\partial u_i'}{\partial t} + U_k \frac{\partial u_i'}{\partial X_k} + u_k' \frac{\partial U_i}{\partial X_k} = -\frac{1}{\rho_e}\frac{\partial p'}{\partial X_i} + \nu_e \Delta u_i' \tag{4.25}$$

or, in the vector form,

$$\nabla \cdot \boldsymbol{u}' = 0, \tag{4.26}$$

$$\frac{\partial \boldsymbol{u}'}{\partial t} + (\boldsymbol{U}\nabla)\boldsymbol{u}' + (\boldsymbol{u}'\nabla)\boldsymbol{U} = -\frac{1}{\rho_e}\nabla p + \nu_e \Delta \boldsymbol{u}'. \tag{4.27}$$

The boundary condition on a rigid surface is $\boldsymbol{u}' = 0$.

Differentiating Eq. (4.25) with respect to x_i, summarizing the result over i and using the continuity equation (4.24, we obtain Eq. (4.13). Therefore the general solution of equations (4.24) and (4.25) will be determined once we set the initial values of velocity perturbations $u_i'(\boldsymbol{x}, 0)$. After solving Eqs. (4.24) and (4.25), we can find conditions under which perturbations will not decay in time. These are the hydrodynamic instability conditions, that is, the conditions for the transformation of the flow from laminar to turbulent.

If the solutions $\boldsymbol{u}(x)$ and $p(x)$, whose stability are investigated, are time-independent, the system of equations (4.26) and (4.27) allows a solution in the form

$$\boldsymbol{u}'(\boldsymbol{X}, t) = e^{-i\omega t} f_\omega(\boldsymbol{X}), \quad p'(\boldsymbol{X}, t) = e^{-i\omega t} g_\omega(\boldsymbol{X}). \tag{4.28}$$

Here ω is the complex frequency, f_ω and g_ω – amplitudes that must be found by solving the eigenvalue problem for the system of linear partial differential equations.

If the coefficients of this system do not depend on some spatial coordinate, the number of unknown variables can be reduced by assuming an exponential dependence of f_ω and g_ω on this coordinate. Thus, when an unperturbed flow depends only on one coordinate X_3, we can write

$$f_\omega(\boldsymbol{X}) = e^{i(k_1 X_1 + k_2 X_2)} \tilde{f}(X_3); \quad g_\omega(\boldsymbol{X}) = e^{i(k_1 X_1 + k_2 X_2)} \tilde{g}(X_3).$$

The eigenvalue problem reduces to a system of ordinary differential equations, and we can find the eigenfrequencies ω. In the flow region is finite, the eigenfrequencies form a discrete set. For the flow to be stable, it is necessary and sufficient that all eigenfrequencies should satisfy the condition $\mathrm{Im}(\omega) < 0$.

Since the Reynolds number enters the equations (4.26) and (4.27) written in the dimensionless form, the eigenfrequencies functionally depend on Re as parameter. Since at $\mathrm{Re} \to 0$ (the rest state) the flow is stable, and at $\mathrm{Re} \to \infty$, it should be unstable, there always exists a Re_{cr} at which the stable flow changes to an unstable one. It means that as Re increases, imaginary parts of some eigenfrequencies should increase and become positive. Because different eigenvalues can change their sign at different values of Re, we should take the smallest of these critical Reynolds numbers as Re_{cr}.

There are papers are devoted to the subject of to hydrodynamic stability that contain solutions of the stability problem for various flows: flow taking place between two rotating cylinders, convective flow in a fluid layer heated from below, plane-parallel flows, flow in pipes, boundary layer flows, and so on. Comparisons with the corresponding experimental data show that theoretical values of Re_{cr} do not always agree with experimental values. For example, the theory of stability of a plane Poiseuille flow gives a noticeably higher value of Re_{cr} than the one obtained in experiments on turbulent flows in a plane channel. A large discrepancy between theoretical and experimental values of Re_{cr} for this and other flows shows that the transition from a laminar flow to a turbulent one may not always be described by the linear perturbation theory. In the above-considered formulation of the problem, perturbations of hydrodynamic quantities were assumed to be small. But in the case of an unstable flow, initially small perturbations can become finite after a while, and then the small perturbation theory will no longer be applicable. Hence, the linear theory of small perturbations is capable of describing the initial stage of turbulence only, and cannot give a complete picture of the process. For finite perturbations, the stability problem reduces to a system of nonlinear equations. This is why the nonlinear theory of the beginning of turbulence is very complex and does not afford a complete solution of the problem as of today.

4.6
The Reynolds Equations

The primary characteristic feature of the turbulent flow is disordered (random) fluctuations of its hydrodynamic parameters. As a result, the dependence of these parameters on spatial coordinates at a given time, as well as the dependence on time at a given spatial point, is highly complex and difficult to handle. Besides, even if we reproduce one and the same flow under the same conditions, hydrodynamic parameters will still assume different values. In practice, one has no choice but to consider a set of similar flows, assuming that hydrodynamic parameters are random variables. This means that in a turbulent flow, an individual (deterministic) description of hydrodynamic fields of velocity, pressure, and so on, is practically impossible, and reliance on statistical methods becomes unavoidable. We can then define the turbulent flow as the one for which there exists a statistical ensemble of similar flows characterized by known probability distributions, with continuous *probability density functions* (PDFs) for the hydrodynamic fields.

In practice, it is quite unnecessary to know all the minutiae of hydrodynamic fields as we are primarily interested in their average characteristics. So we have a powerful reason to employ the averaging methods, which will allow us to operate with smooth and repetitive average values of the flow parameters.

As we stated in Section 1.7, ensemble averaging can be replaced by averaging over the time or by spatial averaging thanks to the ergodic hypothesis. Further on, the validity of the ergodic hypothesis will be assumed by default.

The simplest statistical characteristics of random hydrodynanic fields are their average values such as $\langle u\rangle$, $\langle p\rangle$, and so on. We shall reserve the term "fluctuations" for the deviations of individual values from their averages, for example, $u' = u - \langle u\rangle$, $p' = p - \langle p\rangle$, and so on. Then any hydrodynanic field can be expressed as a sum of the average value and the fluctuation:

$$u = \langle u\rangle + u', \quad p = \langle p\rangle + p', \quad \text{etc.} \tag{4.29}$$

Average values behave in a rather smooth manner, while fluctuations are characterized by intense spatial and temporal "jumps". It is fluctuations that define turbulent inhomogeneities. Note that the scale and period of inhomogeneities can, generally speaking, be arbitrarily small. However, small-scale inhomogeneities must be accompanied by large velocity gradients, which requires high expenditures of energy to overcome friction forces that become quite considerable on such small scales. Thus the existence of microflows on very low scales is almost impossible. This is why turbulent motions should be characterized by minimum scales and minimum periods of inhomogeneities.

For many turbulent flows inside pipes, the characteristic minimum scale of fluctuations ranges from 0.1 to 1 mm. On distances comparable to the minimum scale of fluctuations and on time intervals comparable to the minimum period of fluctuations, all hydrodynamic fields vary slowly and can be described by differentiable functions. Hence, a description of turbulent flows by means of differential equations is quite possible. But direct application of these equations proves to be very difficult and sometimes outright impossible, because hydrodynamic fields in a turbulent flow are non-stationary and depend on initial conditions, so even small perturbations will lead to unstable solutions. Therefore conventional hydrodynamic equations are all but useless for calculating individual hydrodynamic fields. But this does not mean that hydrodynamic equations cannot be used at all. Hydrodynamic equations have proved to be extremely useful for obtaining connections between statistical characteristics of turbulent hydrodynamic fields. The simplest of these connections were first established by Reynolds, who averaged the equations of motion of a viscous incompressible fluid. The resulting equations are known as the Reynolds equations. Before we proceed to average hydrodynamic equations, we must formulate the basic rules for the averaging of hydrodynamic fields; they were first established by Reynolds.

$$\langle f+g\rangle = \langle f\rangle + \langle g\rangle, \tag{4.30}$$

$$\langle af\rangle = a\langle f\rangle \quad \text{if } a = const, \tag{4.31}$$

$$\langle a\rangle = a \quad \text{if } a = const, \tag{4.32}$$

$$\left\langle \frac{\partial f}{\partial s}\right\rangle = \left\langle \frac{\partial f}{\partial s}\right\rangle, \quad s - \text{coordinate or time}, \tag{4.33}$$

$$\langle\langle f\rangle g\rangle = \langle f\rangle\langle g\rangle. \tag{4.34}$$

Eqs. (4.30)–(4.34) are called the Reynolds rules. Taking in consecutive order $g=1$, $g=\langle u\rangle$, and $g=u'=u-\langle u\rangle$, we derive additional rules from Eqs. (4.30)–(4.34):

$$\langle\langle f\rangle\rangle = \langle f\rangle, \quad \langle f'\rangle = \langle f-\langle f\rangle\rangle = 0, \quad \langle\langle f\rangle\langle u\rangle\rangle = \langle f\rangle\langle u\rangle,$$
$$\langle\langle f\rangle u'\rangle = \langle f\rangle\langle u'\rangle = 0, \tag{4.35}$$

Let us proceed to average the Navier–Stokes equations (4.2) and (4.7), taking

$$\boldsymbol{u} = \langle \boldsymbol{u}\rangle + \boldsymbol{u}', \quad p = \langle p\rangle + p', \tag{4.36}$$

and using the continuity equation (4.2) to transform the second term on the left-hand side of Eq. (4.7) to

$$u_k \frac{\partial u_i}{\partial X_k} = \frac{\partial}{\partial X_k}(u_i u_k).$$

Applying the Reynolds rules, we obtain:

$$\frac{\partial\langle u_i\rangle}{\partial X_i} = 0, \tag{4.37}$$

$$\frac{\partial\langle u_i\rangle}{\partial t} + \frac{\partial}{\partial X_k}\left(\langle u_i\rangle\langle u_k\rangle + \langle u_i' u_k'\rangle\right) = -\frac{1}{\rho_e}\frac{\partial\langle p\rangle}{\partial X_i} + \nu_e \Delta\langle u_i\rangle + \langle f_i\rangle. \tag{4.38}$$

Eqs. (4.37)–(4.38) are called the Reynolds equations. The advantage of these equations as compared to Eq. (4.2) and Eq. (4.7) is that they operate with smoothly varying averaged quantities. But at the same time they contain new unknown variables $\langle u_i' u_k'\rangle$ characterizing fluctuational components of velocity; you should think of them as components of a second-rank correlation tensor (see Section 1.7). The appearance of new unknowns is the consequence of nonlinearity of the Navier–Stokes equations.

The physical meaning of the terms with new unknowns becomes apparent if one carries them to the right-hand side of Eq. (4.38) and combines them with the viscous term:

$$\frac{\partial\langle u_i\rangle}{\partial t}+\langle u_k\rangle\frac{\partial\langle u_i\rangle}{\partial X_k}=-\frac{1}{\rho_e}\frac{\partial\langle p\rangle}{\partial X_i}+\frac{\partial}{\partial X_k}\left(\nu_e\frac{\partial\langle u_i\rangle}{\partial X_k}-\langle u_i'u_k'\rangle\right)+\langle f_i\rangle. \tag{4.39}$$

Comparing the resultant equation with Eq. (4.3), one concludes that the stress tensor is not represented by the viscous stress tensor $\sigma_{ij}=2\mu_e E_{ij}$ as in the laminar flow, but by the tensor

$$\tau_{ij}=2\mu_e E_{ij}-\rho_e\left\langle u_i'u_j'\right\rangle=\sigma_{ij}+\tau_{ij}^{(1)}. \tag{4.40}$$

It follows from Eq. (4.40) that turbulence gives rise to additional stresses $\tau_{ij}^{(1)}$, which are induced by turbulent fluctuations. These additional stresses are called the Reynolds stresses. Turbulent stresses are examined using the same method that is commonly applied in hydromechanics. In particular, one can show that the quantities $\rho_e\left\langle u_i'u_j'\right\rangle$ stand for the normal components of turbulent stresses at $i=j$ and for the tangential components at $i\neq j$. In this context, we should mention that the effect of turbulent mixing on the averaged flow is similar to the effect of viscosity, because turbulent fluctuations promote additional momentum transfer from one fluid volume to another in the same way as molecular viscosity forces promote the transport of momentum in the kinetic molecular theory.

When considering the Reynolds stress tensor, we are primarily interested in tensor component describing the transfer of momentum transfer from the flow to the body placed into the flow, because momentum transfer characterizes the friction force (the drag force) acting on the body. Let us look at a simple case: a plane wall $X_3=0$ is placed into a turbulent flow, which is moving along the X_1-axis and parallel to the wall. Then the friction force acting on a unit area of the wall is directed along the X_1-axis and equal to

$$\tau_0=(\langle\sigma_{13}\rangle-\rho_e\langle u_1'u_3'\rangle)_{X_3=0},\quad \langle\sigma_{13}\rangle=\rho_e\nu_e\left(\frac{\partial\langle u_1\rangle}{\partial X_3}+\frac{\partial\langle u_3\rangle}{\partial X_1}\right).$$

The zero relative velocity condition is satisfied at the wall surface, therefore u_i', $\langle u_i'\rangle$, and their derivatives with respect to X_1 are all equal to zero at $X_3=0$ and

$$\tau_0=\rho_e\nu_e\left(\frac{\partial\langle u_1\rangle}{\partial X_3}\right)_{X_3=0}. \tag{4.41}$$

Near the wall, the average flow velocity is directed parallel the wall, and the friction stress is equal to

$$\tau=\rho_e\nu_e\frac{\partial\langle u_1\rangle}{\partial X_3}-\rho_e\langle u_1'u_3'\rangle. \tag{4.42}$$

Suppose that

$$-\rho_e\langle u_1'u_3'\rangle=\rho_e\nu_t\frac{\partial\langle u_1\rangle}{\partial X_3}, \tag{4.43}$$

The factor ν_t has the dimensionality of the kinematic viscosity coefficient (m^2/s) and by analogy is called the coefficient of turbulent viscosity. In contrast to the ordinary (molecular) viscosity coefficient ν_e, the turbulent viscosity coefficient ν_t characterizes statistical properties of fluctuational motion, rather than physical properties of the fluid. In the general case ν_t does not remain constant but varies in space and in time. The turbulent viscosity coefficient is much larger than the molecular viscosity coefficient, since $\nu_t/\nu_e \sim \mathrm{Re} \gg 1$.

By virtue of Eq. (4.43), the friction stress near the wall can be written as

$$\tau = \rho_e(\nu_e + \nu_t)\frac{\partial\langle u_1\rangle}{\partial X_3}, \tag{4.44}$$

while far away from the wall, we have $\nu_t \gg \nu_e$ and $\tau \approx \tau^{(1)} = -\rho_e\langle u_1' u_3'\rangle$.

Hence, the Reynolds equations (4.39) are essentially the equations of conservation of momentum for a turbulent flow, and the Reynolds stresses describe the turbulent transport of momentum.

Similarly, we can derive the equations of conservation of other substances, such as heat and matter. Taking $\vartheta = \langle\vartheta\rangle + \vartheta'$ in Eqs. (4.20)–(4.21) and averaging these equations, we get

$$\frac{\partial\langle\vartheta\rangle}{\partial t} + \frac{\partial}{\partial X_i}\left(\langle u_i\rangle\langle\vartheta\rangle + \langle u_i'\vartheta'\rangle\right) = \chi\Delta\langle\vartheta\rangle$$

or

$$\frac{\partial\langle\vartheta\rangle}{\partial t} + \langle u_i\rangle\frac{\partial\langle\vartheta\rangle}{\partial X_i} = \frac{\partial}{\partial X_i}\left(\chi\frac{\partial\langle\vartheta\rangle}{\partial X_i} - \langle u_i'\vartheta'\rangle\right). \tag{4.45}$$

The last equation is written in the divergent form (4.39), and ϑ can be either temperature or concentration of the passive impurity. In the former case χ has the meaning of thermal diffusivity, and in the latter it is the diffusion coefficient.

The equations of heat and mass transfer have the same structure as the Reynolds equation. In these two equations, there appears an additional flux caused by turbulent fluctuations: the heat flux $-c_p\rho_e\langle\vartheta' u_i'\rangle$ or the mass flux of passive impurity $-\rho_e\langle\vartheta' u_i'\rangle$. By analogy with Eq. (4.43), the additional heat flux can be written as

$$-c_p\rho\langle\vartheta' u_i'\rangle = c_p\rho\chi_t\frac{\partial\langle\vartheta\rangle}{\partial X_i}, \tag{4.46}$$

where ϑ is temperature and χ_t is turbulent thermal diffusivity. By the same token, we write for the mass transfer of passive impurity:

$$-\rho\langle\vartheta' u_i'\rangle = -\rho D_t\frac{\partial\langle\vartheta\rangle}{\partial X_i}, \tag{4.47}$$

where ϑ is the concentration of passive impurity and D_t – the turbulent diffusion coefficient.

We have defined the coefficients of turbulent viscosity ν_t, thermal diffusivity χ_t, and diffusion D_t for the case of one-dimensional flow. For multidimensional flows, these coefficients will be tensor, rather than scalar, quantities.

The appearance of additional terms containing Reynolds stresses $\tau_{ij}^{(1)}$ in the Reynolds equations means that the system of equations will no longer be a closed. In order to obtain a closed system of equations, one has to write additional equations that would describe $\tau_{ij}^{(1)}$.

A general method that aims to derive the necessary equations for the Reynolds stresses has been proposed by Keller and Friedman. However, in each of the newly derived equations, there still appear new unknowns, whose determination, in its turn, requires new equations. The resulting system of equations (the Keller–Friedman chain) becomes infinite because any finite subsystem turns out to be unclosed. Nevertheless, the equations for $\tau_{ij}^{(1)}$ still lead us to some important qualitative conclusions about the properties of turbulent flows.

4.7 The Equation of Turbulent Energy Balance

Kolmogorov was the first to suggest using the energy balance equation in addition to the Reynolds equations. Since the quantity $\langle u_i' u_j' \rangle$ that we are trying to determine is a second order moment, let us employ the following general method that enables us to compile the equations for moments.

Let $u_1, u_2, \ldots, u_N$ be N hydrodynamic fields of the turbulent flow, and $\boldsymbol{X}_1, \boldsymbol{X}_2, \ldots, \boldsymbol{X}_N$ – N points in the volume filled by the fluid. The fields as well as the points might all be different, or some of them might be the same. Let us consider N-th order moment

$$B_{u_1 u_2 \ldots u_N}(\boldsymbol{X}_1, \boldsymbol{X}_2, \ldots, \boldsymbol{X}_N, t) = \langle u_1(\boldsymbol{X}_1, t) u_2(\boldsymbol{X}_2, t) \ldots u_N(\boldsymbol{X}_N, t) \rangle.$$

Differentiating this relation with respect to time and using the property (4.33), one obtains:

$$\begin{aligned} \frac{\partial}{\partial t} B_{u_1 u_2 \ldots u_N}(\boldsymbol{X}_1, \boldsymbol{X}_2, \ldots, \boldsymbol{X}_N, t) &= \left\langle \frac{\partial u_1(\boldsymbol{X}_1, t)}{\partial t} u_2(\boldsymbol{X}_2, t) \ldots u_N(\boldsymbol{X}_N, t) \right\rangle \\ &+ \left\langle u_1(\boldsymbol{X}_1, t) \frac{\partial u_2(\boldsymbol{X}_2, t)}{\partial t} \ldots u_N(\boldsymbol{X}_N, t) \right\rangle \ldots \\ &+ \left\langle u_1(\boldsymbol{X}_1, t) u_2(\boldsymbol{X}_2, t) \ldots \frac{\partial u_N(\boldsymbol{X}_N, t)}{\partial t} \right\rangle. \quad (4.48) \end{aligned}$$

Eliminating the derivatives $\partial u_i(\boldsymbol{X}_j, t)/\partial t$ on the right-hand side with the help of Eq. (4.9), we obtain a balance equation for the moment $B_{u_1 u_2 \ldots u_N}$ in the form of a combination of hydrodynamic fields and spatial coordinates. Let us first apply this method to the unaveraged quantities $\rho\ u_i u_j$. From the self-obvious equality

$$\frac{\partial \rho u_i u_j}{\partial t} = \rho u_i \frac{\partial u_j}{\partial t} + \rho u_j \frac{\partial u_i}{\partial t}$$

and from the momentum equation (4.9) that has been brought to the form

$$\frac{\partial \rho u_i}{\partial t} + \frac{\partial}{\partial X_k}(\rho u_i u_k + p\delta_{ik} - \sigma_{ik}) = F_i,$$

we have just used the continuity equation (4.2)), one obtains:

$$\begin{aligned}&\frac{\partial(\rho u_i u_j)}{\partial t} + \frac{\partial}{\partial X_k}(\rho u_i u_j u_k + (pu_i\delta_{jk} + pu_j\delta_{ik}) - (u_i\sigma_{jk} + u_j\sigma_{ik}))\\ &\quad = (\rho u_i F_j + \rho u_j F_i) + p\left(\frac{\partial u_i}{\partial X_j} + \frac{\partial u_j}{\partial X_i}\right) - \left(\sigma_{ik}\frac{\partial u_j}{\partial X_k} + \sigma_{jk}\frac{\partial u_i}{\partial X_k}\right).\end{aligned} \tag{4.49}$$

If we introduce the density of kinetic energy

$$e_k = r u_i u_i / 2,$$

then at $i = j$, the equation (4.49) turns into an equation for kinetic energy:

$$\frac{\partial e_k}{\partial t} + \frac{\partial}{\partial X_k}(e_k u_k + p u_k - u_i \sigma_{ki}) = p u_k X_k - \rho\varepsilon, \tag{4.50}$$

where

$$\rho\varepsilon = \frac{\mu_e}{2}\sum_{l,m}\left(\frac{\partial\langle u_l\rangle}{\partial X_m} + \frac{\partial\langle u_m\rangle}{\partial X_l}\right)^2$$

is the dissipation of kinetic energy in a unit volume of the fluid in a unit time.

It follows from this equation that the change of kinetic energy occurs due to the following factors: energy transfer via convective flux and via the work performed by pressure forces and molecular forces (second term on the left-hand side); work of the body force (first term on the right-hand side); viscous dissipation of energy, where the energy density ε is given by Eq. (4.18).

The averaged continuity equation (4.37) allows to bring the Reynolds equations to the form

$$\frac{\partial \rho\langle u_i\rangle}{\partial t} + \frac{\partial}{\partial X_k}(\rho\langle u_i u_k\rangle + \rho\langle u_i' u_k'\rangle + \langle p\rangle\delta_{ik} - \langle\delta_{ik}\rangle) = \langle F_i\rangle. \tag{4.51}$$

Application of this method to the moments $\rho\langle u_i u_j\rangle$ yields

$$\begin{aligned}\frac{\partial\rho\langle u_i\rangle\langle u_j\rangle}{\partial t}&+\frac{\partial}{\partial X_k}\left(\rho\langle u_i\rangle\langle u_j\rangle\langle u_k\rangle+\rho\langle u_i'u_k'\rangle\langle u_j\rangle+\rho\left\langle u_j'u_k'\right\rangle\langle u_i\rangle\right.\\&\left.+\left(\langle p\rangle\langle u_i\rangle\delta_{jk}+\langle p\rangle\langle u_j\rangle\delta_{ik}\right)-\left(\langle u_i\rangle\langle\delta_{jk}\rangle+\langle u_j\rangle\langle\delta_{ik}\rangle\right)\right)\\&=\left(\rho\langle u_i\rangle\langle F_j\rangle+\rho\langle u_j\rangle\langle F_i\rangle\right)+\langle p\rangle\left(\frac{\partial\langle u_i\rangle}{\partial X_j}+\frac{\partial\langle u_j\rangle}{\partial X_i}\right)\\&-\left(\langle\delta_{ik}\rangle\frac{\partial\langle u_j\rangle}{\partial X_k}+\langle\delta_{jk}\rangle\frac{\partial\langle u_i\rangle}{\partial X_k}\right)+\left(\rho\langle u_i'u_k'\rangle\frac{\partial\langle u_j\rangle}{\partial X_k}+\rho\left\langle u_j'u_k'\right\rangle\frac{\partial\langle u_i\rangle}{\partial X_k}\right).\end{aligned}\tag{4.52}$$

If we now introduce the density of kinetic energy of the averaged turbulent flow by the relation

$$e_s=\rho\langle u_i\rangle\langle u_i\rangle/2$$

Eq. (4.51) turns into an equation for the density of kinetic energy of the averaged turbulent flow:

$$\begin{aligned}\frac{\partial e_s}{\partial t}&+\frac{\partial}{\partial X_k}\left(e_s\langle u_k\rangle+\rho\langle u_k'u_i'\rangle\langle u_i\rangle+\langle p\rangle\langle u_k\rangle-\langle u_i\rangle\langle\sigma_{ki}\rangle\right)\\&=\rho\langle u_k\rangle\langle F_k\rangle-\rho\varepsilon_s+\rho\langle u_k'u_i'\rangle\frac{\partial\langle u_i\rangle}{\partial X_k},\end{aligned}\tag{4.53}$$

where

$$\varepsilon_s=\frac{1}{\rho}\langle\sigma_{lm}\rangle\frac{\partial\langle u_l\rangle}{\partial X_m}=\frac{\nu_e}{2}\sum_{l,m}\left(\frac{\partial\langle u_l\rangle}{\partial X_m}+\frac{\partial\langle u_m\rangle}{\partial X_l}\right)^2$$

is the specific dissipation of energy of the averaged flow that occurs due to the viscous forces. The physical meaning of the terms in Eq. (4.53) is the same as the meaning of the terms in Eq. (4.50), with the exception of the term $\rho\langle u_k'u_i'\rangle\langle u_i\rangle$, which corresponds to the transport of energy via turbulent viscosity.

The equation for components of the Reynolds stress tensor can be derived by subtracting Eq. (4.52) term by term from the averaged Eq. (4.49):

$$\begin{aligned}\frac{\partial\rho\langle u_i\rangle\langle u_j\rangle}{\partial t}&+\frac{\partial}{\partial X_k}\left(\rho\left\langle u_i'u_j'\right\rangle\langle u_k\rangle+\rho\left\langle u_i'u_k'u_j'\right\rangle+\left(\langle p'u_i'\rangle\delta_{jk}\right.\right.\\&\left.\left.+\left\langle p'u_j'\right\rangle\delta_{ik}\right)-\left(\left\langle u_i'\sigma_{jk}'\right\rangle+\left\langle u_j'\sigma_{ik}'\right\rangle\right)\right)=\rho\left\langle u_i'F_j'\right\rangle\\&+\rho\left\langle u_j'F_i'\right\rangle+\left\langle p'\left(\frac{\partial u_i'}{\partial X_j}+\frac{\partial u_j'}{\partial X_i}\right)\right\rangle-\left\langle\sigma_{ik}'\frac{\partial u_j'}{\partial X_k}+\sigma_{jk}'\frac{\partial u_i'}{\partial X_k}\right\rangle\\&-\left(\rho\langle u_i'u_k'\rangle\frac{\partial\langle u_j\rangle}{\partial X_k}+\rho\left\langle u_j'u_k'\right\rangle\frac{\partial\langle u_i\rangle}{\partial X_k}\right).\end{aligned}\tag{4.54}$$

We see that in addition to the average velocity $\langle u_i \rangle$ and Reynolds stresses $\rho\left\langle u_i' u_j' \right\rangle$, Eq. (4.54) contains new unknowns: third order central moments $\rho\left\langle u_i' u_k' u_j' \right\rangle$; second order moments of fluctuations of velocity and its spatial derivatives appearing in $\left\langle u_i' \sigma_{jk}' \right\rangle$ and $\left\langle \sigma_{jk}' (\partial u_i' / \partial X_k) \right\rangle$; and second order mutual moments of pressure and velocity fields $\left\langle p' u_i' \right\rangle$ and $\left\langle p' (\partial u_i' / \partial X_j) \right\rangle$. The latter moments can be represented as two-point, third order moments of the type $\left\langle u_i'(\boldsymbol{X}, t) u_j'(\boldsymbol{X}', t) u_k'(\boldsymbol{X}', t) \right\rangle$ with the help of Eq. (4.14).

From Eq. (4.54), one can obtain an equation for the average kinetic (turbulent) energy density of fluctuational motion:

$$e_k' = \rho \left\langle u_i' \right\rangle \left\langle u_i' \right\rangle / 2.$$

Putting $i = j$ into Eq. (4.67), one gets:

$$\begin{aligned} &\frac{\partial e_k'}{\partial t} + \frac{\partial}{\partial X_k}\left(e_k' \langle u_k \rangle + \frac{1}{2}\rho \left\langle u_i' u_i' u_k' \right\rangle + \left\langle p' u_k' \right\rangle - \left\langle u_i' \sigma_{ki}' \right\rangle \right) \\ &\quad = \rho \left\langle u_k' F_k' \right\rangle - \rho \langle \varepsilon_k \rangle - \rho \left\langle u_k' u_i' \right\rangle \frac{\partial \langle u_i \rangle}{\partial X_k}, \end{aligned} \tag{4.55}$$

where

$$\langle \varepsilon_k \rangle = \frac{1}{\rho}\left\langle \sigma_{lm}' \frac{\partial u_l}{\partial X_m} \right\rangle = \frac{\nu_e}{2} \sum_{l,m} \left\langle \left(\frac{\partial u_l'}{\partial X_m} + \frac{\partial u_m'}{\partial X_l} \right)^2 \right\rangle$$

is the average specific energy of fluctuational motion under the action of viscous forces.

The terms in Eq. (4.55) have the following physical meaning. The second term on the left-hand side expresses the change of turbulent energy flux density. The four summands in this term represent the contributions from energy transfer by the averaged flow, from turbulent viscosity, from pressure fluctuations, and from molecular viscosity, respectively. The term

$$A = -\rho \left\langle u_k' u_i' \right\rangle \frac{\partial \langle u_i \rangle}{\partial X_k} \tag{4.56}$$

describes energy exchange between the averaged and fluctuational motions.

Thus Eq. (4.55) is the equation of turbulent energy balance. It follows from there that turbulent energy density at a given point inside the flow can change via the following mechanisms: transport of turbulent energy from other regions in the fluid; work performed by external force fluctuations; viscous dissipation of turbulent energy; and finally, transformations of energy of the averaged motion into turbulent energy and vice versa.

One characteristic that we shall be using frequently in the subsequent discussion is the average kinetic energy of fluctuational motion per unit mass of the fluid,

$e_k = e'_k/\rho = \langle u'_i u'_i \rangle /2$. The equation for e_k can easily be obtained from Eq. (4.55):

$$\frac{De_k}{Dt} = -\langle u'_k u'_i \rangle \frac{\partial \langle u_i \rangle}{\partial X_k} - \langle \varepsilon_k \rangle + \frac{\partial}{\partial X_k}\left(-\frac{1}{2}\langle u'_i u'_i u'_k \rangle - \frac{1}{\rho}\langle p' u'_k \rangle \right)$$
$$+ \nu_e \left\langle u'_i \left(\frac{\partial \langle u'_i \rangle}{\partial X_k} + \frac{\partial \langle u'_k \rangle}{\partial X_i} \right) \right\rangle + \langle u'_k F'_k \rangle. \tag{4.57}$$

This equation, in its turn, contains new unknown quantities $\langle u'_i u'_i u'_k \rangle$, $\langle p' u'_k \rangle$, and $\langle \varepsilon_k \rangle$. Therefore a system of equations containing the Reynolds equation (4.51) and either the equation for Reynolds stresses (4.54) or the equation for turbulent energy (4.57) will be unclosed. One can always try to construct new equations for the new unknowns, but the derived system will also be unclosed, because it will contain unknown moments of higher orders. Hence, construction of additional equations for higher-order moments gets us nowhere as we try to obtain a closed system of equations describing the turbulent flow. The Reynolds equations and the equation of turbulent energy balance only allow us to infer the existence of certain connections between different statistical characteristics of turbulence, but they cannot be solved.

The only way out from this situation is to attempt to close the system of equations by making additional assumptions that are based on certain physical considerations and justified by their agreement with experimental data. In other words, we aim to specify the missing connections between statistical characteristics of turbulence irrespective of the available equations.

The "easiest" way to close the system is to simply drop the higher-order moments. It turns out, however, that such a procedure attains its purpose only for relatively small Reynolds numbers that do not present any practical interest. We are interested specifically in the case of great Reynolds numbers, or, to use another term, in the case of fully developed turbulence.

In a few cases, the form of additional connections between statistical characteristics can be guessed from dimensionality considerations; the expressions derived in this way are accurate up to a certain small number of empirical constants. But the dimensionality theory still stops halfway in solving the problem, because the resultant relations contain unknown functions and (or) constants, which then have to be determined experimentally. The total number of these functions and constants can be large, because different functions and constants are needed for different flows (flows in pipes, flows bypassing a solid body, flows in boundary layers, jet flows, etc.)

Yet another closure method uses transport equations to find the characteristics of turbulence such as turbulent energy, turbulent viscosity, and the integral scale of turbulence. This method is the most popular as of today and is widely used in numerical calculations for different turbulent flows.

It is quite natural that one would like to make his task simpler by finding the minimum required number of additional relations, functions, or constants that would applicable all at once to many different flows. Unfortunately, as of today, we are still lacking a universal theory that would describe all kinds of turbulent flows. Inevitably, all the existing turbulent flow models are valid for only one or several types of flows.

Theories of turbulence that encompass relations found empirically or guessed from physical considerations and then proved by experiments (in addition to the available hydrodynamic equations) are referred to as semi-empirical theories.

The existing models of turbulence will be examined in more detail in Section 4.10. But we must first discuss the internal structure of turbulence, and in particular, the underlying concept of isotropic turbulence.

4.8
Isotropic Turbulence

Turbulence is called homogeneous when all hydrodynamic fields are homogeneous random fields (see Section 1.9), and isotropic when all hydrodynamic fields are isotropic random fields (see Section 1.10). Isotropic turbulence is a mathematical idealization, which is suitable for approximate description of some special turbulent flows. In fact, turbulence can be isotropic only when the fluid occupies the entire space. Real flows always have boundaries, and this is where isotropy ends. The case of isotropic turbulence is the simplest one, yet it allows us to establish some distinguishing properties of turbulence. This explains why the concept of isotropic turbulence, which was first introduced by Taylor, has played such a crucial role in the development of the modern theory of statistical turbulence. Later on, Kolmogorov proposed a more general concept of locally isotropic turbulence, which embraced a greater variety of real flows and has since proved itself a powerful tool for the analysis of various turbulent flows.

Consider isotropic turbulence in a viscous incompressible fluid in the absence of external forces. As evidenced by the discussion in Sections 4.5 and 4.7, we are primarily interested in the components of correlation tensors, which, according to the definition of a homogeneous, isotropic random vector field, depend on $\boldsymbol{r}=\boldsymbol{X}'-\boldsymbol{X}$ and t, where $\boldsymbol{X}'$ and $\boldsymbol{X}$ are two arbitrary points in space. For a homogeneous, isotropic random velocity field, the second order correlation tensor

$$B_{ij}(\boldsymbol{r},t)=\left\langle u_i(\boldsymbol{X},t)u_j(\boldsymbol{X}+\boldsymbol{r},t)\right\rangle$$

is expressed in terms of two scalar functions, $B_{LL}(r,t)$ and $B_{NN}(r,t)$:

$$B_{ij}(\boldsymbol{r},t)=\left(B_{LL}(r,t)-B_{NN}(r,t)\right)\frac{r_ir_j}{r^2}+B_{NN}(r,t)\delta_{ij},\tag{4.58}$$

where $B_{LL}(r,t)=\langle u_L(\boldsymbol{X},t)u_L(\boldsymbol{X}+\boldsymbol{r},t)\rangle$ and $B_{NN}(r,t)=\langle u_N(\boldsymbol{X},t)u_N(\boldsymbol{X}+\boldsymbol{r},t)\rangle$ are the longitudinal and transverse correlation functions; u_L and u_N are projections of the velocity vector $\boldsymbol{u}$ onto the directions parallel and perpendicular to $\boldsymbol{r}$, and $r=|\boldsymbol{r}|$. By the same token, the third order correlation tensor

$$B_{ij,k}(\boldsymbol{r},t)=\left\langle u_i(\boldsymbol{x},t)u_j(\boldsymbol{x},t)u_k(\boldsymbol{x}+\boldsymbol{r},t)\right\rangle$$

can be expressed in terms of three scalar functions, $B_{LL,L}(r, t)$, $B_{LN,N}(r, t)$, and $B_{NN,L}(r, t)$:

$$\begin{aligned} B_{ij,k}(r,t) =& (B_{LL,L}(r,t) - 2B_{LN,N}(r,t) - B_{NN,L}(r,t)) \frac{r_i r_j r_k}{r^3} \\ &+ B_{NN,L}(r,t)) \frac{r_k}{r} \delta_{ij} + B_{LN,N}(r,t) \left(\frac{r_i}{r} \delta_{jk} + \frac{r_j}{r} \delta_{ik} \right). \end{aligned} \tag{4.59}$$

The relations (4.58) and (4.59) can be considerably simplified if the velocity field is solenoidal or potential. The former case is true for an incompressible fluid ($\nabla \cdot \boldsymbol{u} = 0$), and the second – for an ideal fluid ($\nabla \times \boldsymbol{u} = 0$). Since the fluid is assumed to be incompressible, we have a solenoidal velocity field. Then the continuity equation leads us to

$$B_{NN}(r,t) = B_{LL}(r,t) + \frac{r}{2} \frac{\partial}{\partial r} (B_{LL}(r,t)), \tag{4.60}$$

which is the so-called Karman equation. One can see that the second order correlation tensor of an isotropic solenoidal vector field can be expressed in terms of a single scalar function.

The third order correlation tensor of an isotropic solenoidal vector field $B_{ij,k}(r, t)$ is also expressed in terms of a single scalar function, because

$$B_{NN,L}(r,t) = -\frac{1}{2} B_{LL,L}(r,t).$$

$$B_{LN,N}(r,t) = \frac{1}{2} B_{LL,L}(r,t) + \frac{r}{4} \frac{\partial}{\partial r} (B_{LL,L}(r,t)). \tag{4.61}$$

We should mention yet another important property of isotropic turbulent vector fields. Any isotropic random vector field $\boldsymbol{u}(\boldsymbol{x})$ can be represented as a sum of two mutually uncorrelated fields, one of which is solenoidal and the other – potential. The corollary is that no scalar isotropic field can correlate with a solenoidal vector field. If we choose pressure to be our scalar field and velocity – our vector field, then this corollary reduces to the statement that

$$B_{Pi}(\boldsymbol{r},t) = B_{PL}(\boldsymbol{r},t) \frac{r_i}{r} = 0, \tag{4.62}$$

where $B_{Pi}(\boldsymbol{r}, t) = \langle p(\boldsymbol{r}, t) u_i(\boldsymbol{r}, t) \rangle$, $B_{PL}(\boldsymbol{r}, t) = \langle p(\boldsymbol{r}, {}_t) u_L(\boldsymbol{r}, t) \rangle$.

It is now easy to derive the dynamic equation for the correlation tensor $B_{ij}(\boldsymbol{r}, t)$. Let us apply the equations (4.9) (with **f** set to zero) to i-th component of velocity, u_i, at the point $\boldsymbol{X}$ and to j-th component of velocity, u'_j, at the point $\boldsymbol{X} + \boldsymbol{r} = \boldsymbol{X}'$. Multiply the first equation by u'_j and the second – by u_i, add both equations together and average the result. The outcome is

$$\frac{\partial\left\langle u_i u_j'\right\rangle}{\partial t}+\frac{\partial\left\langle u_i u_{\mathrm{k}} u_j'\right\rangle}{\partial X_{\mathrm{k}}}+\frac{\partial\left\langle u_i u_j' u_{\mathrm{k}}'\right\rangle}{\partial X_{\mathrm{k}}'}=-\frac{1}{\rho}\left(\frac{\partial\left\langle p u_j'\right\rangle}{\partial X_i}+\frac{\partial\left\langle p' u_i'\right\rangle}{\partial X_j'}\right)$$
$$+\nu_e\left(\frac{\partial^2\left\langle u_i u_j'\right\rangle}{\partial X_{\mathrm{k}}\partial X_{\mathrm{k}}}+\frac{\partial^2\left\langle u_i u_j'\right\rangle}{\partial X_{\mathrm{k}}'\partial X_{\mathrm{k}}'}\right). \tag{4.63}$$

It follows from the homogeneity of turbulence that all two-point moments depend on $\boldsymbol{r}=\boldsymbol{X}'-\boldsymbol{X}$, therefore $\partial/\partial X_{\mathrm{k}}$ and $\partial/\partial X_{\mathrm{k}}'$ are respectively equal to $-\partial/\partial r_{\mathrm{k}}$ and $\partial/\partial r_{\mathrm{k}}$. As a result, Eq. (4.63) reduces to

$$\frac{\partial B_{ij}(r,t)}{\partial t}=\frac{\partial}{\partial r_{\mathrm{k}}}\left(B_{i\mathrm{k},j}(r,t)-B_{i,j\mathrm{k}}(r,t)\right)$$
$$+\frac{1}{\rho}\left(\frac{\partial B_{pj}(r,t)}{\partial r_i}-\frac{\partial B_{ip}(r,t)}{\partial r_j}\right)+2\nu_e\frac{\partial^2 B_{ij}(r,t)}{\partial r_{\mathrm{k}}\partial r_{\mathrm{k}}}. \tag{4.65}$$

From the property of isotropy, one can deduce the relations (4.62), thus establishing that $B_{pj}(\boldsymbol{r},t)=B_{ip}(\boldsymbol{r},t)=0$, and the relations (4.60), (4.61), which mean that the tensors $B_{ij}(\boldsymbol{r},t)$, $B_{i\mathrm{k},j}(\boldsymbol{r},t)$, and $B_{i,j\mathrm{k}}(\boldsymbol{r},t)=B_{j\mathrm{k},i}(-\boldsymbol{r},t)$ can be expressed through scalar functions $B_{LL}(\boldsymbol{r},t)$ and $B_{LL,\mathrm{L}}(r,t)$. After some algebra, we get the Karman–Howarth equation

$$\frac{\partial B_{LL}(\boldsymbol{r},t)}{\partial t}=\left(\frac{\partial}{\partial r}+\frac{4}{r}\right)\left(\partial B_{LL,L}(r,t)+2\nu_e\frac{\partial B_{LL}(\boldsymbol{r}],t)}{\partial r}\right). \tag{4.65}$$

Just like the Reynolds equations, Eq. (4.65) cannot be solved, because it contains two unknowns, $B_{LL}(r,t)$ and $B_{LL,L}(r,t)$.

Consider some important corollaries that follow from the derived equations for correlation functions. These are equations for some functions of r and t, from which one can obtain certain numerical characteristics describing turbulence as a whole (in other words, these characteristics are independent of the distance r between the two points under consideration). To this end, it is sufficient to expand the functions that appear in these equations as a Taylor series over the powers of r and then equate the terms having the same power.

We begin with the Karman–Howarth equation (4.65). The zeroth term of the expansion (i.e., $r=0$) gives us

$$\frac{dB_{LL}(0)}{dt}=10\nu_e\left(\frac{\partial^2 B_{LL}}{\partial r^2}\right)_{r=0}. \tag{4.66}$$

Since $B_{LL}(0)=\langle u^2\rangle$, Eq. (4.66) can be rewritten in the form

$$\frac{d}{dt}\left(\frac{3}{2}\langle u^2\rangle\right)=15\nu_e\langle u^2\rangle f''(0)=-\frac{15\nu_e\langle u^2\rangle}{\lambda_t^2}, \tag{4.67}$$

where u is the velocity component along the X-axis, $f(r) = B_{LL}(\mathrm{r})/B_{LL}(0)$, and $\lambda_t^2 = -1/f''(0)$. Note that $B_{LL}(0) = \langle(\boldsymbol{u}(\boldsymbol{x}))^2\rangle/3$.

Eq. (4.67) represents the balance of energy for isotropic turbulence. It describes the rate of decrease of the average kinetic energy of turbulence due to the action of viscous forces. The parameter λ_t has the dimensionality of a length and is called the Taylor microscale. It can be regarded as the smallest size of eddies, which are responsible for energy dissipation. As far as $d\langle u^2\rangle/dt \sim \langle u^2\rangle/\tau_t$, where τ_t is the characteristic time of hydrodynamic relaxation (the Taylor time microscale), Eq. (4.67) gives $\tau_t \sim \lambda_t^2/10\nu_e$. One can use the Karman equation to express the microscale λ_t through the transverse correlation $\lambda_t^2 = -2/g''(0)$, where $g(r) = B_{NN}(r)/B_{NN}(0)$.

The expressions for Taylor microscales can be brought to the form

$$\left\langle\left(\frac{\partial u}{\partial X}\right)^2\right\rangle = \frac{\langle u^2\rangle}{\lambda_t^2}, \quad \left\langle\left(\frac{\partial u}{\partial Y}\right)^2\right\rangle = 2\frac{\langle u^2\rangle}{\lambda_t^2}. \tag{4.68}$$

The relations (4.68) make it possible to determine λ_t by finding the values of $\langle(\partial u/\partial X)^2\rangle$ and $\langle(\partial u/\partial X)^2\rangle$ from the experiment. Obtaining λ_t and $\langle u^2\rangle$ from independent measurements (they can be conducted, for example, behind the grate in a wind tunnel, at different distances from the grate), we can prove the relation (4.67) and establish the attenuation formula for $\langle u^2\rangle$. The averaged squares of all velocity components decrease with time in accordance with the "5/2 law", namely,

$$\langle u^2\rangle = \frac{C}{(t-t_0)^{5/2}}, \tag{4.69}$$

where t_0 is some arbitrarily chosen initial moment ("initial time reading"), and λ_t^2 increases linearly with time

$$\lambda_t^2 = 4\nu_e(t-t_0). \tag{4.70}$$

Since $B_{LL}(r)$ is an even function of r, while $B_{LL,L}(r)$ is an odd function, both sides of Eq. (4.65) contain only even degrees of r. Equating the coefficients for r^2 in the Taylor expansion, one gets the equation

$$\frac{1}{2}\frac{d}{dt}B''_{LL}(0) = \frac{7}{6}B'''_{LL,L}(0) + \frac{7}{3}\nu_e B^{IV}_{LL}(0). \tag{4.71}$$

This equation is interpreted as the balance equation for a vortex, since the correlation tensor for a vortex is equal to $B_{\omega_i\omega_i}(\boldsymbol{r}) = -\Delta B_{ii}(\boldsymbol{r})$.

If at $r \to \infty$, the quantity $B_{LL}(r)$ goes to zero faster than r^{-5}, then from Eq. (4.65) there follows the relation

$$\int_0^\infty r^4 B_{LL}(r)dr = \Lambda = const, \tag{4.72}$$

which has the form of a conservation law. The quantity Λ is called the Loitsyansky integral (or the Loitsyansky invariant).

Similarly, by using Eq. (4.13) that expresses pressure in terms of velocity, we can study the statistical properties of a scalar hydrodynamic field – pressure, and by using the equations of heat conduction (4.20) and diffusion (4.21), we can study the statistical properties of scalar fields – temperature and concentration (see the end of this section).

In addition to the above-mentioned Taylor microscale, the theory of turbulence introduces four other length scales: longitudinal and transverse differential scales

$$\lambda_1 = \left(-\frac{B_{LL}(0)}{2B''_{LL}(0)}\right)^{1/2}, \quad \lambda_2 = \left(-\frac{B_{NN}(0)}{2B''_{NN}(0)}\right)^{1/2} \tag{4.73}$$

and longitudinal and transverse integral scales

$$L_1 = \frac{1}{B_{LL}(0)}\int_0^\infty B_{LL}(r)dr, \quad L_2 = \frac{1}{B_{NN}(0)}\int_0^\infty B_{NN}(r)dr. \tag{4.74}$$

Comparing the relations (4.74) with the formulas (1.99) and (1,102), one can conclude that integral length scale has the meaning of characteristic correlation length, that is, the average distance that turbulent perturbations can travel. Since correlation between velocities at two different points decreases with increase of the distance between these points, the integral scale is equal by an order of magnitude to the maximum distance between these points at which the velocities still show a noticeable correlation.

A further insight into isotropic turbulence can be gained by examining correlation functions in the wavenumber space. As was mentioned in Section 1.10, spectral representations of random functions have the meaning of superposition of harmonic oscillations for stationary random processes. For an isotropic turbulent field, the spectral representation looks especially simple. Representations for components of the correlation tensor $B_{ij}(\boldsymbol{r}, t)$ and its spectral tensor $F_{ij}(\boldsymbol{k})$ are found from the definitions (1.104) and (1.105):

$$B_{ij}(\boldsymbol{r}, t) = 4\pi\int_0^\infty \frac{\sin(kr)}{kr} F_{ij}(k, t)k^2 dk, \tag{4.75}$$

$$F_{ij}(\boldsymbol{k}, t) = \frac{1}{2\pi^2}\int_0^\infty \frac{\sin(k\boldsymbol{r})}{kr} B_{ij}(r, t)r^2 dr, \tag{4.76}$$

The spectrum $F_{ij}(\boldsymbol{k})$ is symmetric and nonnegative, and its corresponding quadratic form is positive definite. The isotropy condition implies that $F_{ij}(\boldsymbol{k})$ can be represented in the form (4.58), namely,

$$F_{ij}(\boldsymbol{k},t) = (F_{LL}(k,t) - F_{NN}(k,t))\frac{k_i k_j}{k^2} + F_{NN}(r,t)\delta_{ij}, \tag{4.77}$$

where $F_{LL}(k, t)$ and $F_{NN}(k, t)$ are the longitudinal and transverse spectra.

The spectral representation of the average energy is

$$\frac{1}{2}\langle \boldsymbol{u}^2(\boldsymbol{X},t)\rangle = \frac{1}{2}B_{ii}(0,t) = \int_0^\infty E(k,t)dk. \tag{4.78}$$

In the isotropic case, the last relation transforms to

$$E(k,t) = 4\pi k^2 \frac{F_{ij}(k,t)}{2} = 2\pi k^2 (F_{LL}(k,t) + 2F_{NN}(k,t)). \tag{4.79}$$

The conditions of solenoidality and potentiality allow us to simplify the expressions for $E(k, t)$ and $F_{ij}(k, t)$:

$$E(k,t) = \begin{cases} 4\pi k^2 F_{NN}(k,t), & \text{for solenoidal field,} \\ 2\pi k^2 F_{LL}(k,t), & \text{for potential field,} \end{cases} \tag{4.80}$$

$$F_{ij}(k,t) = \begin{cases} \dfrac{E(k)}{4\pi k^2}\left(\delta_{ij} - \dfrac{k_i k_j}{k^2}\right), & \text{for solenoidal field,} \\ \dfrac{E(k)k_i k_j}{2\pi k^4}, & \text{for potential field,} \end{cases} \tag{4.81}$$

For a solenoidal field, the longitudinal B_{LL} and transverse B_{NN} correlation functions are connected with $E(k)$ through the relations

$$\begin{aligned} B_{LL}(r) &= 2\int_0^\infty \left(-\frac{\cos kr}{(kr)^2} + \frac{\sin kr}{(kr)^2}\right) E(k,t)dk, \\ B_{NN}(r) &= 2\int_0^\infty \left(\frac{\sin kr}{kr} + \frac{\cos kr}{(kr)^2} - \frac{\sin kr}{(kr)^3}\right) E(k,t)dk, \end{aligned} \tag{4.82}$$

whereas $E(k)$ is expressed through B_{LL} as

$$E(k) = \frac{1}{\pi}\int_0^\infty (kr\sin kr - k^2 r^2 \cos kr) B_{LL}(r)dr. \tag{4.83}$$

Similar relations exist for the spectrum of a third order correlation tensor of an isotropic field $\boldsymbol{u}(\boldsymbol{x})$:

$$F_{ij,k}(k) = iF_{LN,N}(k)\left(\delta_{jk}\frac{k_i}{k} + \delta_{ik}\frac{k_j}{k} - 2\frac{k_i k_j k_k}{k^3}\right),$$

$$F_{LN,N}(k) = \frac{1}{8\pi^2}\int_0^\infty \left(\sin kr + \frac{3\cos kr}{kr} - \frac{3\sin kr}{(kr)^2}\right) B_{LL,L}(r) r^2 dr.$$

The Karman–Howarth equation (4.65) has the following spectral representation:

$$\frac{\partial F_{NN}(k,t)}{\partial t} = -2kF_{LN,N}(k,t) - 2\nu k^2 F_{NN}(k,t) \tag{4.84}$$

or

$$\frac{\partial E(k,t)}{\partial t} = -8\pi k^3 F_{LN,N}(k,t) - 2\nu k^2 E(k,t) \tag{4.85}$$

Equations (4.84) and (4.85) describe the time rate of change of the spectral distribution of isotropic turbulence energy. The second term on the right-hand side gives energy dissipation due to viscosity. The viscosity-related increase in the dissipation of kinetic energy of a perturbation with the wave number k is proportional to the intensity of this perturbation; $2\nu k^2$ is the proportionality coefficient. Hence, the energy of long-wavelength perturbations (small values of k) decreases under the action of viscosity at much slower rates than the energy of short-wavelength perturbations. The reason for this is that short-wavelength perturbations produce large velocity gradients, and the viscous friction force is proportional to the velocity gradient. The first term on the right-hand side of equations (4.84) and (4.85) describes the energy change of the spectral component of turbulence with the wavenumber k due to nonlinear inertial terms of hydrodynamic equations. This change leads to a redistribution of energy between spectral components without changing the total energy of turbulence. Hence, any change in the total energy of turbulence is caused exclusively by viscous forces, that is,

$$\frac{\partial}{\partial t}\frac{\langle u_i u_j\rangle}{2} = \frac{\partial}{\partial t}\int_0^\infty E(k,t)dt = -2\nu_e\int_0^\infty k^2 E(k,t)dt. \tag{4.86}$$

The first term on the right-hand side of Eq. (4.85) is negative at small values of k and positive at large values of k, therefore turbulent mixing leads to breakup of turbulent perturbations, that is, to energy transfer from large-scale to small-scale components, with energy being spent to overcome viscous friction. Hence viscosity becomes a major factor for small-scale components. This fact will be used in the next section as we examine the inner structure of developed turbulence.

When looking at the inner structure of developed turbulence, we are not as much concerned with correlations between components of velocities at different points $\boldsymbol{X}+\boldsymbol{r}$ and $\boldsymbol{X}$ at a given moment of time (i.e., with components of the tensor B_{ij}) as we are with correlations between components of velocity differences $\Delta_r\mathbf{u} = \boldsymbol{u}(\boldsymbol{X}+\boldsymbol{r}) - \boldsymbol{u}(\boldsymbol{r})$ at these points. For isotropic turbulence, the condition $\langle \Delta_r \boldsymbol{u}\rangle = 0$ must be valid (see Section 1.10). The corresponding symmetric tensor has components

$$b_{ij} = \langle (\boldsymbol{u}_i(\boldsymbol{X}+\boldsymbol{r}) - \boldsymbol{u}_i(\boldsymbol{r}))(\boldsymbol{u}_j(\boldsymbol{X}+\boldsymbol{r}) - \boldsymbol{u}_j(\boldsymbol{r}))\rangle, \tag{4.87}$$

known as the structure functions. For simplicity's sake, we are considering stationary processes only, hence the omission of the time t in Eq. (4.87).

Structure functions for an isotropic field can be written in the form similar to that of Eq. (4.58):

$$b_{ij}(\boldsymbol{r}) = (b_{LL}(r) - b_{NN}(r))\frac{r_i r_j}{r^2} + b_{NN}(r)\delta_{ij}, \tag{4.88}$$

where $b_{LL}(r, t)$ and $b_{NN}(r, t)$ are the longitudinal and transverse structure functions equal to

$$b_{LL}(r,t) = \left\langle (\boldsymbol{u}_L(\boldsymbol{X}+\boldsymbol{r}) - \boldsymbol{u}_L(\boldsymbol{r}))^2 \right\rangle, \quad b_{NN}(r,t) = \left\langle (\boldsymbol{u}_N(\boldsymbol{X}+\boldsymbol{r}) - \boldsymbol{u}_N(\boldsymbol{r}))^2 \right\rangle.$$

The longitudinal and transverse structure functions are connected with the corresponding correlation functions through the relations

$$b_{LL}(r) = 2(B(0) - B_{LL}(r)), \quad b_{NN}(r) = 2(B(0) - B_{NN}(r)). \tag{4.89}$$

Here $B(0) = B_{LL}(0) = B_{NN}(0) = \langle u^2\rangle/3$.

For a solenoidal field $\boldsymbol{u}(\boldsymbol{x})$ (incompressible fluid), the longitudinal and transverse structure functions b_{LL} and b_{NN} are mutually connected through an equation similar to the Karman equation (4.60):

$$b_{NN}(r,t) = b_{LL}(r,t) + \frac{r}{2}\frac{\partial}{\partial r}(b_{LL}(r,t)). \tag{4.90}$$

In addition to the two-point second order moments of velocity difference, b_{ij}, one can introduce two-point third order moments of velocity difference,

$$b_{ijk} = \langle (u_i(\boldsymbol{X}+\boldsymbol{r}) - u_i(\boldsymbol{r}))(u_j(\boldsymbol{X}+\boldsymbol{r}) - u_j(\boldsymbol{r}))(u_k(\boldsymbol{X}+\boldsymbol{r}) - u_k(\boldsymbol{r}))\rangle,$$

which can be expressed through single scalar function $b_{LLL}(r)$ by virtue of the isotropy condition:

$$b_{ijk} = \frac{1}{2}\left(b_{LLL}(r) - r\frac{\partial b_{LLL}(r)}{\partial r}\right)\frac{r_i r_j r_k}{r^3} + \frac{1}{6}\left(b_{LLl}(r) + r\frac{\partial b_{LLl}(r)}{\partial r}\right)$$
$$\left(\frac{r_i}{r}\delta_{jk} + \frac{r_j}{r}\delta_{ik} + \frac{r_k}{r}\delta_{ij}\right).$$

Once again, the system of hydrodynamic equations for an isotropic turbulent flow turns out to be unclosed, which is evident, for example, from Eq. (4.90) that contains two unknown functions $b_{NN}(r, t)$ and $b_{LL}(r, t)$. We have to come up with additional hypotheses and relations in order to close this system.

Our previous analysis for the isotropic vector field can be extended to the case of an isotropic scalar random field, for example, the field of passive impurity concentration. Let $C(\boldsymbol{X}, t)$ be the concentration of a substance in the fluid and $C'(\boldsymbol{X}, t)$ – its fluctuation relative to the average value $\langle C\rangle$. The fields of velocity (a vector field) and concentration (a scalar field) are assumed to be isotropic, so $\langle \boldsymbol{u}\rangle$ and $\langle C\rangle$ are constants. A theoretical examination of the turbulent scalar field can be performed in the same manner as for the vector field in the preceding discussion.

Let $C'_a = C'(\boldsymbol{X}_a, t)$ and $C'_b = C'(\boldsymbol{X}_b, t)$ denote concentration fluctuations at the points $\boldsymbol{X}_a$ and $\boldsymbol{X}_b = \boldsymbol{X}_a + \boldsymbol{r}$ at one and the same instant of time. The correlation of these quantities is $B_{ab} = \langle C'_a C'_b\rangle$ and the corresponding correlation coefficient is $\rho_{cc} = \langle C'_a C'_b\rangle / \langle (C')^2\rangle$. The quantity $\langle (C'_a)^2\rangle = \langle (C'_b)^2\rangle = \langle (C')^2\rangle$ is called the intensity of concentration fluctuations. Just as for the vector field (see Eqs. (4.73)–(4.74)), we can introduce two length scales for the scalar field: differential scale (microscale) λ_c and integral scale (macroscale) L_c:

$$\frac{2}{\lambda_c^2} = -\left(\frac{\partial^2 \rho_{cc}}{\partial r^2}\right)_{r=0}, \quad L_c = \int_0^{\infty} \rho_{cc}(r)dr. \tag{4.91}$$

Having looked at correlations between the values of one and the same scalar quantity C, we may now ask about correlations between C and components of the velocity vector u_i – either at one and the same point or at different points. It turns out that, due to the fact that no scalar isotropic field can correlate with a solenoidal vector field, these correlations are absent, that is, $\langle (u_i)_a C_a\rangle = 0$ and $\langle (u_i)_a C_b\rangle = 0$.

We can also introduce correlations of higher order, for example, third order correlations at two points $\langle (u_i)_a C_a C_b\rangle$ and $\langle (u_i)_a (u_j)_b C_b\rangle$. It is obvious that $\langle (u_i C^2)_a\rangle = 0$ and $\langle (u_i^2)_a C_b^2\rangle = 0$.

The correlation function $B_{ab}(r, t)$ for an isotropic turbulent field satisfies a dynamic equation of the Karman–Howarth type:

$$\frac{\partial B_{ab}}{\partial t} = 2\left(\frac{\partial}{\partial r} + \frac{2}{r}\right)\left(B_{La,b} + D_m \frac{\partial B_{ab}}{\partial r}\right). \tag{4.92}$$

where D_m is the coefficient of molecular diffusion; $B_{La,b} = \langle (u_L)_a C_a C_b\rangle$; $(u_L)_a$ is the velocity component along the vector $\boldsymbol{r}$ connecting the points a and b. This equation is called the Corrsin equation.

The Corrsin equation leads to the dynamic equation for the intensity of concentration fluctuations $\langle (C')^2\rangle$. One can derive it by going to the limit $r \to 0$ and expanding the functions entering Eq. (4.92) in a Taylor series similarly to the derivation of Eq. (4.66).

$$\frac{d\langle (C'2)\rangle}{dt} = -12\frac{D_m}{\lambda_c^2}\langle (C')^2\rangle. \quad (4.93)$$

One can see from Eq. (4.93) that the intensity of concentration fluctuations $\langle (C')^2\rangle$ decreases with time, and furthermore, the characteristic time of this decrease is inversely proportional to the coefficient of molecular diffusion. So, in the final analysis, attenuation of intensity of concentration fluctuations is caused solely by the molecular diffusion, just as attenuation of turbulence is caused solely by the molecular viscosity. Introducing τ_c – the characteristic relaxation time (aka time of micromixing) of the scalar field – through the relation $d\langle (C')^2\rangle/dt \sim \langle (C')^2\rangle/\tau_c$, we readily get $\tau_c \sim \lambda_c^2/12D_m$ from Eq. (4.93).

Scalar fields can be also represented in the spectral form. Let us introduce the spectral representation of correlations B_{ij} according to the formulas (4.75) and (4.76), in which i and j should be replaced by a and b:

$$B_{ab}(r,t) = 4\pi\int_0^\infty \frac{\sin kr}{kr}k^2 F_{ab}(k,t)dk, \quad (4.94)$$

$$F_{ab}(k,t) = \frac{1}{2\pi^2}\int_0^\infty \frac{\sin kr}{kr}k^2 B_{ab}(r,t)dr. \quad (4.95)$$

At $r \to 0$, one gets the spectral representation of concentration fluctuation intensity $\langle (C')^2\rangle$:

$$\left\langle (C')^2\right\rangle = 4\pi\int_0^\infty k^2 F_{ab}(k,t)dk. \quad (4.96)$$

By analogy with Eq. (4.79), we can introduce the function $E_c(k,t) = 4k^2 F_{ab}(k,t)$. The relation (4.96) takes the form

$$\left\langle (C')^2\right\rangle = \int_0^\infty E_c(k,t)dk \quad (4.97)$$

and the relations (4.94)–(4.95) transform to

$$B_{ab}(r,t) = \int_0^\infty \frac{\sin kr}{kr}E_c(k,t)dk, \quad E_c(k,t) = \frac{1}{2\pi}\int_0^\infty kr\sin kr\, B_{ab}(r,t)dr.$$

The differential and integral length scales of the scalar field are expressed as

$$\frac{2}{\lambda_c^2}=\frac{1}{3}\frac{1}{\langle (C')^2\rangle}\int_0^\infty k^2 E_c(k,t)dk, \quad L_c=\frac{\pi}{2}\frac{1}{\langle (C')^2\rangle}\int_0^\infty \frac{E_c(k,t)}{k}dk. \tag{4.98}$$

The spectral representation of the Corrsin equation gives rise to the following dynamic equation for the spectrum $E_c(k, t)$:

$$\frac{\partial E_c(k,t)}{\partial t}=F_c(k,t)-2D_m k^2 E_c(k,t), \tag{4.99}$$

where $F_c(k, t)=-8k^2F_{La,b}(k, t)$; $F_{La,b}(k, t)$ is the spectral representation of $B_{La,b}$ in accordance with Eq. (4.94).

4.9 The Local Structure of Fully Developed Turbulence

The concept of isotropic turbulence introduced in the previous section is a mathematical idealization that has little to do with real turbulent flows. Yet it would be a mistake to think that it has no practical importance. In the present section, we are going to introduce the concept of local isotropic turbulence, which makes it possible to examine the local structure of the turbulent flow with rather simple methods and, above all, has direct application to real turbulent flows at very high Reynolds numbers, that is, at $\mathrm{Re} \gg \mathrm{Re}_{cr}$ [12–15]. We reserve the term "developed turbulent flow" for this turbulent flow regime. It remains to note that it is precisely such flows that present the greatest practical interest in practical applications.

A distinguishing feature of developed turbulence is the presence of fluctuational motions with various amplitudes that get superimposed on the averaged flow described by the velocity U. To describe turbulent fluctuations, one has to specify not only the absolute velocity values, but also the distances at which velocity can change noticeably. Such distances are called "motion scales" or "scales of eddies". The latter notion eludes precise definition, but for all practical purposes, it is acceptable to imagine a region of size λ, within which the turbulent motion is localized. In the subsequent discussion, the term "motion scale" will be understood to refer to just such a region. The most rapid fluctuational motions have the largest motion scales. Their velocities are equal by the order of magnitude to the average flow velocity U and their motion scales – to the characteristic linear scale L of the flow. For example, if the fluid is flowing inside a pipe, then U is the average flow rate velocity and L is the diameter of the pipe. Such fluctuation are called "large-scale". There also exist small-scale fluctuations. Before we define the meaning of "small scale", let us remind you that in principle, the size of fluctuations can be as small as desired – up to the mean free path of a molecule. However, fluctuations that have a very small scale give rise to extremely large velocity gradients, which in their turn invoke strong forces of viscous friction, causing a very rapid decay of such fluctuations. Hence, the size of fluctuations should be bounded from below by some scale λ_0 (we will talk about this motion scale later on). Fluctuations with scales $\lambda \ll L$ are defined as small-scale fluctuations. Small-scale fluctuations of the size $\lambda \sim \lambda_0$ are accompanied by considerable energy dissipation, with subsequent conversion of energy into heat. Finally, there is an intermediate region with scales $\lambda_0 \ll \lambda \ll L$. Hence, the entire spectrum of motion scales can be divided into three regions: the energy region $\lambda \sim L$, the inertia region $\lambda_0 \ll \lambda \ll L$, and the viscous dissipation region $\lambda \leq \lambda_0$. To be sure, this classification is very inexact because it is impossible to establish sharply defined boundaries between these regions.

It turns out that small-scale perturbations in a turbulent flow with a very high Reynolds number can be regarded as isotropic, and it is just this property of developed turbulence that we call local isotropy. This statement is based on the following qualitative model of developed turbulence. According to this scheme, developed turbulence consists of a set of disordered perturbations (eddies) that differ from each other by their scale λ and velocity u_λ. As we gradually increase Re, the fluid flow accomplishes a transition from a laminar to a turbulent flow, and then, at a further increase of Re – to developed turbulence. Perturbations of different scales do not appear at the same moment. First, when Re becomes larger than Re_{cr}, large-scale fluctuations emerge. As Re keeps increasing, these fluctuations give birth to smaller-scale perturbations, transferring to them a part of their kinetic energy. Those perturbations, in their turn, give birth to even smaller perturbations, and so on. Eventually we get the entire spectrum of fluctuations, where each perturbation gets its kinetic energy from its larger-scale "parent". Perturbations can disintegrate because of their instability. Indeed, each perturbation (fluctuation) is characterized by its own Reynolds number $\mathrm{Re}_\lambda = u_\lambda \lambda / \nu_e$. For the largest fluctuations, the Reynolds number is equal by the order of magnitude to the Reynolds number of the bulk flow, and since $\mathrm{Re} \gg \mathrm{Re}_{cr}$, large fluctuations are unstable and disintegrate into smaller-scale fluctuations. The Reynolds number of these newly-generated fluctuations is still too large, so they too disintegrate into smaller fluctuations, and so on. The chain of ever-smaller fluctuations continues until the scale of resulting fluctuations approaches λ_0. This scale corresponds to the Reynolds number $Re_{\lambda_0} \sim 1$ and is called the inner (or Kolmogorov) scale of turbulence. Motions whose scale is λ_0 or less are hydrodynamically stable and do not disintegrate. For such fluctuations, viscous friction forces are essential. The energy of such fluctuations eventually dissipates into heat.

Hence, instability of the averaged motion leads to a continuous flow of energy over the spectrum of fluctuations – from large-scale fluctuations to the fluctuations of minimum scale – with subsequent conversion into heat. In order to for developed turbulence to be sustainable, one has to continuously supply the averaged motion with energy from an external source. It is easy to see that the average specific dissipation energy $\bar{\varepsilon}$ (average amount of energy per unit mass per unit time), is an important parameter characterizing the intensity of developed turbulence.

The averaged fluid flow is generally inhomogeneous, anisotropic, and nonstationary. Because of the random character of energy transfer from large-scale motions to small-scale ones, the orientating influence of the averaged flow will have less and less effect on statistical characteristics of fluctuations as the scale decreases. It is therefore quite natural to assume that in the case of developed turbulence, all perturbations except the largest ones are isotropic. The change of the average flow velocity $\langle \boldsymbol{u} \rangle = \boldsymbol{U}$ with distance becomes noticeable only for distances of the order L. The distance has to be that large in order for inhomogeneity to affect the average flow velocity. Therefore inhomogeneity is only important for large-scale fluctuations, but does not affect small-scale fluctuations. Hence, the second assumption boils down to that of statistical homogeneity of small-scale fluctuations. As the fluctuation scale λ decreases, so does its characteristic period $t_\lambda = \lambda/u_\lambda$. For small-scale fluctuations it becomes much shorter than the characteristic time $t_L = L/U$ during which the averaged flow remains non-stationary. In other words, the change of the average velocity that is responsible for non-stationary character of the flow takes much longer than the change of statistical characteristics of small-scale fluctuations. Therefore small-scale perturbations can be regarded as stationary, or, more precisely, quasi-stationary. Recall that a quasi-stationary flow defined as a flow whose parameters do not explicitly depend on time, while the flow itself does change with time because of its dependence on the integral characteristics of the flow.

Hence, the outlined mechanism of developed turbulence leads us to the logical assumption that the statistical regime of small-scale fluctuations (i.e., the ones with length scale $\lambda \ll L$ and time scale $t_\lambda \ll T_L$) will be stationary, homogeneous, and isotropic over sufficiently small spacetime regions. This assumption forms the basis of the theory of local isotropic turbulence, which was first formulated by Kolmogorov. Though this assumption cannot be proved rigorously, many functional dependences that follow from the theory of local isotropic turbulence have been confirmed by numerous experiments.

We shall now proceed to determine the general qualitative characteristics of developed turbulent flow, keeping in mind what we have just said about the pattern of developed turbulence and using some dimensionality consideration. We begin by considering large-scale fluctuations. In accordance with the preceding discussion, large-scale fluctuations are characterized by the following parameters: characteristic external integral length scale, equal by the order of magnitude to the characteristic length scale of the averaged flow L; characteristic velocity change ΔU of the most rapid fluctuations on the distance of equal to the scale of fluctuations $\lambda \sim L$ (ΔU has the same order of magnitude as U); specific dissipation of energy $\bar{\varepsilon}$, equal to

$$\bar{\varepsilon} = \frac{\varepsilon}{\rho_e} = \frac{1}{2}\frac{\mu_e}{\rho_e}\sum_{i,j}\left\langle\left(\frac{\partial u_i}{\partial X_j} + \frac{\partial u_j}{\partial X_i}\right)^2\right\rangle, \tag{4.100}$$

and fluid density ρ_e. Since for large-scale fluctuations, the Reynolds number is large, $\mathrm{Re} \gg 1$, the coefficient of molecular viscosity μ_e is not included in the list of characteristic parameters. Nevertheless, energy dissipation does take place, and by analogy with the formula (4.100), it should be characterized by the coefficient of turbulent viscosity μ_t. Since the expression inside the brackets in Eq. (4.100) has the same order of magnitude as $(\Delta U)^2/L^2$, we have

$$\varepsilon \sim \mu_t \frac{(\Delta U)^2}{L^2}. \tag{4.101}$$

On the other hand, in view of dimensionality considerations, the quantities ε and $\bar{\varepsilon}$ should be expressed through dimensional parameters L, ΔU, and ρ. Therefore,

$$\varepsilon \sim \frac{\rho(\Delta U)^3}{L}, \quad \bar{\varepsilon} \sim \frac{(\Delta U)^3}{L}. \tag{4.102}$$

Thus Eqs. (4.101)–(4.102) give us the dynamic and kinematic turbulent viscosities:

$$\mu_t \sim \rho \Delta U L. \tag{4.103}$$

$$\nu_t = \frac{\mu_t}{\rho} = \Delta U L. \tag{4.104}$$

The ratio of molecular and turbulent viscosities is equal to

$$\frac{\nu_e}{\nu_t} \sim \frac{\nu_e}{\Delta U L} \sim \frac{1}{\mathrm{Re}} \ll 1. \tag{4.105}$$

So the coefficient of turbulent viscosity is much larger than the coefficient of molecular viscosity.

The pressure change is approximately (i.e., by the order of magnitude) equal to

$$\Delta p \sim \rho(\Delta U)^2. \tag{4.106}$$

Let us go on to small-scale fluctuations with the scale $\lambda \ll L$. We begin with the inertia region $\lambda_0 \ll \lambda \ll L$ where the motion can be considered as non-viscous. The velocity u_λ of fluctuations having the scale λ does not depend on μ_e or on the external parameters L and ΔU, because $\lambda \ll L$. Therefore u_λ can depend only on ρ, λ, and $\bar{\varepsilon}$ (or ε). The only combination of these quantities that has the dimensionality of velocity is $(\bar{\varepsilon}\lambda)^{1/3} = (\varepsilon\lambda/\rho)^{1/3}$. Substituting ε from Eq. (4.102) into this formula, we get

$$u_\lambda \sim (\bar{\varepsilon}\lambda)^{1/3} = \Delta U \left(\frac{\lambda}{L}\right)^{1/3}. \tag{4.107}$$

We see from Eq. (4.107) that the change of fluctuation velocity on a small distance λ is proportional to $\lambda^{1/3}$. This principle is known as the Kolmogorov–Obukhov law. It can be represented in a spectral form ("spectral" here refers to the spatial spectrum) by assigning to each fluctuation its wave number $k \sim 1/\lambda$ instead of λ and the kinetic energy $E(k)dk$ per unit mass contained in fluctuations with wave numbers in the interval $(k, k + dk)$. Since the dimensionality of $E(k)$ is $\mathrm{m^3/s^2}$, we should compose from the parameters $\bar{\varepsilon}$ and k a combination that would have this dimensionality:

$$E(k) \sim \bar{\varepsilon}^{2/3} k^{-5/3}. \tag{4.108}$$

Integrating (4.10) over k from k to ∞, one gets the total kinetic energy contained within fluctuations whose scale is $\leq \lambda$:

$$\int_k^\infty E(k)dk \sim \frac{\bar{\varepsilon}^{2/3}}{k} \sim (\bar{\varepsilon}\lambda)^{2/3} \sim u_\lambda^2.$$

Then u_λ^2 is equal by the order of magnitude to the total kinetic energy contained within such fluctuations. If we define the coefficient of turbulent viscosity as $\nu_t \sim \lambda u_\lambda$ by analogy with the formula (4.104), the relation (4.107) will take the form

$$\bar{\varepsilon} \sim \frac{u_\lambda^3}{\lambda} \sim \nu_t \left(\frac{u_\lambda}{\lambda}\right)^2. \tag{4.109}$$

Let us introduce the characteristic period of fluctuations, $t_\lambda = \lambda/u_\lambda$ and determine the order of velocity change Δu_t at a given point in space during the time t_λ that is small compared to the characteristic external time $t_L \sim L/U$. The presence of the averaged flow leads to conclude that after the time t_λ any arbitrary point in space will be filled with the fluid that initially was separated from this point by the distance Ut_λ. Therefore Δu_t can be derived from the formula (4.131), in which λ should be replaced by Ut_λ:

$$\Delta u_t \sim (\bar{\varepsilon} U t_\lambda)^{1/3}. \tag{4.110}$$

One should distinguish the quantity Δu_t from the change of velocity $\Delta u_t'$ of a given volume element of the fluid (fluid particle) moving in space. Since the latter depends only on the parameters $\bar{\varepsilon}$ and t_λ, we must compose a combination of this parameters that has the dimensionality of velocity dimension:

$$\Delta u_t' \sim (\bar{\varepsilon} t_\lambda)^{1/2}. \tag{4.111}$$

It is readily seen that the change of velocity of a moving fluid particle is proportional to $\Delta u_t \sim t_\lambda^{1/2}$, while the change of velocity of the fluid at a given point in space obeys another law: $\Delta u_t \sim t_\lambda^{1/3}$, so when $t_\lambda \ll T$, we have $\Delta u_t' \ll \Delta u_t$. Now the formulas (4.107) and (4.110) can be represented as

$$\frac{u_\lambda}{\Delta U} \sim \left(\frac{\lambda}{L}\right)^{1/3}, \quad \frac{u_t}{\Delta U} \sim \left(\frac{t_\lambda}{T}\right)^{1/3}. \tag{4.112}$$

The form of these relations shows that characteristics of small-scale fluctuations in different developed turbulent flows differ from each other only by their length and velocity (or length and time) scales. This statement forms the essence of the self-similarity property of the local isotropic turbulence.

Let us determine the distance λ_0 at which viscous effects become significant. As we noted earlier, this distance corresponds to the local Reynolds number

$\mathrm{Re}_\lambda = u_\lambda \lambda / \nu_e \sim 1$. Substituting the relation (4.197), we write

$$\mathrm{Re}_\lambda = \frac{\Delta U \lambda^{4/3}}{\nu_e L^{1/3}} \sim \mathrm{Re}\left(\frac{\lambda}{L}\right)^{4/3} \sim 1, \tag{4.113}$$

where Re is the Reynolds number of the average flow. This condition yields λ_0:

$$\lambda_0 \sim \frac{L}{\mathrm{Re}^{3/4}} \sim \left(\frac{\nu_e^3}{\bar{\varepsilon}}\right)^{1/4}. \tag{4.114}$$

The characteristic velocity and characteristic time of such fluctuations are obtained from Eq. (4.107):

$$u_{\lambda_0} \sim \frac{\Delta U}{\mathrm{Re}^{1/4}} = \lambda_0 \left(\frac{\bar{\varepsilon}}{\nu_e}\right)^{1/2}, \quad \tau_{\lambda_0} = \frac{\lambda_0}{u_{\lambda_0}} \sim \left(\frac{\nu}{\bar{\varepsilon}}\right)^{1/2}. \tag{4.115}$$

The scales λ_0 and τ_{λ_0} are respectively known as the Kolmogorov (or inner) spatial and temporal microscales. The values of λ_0 and u_{λ_0} decrease with increase of the Reynolds number of the average flow.

At $\lambda \leq \lambda_0$, the motion of the fluid has viscous character. Turbulent fluctuations do not vanish suddenly; instead, they gradually decay subject to viscous forces. Since velocity changes rather smoothly in this region, it can be expanded in Taylor series over the powers of λ. Let us keep only the first term of the series $u_\lambda \sim const\ \lambda$ and determine the constant from the condition $u_\lambda \sim u_{\lambda_0}$ at $\lambda \sim \lambda_0$. Then

$$u_\lambda \sim \frac{u_{\lambda_0}}{\lambda_0}\lambda \sim \frac{\Delta U}{L}\lambda \mathrm{Re}^{1/2}. \tag{4.116}$$

Scales of turbulent fluctuations are function as their spatial characteristics. In addition to those, we may consider the time characteristics of fluctuations, namely, the frequencies ω_λ. The whole frequency spectrum can be divided into three intervals. The lower end of the spectrum, $\omega_L \sim U/L$, corresponds to the energy region; the upper end, $\omega_{\lambda_0} \sim U/\lambda_0 \sim URe^{3/4}/L$, corresponds to the dissipation region; and the intermediate interval $\omega_L \ll \omega_\lambda \ll \omega_{\lambda_0}$ – to the inertia region. The inequality $\omega_\lambda \gg \omega_L$ means that the external (average) flow can be considered as stationary with respect to the local properties of small-scale fluctuations.

Energy distribution over the frequency spectrum in the inertia region is derived from Eq. (4.108) by replacing k with ω_λ / U:

$$E(\omega) \sim (U\bar{\varepsilon})^{2/3}\omega_\lambda^{-5/3}. \tag{4.117}$$

The frequency ω_λ defines the repetition period of velocity at a fixed point in space. Together with ω_λ, we can introduce another frequency ω'_λ, which stands for the repetition period of velocity of a chosen fluid particle. The distribution of energy over the frequency spectrum for such particles does not depend on U but depends only on $\bar{\varepsilon}$ and ω_λ. We conclude from dimensionality considerations that

$$E(\omega'_\lambda) \sim \frac{\bar{\varepsilon}}{\omega'_\lambda}. \tag{4.118}$$

We now apply the results obtained for small-scale fluctuations to estimate the velocity of inertialess particles buoyant in the fluid. Turbulent mixing causes particles to gradually move away from each other. Consider two particles such that the initial interparticle distance does not exceed the size of fluctuations from the inertia region. We make this requirement because otherwise large fluctuations would just transport the two particles without changing the interparticle distance. Our assumption allows us to find the rate of change of interparticle distance δ from the equation

$$\frac{d\delta}{dt} \sim u_\lambda \sim (\bar{\varepsilon}\delta)^{1/3}. \tag{4.119}$$

By solving this equation at a given initial value of the interparticle distance δ_0, we find the time it takes for the two particles to move away from each other so that the gap between them reaches the value δ_1. In the limiting case $\delta_1 \gg \delta_0$ this time is

$$t \sim \frac{\delta_1^{2/3}}{\bar{\varepsilon}^{1/3}}. \tag{4.120}$$

Now consider the correlations of velocity differences of two neighboring particles at a fixed instant of time. These correlations were introduced as structure functions in Section 4.8. Even the formula (4.107) gives a qualitative correlation of velocities at two points separated by a distance $\lambda \ll L$, in other words, it provides a connection between velocity values at two neighboring points. Components of the correlation tensor $b_{ik}(\mathbf{r}, t)$ serve as quantitative characteristics of this correlation. In an isotropic vector field, these components depend on two scalar functions – longitudinal and transverse functions $b_{NN}(r, t)$ and $b_{LL}(r, t)$, where $r = |\mathbf{r}_2 - \mathbf{r}_1|$ is the length of the radius vector between the points $\mathbf{r}_1$ and $\mathbf{r}_2$. In the case of local isotropic turbulence, we have $\lambda_0 \leq r \ll L$.

The change of velocity at small distances is caused by small-scale fluctuations and is independent of the average flow. Therefore our analysis of correlation and structure functions can be simplified if we assume that isotropy and homogeneity take place not only at small scales, but at large scales as well. Then the average velocity can be taken to be zero (see Section 1.10), and we can take advantage of the relations between the functions b_{LL} and b_{NN} that have been established in Section 4.8.

Because of Eq. (4.107), the difference of velocities over the distance r in the inertia region is proportional to $r^{1/3}$, therefore b_{LL} and b_{NN} are proportional to $r^{2/3}$. In other words, in any turbulent flow with a sufficiently high Reynolds number, the root-mean-square value of the difference of velocities at two points separated by the distance r (where r is neither too small nor too large) should be proportional to $r^{2/3}$.

This law, which was first established by Kolmogorov, is one of the most important laws describing turbulent flows and is called "the law of two thirds".

Switching to the spectral form, we can formulate a similar law for the energy spectrum:

$$E(k) \sim \bar{\varepsilon}^{2/3} k^{-5/3}, \tag{4.121}$$

which is called "the law of five thirds".

Let us now obtain the connection between b_{LL} and b_{NN} in the inertia region. First we transform Eq. (4.90) to the form

$$b_{NN} = \frac{1}{2r}\frac{d}{dr}(r^2 b_{LL}). \tag{4.122}$$

Recalling that both b_{NN} and b_{LL} are proportional to $r^{2/3}$, we can write

$$b_{NN} = 4b_{LL}/3, \quad (\lambda_0 \ll r < \ll L). \tag{4.123}$$

In the dissipation region ($\lambda \le \lambda_0$), the velocity difference at two neighboring points is proportional to r, as follows from Eq. (4.116). Then b_{LL} and b_{NN} are proportional to r^2 and the formula (4.122) reduces to

$$b_{NN} = 2b_{LL}, \quad (r \le \lambda_0). \tag{4.124}$$

The longitudinal and transverse functions b_{LL} and b_{NN} for small-scale fluctuations can be expressed in terms of specific dissipation of energy:

$$b_{NN} \sim \frac{2\bar{\varepsilon}}{15\nu_e} r^2, \quad b_{LL} \sim \frac{\bar{\varepsilon}}{15\nu_e} r^2. \tag{4.125}$$

The above-considered case corresponds to the situation when the average fluid flow is absent, for example, the fluid has been subjected to intensive shaking and then left alone. Such motion decays with time, and small-scale fluctuations decay in accordance with the power law

$$u_\lambda \sim t^{-5/4}. \tag{4.126}$$

The adduced statistical characteristics have been examined using only considerations of similarity and dimensionality, which do not invoke any hydrodynamic equations. The main conclusion is that the statistical regime of small-scale components of turbulence at high Reynolds numbers is quite independent from the peculiarities of the macroscopic structure of the flow, which can affect only the value of $\bar{\varepsilon}$. Therefore dynamic equations for the characteristics of locally isotropic turbulence do not depend on the character of large-scale motions, and it is sufficient to consider the case of isotropic turbulence in unbounded space and find the connections between its local characteristics. The obtained characteristics will then be the same for all turbulent flows sharing the same values of $\bar{\varepsilon}$ and ν_e if $Re \gg 1$. Hence, all the relations given above are universal for any locally isotropic turbulent flow.

Note that in the case of local isotropy the system of dynamic equations is also unclosed, and we need additional hypothesis and relations to close such a system.

Our analysis of the local structure of the velocity vector field can be repeated for a scalar field of passive impurity concentration $C(\boldsymbol{X}, t)$. Consider a developed turbulent flow of some fluid containing a passive impurity; the impurity does not influence the turbulent flow of the carrier fluid. An intense mixing of fluid volumes with different impurity concentrations occurs in a developed turbulent flow. Under the action of fluctuations with different scales, there occurs mixing of both small volumes (microvolumes) and relatively large volumes (macrovolumes or, to use a different term, moles). As was shown earlier, small-scale perturbations in a developed turbulent flow can be considered stationary and isotropic, that is, locally isotropic. It is natural to expect that perturbations of the field of concentrations in a small regions or space will also be stationary and isotropic, in other words, that the scalar field $C(\boldsymbol{X}, t)$ in such regions will be scalar isotropic.

Gradual disintegration of fluctuations – starting from the largest-scale ones (whose size is equal by the order of magnitude to the characteristic linear size L of the flow region) and all the way down to the Kolmogorov microscale λ_0 – is the underlying process responsible for the formation of velocity fluctuation spectrum. Of all characteristics of large-scale motions, only specific energy dissipation $\bar{\varepsilon}$ has an effect on small-scale motions. The same reasoning can be applied to the field of concentrations by replacing the Reynolds number $Re = \Delta U\, L/\nu_e$ with the diffusion Peclet number $Pe_D = L\delta U/D_m$, assuming $Pe_D \gg 1$ and, of course, keeping the condition $Re \gg 1$. Here L is the characteristic linear scale of change of average concentration $\langle C(\boldsymbol{X})\rangle$; δU is the change of average velocity on the distance L; D_m is the coefficient of molecular diffusion. If $L > \mathrm{L}$, we should take ΔU as our δU.

Because of their instability, large-scale fluctuations of concentration will rise to smaller and smaller fluctuations – all the way down to the minimum fluctuation, which has the inner concentration scale $\lambda_c^{(0)}$. Using the same reasoning as for the vector field, we arrive at the statement that in a spatial region of scale $\lambda_c \ll \mathrm{L}$, the field of concentrations will be locally isotropic so that the average concentration $\langle C\rangle$ can be considered constant. The degree of concentration inhomogeneity in these regions is given by the parameter characterizing the change of concentration fluctuations $C' = C - \langle C\rangle$. The meaning of this parameter is analogous to specific energy dissipation $\bar{\varepsilon}$, which is determined by the velocity gradient $\nabla \boldsymbol{u} = \partial u_i/\partial X_j$ (see Eq. (4.100)) rather than velocity fluctuations. Therefore it is quite natural to assume that a quantity similar to (4.100), namely,

$$\langle N\rangle = D_m\left\langle (\nabla C')^2\right\rangle. \tag{4.127}$$

will serve as a measure of concentration inhomogeneity.

This parameter is called dissipation of concentration inhomogeneity. Since we have $\langle C\rangle = const$ for small-scale ($\lambda_c \ll L$) fluctuations of concentration, the formula

(4.128) can be rewritten as

$$\langle N \rangle = D_m \left\langle (\nabla C)^2 \right\rangle. \tag{4.128}$$

As far as there are two characteristic length scales L and L, let us introduce $L_0 = \min(\mathrm{L}, L)$ and divide the entire spectrum of concentration fluctuations into two intervals: the interval of large-scale fluctuations with $\lambda_c \sim L_0$ and the interval of small-scale fluctuations with $\lambda_c \ll L_0$. For the first interval, the characteristic quantities are the length scale L_0, the change of average velocity $\Delta_{L_0} U$, and the change of average concentration $\Delta_{L_0}\langle C \rangle$. We can build a combination having the dimensionality of $\langle N \rangle$ from these parameters:

$$\langle N \rangle \sim \frac{\Delta_{L_0} U (\Delta_{L_0} \langle C \rangle)}{L_0}. \tag{4.129}$$

In practice, characteristic length scales L and L are equal by the order of magnitude, that is, $L_0 = L \sim \mathrm{L}$. Therefore we can introduce the coefficient of turbulent diffusion D_t by analogy with the formula (4.101):

$$\langle N \rangle = D_t \left(\frac{\Delta \langle C \rangle}{L} \right)^2. \tag{4.130}$$

Comparing the relations (4.129), (4.131) with (4.101), (4.104) and keeping in mind that $L \sim \mathrm{L}$, we get:

$$D_t \sim L \Delta U \sim \nu_t. \tag{4.131}$$

Thus the coefficients of turbulent diffusion and turbulent viscosity have the same order of magnitude.

Now consider the small-scale interval $\lambda_c \ll L_0$. The velocity field of small-scale perturbations is characterized by two dimensional parameters: specific dissipation of energy $\bar{\varepsilon}$ and kinematic viscosity coefficient ν_e. When examining the consideration field in this region, we should bring in two more parameters – the coefficient of molecular diffusion D_m, which enters the diffusion equation for the impurity,

$$\frac{\partial C}{\partial t} + u_{\mathrm{k}} \frac{\partial C}{\partial X_{\mathrm{k}}} = D_m \Delta C \tag{4.132}$$

and the dissipation of concentration inhomogeneity $\langle N \rangle$. The ratio between the convective term (second summand on the left-hand side) and the diffusional term (right-hand side) in Eq. (4.132) is equal by the order of magnitude to the diffusion Peclet number $\mathrm{Pe}_D = \Delta_{L_0} L_0 / D_m$. Molecular diffusion plays a considerable role only for $\mathrm{Pe}_D < 1$. For the averaged concentration, the Peclet number is usually $\mathrm{Pe}_D \gg 1$, so the effect of molecular diffusion is negligible for large-scale perturbations of concentration, and turbulent diffusion emerges as the main mechanism behind macroscopic mixing of regions with different impurity concentrations. Since $\mathrm{Pe}_D \sim \lambda_c$, a smaller scale of concentration fluctuations means a smaller Pe_D. When $\lambda_c = \lambda_c^{(0)}$, the Peclet number is $\mathrm{Pe}_D = 1$, and when $\lambda_c \leq \lambda_c^{(0)}$, the inequality $\mathrm{Pe}_D \leq 1$ is satisfied. Thus the interval of small-scale concentration fluctuations contains two subintervals: the convective interval $\lambda_c^{(0)} \ll \lambda_c \ll < L$, for which $\mathrm{Pe}_D \gg 1$, and the dissipation interval $\lambda_c \leq \lambda_c^{(0)}$, for which $\mathrm{Pe} \leq 1$.

In the convective interval, molecular diffusion does not play any noticeable role, therefore the parameter D_m does not figure among its governing parameters. The governing parameters in this interval are ε, ν_e, and $\langle N \rangle$. In the dissipative interval, where molecular diffusion plays a noticeable role, the governing parameters will include $\bar{\varepsilon}$, ν_e, D_m, and $\langle N \rangle$. Two combinations with the dimensionality of length can be built from these four parameters: the inner scale of turbulence (Kolmogorov scale)

$$\lambda_0 = \left(\frac{\nu_e^3}{\bar{\varepsilon}}\right)^{1/4} = \frac{L}{\mathrm{Re}^{3/4}}$$

and the inner scale of diffusion (Batchelor scale)

$$\lambda_c^{(0)} = \lambda_b = \left(\frac{\lambda_e D_m^2}{\bar{\varepsilon}}\right)^{1/4} = \lambda_0 Sc^{-1/2}, \tag{4.133}$$

where $\mathrm{Sc} = \lambda_e / D_m$ is the Schmidt number. It should be noted that for fluids, $\mathrm{Sc} \sim 10$, whereas for infinitely dilute solutions, $\mathrm{Sc} \sim 10^3$.

By its physical meaning, the convective interval should comprise those concentration scales for which molecular diffusion is negligibly small as compared to convection. But the presence of two quantities (ν_e and D_m) of the same dimensionality leads to the appearance of a new dimensionless number – the Schmidt number Sc – and to the dependence of the lower end of the interval upon Sc in accordance with Eq. (4.133). Therefore the condition $\lambda \gg \lambda_c^{(0)}$ alone does not guarantee that the fluctuation belongs to the convective interval. To determine where the intervals where convection or molecular diffusion dominates, one should compare the transport coefficients ν_e and D_m. If they are of the same order of magnitude, then $\mathrm{Sc} \sim 1$ and the length scales λ_0 and λ_b are roughly the same. The cases of $\mathrm{Sc} > 1$ and $\mathrm{Sc} < 1$ require additional study. As long as $\mathrm{Sc} \gg 1$, we have $\lambda_b \ll \lambda_0$, in other words, there exists within the viscous interval a visco-diffusional interval where molecular diffusion plays a significant role.

In conclusion, we shall give an approximate expression for the structure function of the concentration field $d_{cc}(r) = \langle [C(\boldsymbol{X} + \boldsymbol{r}) - C(\boldsymbol{X})]^2 \rangle$. For small-scale fluctuations of concentration, the dimensionality theory suggests the expression

$$d_{cc}(r) = \langle N \rangle (\bar{\varepsilon})^{-1/2} D_m^{1/2} F\left(\frac{r}{\lambda_b}, \frac{\nu_e}{D_m}\right). \tag{4.134}$$

In the dissipation region, at $r \ll \lambda_b$, there exists the following representation:

$$d_{cc}(r) \approx \frac{\langle N \rangle}{3D_m} r^2. \tag{4.135}$$

In convective interval, at $L \gg r \gg \lambda_b$, the structure function is given by

$$d_{cc}(r) \approx \frac{\langle N \rangle}{3(\bar{\varepsilon})^{1/3}} r^{2/3}. \tag{4.136}$$

The formula (4.136) is called "the law of two thirds" for the concentration field. The spectral "law of five thirds" for a local isotropic concentration field has the form

$$E_c(\mathrm{k}) \approx \frac{\langle N \rangle}{(\bar{\varepsilon})^{1/3} \mathrm{k}^{5/3}}. \tag{4.137}$$

4.10 Turbulent Flow Models

It was shown in Section 9.6 that the system of Reynolds equations describing a turbulent fluid flow is unclosed, because the number of unknowns is greater than the number of equations. Attempts to close the system by adding equations for higher-order moments were unsuccessful because those additional equations contain new moments of higher order. Therefore neither the Reynolds equations on their own, nor a system of Reynolds equations plus equations for higher-order moments (e.g., the energy equation discussed in Section 4.7), nor the simplified equations for isotropic (see Section 4.8) or locally isotropic turbulence (see Section 4.9) can be solved. All they can do for us is to establish certain connections between different statistical characteristics of turbulence.

There exist several possible ways to close the system of Reynolds equations. The first way is to use experimental data to determine the functional connections between moments of some definite order and the lower-order moments. A second way is to deduce these connections from simple hypotheses that are well justified on physical grounds and are accurate up to some empirical constants. This method lies in the basis of all semi-empirical theories of turbulence. Finally, the third and currently the most widespread method is based on the use of transport equations for some characteristics of turbulence.

It should be emphasized that there are no universal relations that would be applicable to all turbulent flows. Each of the existing approximations is suitable only for some type of flows, for example, flows in tubes, boundary layer flows, jet flows, flows past a body, and so forth.

The present section offers a review of several models from which one can derive additional equations and thus close the system.

4.10.1
Semi-empirical Theories of Turbulence

The failure of the Reynolds equations (4.37), (4.39) to form a closed system is explained by the presence of new unknowns – correlations of velocity fluctuations that appear in the Reynolds stresses $\tau_{ij}^{(1)}$. The simplest way to close the Reynolds equations is to establish connections between the Reynolds stresses and the average hydrodynamic fields. Such methods are called local equilibrium algebraic methods and the corresponding relations are said to be of the gradient type.

The methods based on approximating the Reynolds stresses with the help of parameters determined by the average velocity profile in the given cross section are well-developed and widely used for calculating different flows. The range of application of these methods is limited to the turbulent flows whose turbulence characteristics in a given cross section do not depend on their distributions in the preceding cross sections. The main difficulty is to find the range of applicability of these methods. Some of the resulting connections are adduced below.

1. *The Boussinesq model*:

For simplicity's sake, we consider a stationary fluid flow in a flat channel in the absence of external forces ($f_i = 0$). The average velocity has one component $\langle u_x \rangle = U$ parallel to the channel wall and depending only on the transverse coordinate Z. Suppose that $u_x = U + u'$, $u_z = w'$, where u' and w' are fluctuations of the longitudinal and transverse velocity components. Then Eqs. (4.39) take the form

$$\frac{\partial \tau}{\partial z} = \frac{\partial \langle p \rangle}{\partial X}, \quad \rho_e \frac{\partial \langle w'^2 \rangle}{\partial Z} = -\frac{\partial \langle p \rangle}{\partial Z}, \tag{4.138}$$

where the stress τ is equal to

$$\tau = \rho_e \nu_e \frac{d\langle u \rangle}{dZ} - \rho \langle u'w' \rangle. \tag{4.139}$$

A new unknown function $\tau' = -\rho_e \langle u'w' \rangle$ appears in the equations. Thus, in order to close the set of equations (4.138) and (4.139), it is sufficient to express τ' through $U(Z)$. The Boussinesq hypothesis states that the following equality is valid:

$$-\rho_e \langle u'w' \rangle = \rho_e \nu_t \frac{dU}{dZ}, \tag{4.140}$$

where ν_t is a quantity with the dimensionality of viscosity; it is called the turbulent viscosity coefficient.

Strictly speaking, the equality (4.140) does not constitute a closure relation, because in order to determine the new unknown ν_t, one needs to have experimental data or to formulate a supplementary hypotheses. The simplest way out is to take $\nu_t = const$. But then, in a notable analogy with Eq. (4.44), the introduction of ν_t will be tantamount to replacing the fluid viscosity ν_e with $\nu_e + \nu_t$, and the Reynolds equations will be equivalent to equations for the laminar flow with a new viscosity coefficient. In this case the obtained velocity profile will be a parabolic Poiseuille profile, while it is well known that such turbulent flows have a logarithmic, rather than parabolic, velocity profile. We conclude that ν_t cannot be a constant; instead, it should be a function of Z.

Let us estimate the form of this function by using the dimensionality theory. Consider the flow near a flat wall. Let ΔU be the characteristic variation of velocity at the distance Z from the wall. Since no characteristic linear size has been assigned to our flow, we shall take Z as the characteristic linear size. The two governed parameters ΔU and Z can form the quantity

$$\nu_t = \Delta U Z \tag{4.141}$$

that has the dimensionality of viscosity. From other hand, in the vicinity of the wall, we can assume $\Delta U \sim Z dU/dZ$. Then the friction force per unit area of the wall is

$$\tau_f = \mu_t \frac{dU}{dZ} = \rho_e \nu_t \frac{dU}{dZ} \sim \rho_e Z^2 \left(\frac{dU}{dZ}\right)^2, \tag{4.142}$$

from which there follows

$$\frac{dU}{dZ} \sim \left(\frac{\tau_f}{\rho_e}\right)^{1/2} \frac{1}{Z}.$$

Since the value of τ_f at the wall has to be constant, we have

$$U \sim \left(\frac{\tau_f}{\rho_e}\right)^{1/2} \ln Z. \tag{4.143}$$

Thus, a simple estimation shows that ν_t decays linearly as we get closer to the wall and that the longitudinal velocity profile has a logarithmic form. Actually, the structure of the flow in the vicinity of the wall is more complex. A detailed analysis of the flow structure, which takes into account the transformation of the turbulent boundary layer into a viscous boundary layer, shows that in fact, in the region adjacent to the wall, ν_t decays much faster (never slower than z^3).

In spite of the fact that the assumption $\nu_t = const$ is inadmissible for turbulent flow inside pipes, there are some flows for which this simple model is acceptable, such as, for example, turbulent jet flows and turbulent flows in the open atmosphere. For such flows, ν_t should be considered as a parameter that varies for different flows and is determined from experiments.

We should also mention that a similar model can be applied to the problems involving heat or passive impurity propagation in a turbulent flow once we introduce the coefficient of turbulent thermal diffusion χ_t and the diffusion coefficient D_t (see Eq. (4.46) and Eq. (4.47).

2. *The Prandtl model*:

The model proposed by Prandtl is based on the concept of mixing length. Prandtl attributes a physical meaning to the quantity L in Eq. (4.104), taking his hint from the analogy between the turbulent flow and the random molecular motion in the kinetic theory of gases. According to this theory, viscosity ν_e is defined by the same formula where ΔU is the average velocity and L is the mean free path of molecules. The velocity of molecular motion is certainly much greater then the average velocity of a turbulent fluid flow, whereas the mean free path of molecules is much shorter than the scale of fluctuations, so that the product of these quantities gives the difference between ν_e and ν_t that is in good agreement with the formula (4.105).

Similarly to the exchange of momentum between molecules, in a turbulent momentum exchange, a finite fluid volume leaving the layer separated from the given layer by a certain distance conserves its average momentum until it reaches the given layer. There it mixes with the ambient fluid, transferring the entire momentum difference to this fluid. The average distance between the initial layer, from where the volume has started its journey, and the destination layer, where it mixes with the ambient fluid, is called the mixing length. That is why Prandtl's theory is sometimes called the mixing length theory.

As in the previous model, we shall consider a plane-parallel flow with the average velocity U along the X-axis. The Z-axis is perpendicular to the X-axis and is pointing in the upward direction. The adopted model says that the volumes coming from the lower layer $Z-l'$ and from the upper layer $Z+l'$ will reach the layer Z. If the mixing of the arriving volumes with the ambient fluid happens instantaneously, the volumes bring to the layer Z the same momentum which they held initially while inside the layers $Z-l'$ and $Z+l'$. Such an exchange will lead to the emergence of fluctuations of the transverse velocity w' which by their order of magnitude are equal to

$$w' \sim U(Z \mp l') - U(Z) \sim \mp l' \frac{dU(Z)}{dZ}. \tag{4.144}$$

We may now determine the friction force per unit area exerted on the layer Z by the upper and lower layers. If we designate the momentum from the upper layer as positive and the one from the lower layer – as negative, then

$$\tau_f = \rho_e \langle w'(U(Z \mp l') - U(Z)) \rangle \sim \mp \rho_e \langle w' l' \rangle \frac{dU(Z)}{dZ}.$$

Plugging in the expression (4.144) for w' and designating $l^2 = \langle l'^2 \rangle$, we obtain

$$\tau_f = \mp \rho_e l^2 \left(\frac{dU(Z)}{dZ} \right)^2.$$

Since τ_f should be a positive quantity, the latter formula can be represented as

$$\tau_f = \rho_e l^2 \left|\frac{dU(Z)}{dZ}\right| \frac{dU(Z)}{dZ}. \tag{4.145}$$

Now, similarly to Eq. (4.42), if we take $\tau_t = \rho\nu_t \, dU/dZ$ and use the relation (4.145), we have

$$\nu_t = l^2 \left|\frac{dU(Z)}{dZ}\right|. \tag{4.146}$$

For the problem of a flow in a plane channel near the wall, we can take $l \sim Z$. Then, using the formulas (4.145) and (4.146), we obtain a logarithmic velocity profile near the wall but outside the viscous sublayer. However, in the region close to the symmetry axis of the channel, the obtained approach is unacceptable. In the latter case, it is better to take $l = const$.

In contrast to the Boussinesq model, the unknown parameter in the Prandtl model is the mixing length l, which depends on coordinates and must be obtained experimentally for any specific flow. Prandtl's model, as well as Boussinesq's one, is not applicable to all turbulent flows.

3. *The Taylor model*:

Taylor has suggested his model, known as the theory of eddy transport, in an attempt to properly account for the influence of pressure fluctuations on fluid particles. The theory is similar to Prandtl's in that it also uses the concept of mixing length, but unlike Prandtl, Taylor considers the mixing layer for the velocity vortex, and not for the momentum vortex.

Consider a two-dimensional flow with the average velocity $\langle \boldsymbol{u} \rangle = (U, W)$ in the (X, Z) plane. In this flow, the average vortex has only one component:

$$\langle \omega_y \rangle = \Omega = \frac{\partial U}{\partial Z} - \frac{\partial W}{\partial X}.$$

Let the average flow be parallel to the X-axis. Then $\langle \boldsymbol{u} \rangle = (X, 0)$ and the turbulent component of stress $\tau^{(1)} = \rho_e \langle u'w' \rangle$ obeys the momentum equation

$$\begin{aligned}\frac{\partial \tau}{\partial Z} &= -\frac{\partial}{\partial Z}\langle u'w' \rangle = -\left(\left\langle u'\frac{\partial w'}{\partial Z}\right\rangle + \left\langle w'\frac{\partial u'}{\partial Z}\right\rangle\right) \\ &= -\left\langle w'\left(\frac{\partial u'}{\partial Z} - \frac{\partial w'}{\partial Z}\right)\right\rangle + \frac{1}{2}\frac{\partial}{\partial X}(\langle u'2 \rangle - \langle w'2 \rangle).\end{aligned} \tag{4.147}$$

When deriving this equation, we used the continuity equation for velocity fluctuations,

$$\frac{\partial u'}{\partial Z} + \frac{\partial w'}{\partial Z} = 0.$$

Let the flow be uniform along the X-axis. Then all derivatives with respect to X are equal to zero and the equation (4.147) transforms to

$$\frac{\partial \tau}{\partial Z} = -\rho_e \left\langle w' \omega_y' \right\rangle. \tag{4.148}$$

We now introduce the mixing length l_1' for the velocity vortex through the relation similar to Eq. (4.144):

$$\omega_y' = l_1' \frac{\partial \Omega}{\partial Z}, \tag{4.149}$$

where $\Omega = \langle \omega_y \rangle = dU/dZ$. We also retain the expression (4.144) for transverse velocity fluctuations w':

$$w' = l' \frac{\partial U}{\partial Z}. \tag{4.150}$$

Then

$$\frac{\partial \tau}{\partial Z} = -\rho_e \left\langle w' \omega_y' \right\rangle = -\rho_e \langle l_1' l' \rangle \frac{\partial U}{\partial Z} \frac{\partial \Omega}{\partial Z} = \rho_e l_1^2 \frac{\partial U}{\partial Z} \frac{\partial^2 U}{\partial Z^2}, \tag{4.151}$$

where $l_1 = \left(-\langle l_1' l' \rangle\right)^{1/2}$ is the characteristic length that plays the same role in the Taylor model that l plays in the Prandtl model.

Taking $l_1 = const$, we get from Eq. (4.151)

$$\tau_f = \frac{1}{2} \rho_e l_1^2 \left(\frac{dU}{dZ}\right)^2. \tag{4.152}$$

This expression coincides with the formula (4.145) in the Prandtl theory if we take $l_1 = \sqrt{2}l$. In all other respects, the Taylor theory is different from Prandtl theory, and makes different predictions. For example, the velocity profile for the channel flow predicted by the Taylor theory is in good agreement with experimental data all the way to the central axis of the channel, in a stark contrast with predictions of the Prandtl theory.

As all other semi-empirical models, the Taylor model does not solve the closure problem entirely, because it reduces to a single empirical parameter – the vortex mixing length l_1. The main shortcoming of the Taylor theory is it limited range of application – it is suitable only for two-dimensional problems.

Finally, it is necessary to make the following note. The coefficient of turbulent viscosity ν_t in the Boussinesq model and the mixing lengths l and l_1 in the Prandtl and Taylor models have been introduced purely formally (albeit with some supporting physical rationalizations) for plane-parallel fluid flows in an attempt to describe the simplest fluid flows – in pipes, channels, and boundary layers. We still need to show now how these parameters can be introduced in the general case of an arbitrary spatial flow.

Let us suppose that turbulence emerges as a result of transition of a part of the average flow energy into small scale perturbations. Then, according to the energy balance equation (4.55), the inequality $A > 0$ should hold, where A is a term in the energy equation (see Eq. (4.56) describing the exchange of energy between the averaged motion and the fluctuational motion. Indeed, the condition $A > 0$ means that the turbulent energy density e_k at a given point increases at the expense of energy of the averaged flow. Then all statistical characteristics of turbulence, including the Reynolds stresses, should depend on the field of the average velocity.

The Reynolds stresses $\tau_{ij}^{(1)}$ play the same role with respect to the averaged motion as viscous forces with respect to the laminar flow. Therefore, when deformation of fluid particles is not taken into account, the averaged flow is similar to the motion of a rigid body, and the Reynolds stresses are pointing along the normal to any surface element arbitrarily selected within the fluid. Then the tensor $\rho\langle u_i' u_i' \rangle$ is isotropic and can be represented as a spherical component of the rate-of-strain tensor (see Eq. (4.4)):

$$\rho\left\langle u_i' u_j' \right\rangle = c\delta_{ij}, \quad c = \frac{1}{3}\rho\langle u_k' u_k' \rangle = \frac{2}{3}\rho e_k. \tag{4.153}$$

The turbulent energy ρe_k is similar to $-p$ in the incompressible liquid law (4.4).

In the general case that takes into account the deformation of fluid particles, the stresses $\tau_{ij} = \rho\left\langle u_i' u_j' \right\rangle$ depend on the derivatives of the average velocity with respect to coordinates. Since the tensor $\tau_{ij}^{(1)}$ is symmetrical, it depends on the rate-of-strain tensor E_{ij} (see Eq. (4.5)). In the case of small deformations, this dependence is linear and the proportionality coefficient has the meaning of turbulent viscosity coefficient, analogously to the Navier–Stokes law.

Let us now dwell on the analogy with the kinetic theory of gases. This theory holds that the coefficient of molecular viscosity is equal to $\nu_e \sim u_m l_m$, where u_m and l_m are, respectively, the average velocity and the mean free path of molecules. Suppose that a similar relation is true for the turbulent motion, with the root-mean-square value of velocity fluctuation functioning as our u_m, and the integral scale of turbulence – as l_m. In the Prandtl theory, this scale is the mixing length, which has the order of the integral scale of turbulence and, as we noted in Section 4.8, has the meaning of the average distance that turbulent fluctuations can travel. As we are concerned with spatial motions, the turbulence will be characterized by different scales assigned to different directions. The set of scales l_{ij} will then form a symmetric scale tensor. Now, using this tensor and taking advantage of the symmetry of the tensor $\rho\left\langle u_i' u_j' \right\rangle$, we can assume

$$\tau_{ij}^{(1)} = \rho\left\langle u_i' u_j' \right\rangle = \frac{2}{3}\rho e_k \delta_{ij} - \rho\sqrt{e_k}(l_{ik} E_{kj} + l_{jk} E_{ki}). \tag{4.154}$$

This formula was first suggested by Monin. It can be thought of as a generalization of the Boussinesq and Prandtl models.

Sometimes, in the first approximation, we can take

$$l_{ij} = l\delta_{ij}.$$

Then (4.154) takes the form

$$\tau_{ij}^{(1)} = \frac{2}{3}\rho e_k \delta_{ij} - \rho l \sqrt{e_k} E_{ij}.$$

Defining the coefficient of turbulent viscosity as

$$\nu_t = l\sqrt{e_k},$$

one gets

$$\tau_{ij}^{(1)} = \frac{2}{3}\rho e_k \delta_{ij} - 2\rho \nu_t E_{ij}. \qquad (4.155)$$

4.10.2
The Use of Transport Equations

Following the papers [16–18], we make a brief review of several models of turbulence that are based on transport equations for various statistical characteristics of turbulent flows. In these models, the minimum possible number of parameters that can be used to describe turbulence is equal to three. The turbulent stress, the energy of turbulence, and a third parameter which, when combined with the energy of turbulence, would result in a quantity having the dimensionality of length, is the most common choice of parameters. These models have been tested for problems that involve flows in channels and boundary layers. The most influential publications that shaped the development of this models are [19–22].

Numerous models tailored for different types of turbulent flows have been proposed as of today. All of them can be classified as one-, two-, and three- parametric according to the number of transport equations employed. If a model contains less then three transport equations, it means that this model includes some algebraic relations between various characteristics of turbulence. An increase of the number of transport equations complicates the problem considerably, as we are faced with the necessity to measure the constants involved and determine the range of applicability of the model. The models listed below, as well as the corresponding equations, are written in the approximation of a stationary plane boundary layer for a homogeneous incompressible fluid ($\rho = const$). The velocity $\boldsymbol{u}$ has two components, $u_1 = u$ and $u_2 = v$ along the X и Y coordinate axes.

1. *One-parametric models*:

These models use one equation for the turbulence energy $e_k = 0.5\sum\langle u_i'2\rangle$, for the Reynolds stress (shear stress) $\tau' = \tau_{12}^{(1)}/\rho = -\langle u'v'\rangle$, or for the turbulent viscosity ν_t.

a. Kolmogorov [19] was the first to suggest to use the equation for the turbulence energy. In the stationary plane boundary layer approximation, this equation has the following form:

$$\frac{De_k}{Dt} = \frac{\partial}{\partial Y}\left(D_E \frac{\partial e_k}{\partial Y}\right) + \tau' \frac{\partial \langle u \rangle}{\partial Y} - \bar{\varepsilon}, \tag{4.156}$$

where $D/D_t = \langle u \rangle \partial/\partial X + \langle v \rangle \partial/\partial Y$; $\bar{\varepsilon} = \nu_e \langle (\partial u_i'/\partial X_k)(\partial u_i'/\partial X_k) \rangle$ is the specific dissipation of energy; and D_E is the effective diffusion coefficient. The respective terms on the right-hand side of Eq. (4.156) describe the processes of diffusion, production, and dissipation of energy. The diffusional term is written in the gradient form, with the effective diffusion coefficient D_E. The parameters $\bar{\varepsilon}$ and D_E entering this equation were defined by Kolmogorov based on dimensionality considerations:

$$\bar{\varepsilon} = C_E L^{-1} e_k^{3/2}, \quad D_E = C_D L \sqrt{e_k}. \tag{4.157}$$

Making a correction to account for the molecular viscosity, we rewrite Eq. (4.157) as

$$\bar{\varepsilon} = C_E L^{-1} e_k^{3/2} + C_E' \nu_e e_k / L^2, \quad D_E = C_D L \sqrt{e_k} + \nu_e. \tag{4.158}$$

Here L is the integral scale of turbulence, and C_E, C_E', and C_D are empirical constants.

A shortcoming of this model is the need to specify the integral scale of turbulence L, which depends on the flow pattern.

b. Transport equation for the Reynolds stress $\tau' = -\langle u'v' \rangle$ is derived in [23]. For a plane-parallel flow of an incompressible fluid in a boundary layer, it has the following form:

$$\frac{D\tau'}{Dt} = a_1 \tau' \frac{\partial \langle u \rangle}{\partial Y} - a_1 \frac{\tau^{3/2}}{L} - a_1 \sqrt{\tau_m} - \frac{\partial}{\partial Y}(G\tau'). \tag{4.159}$$

The presence of the empirical functions L and G together with the empirical constants a_1 and τ_m is the main shortcoming of this model.

It should be noted that, in contrast to the model a), where the transport equation (4.156) is parabolic because of the first term on the right-hand side, Eq. (4.159) is hyperbolic.

c. Transport equation for turbulent viscosity ν_t was first proposed in [24] and later specified in more detail in [25]. For a plane boundary layer of incompressible fluid, it has the form

$$\frac{D\nu_t}{Dt} = \alpha \nu_t \left| \frac{\partial \langle u \rangle}{\partial Y} \right| - \gamma \frac{\nu_t(\nu + \beta \nu_t)}{s^2} + \frac{\partial}{\partial Y}\left((\nu_e + \kappa \nu_t) \frac{\partial \nu_t}{\partial Y} \right), \tag{4.160}$$

where s is the minimum distance from the wall; α, β, γ, and κ are empirical constants.

One-parametric models of turbulence use the transport equation to determine only one of the quantities characterizing the turbulent flow. As a rule, this quantity is either the energy of turbulence e_k (see Eq. (4.156) or the turbulent viscosity ν_t (see Eq. (4.160). Somewhat less common is the one-parametric model that uses the transport equation (4.159) for the Reynolds stress.

A serious disadvantage of one-parametric models is the necessity to specify the scale of turbulence L, which is not known beforehand and cannot be determined without additional hypotheses dependent on the type of the flow. For simple flows, the scale of turbulence can be determined through the governing parameters at a given point inside the flow. Thus, for a flow near a plane wall, L can be taken proportional to the distance between the given point and the wall, whereas for jet flows, L is usually taken proportional to the width of the jet. In complex turbulent flows, it is impossible to express L in terms of the governing parameters of the flow at a given point. Besides, in such flows the scale of turbulence, as well as other governing parameters, usually depends not only on their values at a given point but also on the entire prehistory of the flow (for example, on the conditions at the channel entrance plus the boundary conditions).

d. Transport equation for the scale of turbulence L was proposed in [22], where it was used to calculate the shape of the turbulent boundary layer on a flat plate. The equation for $F = L^2/2$ has the form

$$u\frac{\partial F}{\partial X}+w\frac{\partial F}{\partial Y}=\nu_e\frac{\partial^2 F}{\partial Y^2}-C_L\frac{\nu_t}{e_k}\left(\frac{\partial u}{\partial Y}\right)^2 F+C_F\left|1-\frac{2F}{s^2}\varphi\left(\frac{2F}{s^2}\right)\right|\frac{F^2}{e_k}\varepsilon_E,$$
$$w=v+0.5\nu_e\frac{\partial \ln F}{\partial Y}-(\nu_e+D_E)\frac{\partial \ln e_k}{\partial Y},\qquad (4.161)$$

where e_k is the energy of turbulence; s – distance from the plate; $\varepsilon = \varphi e_k^{3/2}/L$, C_L, C_F, and φ_1 – empirical constants.

2. *Two-parametric models*:

Turbulence models that use two transport equations to determine the characteristics of turbulence are called two-parametric models. The majority of such models involve a transport equations for the energy of turbulence e_k and a transport equation for the specific energy dissipation $\bar{\varepsilon} = \nu_e\langle(\partial u_i'/\partial X_k)(\partial u_i'/\partial X_k)\rangle$ or for the function $F = e_k^m L^n$.

The first two-parametric model was proposed in [19]. This work considers transport equations for e_k and the combination $\sqrt{e_k}/L$. The Reynolds stress is determined by the relation

$$\tau' = -\langle u'v'\rangle = C\sqrt{e_k}L\frac{\partial\langle u\rangle}{\partial Y}, \qquad (4.162)$$

where C is constant.

Paper [26] was the first to use a transport equation for $\bar{\varepsilon}$ in a two-parametric model. As of today, the most popular two-parametric models are ones that describe turbulent flows by two transport equations for the functions e_k and $\bar{\varepsilon}$. In many publications the energy of turbulence is denoted by k, hence the commonly used term "$k-\varepsilon$ models".

Consider two models: the $e-F$ model and the $k-\varepsilon$ model.

a. In addition to Eq. (4.156) for turbulent energy e_k, the two-parametric $e-F$ model contains the following equation for the function $F = e_k^m L^n$:

$$\frac{DF}{Dt} = \frac{\partial}{\partial Y}\left(D_F \frac{\partial F}{\partial Y}\right) - (C\sqrt{e_k}L + C_1 \nu_e)\frac{e_k}{L^2} + \gamma F \frac{\partial \langle u \rangle}{\partial Y} + \Psi, \tag{4.163}$$

where $D_F = a_F\sqrt{e_k}L + \alpha_F \nu$; $L = (e_k^m/F)^{1/n}$; a_F and α_F are empirical constants; γ is a function of τ', e_k, and $\partial\langle u\rangle/\partial Y$; Ψ is a function that depends on the sign of n: at $n < 0$, it is zero, while at $n > 0$, a special form of Ψ is required. The reason for such behavior of Ψ is that the first term on the right-hand side of Eq. (4.163) describes a diffusional process characterized by the diffusion coefficient D_F. For positive D_F, the maximum value of F must decrease with time, which happens only at $n < 0$. Therefore the case $n > 0$ requires the presence of such a function Ψ that D_F would not change its sign.
Accordingly, all $e-F$ models fall into two categories: those with $n > 0$ and those with $n < 0$. The models with $n > 0$ use the function $e_k L$ or $\sqrt{e_k}L$ for F. The models with $n < 0$, on the other hand, use one of the functions $\sqrt{e_k}/L$, e_r/L^2, and $e_k^{3/2}/L$.

b. $k-\varepsilon$ models use the transport equation for specific energy dissipation $\bar{\varepsilon}$ (see Eq. (4.100). In the plane boundary layer approximation, this equation has the form

$$\frac{D\bar{\varepsilon}}{Dt} = \frac{\partial}{\partial Y}\left(D_\varepsilon \frac{\partial \bar{\varepsilon}}{\partial Y}\right) - C_1 f_1 \frac{\bar{\varepsilon}^2}{e_k} + C_2 f_2 \frac{\bar{\varepsilon}}{e_k}\frac{\partial \langle u \rangle}{\partial Y} + \Psi. \tag{4.164}$$

Here $D_\varepsilon = a_\varepsilon\sqrt{e_k}L + \alpha_\varepsilon \nu_e$; C_1, C_2, a_ε, and α_ε are constants; f_1, f_2, and Ψ are functions that depend on the governing parameters.

Eq. (4.164) has some peculiarities, one of which has to do with the behavior of $\bar{\varepsilon}$ near the wall.

Two-parametric models rely on the following relation to determine the Reynolds stress:

$$\tau' = -\langle u'v' \rangle = C_\mu f_\mu(\mathrm{Re})\sqrt{e_k}L\frac{\partial \langle u \rangle}{\partial Y}. \tag{4.165}$$

The scale of turbulence L that appears in this formula is defined differently in different models. The purpose of the function $f_\mu(\mathrm{Re})$ is to describe the effect of viscosity on τ'. This dependence is not universal but varies with the type of the flow.

3. *Three-parametric models*:

A distinguishing feature of three-parametric models that sets them apart from other models is that transport equations are written for all characteristics of turbulence that are employed by the model. Instead of introducing turbulent viscosity to find the Reynolds stress, these models rely on the corresponding transport equation whose structure is similar to that of the transport equation for the energy of turbulence. Models of this type are sometimes called Reynolds stress models.

Three-parametric models include transport equations for the shear stress $\tau' = -\langle u'v' \rangle$, for the energy of turbulence $e_k = 0.5 \sum \langle u_i' 2 \rangle$, and for the parameter $F = e_k^m L^n$.

As far as equations for all three characteristics of turbulence have identical structure, they can be written in the general form

$$\frac{D\Phi}{Dt} = -D_\Phi \frac{\Phi}{L^2} + \gamma_\Phi \Gamma_\Phi \frac{\partial \langle u \rangle}{\partial Y} + \frac{\partial}{\partial Y}\left(D_\Phi^* \frac{\partial \langle u \rangle}{\partial Y} \right), \tag{4.166}$$

where Φ successively assumes the values e_k, τ', and $F = e_k^m L^n$. The values of D_Φ, γ_Φ, Γ_Φ, and D_Φ^* are different for each of these equations:

$$D_E = \alpha_E \sqrt{e_k} L + \beta_E \nu; \quad \Gamma_E = \tau'; \quad \gamma_E = 1; \quad D_\tau = \alpha_\tau \sqrt{e_k} L + \beta_\tau \nu_e;$$
$$\Gamma_\tau = e_k; \quad \gamma_\tau = m \frac{\tau}{E_\tau} - n\gamma_F' sign\left(\frac{\partial \langle u \rangle}{\partial Y} \right),$$

where α_E, β_E, α_τ, β_τ, γ_τ, and γ_F' are constants.

A number of current publications present theoretical results for stationary turbulent flows obtained from three-parametric models. The experience with three-parametric models (as well as other, more sophisticated models) indicates that as we increase the number of differential equations (e.g., by using equations for third-order moments), we have to deal with ever-increasing number of empirical constants without any gain in accuracy or versatility of the models.

Finally, it should be noted that the division of all existing models of turbulence into semi-empirical models and models that employ transport equations for the characteristics of turbulence is really a matter of convention because models of the second type also rely on experiments to garner the functions and constants involved and to determine the range of applicability. In this sense, all models are semi-empirical.

4.11 Use of the Characteristic Functional in the Theory of Turbulence

If the flow is laminar, hydrodynamic equations uniquely determine the values of all hydrodynamic characteristics of the flow at any future instant of time, given the initial and boundary values of hydrodynamic fields. In the case of a turbulent flow, initial values of hydrodynamic fields also determine their future values in a unique way. But, in contrast to the laminar flow, these future values prove extremely sensitive to random uncontrollable perturbations of initial and boundary conditions. Besides, they have such complicated and tangled form that it is quite futile to attempt a rigorous derivation by solving the corresponding differential equations. Only probability distributions of hydrodynamic fields might be of any interest to us, but not the exact values of these fields. This is why hydrodynamic equations for turbulent flows are used to study the corresponding probability distributions and the statistical characteristics of turbulence that follow from these distributions.

In the previous sections, we showed how to determine statistical characteristics of turbulence – average values and correlation functions of the velocity field – from hydrodynamic equations and algebraic relations. In this section, we are going to show how to get a complete statistical description of the velocity field from the probability distributions [7].

The fields of hydrodynamic parameters (velocity, pressure, temperature) in a turbulent flow are random fields, each field having its own probability density distribution (see Sections 1.2 and 1.7). For an incompressible fluid, the problem reduces to examining the velocity field.

Consider the probability density distribution of velocity components $u_k(\boldsymbol{X}, t)$, $k = 1, 2, 3$, assuming that the fields $u_k(\boldsymbol{X}, t)$ are random fields, that is, the values $u_k(\boldsymbol{X}, t) = u_k(\boldsymbol{M})$ at any fixed point $\boldsymbol{M} = (\boldsymbol{X}, t)$ in spacetime are random quantities. Hence to each pair value $(\boldsymbol{X}, t)$ there should correspond some probability density distribution $p(u_k)$ dependent on the point M.

Recall that the probability density distribution $p(u_k)$ is defined by the following equality:

$$P\{u_k < u_k(\boldsymbol{X}, t) < u_k + du_v\} = p(u_k)du_k. \tag{4.167}$$

Such a distribution is called one-dimensional. Now, consider two spacetime points $M^{(1)} = (X^{(1)}, t^{(1)})$, $M^{(2)} = (X^{(2)}, t^{(2)})$ and the values of velocity u_k at these points, $u_k(M^{(1)}) = u_k^{(1)}$, $u_k(M^{(2)}) = u_k^{(2)}$. Then the two-dimensional probability density distribution $p(u_k^{(1)}, u_k^{(2)})$ can be defined by a relation similar to Eq. (4.167):

$$\begin{aligned} &P\left\{u_k^{(1)} < u_k(M_1) < u_k^{(1)} + du_k^{(1)};\quad u_k^{(2)} < u_k(M_2) < u_k^{(2)} + du_k^{(2)}\right\} \\ &= p(u_k^{(1)}, u_k^{(2)})du_k^{(1)}du_k^{(2)}. \end{aligned} \tag{4.168}$$

Finally, for a system of N arbitrarily chosen spacetime points $M^{(1)} = (X^{(1)}, t^{(1)})$, $M^{(2)} = (X^{(2)}, t^{(2)}), \ldots, M^{(N)} = (X^{(N)}, t^{(N)})$ and the corresponding values $u_k(M^{(j)}) = u_k^{(j)}$ of the velocity field u_k at these points, we can introduce the N-dimensional probability density distribution $p(u_k^{(1)}, u_k^{(1)}, \ldots, u_k^{(N)})$ according to the equality

$$\begin{aligned} &P\left\{u_k^{(1)} < u_k(M^{(1)}) < u_k^{(1)} + du_k^{(1)}, u_k^{(2)} < u_k(M^{(2)}) < u_k^{(2)} + du_k^{(2)}, \ldots,\right. \\ &\left. u_k^{(N)} < u_k(M^{(N)}) < u_k^{(N)} + du_k^{(N)}\right\} = p(u_k^{(1)}, u_k^{(2)}, \ldots, u_k^{(N)})du_k^{(1)}du_k^{(2)}\ldots u_k^{(N)}. \end{aligned} \tag{4.169}$$

This N-dimensional probability density distribution satisfies conditions 1–5 of Section 1.5.

It was shown in Section 1.8 that instead of $p(u_k^{(1)}, u_k^{(1)}, \ldots, u_k^{(N)})$, it is convenient to consider its N-dimensional Fourier transform

$$\varphi(\rho_k^{(1)}, \rho_k^{(2)}, \ldots, \rho_k^{(N)}) = \int_{-\infty}^{\infty}\int_{-\infty}^{\infty} \cdots \int_{-\infty}^{\infty} \exp\left\{ i\sum_{j=1}^{N} \rho_k^{(j)} u_k^{(j)} \right\} \times p(u_k^{(1)}, u_k^{(1)}, \ldots, u_k^{(N)}) du_k^{(1)} du_k^{(2)} \ldots u_k^{(N)}. \tag{4.170}$$

Introducing vectors $\boldsymbol{\rho}_k = (\rho_k^{(1)}, \rho_k^{(2)}, \ldots, \rho_k^{(N)})$, $\boldsymbol{u}_k = (u_k^{(1)}, u_k^{(2)}, \ldots, u_k^{(N)})$ and using the averaging rule (1.27), we can rewrite the relation (4.170) in a compact form:

$$\varphi(\boldsymbol{\rho}) = \int_{-\infty}^{\infty} \exp(i\boldsymbol{\rho}_k \cdot \boldsymbol{u}_k)\, \boldsymbol{p}(\boldsymbol{u}_k) d\boldsymbol{u}_k = \langle \exp(i\boldsymbol{\rho}_k \cdot \boldsymbol{u}_k) \rangle. \tag{4.171}$$

The function $\varphi(\boldsymbol{\rho})$ is called the characteristic function. We outlined its properties in Section (1.5). The most important of them are

$$\varphi(\mathbf{0}) = \mathbf{1}, \tag{4.172}$$

which follows from the normalization condition for probability density, and

$$\varphi(\rho_k^{(1)}, \rho_k^{(2)}, \ldots, \rho_k^{(n)}) = \varphi(\rho_k^{(1)}, \rho_k^{(2)}, \ldots, \rho_k^{(n)}, 0, 0, \ldots, 0), \tag{4.173}$$

where $n < N$ and number of zeros equals $N - n$. Taking the inverse Fourier transform

$$p(\boldsymbol{u}) = \frac{1}{(2\pi)^N} \int_{-\infty}^{\infty} \exp(-i\boldsymbol{\rho}_k \cdot \boldsymbol{u}_k) \varphi(\boldsymbol{\rho}_k) d\boldsymbol{\rho}_k, \tag{4.174}$$

of a given characteristic function, one can find the probability density distribution. The problem of finding the probability density distribution of velocity thus reduces to that of finding the characteristic function. In particular, using the property (4.173) and the relation (4.174), one can find the probability density distribution if the number of dimensions is smaller than N.

The shortcoming of the characteristic function is that its range of application is limited a discrete system of points. For a continuous domain of points, it must be replaced with the characteristic functional $\rho(\boldsymbol{X})$ (see Section 1.15).

For a random function $u_k(X)$ of a single variable X (one-dimensional flow) defined on a finite interval (a, b), the characteristic functional is defined as (see Section 1.15)

$$\Phi[\rho_k(X)] = \left\langle \exp\left\{ i\int_b^a \rho_k(X)u_k(X)dx \right\}\right\rangle. \quad (4.175)$$

If the characteristic functional $\Phi[\rho_k(X)]$ is given, then, setting in Eq. (4.175)

$$\rho_k(X) = \sum_{j=1}^{N} \rho_k^{(j)}\delta(X-X^{(j)})$$

and recalling the properties of the delta function (see Section 1.3), we find that

$$\begin{aligned}\Phi[\rho_k(X)] &= \exp\left\langle \left\{ i\int_b^a \rho_i^{(k)}\delta(X-X^{(k)})u_i(X)dx \right\}\right\rangle \\ &= \exp\left\langle i\sum_{j=1}^{N} \rho_k^{(j)}u_k^{(j)}dX \right\rangle = \varphi(\rho_k^{(1)}, \rho_k^{(2)}, \ldots, \rho_k^{(N)}).\end{aligned} \quad (4.176)$$

So, the knowledge of the characteristic functional enables to find the multidimensional characteristic functional of a random function for any discrete system of points.

Putting $\rho_k(X) = 0$ into Eq. (4.175) and using the property (4.172), one gets the following important property of the characteristic functional:

$$\Phi[\rho_k(X)]_{\rho_k(X)\equiv 0} = 1. \quad (4.177)$$

Similarly, one can find the characteristic functional of a random function $u_k(\boldsymbol{X}, t)$ that depends on coordinates $\boldsymbol{X}(X_1, X_2, X_3)$ and time t:

$$\Phi[\rho_k(\boldsymbol{X}, t)] = \left\langle \exp\left\{ i\int \int_{-\infty}^{\infty} \int_{-\infty}^{\infty} \int_{-\infty}^{\infty} \rho_k(\boldsymbol{X}, t)u_k(\boldsymbol{X}, t)d\boldsymbol{X}\,dt \right\}\right\rangle, \quad (4.178)$$

where $d\boldsymbol{X} = dX_1, dX_2dX_3$.

For several statistically correlated functions such as, for example, three components of velocity $\boldsymbol{u}(\boldsymbol{X}, t) = \{u_1(\boldsymbol{X}, t), u_2(\boldsymbol{X}, t), u_3(\boldsymbol{X}, t)\}$, we can introduce the characteristic functional

$$\begin{aligned}\Phi[\rho(\boldsymbol{X}, t)] &= \Phi[\rho_1(\boldsymbol{X}, t), \rho_2(\boldsymbol{X}, t), \rho_3(\boldsymbol{X}, t)] \\ &= \left\langle \exp\left\{ i\int \int_{-\infty}^{\infty} \int_{-\infty}^{\infty} \int_{-\infty}^{\infty} \sum_{k=1}^{3} \rho_k(\boldsymbol{X}, t)u_k(\boldsymbol{X}, t)d\boldsymbol{X}dt \right\}\right\rangle.\end{aligned} \quad (4.179)$$

Let us now consider a system of points $(X^{(1)}, t^{(1)})$, $(X^{(2)}, t^{(2)})$, ..., $(X^{(N)}, t^{(N)})$ and plug into Eq. (4.179) the expression

$$\rho_k(X,t) = \sum_{j=1}^{N} \rho_k^{(j)} \delta(X - X^{(j)}) \delta(t - t^{(j)}).$$

The resultant values of the characteristic functional will coincide with the characteristic functions of probability density distributions for the values $u^{(k)} = u(X^{(k)}, t^{(k)})$ of the velocity field $u(X, t)$ at a finite set of spacetime points. One can see that all finite-dimensional probability density distributions of a field $u(X, t)$ are uniquely determined by the values of the characteristic functional.

A functional of the type (4.178) is called a spacetime characteristic functional. A less comprehensive but more simple statistical description of the random field is provided by the spatial characteristic functional

$$\Phi[\rho(X), t] = \Phi[\rho_1(X), \rho_2(X), \rho_3(X)] = \left\langle \exp\left\{ i \int_{-\infty}^{\infty}\int_{-\infty}^{\infty}\int_{-\infty}^{\infty} \sum_{k=1}^{3} \rho_k(X) u_k(X) dX \right\} \right\rangle. \tag{4.180}$$

This characteristic functional contains a complete statistical description of the velocity field $u(X, t)$ at a fixed instant of time t but it cannot be used to calculate joint statistical characteristics of the velocity field at different instants of time. For simplicity''s sake, we shall consider only spatial characteristic functionals.

The knowledge of the characteristic functional enables us to find all statistical characteristics of the random field $u(X, t)$. In order to determine them, we have to use variational (functional) derivatives (see Section 1.14).

Recall that for a functional

$$\Phi[\rho(X)] = \int_a^b\int_a^b u(X_1, X_2)\rho(X_1)\rho(X_2) dX_1 dX_2,$$

the first functional derivative is equal to (see Eq. 1.174)

$$\frac{\delta\Phi[\rho(X)]}{\delta\rho(X_1)} = \int_a^b (u(X_2, X_1) + u(X_1, X_2))\rho(X_2) dX_2.$$

It depends on the point X_1, which plays the role of a parameter.

The second functional derivative of this functional is

$$\frac{\delta^2\Phi[\rho(X)]}{\delta\rho(X_1)\delta\rho(X_2)} = u(X_1, X_2) + u(X_2, X_1).$$

It depends on two points X_1 and X_2. For a functional of a more general form, we should define (if it exists) the N-th functional derivative

$$\frac{\delta^N \Phi[\rho(X)]}{\delta\rho(X_1)\delta\rho(X_2)\ldots\delta\rho(X_N)},$$

which depends on N points X_1, X_2, ..., X_N as on parameters.

Let us apply the rule of functional differentiation to the characteristic functional (4.175). Since the operations of averaging and differentiation are interchangeable, we have

$$\frac{\delta\Phi[\rho_k(X)]}{\delta\rho_k(X^{(1)})} = i\left\langle u_k(X^{(1)})\exp\left\{i\int_a^b \rho_k(X)u_k(X)dX\right\}\right\rangle, \tag{4.181}$$

$$\ldots\ldots\ldots\ldots\ldots\ldots\ldots\ldots\ldots\ldots\ldots\ldots\ldots,$$

$$\frac{\delta^N \Phi[\rho_k(X)]}{\delta\rho_k(X_1^{(1)})\delta\rho_k(X_2^{(2)})\ldots\delta\rho_k(X_N^{(N)})} = i^n\left\langle u_k(X^{(1)})u_k(X^{(2)})\ldots u_k(X^{(N)})\exp\left\{i\int_a^b \rho_k(X)u_k(X)dX\right\}\right\rangle. \tag{4.182}$$

Eq. (4.182) together with the general expression (1.53) yields the N-point moment of the random field $u_i(X)$:

$$B_{u_k u_k \ldots u_k} = \left\langle u_k(X^{(1)})u_k(X^{(2)})\ldots u_k(X^{(N)})\right\rangle = \left\langle u_k^{(1)} u_k^{(2)}\ldots u_k^{(N)}\right\rangle = (-i)^N \left.\frac{\delta^N \Phi[\rho_k(X)]}{\delta\rho_k(X_1^{(1)})\delta\rho_k(X_2^{(2)})\ldots\delta\rho_k(X_N^{(N)})}\right|_{\rho_i(x)=0}. \tag{4.183}$$

For a random field of n-dimensional velocity $\boldsymbol{u} = (u_1(\boldsymbol{X}),\ u_2(\boldsymbol{X}),\ \ldots,\ u_n(\boldsymbol{X}))$, the formulas (4.179) and (4.183) give us the following expressions for multi-point moments of a random multidimensional velocity field:

$$B_{k_1 k_2 \ldots k_n} = \boldsymbol{X}^{(1)}, \boldsymbol{X}^{(2)}, \ldots, \boldsymbol{X}^{(n)} = \left\langle u_{k_1}(\boldsymbol{X}^{(1)}), u_{k_2}(\boldsymbol{X}^{(2)})\ldots u_{k_n}(\boldsymbol{X}^{(N)})\right\rangle = (-i)^n \left.\frac{\delta^n \Phi[\rho_k(\boldsymbol{X})]}{\delta\rho_{k_1}(\boldsymbol{X}_1^{(1)})\delta\rho_{k_2}(\boldsymbol{X}_2^{(2)})\ldots\delta\rho_{k_n}(\boldsymbol{X}_n^{(n)})}\right|_{\rho(x)\,\equiv 0.} \tag{4.184}$$

A characteristic functional that has functional derivatives of all orders can be expanded in a Taylor series:

$$\Phi[\rho(\boldsymbol{X})] = \Phi[0] + i\sum_{k=1}^{N}\int \frac{\delta\Phi[0]}{\delta\rho_k(X_1^{(1)})}\rho_k(\boldsymbol{X}^{(1)})d\boldsymbol{X}^{(1)} + \sum_{k=1}^{N}\sum_{l=1}^{N}\int \frac{\delta\Phi[0]}{\delta\rho_k(\boldsymbol{X}^{(1)})\delta\rho_l(X^{(2)})}\rho_k(\boldsymbol{X}^{(1)})\rho_l(\boldsymbol{X}^{(2)})d\boldsymbol{X}^{(1)}d\boldsymbol{X}^{(2)} + \ldots. \tag{4.185}$$

Let us employ the expressions (4.184) and (4.185) to expand the characteristic functional of a random velocity field into a series with respect to correlation functions:

$$\begin{aligned}\Phi[\rho(\boldsymbol{X})] &= 1 + i\sum_{k=1}^{N}\int\left\langle u_k(\boldsymbol{X}^{(1)})\right\rangle \rho_k(\boldsymbol{X}^{(1)})d\boldsymbol{X}^{(1)} \\ &-\sum_{k=1}^{N}\sum_{l=1}^{N}\int\int B_{kl}(\boldsymbol{X}^{(1)}, X^{(2)})\rho_k(\boldsymbol{X}^{(1)})\rho_l(\boldsymbol{X}^{(2)})d\boldsymbol{X}^{(1)}d\boldsymbol{X}^{(2)} \\ &+\ldots + i^n\sum_{k_1=1}^{N}\ldots\sum_{k_n=1}^{N}\int\ldots\int B_{k_1k_2\ldots k_n}(\boldsymbol{X}^{(1)}, X^{(2)},\ldots, X^{(n)}) \\ &\times \rho_{k_1}(\boldsymbol{X}^{(1)})\rho_{k_2}(\boldsymbol{X}^{(2)})\ldots\rho_{k_n}(\boldsymbol{X}^{(n)})d\boldsymbol{X}^{(1)}d\boldsymbol{X}^{(2)}\ldots d\boldsymbol{X}^{(n)}.\end{aligned} \tag{4.186}$$

Expanding the logarithm of the characteristic functional or the cumulant-generating function $\psi = \ln\,\varphi$ (see Section 1.6) in a Taylor series, we can represent the series (4.186) in a somewhat different form

$$\begin{aligned}\Phi[\rho(X)] &= \exp\left\{1\sum_{k=1}^{N}\int\left\langle u_k(\boldsymbol{X}^{(1)})\right\rangle \rho_k(\boldsymbol{X}^{(1)})d\boldsymbol{X}^{(1)}\right. \\ &-\sum_{k=1}^{N}\sum_{l=1}^{N}\int\int b_{kl}(\boldsymbol{X}^{(1)}, X^{(2)})\rho_k(\boldsymbol{X}^{(1)})\rho_l(\boldsymbol{X}^{(2)})d\boldsymbol{X}^{(1)}d\boldsymbol{X}^{(2)} + \ldots + \\ &+ i^n\sum_{k_1=1}^{N}\ldots\sum_{k_n=1}^{N}\int\ldots\int S_{k_1k_2\ldots k_n}(\boldsymbol{X}^{(1)}, X^{(2)},\ldots, X^{(n)}) \\ &\left.\times \rho_{k_1}(\boldsymbol{X}^{(1)})\rho_{k_2}(\boldsymbol{X}^{(2)})\ldots\rho_{k_n}(\boldsymbol{X}^{(n)})d\boldsymbol{X}^{(1)}dX^{(2)}\ldots d\boldsymbol{X}^{(n)}\right\},\end{aligned} \tag{4.187}$$

where b_{ij} are central moments (correlations of fluctuations) and $S_{i_1\ldots i_n}$ are cumulants, or semi-invariants.

If the random field is Gaussian, then cumulants of the second order vanish (see Section 1.8). When considering a turbulent fluid flow or the behavior of particles in a turbulent flow, we assume that the distribution of statistical characteristics is Gaussian or close to Gaussian. Such an assumption considerably simplifies the analysis. In particular, it leaves only the first two terms in the series (4.187). In this sense, the expansion (4.187) is more preferable as compared to the expansion (4.186).

Our next task is to find the characteristic functional of the velocity field in a turbulent flow. In an incompressible uniform fluid, the density is $\rho = const$, whereas the pressure field $p(\boldsymbol{X}, t)$ obeys the equation (4.13) and can, in principle, be expressed through the velocity field $\boldsymbol{u}(\boldsymbol{X}, t)$. Therefore the velocity field gives a complete hydrodynamic description for the case of a uniform, incompressible fluid flow, whereas the characteristic functional of the velocity field gives a complete statistical description of both the velocity field and the turbulent fluid flow.

It was shown earlier that the moments of the velocity field can be determined from the characteristic functional (see Eq. (4.184)). On the other hand, these moments satisfy the infinite system of equations following from the Navier–Stokes hydrodynamic equations (the Keller–Friedman chain). Therefore the characteristic functional should also satisfy some equation containing functional derivatives. Let us derive this equation for a spatial characteristic functional $\Phi[\rho(\boldsymbol{X}),\ t]$.

Denote the scalar product of two functions $\boldsymbol{\vartheta}(\boldsymbol{M})$ and $\boldsymbol{u}(\boldsymbol{M})$ of the point $\boldsymbol{M}=\boldsymbol{X}$ or $\boldsymbol{M}=(\boldsymbol{X},\ t)$ in the functional space as

$$(\boldsymbol{\vartheta}\cdot\boldsymbol{u}) = \int \boldsymbol{\vartheta}(\boldsymbol{M})\boldsymbol{u}(\boldsymbol{M})d\boldsymbol{M}, \tag{4.188}$$

where $d\boldsymbol{M}=d\boldsymbol{X}$ for $\boldsymbol{M}=\boldsymbol{X}$ or $d\boldsymbol{M}=d\boldsymbol{X}\,dt$ for $\boldsymbol{M}=(\boldsymbol{X},\ t)$.

The definition (4.175) of a spatial characteristic functional can then be represented as

$$\Phi[\boldsymbol{\vartheta}(\boldsymbol{X}),t] = \exp(i(\boldsymbol{\vartheta}\cdot\boldsymbol{u})). \tag{4.189}$$

Let us denote the functional derivative operator by $D_{\mathsf{k}}(M)=\delta/\delta\vartheta_{\mathsf{k}}(M)$. Then the first and second functional derivatives of the characteristic functional can be expressed as

$$D_{\mathsf{k}}(M)\Phi = i\left\langle u_{\mathsf{k}}^{(M)}\exp(i(\boldsymbol{\vartheta}\cdot\boldsymbol{u}))\right\rangle, \tag{4.190}$$

$$D_{\mathsf{k}}(M_1)D_l(M_2)\Phi = i\left\langle u_{\mathsf{k}}^{(M_1)}u_l^{(M_2)}\exp(i(\boldsymbol{\vartheta}\cdot\boldsymbol{u}))\right\rangle, \tag{4.191}$$

Now, we use the Navier–Stokes equations. We start with the continuity equation for an incompressible fluid

$$\frac{\partial u_{\mathsf{k}}}{\partial X_{\mathsf{k}}} = 0. \tag{4.192}$$

This is condition that must be valid in order for the velocity field to be of non-divergent (solenoidal). Since time t does not enter this equation, the difference between spatial functional and spacetime functionals is not important and any conclusion derived from the continuity equation will be true for both functionals.

Suppose the fluid occupies some volume V bounded by a solid surface S. The impermeability condition $u_n|_s=0$ must be satisfied on the surface, where u_n is the projection of velocity upon the outward normal to S. Consider the gradient of a scalar function $\Delta f(\boldsymbol{X})$. From the definition (4.188), the continuity equation, and the divergence theorem, there follows

$$(\nabla f \cdot \boldsymbol{u}) = \int_V u_k \frac{\partial f}{\partial X_k} dV = \int_V \frac{\partial (f u_k)}{\partial X_k} dV = -\int_S u_n f \, dS = 0.$$

Then for any vector function $\boldsymbol{\vartheta}(\mathbf{x})$there should hold

$$(\{\boldsymbol{\vartheta}(\boldsymbol{X}) + \nabla f\} \cdot \boldsymbol{u}) = (\boldsymbol{\vartheta} \cdot \boldsymbol{u}).$$

We conclude from Eq. (4.189) that

$$\Phi[\boldsymbol{\vartheta}(\boldsymbol{X}) + \nabla f] = \Phi[\boldsymbol{\vartheta}(\boldsymbol{X})].$$

Differentiating both sides of Eq. (4.189) with respect to X_k and using the fact that $\exp(i(\vartheta \cdot \boldsymbol{u}))$ does not depend on $\boldsymbol{X}$, we then sum the resultant expression result over k and use continuity equation to get

$$\frac{\partial}{\partial X_k}(D_k(X)\Phi) = i\left\langle \frac{\partial u_k}{\partial X_k} \exp(i(\boldsymbol{\vartheta} \cdot \boldsymbol{u}))\right\rangle = 0.$$

Consequently, the characteristic functional obeys the equation

$$\frac{\partial}{\partial X_i}(D_i(\boldsymbol{X})\Phi) = 0. \tag{4.193}$$

In order to derive a dynamic equation for the characteristic functional, one should take the hydrodynamic equation of fluid motion (4.6) and use the continuity equation to bring it to the form

$$\frac{\partial \boldsymbol{u}}{\partial t} = -\frac{\partial \boldsymbol{u} u_i}{\partial X_i} - \frac{1}{\rho}\nabla p + \nu_e \Delta u. \tag{4.194}$$

Differentiating Eq. (4.189) with respect to t,

$$\frac{\partial \Phi}{\partial t} = i\left\langle \boldsymbol{\vartheta} \cdot \frac{\partial \boldsymbol{u}}{\partial t} \exp(i(\boldsymbol{\vartheta} \cdot \boldsymbol{u}))\right\rangle$$

and taking the expression (4.184) instead of $\partial \mathbf{u}/\partial t$, we obtain:

$$\frac{\partial \Phi}{\partial t} = i\boldsymbol{\vartheta} \cdot \left\{ -\frac{\partial}{\partial X_i}\left\langle \boldsymbol{u} u_i e^{i(\boldsymbol{\vartheta} \cdot \boldsymbol{u})}\right\rangle - \nabla\left\langle \frac{p}{\rho} e^{i(\boldsymbol{\vartheta} \cdot \boldsymbol{u})}\right\rangle + \nu_e \Delta\left\langle \boldsymbol{u} e^{i(\boldsymbol{\vartheta} \cdot \boldsymbol{u})}\right\rangle\right\}.$$

Let us denote

$$\Pi = \frac{1}{\rho}\left\langle p e^{i(\boldsymbol{\vartheta} \cdot \boldsymbol{u})}\right\rangle = \Pi[\boldsymbol{\vartheta}(\boldsymbol{X}); \boldsymbol{X}, t]$$

and use Eqs. (4.190)–(4.191) to get

$$\frac{\partial \Phi}{\partial t} = \boldsymbol{\vartheta} \cdot \left\{ i \frac{\partial \boldsymbol{D} D_i \Phi}{\partial X_i} + \nu_e \Delta \boldsymbol{D} \Phi - i \nabla \Pi \right\}, \tag{4.195}$$

where $\boldsymbol{D} = (D_1, D_2, D_3)$ is the vector operator of functional differentiation.

The equations (4.193) and (4.195) form a system of two equations for two unknown functionals $\Phi[\boldsymbol{\vartheta}(\boldsymbol{X}), t]$ and $\prod[\boldsymbol{\vartheta}(\boldsymbol{X}); \boldsymbol{X}, t]$. Eliminating the last functional, we get

$$\nabla \cdot \left\{ i \frac{\partial \boldsymbol{D} D_i \Phi}{\partial X_i} + \nu_e \Delta \boldsymbol{D} \Phi - i \nabla \Pi \right\} = \nabla \cdot \left\{ i \left\langle \frac{\partial u}{\partial t} \exp(i(\boldsymbol{\vartheta} \cdot \boldsymbol{u})) \right\rangle \right\}$$
$$= i \left\langle \frac{\partial (\nabla \cdot \boldsymbol{u})}{\partial t} \exp(i(\boldsymbol{\vartheta} \cdot \boldsymbol{u})) \right\rangle = 0.$$

Let us rearranging the left-hand side of this equation:

$$\nabla \cdot \left\{ i \frac{\partial \boldsymbol{D} D_i \Phi}{\partial X_i} + \nu_e \Delta \boldsymbol{D} \Phi - i \nabla \Pi \right\} = \frac{\partial^2 (D_i D_j \Phi)}{\partial X_i \partial X_j} + \nu_e \Delta \frac{\partial (D_i \Phi)}{\partial X_i} - \Delta \Pi.$$

Recall that $\partial(D_i \Phi)/\partial X_i$ due to Eq. (4.193). Then

$$\Delta \Pi = \frac{\partial^2 (D_i D_j \Phi)}{\partial X_i \partial X_j}. \tag{4.196}$$

This equation is similar to the equation (4.13) for pressure. It can, in principle, be resolved with respect to Π (see Eq. (4.14). Let us write the general solution in the operator form

$$\Pi = \Delta^{-1} \left\{ \frac{\partial^2 (D_i D_j \Phi)}{\partial X_i \partial X_j} \right\}. \tag{4.197}$$

Substituting the relation (4.197) into Eq. (4.195), we finally obtain

$$\frac{\partial \Phi}{\partial t} = \boldsymbol{\vartheta} \cdot \left\{ i \frac{\partial \boldsymbol{D} D_i \Phi}{\partial X_i} + \nu_e \Delta \boldsymbol{D} \Phi - i \nabla \Delta^{-1} \frac{\partial^2 (D_i D_j \Phi)}{\partial X_i \partial X_j} \right\}. \tag{4.198}$$

This equation for the characteristic functional $\Phi[\boldsymbol{\vartheta}(\boldsymbol{X}), t]$ was first derived in [27]; it is known as the Hopf equation. It gives us the characteristic functional if the initial value $\Phi[\boldsymbol{\vartheta}(\boldsymbol{X}), t_0] = \Phi_0[\boldsymbol{\vartheta}(\boldsymbol{X})]$ is known.

One distinguishing feature of the Hopf equation is its linearity. Although hydrodynamic equations of the turbulent flow are nonlinear, the problem of statistical dynamics of the turbulent flow turns out to be linear. However, this advantage (as compared to hydrodynamic equations) is negated by mathematical difficulties associated with the Hopf equation, which is a linear equation in functional derivatives.

The knowledge of the characteristic functional enables us to determine all statistical characteristics of turbulence. Therefore the initial value problem that includes the Hopf equation and the corresponding initial condition turns out to be a closed (and the most compact) formulation of the turbulence problem – the problem of finding the statistical characteristics of turbulence from the given statistical characteristics of the initial velocity field. Yet, no specific results have been obtained thus far, because we still don't have a mathematical theory of equations in functional derivative.

4.12 Intermittency in a Turbulent Flow

Analysis of the local structure of turbulence (see Section 4.9) shows that any complicated spatial velocity distribution in a turbulent flow can be represented as a superposition (spectrum) of harmonic fluctuations. The wavelength of large-scale fluctuations is comparable to the characteristic linear size of the flow region. The wavelength of smallest-scale fluctuations is far smaller than the characteristic linear size and decreases with the increase of Re. Thus in the spectrum of turbulence there are a lot of oscillations whose wavelength varies in a wide range.

Large-scale oscillations characterize the energy of turbulence, whereas small-scale oscillations characterize its dissipation, which turns out to be significant for all values of Re.

Since large-scale fluctuations are practically independent of viscosity, they are unstable. Their disintegration gives rise to fluctuations that have a smaller spatial scale and, accordingly, a smaller Reynolds number Re. This process continues until it produces oscillations whose scale is so small that the Reynolds number goes to $Re \sim 1$.

The presence of multiple scales characterizing the process of turbulent transport leads to self-similarity of turbulent flows with respect to the Reynolds number. It means that for all quantities (velocity, pressure, concentration of passive impurities, and so forth), their average values that are determined by large-scale velocity fluctuations do not depend on Re when the latter goes to infinity. The validity of this assertion is confirmed by experiments. But at $Re \to \infty$, the size of small-scale oscillations goes to $\lambda \to 0$. Numerical solution of equations for the moments that follow from the Navier–Stokes equations presents a challenge in this case; so does obtaining experimental data. The difficulty lies in the fact that numerical solutions, as well as experiments, can only give quantities that have been averaged over a spacetime region. From the practical viewpoint, it means that it is necessary to carry out a set of numerical calculations or experiments at different values of Re and λ, and then extrapolate the obtained results to the region where $Re = \infty$ and $\lambda = 0$. The existing experimental data based on the study of intermittency in turbulent flows shows that such extrapolation is impossible in principle.

Of crucial importance in the theory of turbulence is the assumption that the average turbulent energy dissipation

$$\bar{\varepsilon} = \frac{1}{2}\nu_e \sum_{i,j}\left\langle\left(\frac{\partial u_i}{\partial X_j} + \frac{\partial u_j}{\partial X_i}\right)^2\right\rangle = \nu_e\left\langle|\omega|^2\right\rangle + 2\nu_e\left\langle\frac{\partial u_i u_j}{\partial X_i \partial X_j}\right\rangle \tag{4.199}$$

and the average scalar dissipation of concentration inhomogeneity

$$\bar{N} = D_m\left\langle\left(\frac{\partial C}{\partial X}\right)^2\right\rangle = D_m(\nabla C)^2, \tag{4.200}$$

(where $\omega = \nabla \times \boldsymbol{u}$ is vorticity, D_m – coefficient of molecular diffusion, C – passive impurity concentration), both of which depend on the gradients of velocity and concentration, tend to finite but nonzero limits as $\mathrm{Re} \to \infty$, and, moreover, these limiting values are independent of molecular transport coefficients. The assumption about the existence of finite limits for the quantities $\bar{\varepsilon}$ and $\bar{N}$ at $\mathrm{Re} \to \infty$ is supported by experiments.

The quantity $\bar{\varepsilon}$ characterizes the decrease of turbulent energy due to viscosity, and N describes the rate at which (unaveraged) concentration inhomogeneities get "smoothed out" via molecular diffusion. The last statement means that scalar dissipation of concentration inhomogeneities characterizes the mixing rate of passive impurity all the way to the molecular level (this should not be confused with mixing of matter on a macrolevel, that is, on a scale equal to the linear size of the considered problem).

It should be noted that a similar situation takes place for laminar flows at very high Reynolds numbers. Examples include flows in a boundary layer with zero pressure gradient, and flows in the mixing layer between two plane-parallel streams. From the solutions of the corresponding problems (e.g. the Blasius solution) that exist in literature, we see that in both cases, increase of the Reynolds number Re leads to a reduced thickness of the boundary layer, increased velocity and concentration gradients, and constancy of $\bar{\varepsilon}$ and $\bar{N}$.

In laminar flows, when we go to the limit $\mathrm{Re} \to \infty$, there appears a small parameter in the highest derivative in the Navier–Stokes equations. To avoid singularities, one should bring into consideration a thin boundary layer whose thickness tends to zero when $\mathrm{Re} \to \infty$ and outside of which molecular transport is negligible (that is, $\bar{\varepsilon} = \bar{N} = 0$), and the flow is described by the Euler equations.

A similar structure is observed in a turbulent flow. Dissipation processes also take place in narrow regions. The peculiarity of turbulent flows distinguishing them from laminar ones is that these regions move randomly in space, and the values of $\bar{\varepsilon}$ and $\bar{N}$, generally speaking, depend on the Reynolds number. This phenomenon, first discovered by Corrsin, is called intermittency.

So, the essence of the intermittency phenomenon is an utterly irregular distribution of velocity and passive impurity concentration gradients in turbulent flows, when regions of very small gradients (non-turbulent fluid) alternate irregularly with regions of very high gradients (turbulent fluid).

Intermittency in turbulent flows at large Reynolds numbers plays an essential role in our picture of the inner structure of turbulence and of the main features chemical reactions in turbulent flows. In particular, intermittency exerts a profound effect on the shape of probability density distribution of hydrodynamic and concentration characteristics in a turbulent flow.

One distinguishes two types of intermittency: inner and external. To illustrate these notions, consider the outflow of a fume from a smoke stack of an electric power station. Observing the smoke plume, we notice that there is a clearly visible, curved, undulating interface, which the smoke never crosses. A similar picture can be observed not only in a stream of "foreign" gas injected into another gas, but also a boundary layer, or in a trail left behind the body by the fluid that bypasses it. Measurements show that beyond the interface, dissipation of energy is practically zero. Accordingly, spatial distribution of energy dissipation turns out to be very nonuniform. Regions with $\bar{\varepsilon} > 0$ alternate with regions where $\bar{\varepsilon} = 0$. Undulation of boundaries of fluid trails, jets, and boundary layers is usually referred to as external intermittency.

But even within the flow regions, spatial distributions of energy dissipation and concentration dissipation are also highly nonuniform. Regions with intense fluctuations of velocity gradients and concentration gradients alternate with regions where such fluctuations are nearly absent. This phenomenon was first discovered by Batchelor and Townsend and was called inner intermittency.

The distinguishing feature of intermittency is that while the velocity gradient varies greatly, velocity itself does not change much. The explanation is that because of pressure fluctuations, velocity fluctuations are observed over the entire flow region, and one cannot base his study of intermittency on observations of the velocity field. Therefore the most commonly used method of research is the analysis of the velocity gradient field, that is, the analysis of energy dissipation.

We often say that regions of small velocity gradients and impurity concentration gradients are filled with a non-turbulent fluid, and regions of large gradients – with a turbulent fluid. Sometimes we use the term "potential fluid" instead of "non-turbulent fluid". Because of this, a turbulent fluid is sometimes referred to as a "rotating fluid" and a non-turbulent one – as a "non-rotating fluid".

It is well-established by now that intermittency is a typical phenomenon for all turbulent flows. The possibility of intermittency greatly complicates theoretical studies of turbulent flows, because the frequently-used assumptions of homogeneity and isotropy of turbulence are often violated.

The possibility to discern turbulent and non-turbulent fluids as they pass through a given point in space has led to the development of an experimental method called the conditional sampling method. Measurement of flow characteristics at some spatial point allows us to determine the probability density distribution while taking intermittency into account.

In order to understand how intermittency affects the probability density distribution of random quantities, consider a two-dimensional jet of a foreign gas injected into a co-current stream of another gas, for example, air. Let $C(\boldsymbol{X}, t)$ be the mass fraction of injected gas. Let us perform measurements of concentration as a function of time in a fixed point $\boldsymbol{X}$ where intermittency is observed. If we plot time t on the horizontal axis and $C(\boldsymbol{X}, t)$ on the vertical axis, then as the jet passes through the point $\boldsymbol{X}$, the time distribution of concentration is a step function: $C \leq C_\varepsilon$ when the fluid is non-turbulent, that is, at $0 < t < t_1$; and $C > C_\varepsilon$ when the fluid is turbulent, that is, at $t_1 < t < t_2$. This temporal distribution will be repeated again after a while . Measurement data allow us to define the so-called intermittency function

$$\Pi(\boldsymbol{X}, t) = \begin{cases} 0 & \text{at} \quad C < C_\varepsilon, \\ 1 & \text{at} \quad C > C_\varepsilon. \end{cases} \tag{4.201}$$

Mathematically, intermittency is defined as the average value of the intermittency function $\langle \Pi(\boldsymbol{X}) \rangle$ for stationary flows. It is equal to the fraction of time during which the flow is turbulent at the point under consideration. The intermittency function is important as the underlying concept behind the method of conditional sampling.

The correlation of $\Pi(\boldsymbol{X}, t)$ with some particular variable, for example, velocity $u_i(\boldsymbol{X}, t)$, provides the average value of the velocity component u_i during the time interval when only turbulent flow exists at the given point.

Probability density distribution $p(C, \boldsymbol{X})$ is characterized by a peak in the vicinity of $C = 0$, by the effect of measurement error on the peak's structure, and by the appearance of negative values as a result of these errors. For the purposes of theoretical analysis, this distribution is simulated by the delta function $\delta(C)$ with the amplitude $1 - \langle \Pi(\boldsymbol{X}) \rangle$.

So, at any point where intermittency takes place, joint probability density distribution of velocity $\boldsymbol{u}$ and concentration C can be represented as a sum of two components,

$$p(\boldsymbol{u}, C, \boldsymbol{X}) = (1 - \langle \Pi(\boldsymbol{X}) \rangle)\, p_c(\boldsymbol{u}, \boldsymbol{X}) + \langle \Pi(\boldsymbol{X}) \rangle\, p_c(\boldsymbol{u}, C, \boldsymbol{X}). \tag{4.202}$$

The first component corresponds to the external flow, and since the probability to observe the value $C = 1$ of a random quantity (concentration of impurity) in a non-turbulent fluid is small, one can take $p_c(\boldsymbol{u}, \boldsymbol{X}) = \delta(C)$. The second component corresponds to the mixing layer with the probability density distribution $p_c(\boldsymbol{u}, C, \boldsymbol{X})$. Accordingly, we use the terms "non-turbulent" and "turbulent" for these two components.

The expression (4.202) makes it possible to determine the correlation of flow parameters while taking intermittency into account. Thus, the correlation of velocity components is given by

$$\langle u_i u_j \rangle = (1 - \langle \Pi(\boldsymbol{X}) \rangle)(\langle u_i u_j \rangle)_0 + \langle \Pi(\boldsymbol{X}) \rangle (\langle u_i u_j \rangle)_1, \tag{4.203}$$

where $(\langle u_i u_j \rangle)_0$ and $(\langle u_i u_j \rangle)_1$ are the correlations in the external and internal flows.

It must be emphasized that the formula (4.203) is valid in the limiting case of $\mathrm{Re} = \infty$. But it is also used for approximate description of flows (with intermittency taken into account) at large but finite values of the Reynolds number.

The problem of intermittency is discussed in more detailed in [10].

References

1 Prandtl, L. (1956) *Führer durch die Strömungslehre*, Braunschweig.

2 Batchelor, G.K. (1953) *The Theory of Homogeneous Turbulence*, Cambridge University Press, Cambridge.

3 Townsend, A.A. (1976) *The Structure of Turbulent Shear Flow*, Cambridge University Press,Cambridge.

4 Levich, V.G. (1962) *Physicochemical Hydrodynamics*, Prentice–Hall, Englewood Cliffs, N.J.

5 Abramovitch, G.N. (1960) *The Theory of Turbulent Jets*, Physmatgis, Moscow (in Russian).

6 Hinze, J.O. (1959) *Turbulence*, McGraw–Hill, New York.

7 Monin, A.S. and Yaglom, A.M. (1971) *Statistical Fluid Mechanics: Mechanics of Turbulence*, **1**, MIT Press, Cambridge; MA. Monin, A.S. and Yaglom, A.M. (1975) *Statistical Fluid Mechanics: Mechanics of Turbulence*, **2**, MIT Press, Cambridge, MA.

8 Loitsyansky, L.G. (1970) *Mechanics of Fluid and Gas*, Nauka, Moscow (in Russian).

9 Launder, B.E. and Spalding, D.B. (1972) *Mathematical Models of Turbulence*, Academic Press, London & New York.

10 Kuznetsov, Y.R. and Sabel'nikov, V.A. (1990) *Turbulence and Combustion*, Hemisphere, New York.

11 Landau, L.D. and Lifshitz, E.M. (1987) *Fluid Mechanics*, Pergamon Press, Oxford.

12 Kolmogorov, A.N. (1941) Local Structure of Turbulence in Incompressible Fluids at Very High Reynolds Numbers. *Dokl. Akad. Nauk SSSR*, **30**, (4), 299–303 (in Russian).

13 Kolmogorov, A.N. (1941) Energy Dissipation in a Locally Isotropic Turbulence *Dokl. Akad. Nauk USSR*, **32** (1), 19–21 (in Russian).

14 Kolmogorov, A.N. (1941) About Decay of Isotropic Turbulence in an Incompressible Viscous Fluid. *Dokl. Akad. Nauk SSSR*, **31** (6), 538–541 (in Russian).

15 Taylor, G.I. (1935) Statistical Theory of Turbulence. I–IV, *Proc. Roy. Soc. A*, **151**, (874), 421–478.

16 Ginevski, A.S., Ioselevitch, V.A., Kolesnikov, A.V., Lapin, J.V., Pilipenko, V.N. and Sekundov, A.N. (1978) Methods of Calculation of Turbulent Boundary Layers Itogi Nauki i Techniki. VINITI. *Mechanics of Fluid, and Gas Series*. **11**, 155–304 (in Russian).

17 Lushik, V.G., Pavelev, A.A. and Jakubenko, A.E. (1988) Transport Equations for Turbulent Characteristics: Models and Results of Calculations Itogi Nauki i Techniki. VINITI. *Mechanics of Fluid, and Gas Series*, **22**, 3–61 (in Russian).

18 Lushik, V.G., Pavelev, A.A. and Jakubenko, A.E. (1994) Turbulent Flows. Models and Numerical Studies. *Fluid Dyn.* (4), 4–27 (in Russian).

19 Kolmogorov, A.N. (1942) Turbulent Flow Equation for an Incompressible Fluid. *Izvestia Acad. Sci. USS, Phys Series*, **6**, (1–2), 56–58 (in Russian).

20 Rotta, J.C. (1951) Statistische Theorie nichthomogener Turbulenz, **Z. Phys.**,

B. 129, N. 5, S. 547–572, B. 131, N. 1, S. 51–77.

21 Glushko, G.S. (1965) Turbulent Boundary Layer on a Flat Plate in an Incompressible Fluid. *Izvestia Acad. Nauk. SSSR, Fluid Dyn. Series,* (4), 13–23 (in Russian).

22 Glushko, G.S. (1970) Differential Equation for the Scale of Turbulence and Calculation of the Turbulent Boundary Layer on a Flat Plate, Turbulent Flows: collected articles, pp. 37–44. Moscow, Nauka (in Russian).

23 Bradshaw, P., Ferris, D.H. and Atwell, N.P. (1967) Calculation of Boundary Layer Development using the Turbulent Energy Equation. *J. Fluid Mech.,* **28** (3), 593–616.

24 Kovasznay, L.S.G. (1967) Structure of the Turbulent Boundary Layer. *Phys. Fluids,* **10** (9), 25–30.

25 Sekundov, A.N. (1971) Use of Differential Equations for Turbulent Viscosity in the Analysis of Plane Flows with no Self-Similarity. *Izvestia Acad. Nauk SSSR, Fluid Dyn. Series,* (5),114–127 (in Russian).

26 Davidov, B.I. (1961) About Statistical Dynamics of Incompressible Turbulent Fluids. *Dokl. Akad. Nauk USSR,* **136** (1), 47–50 (in Russian).

27 Hopf, E. (1952) Statistical Hydromechanics and functional Calculus. *J. Rat. Mech. Anal.,* **1** (1), 87–123.

5
Particle Motion in a Turbulent Flow

5.1
The Eulerian and Lagrangian Approaches to the Description of Fluid Flow and Particle Motion

In hydrodynamics, there are two approaches to the description of motion of a fluid and the particles suspended in the fluid – Eulerian and Lagrangian [1,2].

Eulerian Approach We are aiming to trace what happens at a given spatial point $\boldsymbol{X}$ at different instants of time t or what happens at different spatial points $\boldsymbol{X}$ at a given instant of time t. Coordinates of a point, $\boldsymbol{X} = (X_1, X_2, X_3)$, depend on the choice of coordinate system. We choose a fixed frame of reference (X_i), $(i = 1, 2, 3)$, for example, a coordinate system attached to a motionless boundary of the flow region. For each point $\boldsymbol{X}$ of the flow region V, we can see different points of the continuum passing through $\boldsymbol{X}$. Motion of the continuum is said to be known when velocity, pressure, temperature, and other parameters are known functions of (X_i) and t. If t is fixed while (X_i) can vary, these functions give a spatial distribution of hydrodynamic parameters at this instant of time. If, on the contrary, (X_i) are fixed while t can change, these dependences give the time evolution of parameters at this spatial point. This formulation of the problem motion of a continuous medium is the essence of the Eulerian approach. Coordinates (X_i), time t, and hydrodynamic parameters given as functions of (X_i) and t are called Eulerian variables.

The description of turbulence presented in Chapter 5 was based on the Eulerian approach. In this paradigm, the flow of an incompressible fluid is completely characterized by the velocity field $\boldsymbol{u}(\boldsymbol{X}, t)$ that provides values of the velocity vector for all points $\boldsymbol{X}$ at different instances of time t. Hydrodynamic equations, in which pressure can be eliminated, provide the values of $\boldsymbol{u}(\boldsymbol{X}, t)$ at any instance of time t when the initial velocity $\boldsymbol{u}(\boldsymbol{X}, t_0) = \boldsymbol{u}_0(\boldsymbol{X})$ and the appropriate boundary conditions are specified.

A serious disadvantage of the Eulerian approach is that it does not allow to trace the motion of small individual particles suspended in the fluid. Since spatial position of such particles varies in time as they move with the carrier fluid, their Eulerian coordinates (X_i) also vary in time.

Statistical Microhydrodynamics. Emmanuil G. Sinaiski and Leonid Zaichik

ISBN: 978-3-527-40656-2

Lagrangian Approach The continuous medium is regarded as a continuous set of elementary volumes that are negligibly small compared to the total volume occupied by the fluid and at the same time large in comparison with the size of molecules. Smallness of elementary volumes makes it possible to consider them as material points (fluid particles) moving relative to a fixed coordinate system $\{X_i\}$. If the coordinates of a fluid particle (relative to a chosen coordinate system) change with time according to

$$X_i = f_i(t), \quad (i = 1, 2, 3), \tag{5.1}$$

we say that the point is moving relative to a fixed coordinate system $\{X_i\}$ and that Eq. (5.1) provides the law of its motion.

However, it is insufficient to give the equation of motion relative to the coordinate system $\{X_i\}$ for each fluid particle in the continuum in the form (5.1). One also has to individualize fluid particles that are identical from the geometrical viewpoint. The way to make the particles physically distinguishable is to notice that they have different spatial positions, that is, different coordinates, at some initial instant t_0. Let the position of a point at the instant t_0 be known and specified by the coordinates ξ_j. Then the law of motion (5.1) will depend not only on t but also on ξ_j:

$$X_i = f_i(t, \xi_1, \xi_2, \xi_3), \quad (i = 1, 2, 3), \tag{5.2}$$

If all ξ_j are fixed and t varies, then (5.2) becomes the equation of motion for a given point of the continuum that had initial coordinates $X_i = \xi_i$. If, on the other hand, t is fixed and ξ_j varies, then Eq. (5.2) gives the spatial distribution of various points of the continuum at a given instant of time t. Finally, when both ξ_j and t vary, the relation (5.2) gives the law of motion for the whole volume of the continuum.

Variables ξ_j individualizing specific points of the continuum, grouped together with the time variable t, are called Lagrangian variables, and Eq. (5.2) is called the equation of motion of the continuum.

If functions (5.2) and their partial derivatives are continuous and one-to-one functions at each instant of time, then

$$\Delta = |\partial X_i / \partial \xi_j| \neq 0$$

and Eq. (5.2) allows us to express ξ_j through X_i:

$$\xi_j = \xi_j(X_i, t), \tag{5.3}$$

functions $\xi_j(X_i, t)$ being continuous. Jakobian Δ has the meaning of the ratio of volumes of fluid elements as a result of transformation (5.2). For incompressible fluids, $\Delta = 1$.

The set of values X_i specifies the spatial domain V occupied by the continuum at the moment t. Since $\xi_j \in V_0$ are coordinates of some point M of the domain V_0 at the initial time $t = t_0$, while $X_i \in V$ are coordinates of the same point, which now belongs to a domain V at the time $t > t_0$, the relation (5.3) can be thought of as a one-to-one and continuous mapping of domains V and V_0. It is a well-known topological property of this class of transformations that a volume is mapped onto a volume, a surface – onto a surface, a line – onto a line, and a closed line – onto a closed line.

In addition to the fixed frame of reference $\{X_i\}$, one can introduce a moving frame $\{\xi_i\}$ attached to the fluid particle (an elementary volume around the point ξ_i). In the mechanics of continuum, this frame is considered as a frame that is "frozen" into the elementary volume. As time goes on, it moves and continuously deforms together with the volume. Therefore Lagrangian coordinates can be regarded as an alternative set of coordinates for the same spatial points, and the coordinate system $\{\xi_i\}$ – as a moving, deformable, curvilinear coordinate system. The points that form an elementary volume of the continuum move relative to the fixed coordinate system $\{X_i\}$, but they are at rest relative to coordinate system $\{\xi_i\}$. Use of coordinates ξ_i and t as independent variables is the essence of the Lagrangian approach to the study of continuum motion.

Comparing the two approaches, one comes to the following conclusion. Eulerian approach implies that one is interested in the change of parameters such as velocity, temperature, concentration, and so forth, at a given point visited by a continuous succession of different points of the medium. Lagrangian approach means that one is interested in the same parameters at a given individual point that is moving with the continuum. From the mathematical viewpoint, the two approaches differ in that in the former, the variables are the coordinates of spatial points X_i, while in the latter – the parameters ξ_i (initial coordinates of fluid particles) individualizing different points of the continuum (plus the time variable t that is present in both approaches).

Consider now how the parameters can be determined using either Eulerian or Lagrangian approach. Take a point of the continuum, whose radius vector relative to a motionless coordinate system at the time t is $\boldsymbol{X}$. It depends on ξ_i and t. Lagrangian approach requires the values of ξ_i to be fixed, so the velocity of this point is equal to

$$\boldsymbol{V} = \left(\frac{\partial \boldsymbol{X}(\boldsymbol{\xi}, t)}{\partial t}\right)_{\xi_j} \tag{5.4}$$

or, in the component form,

$$V_i(\xi, t) = \left(\frac{\partial X_i(\boldsymbol{\xi}_j, t)}{\partial t}\right)_{\xi_j},$$

where $V_i(\xi, t) = V_i(\xi_1, \xi_2, \xi_3, t)$. The other parameters change with time in the same manner. The derivative (5.4) characterizes the change of parameters at a given point in the continuum; it is called substantial, material, or total derivative and is designated as D/Dt or d/dt. In the Eulerian approach, the velocity $\boldsymbol{u}$ and other parameters depend on X_1, X_2, X_3 and t and are determined from hydrodynamic equations. In the Lagrangian approach these parameters can also be determined from hydrodynamic equations, if we write these equations in Lagrangian, rather than Eulerian, variables.

But this transformation leads to cumbersome expressions, and solution of the equations thus obtained involves considerable mathematical difficulties. For this reason, one usually makes use of hydrodynamic equations in Eulerian variables.

If the distribution of a parameter, for example, of temperature ϑ, is given both as $\vartheta = \vartheta(\xi, t)$ (the Lagrangian approach) and $\vartheta = \vartheta(\boldsymbol{X}, t)$ (the Eulerian approach), the substantial derivative follows from Eq. (5.2):

$$\left(\frac{\partial\vartheta}{\partial t}\right)_{\xi_i} = \frac{D\vartheta}{Dt} = \left(\frac{\partial\vartheta}{\partial t}\right)_{X_i} + V_i\frac{\partial\vartheta}{\partial X_i}. \tag{5.5}$$

Let us see how one can make a transition from Eulerian to Lagrangian variables. Let the velocity distribution $u_i = u_i(\boldsymbol{X}, t)$ in Eulerian variables be known, for instance, from the solution of the Navier–Stokes equations. Then from Eq. (5.4) there follows a set of ordinary differential equations

$$\frac{\partial X_i}{\partial t} = u_i(\boldsymbol{X}, t), \quad (i = 1, 2, 3) \tag{5.6}$$

and initial conditions

$$\boldsymbol{X}_i(t_0) = \xi_i. \tag{5.7}$$

These equations describe trajectories $X_i = X_i(\xi_1, \xi_2, \xi_3, t)$ of continuum points whose initial coordinates are equal to ξ_i. Parameters ξ_i individualize these points and can be regarded as Lagrangian variables.

The relations $X_i = X_i(\xi_1, \xi_2, \xi_3, t)$ allow a transition from Eulerian variables X_i to Lagrangian variables ξ_i in all dependencies involving hydrodynamic parameters.

Hereafter, Eulerian velocity and coordinates will be denoted through $\boldsymbol{u}(\boldsymbol{X}, t)$ and $\boldsymbol{X} = (X_1, X_2, X_3)$, and Lagrangian velocity and coordinates – through $\boldsymbol{V} = \boldsymbol{V}(\boldsymbol{\xi}, t)$ and $\boldsymbol{\xi} = (\boldsymbol{\xi}_1, \boldsymbol{\xi}_2, \boldsymbol{\xi}_3)$. The relations (5.2) and (5.3) can then be represented in the form

$$\boldsymbol{X} = \boldsymbol{X}(\boldsymbol{\xi}, t), \quad \boldsymbol{\xi} = \boldsymbol{\xi}(\boldsymbol{X}, t), \tag{5.8}$$

with the initial condition

$$\boldsymbol{\xi} = \boldsymbol{X}(\boldsymbol{\xi}, t_0). \tag{5.9}$$

Eulerian $\boldsymbol{u}(\boldsymbol{X},t)$ and Lagrangian $\boldsymbol{V}(\boldsymbol{\xi}, t)$ velocities are related to each other by

$$V(\boldsymbol{\xi}, t) = \boldsymbol{u}(\boldsymbol{X}(\boldsymbol{\xi}, t), t). \tag{5.10}$$

So, the expression (5.4) can be interpreted as the relation between Eulerian and Lagrangian coordinates:

$$\frac{\partial \boldsymbol{X}(\boldsymbol{\xi}, t)}{\partial t} = \boldsymbol{u}(\boldsymbol{X}(\boldsymbol{\xi}, t), t). \tag{5.11}$$

Another way to establish connection between Eulerian and Lagrangian coordinates is based on writing the conservation condition for some continuum characteristic Ψ, say, fluid particle mass. In the Lagrangian approach, it depends only on Lagrangian coordinates individualizing the point. Therefore $\Psi = \Psi(\xi)$. In Eulerian variables, on the other hand, its value would vary in time at a fixed spatial point. Using both ways of writing one and the same characteristic that is conserved during the motion of the continuum, we obtain

$$\Psi(\xi) = \psi(\boldsymbol{X}(\xi, t), t). \tag{5.12}$$

Eq. (5.12) provides an alternative relation between Eulerian and Lagrangian coordinates.

The condition of conservation of Ψ means that

$$\frac{\partial \Psi(\xi)}{\partial t} = \frac{D\psi(\boldsymbol{X}(\xi, t), t)}{Dt} = \frac{\partial \psi}{\partial t} + u_i \frac{\partial \psi}{\partial X_i} = \frac{\partial \psi}{\partial t} + \frac{\partial (u_i \psi)}{\partial X_i} = 0.$$

In deriving this equation, we have used the relation (5.7) for the substantial derivative as well as the continuity equation $\partial u_i / \partial X_i = 0$.

Hence the Eulerian field of any conserved characteristic of the continuum satisfies the following transport equation:

$$\frac{\partial \psi}{\partial t} + \frac{\partial (u_i \psi)}{\partial X_i} = 0. \tag{5.13}$$

One example of conserved quantity is the unit mass concentrated initially at the point ξ_0. Then Ψ is given by $\Psi = \delta(\xi - \xi_0)$, or, in Eulerian variables,

$$\Psi = \psi(\boldsymbol{X}, t) = \delta(\boldsymbol{X} - \boldsymbol{X}(\xi_0, t)). \tag{5.14}$$

This function obeys Eq. (5.13) with the initial condition $\psi(\boldsymbol{X},0) = \delta(\boldsymbol{X} - \xi_0)$ (see Section 1.12).

When considering turbulent flow in the previous chapter, we assumed the Eulerian velocity field $\boldsymbol{u}(\boldsymbol{X}, t)$ to be a random function at a fixed spatial point $\boldsymbol{X}$. Then, owing to the relation (5.12), the Lagrangian velocity $\boldsymbol{V}(\xi, t)$ should be equal to the value of the random function $\boldsymbol{u}(\boldsymbol{X}, t)$, but, in contrast to the Eulerian velocity field, the point $\boldsymbol{X}$ is a random point $\boldsymbol{X}(\xi, t)$ that corresponds to the random position of a continuum point at the time t, given the initial position $\boldsymbol{X} = \xi$ of this point.

It should be noted that all we said thus far refers not only to fluid particles of the carrier fluid, but also to particles of a foreign medium suspended in the fluid, provided that the size of particles and their volume concentration are sufficiently small so that their inertia and the hindered character of motion, that is, the effect of particle interactions, can be neglected, and that they move with the velocity of the carrier fluid.

5.2 Lagrangian Statistical Characteristics of Turbulence

In a turbulent flow, both Eulerian and Lagrangian velocity fields are random fields. This means that coordinates of a fluid particle will also be random fields, since from Eq. (5.4) there follows

$$X(\xi, t) = \xi + \int_{t_0}^{t} V(\xi, t)dt. \tag{5.15}$$

If we consider a finite set of n fluid particles of the continuum initially located at $\xi_1, \xi_2, \ldots, \xi_n$, then there should exist a multidimensional joint probability density distribution of coordinates X and velocities V of these particles at different instants of time $t_1, t_2, \ldots, t_m$. These distributions are functions of $3n+m$ variables, which include the coordinates of Lagrangian variables $\xi_1, \xi_2, \ldots, \xi_n$ and the times $t_1, t_2, \ldots, t_m$:

$$p(X, V|\xi_1, \xi_2, \ldots, \xi_n; t_1, t_2, \ldots, t_m). \tag{5.16}$$

Multidimensional probability density distributions (5.16) are the basic Lagrangian statistical characteristics of turbulence.

It should be noted that statistical characteristics of the type (5.16) depend only on Lagrangian variables. In addition to distributions (5.16), there can also exist probability density distributions for a set of quantities, some of which are Lagrangian and the others are Eulerian, for example,

$$p(X, V, u|\xi_1, \xi_2, \ldots, \xi_n; X_1, X_2, \ldots, X_n; t_1, t_2, \ldots, t_m).$$

Some general relations exist between different Lagrangian statistical characteristics of turbulence. We shall list the most important of them. The first relation is a consequence of the transport equation (5.13). Since expression (5.14) is a solution of the transport equation, substituting it into Eq. (5.13) and noting that, according to Eq. (5.11),

$$u_i(X, t)\delta(X - X(\xi, t)) = u_i(X(\xi, t), t)\delta(X - X(\xi, t)) = V_i(\xi, t)\delta(X - X(\xi, t)),$$

we obtain

$$\frac{\partial\delta(X - X(\xi, t))}{\partial t} + \frac{\partial V_i(\xi, t)\delta(X - X(\xi, t))}{\partial X_i} = 0. \tag{5.17}$$

Recalling the definition of averaging (see (1.26)) and in view of Eq. (5.16), we can write

$$\langle\delta(X - X(\xi, t))\rangle = p(X|\xi, t),$$

$$\langle V_i(\xi, t)\delta(X - X(\xi, t))\rangle = \int V_i(\xi, t)\, p(X, V|\xi, t)dV.$$

The operation of averaging of Eq. (5.17) results in a statistical analog of the transport equation (5.13):

$$\frac{\partial p(\boldsymbol{X}|\boldsymbol{\xi},t)}{\partial t}+\frac{\partial}{\partial X_i}\int V_i(\boldsymbol{\xi},t)\,p(\boldsymbol{X},\boldsymbol{V}|\boldsymbol{\xi},t)d\boldsymbol{V}=0. \tag{5.18}$$

As was noted earlier, Lagrangian velocity $\boldsymbol{V}(\boldsymbol{\xi}, t)$ is the value of the random function $\boldsymbol{u}(\boldsymbol{X}, t)$ at a random point $\boldsymbol{X}(\boldsymbol{\xi}, t)$. Introduce now the random function $\boldsymbol{V}(\boldsymbol{X}_1, \boldsymbol{\xi}, t)$, which has the meaning of velocity of those fluid particles that were located at $\boldsymbol{X}=\boldsymbol{\xi}$ at the initial moment $t=t_0$ and were then found at the fixed point $\boldsymbol{X}_1=\boldsymbol{X}(\boldsymbol{\xi},t)$ at the later moment $t>t_0$. Such a function corresponds to the probability density distribution $p(\boldsymbol{V}|\boldsymbol{X}_1, \boldsymbol{\xi}, t)$ of the quantity $\boldsymbol{V}(\boldsymbol{\xi}, t)$ under the condition $\boldsymbol{X}(\boldsymbol{\xi}, t)=\boldsymbol{X}_1$.

In the general case, the probability density of the conditional distribution for n fluid particles characterized by Lagrangian velocities $\boldsymbol{V}(\boldsymbol{\xi}_1, t_1), \boldsymbol{V}(\boldsymbol{\xi}_2,t_2), \ldots, \boldsymbol{V}(\boldsymbol{\xi}_n,t_n)$ at different instants of time can be written as

$$\begin{aligned}
&p(\boldsymbol{V}_1,\boldsymbol{V}_2,\ldots,\boldsymbol{V}_n|\boldsymbol{\xi}_1,\boldsymbol{\xi}_2,\ldots,\boldsymbol{\xi}_n;t_1,t_2,\ldots,t_n)\\
&=\int\!\!\int\cdots\int p(\boldsymbol{V}_1,\boldsymbol{V}_2,\ldots,\boldsymbol{V}_n|\boldsymbol{X}_1,\boldsymbol{X}_2,\ldots,\boldsymbol{X}_n;t_1,t_2,\ldots,t_n)\\
&\quad\times p(\boldsymbol{X}_1,\boldsymbol{X}_2,\ldots,\boldsymbol{X}_n|\boldsymbol{\xi}_1,\boldsymbol{\xi}_2,\ldots,\boldsymbol{\xi}_n;t_1,t_2,\ldots,t_n)d\boldsymbol{X}_1d\boldsymbol{X}_2\ldots d\boldsymbol{X}_n.
\end{aligned} \tag{5.19}$$

This formula follows from the theorem of total probability, which states that the probability for n fluid particles located initially at the points $\boldsymbol{\xi}_1, \boldsymbol{\xi}_2, \ldots, \boldsymbol{\xi}_n$ to have velocities $\boldsymbol{V}(\boldsymbol{\xi}_1, t_1), \boldsymbol{V}(\boldsymbol{\xi}_2, t_2), \ldots, \boldsymbol{V}(\boldsymbol{\xi}_n, t_n)$ at the instants of time $t_1, t_2, \ldots, t_m$ is equal to the product of the conditional probability density of these velocities (under the condition that coordinates of these particles at the corresponding instants of time assume the values $\boldsymbol{X}(\boldsymbol{\xi}_1, t_1)=\boldsymbol{X}_1, \boldsymbol{X}(\boldsymbol{\xi}_2, t_2)=\boldsymbol{X}_2, \ldots, \boldsymbol{X}(\boldsymbol{\xi}_n, t_n)=\boldsymbol{X}_n$) and the joint probability density distribution of Lagrangian random quantities $\boldsymbol{X}(\boldsymbol{\xi}_1, t_1), \boldsymbol{X}(\boldsymbol{\xi}_2, t_2), \ldots, \boldsymbol{X}(\boldsymbol{\xi}_n, t_n)$, integrated over all possible values of $\boldsymbol{X}_1, \boldsymbol{X}_2, \ldots, \boldsymbol{X}_n$. In the special case $\boldsymbol{\xi}_1=\boldsymbol{\xi}_2=\ldots=\boldsymbol{\xi}_n=\boldsymbol{\xi}$ this formula contains probability the density distribution of coordinates and velocities of one and the same particle at different instants of time:

$$\begin{aligned}
&p(\boldsymbol{V}_1,\boldsymbol{V}_2,\ldots,\boldsymbol{V}_n|\boldsymbol{\xi};t_1,t_2,\ldots,t_n)\\
&=\int\!\!\int\cdots\int p(\boldsymbol{V}_1,\boldsymbol{V}_2,\ldots,\boldsymbol{V}_n|\boldsymbol{X}_1,\boldsymbol{X}_2,\ldots,\boldsymbol{X}_n;t_1,t_2,\ldots,t_n)\\
&\quad\times p(\boldsymbol{X}_1,\boldsymbol{X}_2,\ldots,\boldsymbol{X}_n|\boldsymbol{\xi};t_1,t_2,\ldots,t_n)d\boldsymbol{X}_1d\boldsymbol{X}_2\ldots d\boldsymbol{X}_n.
\end{aligned} \tag{5.20}$$

In the case of $n=1$ (one fluid particle), we have

$$p(\boldsymbol{V}|\boldsymbol{\xi},t)=\int p(\boldsymbol{V}|\boldsymbol{X},t)\,p(\boldsymbol{X}|\boldsymbol{\xi},t)d\boldsymbol{X}. \tag{5.21}$$

It follows from the last formula that the mean value of Lagrangian velocity $\langle \boldsymbol{V}(\boldsymbol{\xi}, t)\rangle$ may be represented as

$$\langle \boldsymbol{V}(\boldsymbol{\xi}, t)\rangle = \int \langle \boldsymbol{V}(\boldsymbol{X}, \boldsymbol{\xi}, t)\rangle \, p(\boldsymbol{X}|\boldsymbol{\xi}, t) d\boldsymbol{X}. \tag{5.22}$$

After a certain amount of time, the fluid particle forgets the past. This time has the meaning of Lagrangian correlation time $T^{(L)}$,

$$T^{(L)} = \int_{t_0}^{\infty} \frac{\langle V_i(\boldsymbol{\xi}, t) V_i(\boldsymbol{\xi}, t_0)\rangle}{\langle \boldsymbol{V}^2(\boldsymbol{\xi}, t)\rangle \langle \boldsymbol{V}^2(\boldsymbol{\xi}, t_0)\rangle^{1/2}} dt \equiv \int \Psi^{(L)}(t) dt, \tag{5.23}$$

where $\Psi^{(L)}(t)$ is the Lagrangian autocorrelation function.

Hence, it should be anticipated that at the time $t > T^{(L)}$, the probability density distribution of Lagrangian velocity $\boldsymbol{V}(\boldsymbol{X}, \boldsymbol{\xi}, t)$ will, for all practical purposes, be independent of ξ, so at these values of t the random quantity $\boldsymbol{V}(\boldsymbol{X}, \boldsymbol{\xi}, t)$ can be taken equal to the Eulerian velocity $\boldsymbol{u}(\boldsymbol{X}, t)$. Then the probability density distribution $p(\boldsymbol{V}|\boldsymbol{X}, \boldsymbol{\xi}, t)$ will be identically equal to the probability density distribution of $\boldsymbol{V} = \boldsymbol{u}(\boldsymbol{X}, t)$, that is, of the Eulerian velocity field at a fixed spacetime point $(\boldsymbol{X}, t)$. The formula (5.22) will then take the form

$$\langle \boldsymbol{V}(\boldsymbol{\xi}, t)\rangle = \int \langle u(\boldsymbol{X}, t)\rangle \, p(\boldsymbol{X}|\boldsymbol{\xi}, t) d\boldsymbol{X}. \tag{5.24}$$

In the case of $\boldsymbol{n} = 2$, the relation (5.20) at $t = t_0$ and $t_2 = t$ gives the joint probability density distribution of the quantities $\boldsymbol{V}_1 = \boldsymbol{V}(\boldsymbol{\xi}, t_0)$ and $\boldsymbol{V}_2 = \boldsymbol{V}(\boldsymbol{\xi}, t)$ at two successive instants of time:

$$p(\boldsymbol{V}_1, \boldsymbol{V}_2|\boldsymbol{\xi}, t_0, t) = \int p(\boldsymbol{V}_1, \boldsymbol{V}_2|\boldsymbol{X}; \boldsymbol{\xi}, t_0, t) \, p(\boldsymbol{X}|\boldsymbol{\xi}, t) d\boldsymbol{X}. \tag{5.25}$$

It should be kept in mind that $\boldsymbol{V}_1 = \boldsymbol{V}(\boldsymbol{\xi}, t_0) = \boldsymbol{u}(\boldsymbol{\xi}, t_0)$.

From the formula (5.25) one can derive the expression for the Lagrangian correlation function $\langle \boldsymbol{V}_i \boldsymbol{V}_j\rangle$ – second-order moment for the components of the vector $\boldsymbol{V}$:

$$\langle V_i(\boldsymbol{\xi}, t_0) V_j(\boldsymbol{\xi}, t)\rangle = \int \langle V_i(\boldsymbol{\xi}, t_0) V_j(\boldsymbol{X}, \boldsymbol{\xi}, t)\rangle \, p(\boldsymbol{X}|\boldsymbol{\xi}, t) d\boldsymbol{X}. \tag{5.26}$$

In the more general case, the Lagrangian correlation function for the components of the vector $\boldsymbol{V}$ at different instants of time could be derived by using the formulas (5.20) and (5.25):

$$\langle V_i(\boldsymbol{\xi}, t_1) V_j(\boldsymbol{\xi}, t_2)\rangle = \int \langle V_i(\boldsymbol{X}_1, \boldsymbol{\xi}, t_1) V_j(\boldsymbol{X}_1, \boldsymbol{X}_2, \boldsymbol{\xi}, t_1, t_2)\rangle$$

$$\times \, p(\boldsymbol{X}_1, \boldsymbol{X}_2|\boldsymbol{\xi}, t_1, t_2) d\boldsymbol{X}_1 d\boldsymbol{X}_2. \tag{5.27}$$

where $V(X_1, X_2, \xi, t_1, t_2)$ is the random Lagrangian velocity of a fluid particle initially located at the point $X = \xi$ and found at fixed points X_1 and X_2 at the instants of time t_1 and t_2. For sufficiently large values of t_1, one can take $V(X_1, \xi, t_1) = u(X_1, t_1)$, whereas for $t_2 \gg t_1$, it is safe to assume $V(X_1, X_2, \xi, t_1, t_2) = u(X_2, t_2)$.

The motion of a continuum point (fluid particle) being initially at the space point ξ has been hitherto described by the vector $X(\xi, t)$, which gives the random position of this particle at the moment t. This enabled us to derive those statistical characteristics of particle motion that are given by the relations (5.17)–(5.27). Consider now the other statistical characteristics of particle motion. A distinctive feature of the behavior of particles suspended in a turbulent flow is that the distance between particles changes in time. Section 4.9 has shown that two particles that were initially close together tend to separate as time goes on. The time it takes for the interparticle distance to reach a given value is estimated by Eq. (4.120). Therefore it makes sense to consider the displacement of particles from their initial position. First, let us consider the displacement of a single fluid particle.

Instead of the vector $X(\xi, t)$, we shall introduce the vector $Y(\tau)$ of particle displacement from the initial position over the time interval τ:

$$Y(\tau) = X(\xi, t_0 + \tau) - \xi = \int_{t_0}^{t_0+\tau} V(\xi, t)dt. \qquad (5.28)$$

In order to determine statistical characteristics of the random displacement vector $Y(\tau)$, it is necessary to specify the probability density distribution $p(Y|\tau, \xi, t_0)$. Let us start with the case when the time interval t is small compared to the Lagrangian correlation time $T^{(L)}$, that is, $\tau \ll T^{(L)}$. In such a short time, the Lagrangian velocity will be practically unchanged, so Eq. (5.28) takes the form

$$Y(\tau) \approx \tau V(\xi, t_0) = \tau u(\xi, t_0). \qquad (5.29)$$

Consider now the probability density distribution $p(u|\xi, t_0)$ of the Eulerian velocity $(u|\xi, t_0)$ at a fixed point ξ at the instant of time $t = t_0$. Recalling that the vectors Y, u are three-dimensional and using the normalization condition for the probability density together with Eq. (5.29), we obtain

$$p(Y|\tau; \xi, t_0) = \tau^{-3} p(u|\xi, t_0) = \tau^{-3} p\left(\frac{Y}{\tau}|\xi, t_0\right). \qquad (5.30)$$

A large body of experiments shows that in a steady turbulent flow, the distribution $p(u|\xi, t_0)$ and thus the distribution $p(Y|\tau, \xi, t_0)$ is close to normal (Gaussian) at small values of τ, that is, for $\tau \ll T^{(L)}$, and at relatively large values of τ, that is, for $\tau \gg T^{(L)}$. The close resemblance of both particle displacement distributions to a normal one at $\tau \ll T^{(L)}$ and $\tau \gg T^{(L)}$ makes it reasonable to hypothesize that for the intermediate values of τ, the particle displacement distribution will be normal as well. The assumption implies the absence of any radical rearrangement of the distribution (from the normal one at small τ and again to a normal distribution at large τ) in this intermediate region. The main advantage of this assumption is that it considerably simplifies the derivation of statistical characteristics of particle motion in a turbulent flow.

Let us now consider the statistical characteristics of particle displacement. Because of the assumption we made, it will not be very different from Gaussian, so it is sufficient to determine only two first moments (see Section 1.8). In view of Eq. (5.29), the mean value of the vector $\mathbf{Y}(\tau)$ is

$$\langle \mathbf{Y}(\tau)\rangle = \int_{t_0}^{t_0+\tau} \langle \mathbf{V}(\boldsymbol{\xi}, t)\rangle dt. \tag{5.31}$$

For small values of τ, we obtain from Eq. (5.31)

$$\langle \mathbf{Y}(\tau)\rangle \approx \boldsymbol{U}\tau, \tag{5.32}$$

where $\boldsymbol{U}$ is the average flow velocity.

Let us introduce the fluctuations of displacement $\mathbf{Y}'$ and particle velocity $\mathbf{V}'$:

$$\mathbf{Y}'(\tau) = \mathbf{Y}(\tau) - \langle \mathbf{Y}(\tau)\rangle, \quad \mathbf{V}'(\boldsymbol{\xi}, t) = \mathbf{V}(\boldsymbol{\xi}, t) - \langle \mathbf{V}(\boldsymbol{\xi}, t)\rangle.$$

Due to Eq. (5.31),

$$\langle \mathbf{Y}'(\tau)\rangle = \int_{t_0}^{t_0+\tau} \mathbf{V}(\boldsymbol{\xi}, t)\langle \mathbf{V}'(\boldsymbol{\xi}, t)\rangle dt. \tag{5.33}$$

The second central moments for components of the vector $\mathbf{Y}$ at one and the same instant of time τ are components of the fluctuation correlation tensor, which is called the dispersion tensor of fluid particle displacements. Hitherto it has been denoted as b_{ij} (see Section 1.6). Since it is associated with the diffusion tensor in a turbulent flow, let us denote it further on as d_{ij}:

$$d_{ij}(\tau) = \langle Y_i'(\tau) Y_j'(\tau)\rangle = \int_{t_0}^{t_0+\tau}\int_{t_0}^{t_0+\tau} \langle V_i'(\boldsymbol{\xi}, t_1) V_j'(\boldsymbol{\xi}, t_2)\rangle dt_1 dt_2. \tag{5.34}$$

The assumption that the displacement of a fluid particle is distributed in accordance with the Gaussian law means that first two moments characterize this distribution completely.

At small values of τ, we find from Eq. (5.34) while taking into account Eq. (5.29)

$$d_{ij}(\tau) \approx \langle u_i'(\boldsymbol{\xi}, t_0) u_j'(\boldsymbol{\xi}, t_0)\rangle \tau^2 = b_{ij}^{(0)}\tau^2. \tag{5.35}$$

Coefficients $b_{ij}^{(0)} = \langle u'_i(\xi, t_0) u'_j(\xi, t_0) \rangle$ are the correlation functions of fluctuations. In a steady turbulent flow, they depend only on and under the additional condition of homogeneity of Eulerian velocity they are constant.

Consider the case of stationary homogeneous turbulence. Then all hydrodynamic fields are homogeneous random fields and stationary random functions. The mean velocity $\langle \boldsymbol{u} \rangle$ is constant in space and in time (see Section 1.9). In accordance with Eq. (5.10) and Eq. (5.32), we have

$$\langle \boldsymbol{V}(\xi, t) \rangle = \langle \boldsymbol{u}(\boldsymbol{X}(\xi, t)) \rangle = \langle \boldsymbol{u} \rangle, \quad \langle \boldsymbol{Y}(\tau) \rangle = \langle \boldsymbol{u} \rangle \tau. \tag{5.36}$$

The fluctuation velocity $\boldsymbol{V}'(\boldsymbol{X}, t)$of a fluid particle will possess the same statistical characteristics for all $\boldsymbol{X}$, in other words, it will be a stationary random function. The following equation holds:

$$\langle V'_i(\xi, t_1) V'_j(\xi, t_2) \rangle = b_{ij}^{(L)}(t_2 - t_1) = (\langle u'_i 2 \rangle \langle u'_j 2 \rangle)^{1/2} \Psi_{ij}^{(L)}(t_2 - t_1), \tag{5.37}$$

where $\Psi_{ij}^{(L)} = \frac{b_{ij}^{(L)}}{(\langle u'_i 2 \rangle)(\langle u'_j 2 \rangle)^{1/2}} = \frac{\langle u'_i u'_j \rangle}{(\langle u'_i 2 \rangle)(\langle u'_j 2 \rangle)^{1/2}}$ are correlation coefficients.

Here we have used the superscript symbol (L) to indicate that the corresponding function is a Lagrangian correlation function.

Changing the variables in Eq. (5.37) according to $s = (t_2 - t_1)$, $t = (t_1 + t_2)/2$ and substituting the result into Eq. (5.34), one gets

$$\begin{aligned} d_{ij}(\tau) &= \int_0^{\tau} \int_{t_0 + \frac{s}{2}}^{t_0 + \tau - \frac{s}{2}} (b_{ij}^{(L)}(s) + b_{ji}^{(L)}(s)) dt ds \\ &= (\langle u'_i 2 \rangle \langle u'_j 2 \rangle)^{1/2} \int_0^{\tau} \int_{t_0 + \frac{s}{2}}^{t_0 + \tau - \frac{s}{2}} (\Psi_{ij}^{(L)}(s) + \Psi_{ji}^{(L)}(s)) dt ds \\ &= (\langle u'_i 2 \rangle \langle u'_j 2 \rangle)^{1/2} \int_0^{\tau} (\tau - s) [\Psi_{ij}^{(L)}(s) + \Psi_{ji}^{(L)}(s)] ds. \end{aligned} \tag{5.38}$$

The special case $i = j$ yields

$$d_{ii}(\tau) = 2 \int_0^{\tau} (\tau - s) b_{ii}^{(L)}(s) ds = 2 \langle u'_i 2 \rangle \int_0^{\tau} (\tau - s) \Psi_{ii}^{(L)}(s) ds. \tag{5.39}$$

Lagrangian correlation coefficients $\Psi_{ii}^{(L)}(s)$ have the property that $\Psi_{ii}^{(L)}(s) \to 0$ at $s \to \infty$, in agreement with our intuitive understanding that the correlation of Lagrangian velocities of a fluid particle at two different moments $t_2 - t_1 >> T_i^{(L)}$ should not be noticeable. With this in mind, we introduce the characteristic time

$$T_i^{(L)} = \int_{\infty}^{\tau} \Psi_{ii}^{(L)}(s)ds, \tag{5.40}$$

which is simply the time that must pass before we get $\Psi_{ii}^{(L)}(s) \approx 0$. We can take as our Lagrangian correlation time $T^{(L)}$ either the maximum or the average time value from the set $T_i^{(L)}$. Note that this time is of the same order of magnitude as the Lagrangian correlation time introduced earlier by the formula (5.23).

In many cases, especially for large Reynolds numbers, the function $\Psi_{ii}^{(L)}$ can be approximated by an exponential dependence:

$$\Psi_{ii}^{(L)} \approx \exp(-t/T_i^{(L)}). \tag{5.41}$$

Although this is not exactly the right function to describe the shape of the correlation curve, especially at $t \to 0$ and $t \to \infty$, its use leads to satisfactory outcomes in many practical problems.

For $\tau >> T_i^{(L)}$, the expression (5.39) can be replaced by the asymptotic relation

$$d_{ii}(\tau) = 2\langle u_i' 2\rangle \int_0^{\infty} (\tau - s)\Psi_{ii}^{(L)}(s)ds = 2\langle u_i' 2\rangle (T_i^{(L)}\tau - s_i), \tag{5.42}$$

where $S_i = \int_0^{\infty} s\Psi_{ii}^{(L)}(s)ds$ (assuming, of course, that this integral exists).

Since $\tau >> T_i^{(L)}$, we have

$$d_{ii}(\tau) \approx 2\langle u_i' 2\rangle T_i^{(L)}\tau. \tag{5.43}$$

If $i \neq j$, we can still get a similar expression by introducing (instead of the characteristic time (5.40))

$$T_{ij}^{(L)} = \int_0^{\infty} [\Psi_{ij}^{(L)}(s) + \Psi_{ji}^{(L)}(s)]ds. \tag{5.44}$$

For large values of τ, the dispersion of the fluid particle displacement equals

$$d_{ij}(t) \approx (\langle u_i' 2\rangle \langle u_j' 2\rangle)^{1/2} T_{ij}^{(L)}\tau \tag{5.45}$$

whereas for small values of τ, in accordance with Eq. (5.35), it is equal to

$$d_{ij}(\tau) \approx (\langle u_i' 2\rangle \langle u_j' 2\rangle)^{1/2} b_{ij}^{(0)}\tau^2. \tag{5.46}$$

For the intermediate values of τ, the dependence $d_{ij}(\tau)$ is more complex because it also depends on the form of $\Psi_{ij}^{(L)}$.

Hence, the dispersion of the fluid particle's displacement in a turbulent flow is proportional to , given a sufficiently long time $\tau \gg T^{(L)}$, whereas for short $\tau \ll T^{(L)}$, the dispersion is proportional to τ^2.

We must mention the analogy between the obtained result and the laws (3.30) and (3.32) that describe Brownian diffusion. As far as particle displacements are concerned, the difference between the Brownian and turbulent diffusion consists in the form of the proportionally coefficients, as well as in different characteristic times.

The above-considered model of homogeneous, stationary turbulence is an idealization of real turbulence, therefore any application of the obtained results to spatial flows should be questioned. For example, the formula (5.42) can be used only if the homogeneity condition in the direction of the X_i-axis holds. One particular case when this condition is satisfied is that of a steady turbulent flow in a long pipe whose symmetry axis coincides with the X_i-axis of the coordinate system. Let $Y_i(\tau)$ denote the component of the fluid particle's displacement vector $\mathbf{Y}$ along the X_i-axis. Then the corresponding Lagrangian velocity equals

$$V_i(\boldsymbol{\xi}, t_0 + \tau) = \frac{dY_i(\tau)}{d\tau}. \tag{5.47}$$

Denote through $\bar{U}$ the mean-flow-rate velocity of the fluid in the direction of the X_i-axis and assume that after a sufficiently long time , the initial position of the fluid particle will not have any noticeable effect on its motion and, in particular, on its Lagrangian velocity. Then

$$V_i(\boldsymbol{\xi}, t_0 + \tau) \approx \bar{U} \tag{5.48}$$

and one can see from Eq. (5.47) and Eq. (5.48) that at large values of τ, the following equality for the average displacement of the fluid particle along the $\mathbf{X}_i$-axis is valid:

$$\langle V_i(\tau)\rangle \approx \bar{U}\tau. \tag{5.49}$$

The relation (5.48) yields the root-mean-square displacement of the fluid particle along the X_i-axis:

$$\langle [Y_i(\tau) - \langle Y_i(\tau)\rangle]^2\rangle \approx 2\langle u_i'2\rangle T_i^{(L)}\tau. \tag{5.50}$$

The values of $\langle u_i'2\rangle$ and $T_i^{(L)}$ depend on the pipe radius R and on the dynamic velocity $u^* = \sqrt{\tau_0/\rho_e}$, so that

$$\langle u_i'2\rangle T_i^{(L)} = cRu^*$$

where c is a universal constant, τ_0 – the frictional stress at the wall, ρ_e – the fluid density.

Another example is the problem of particle motion in an unbounded turbulent shear flow with constant gradient of the average velocity γ, where the Eulerian field of velocity fluctuations $\boldsymbol{u}'(\boldsymbol{X},\ t)$ is stationary and statistically homogeneous and the average velocity $\langle \boldsymbol{u}(\boldsymbol{X})\rangle$ does not change in time and has a linear dependence on spatial coordinates. Let the average velocity be directed along the X_i-axis, so that

$$\langle u_1\rangle = \dot{\gamma}X_3, \quad \langle u_2\rangle = \langle u_3\rangle = 0.$$

Consider the motion of a fluid particle that was initially ($t=0$) located at the point $\xi=0$. At the moment t the particle is located at a point with coordinates $X_i(t)$ and has Lagrangian velocity $\boldsymbol{V}(t)$. Then, in accordance with the formula (5.10), we have

$$\boldsymbol{V}(t) = \langle \boldsymbol{V}(t)\rangle + \boldsymbol{V}'(t) = \langle \boldsymbol{u}(\boldsymbol{X}(t))\rangle + \boldsymbol{u}'(\boldsymbol{X}(t), t)$$

or, in the component form,

$$V_1(t) = \dot{\gamma}X_3(t) + V_1'(t), \quad V_2(t) = V_2'(t), \quad V_3(t) = V_3'(t).$$

Since $\boldsymbol{V}=d\boldsymbol{X}/dt$, the last equations lead us to write

$$X_1(t) = \int_0^t [\dot{\gamma}X_3(t) + V_1'(t)]dt, \quad X_2(t) = \int_0^t V_2'(t)dt, \quad X_3(t) = \int_0^t V_3'(t)dt.$$

Since $\boldsymbol{u}'\langle \boldsymbol{X},t\rangle = 0$, we have $\langle \boldsymbol{V}'(t)\rangle = 0$ and $\langle \boldsymbol{X}(t)\rangle = 0$. The field $\boldsymbol{u}'(\boldsymbol{X},t)$ is homogeneous and stationary, so $\boldsymbol{V}'(\boldsymbol{X},t)$ is also a homogeneous and stationary function, and its correlation tensor has the form

$$V_i'(t_1)V_j'(t_2) = b_{ij}(t_1-t_2).$$

Let us now find the dispersion tensor $d_{ij}(\tau) = \langle X_i(\tau)X_j(\tau)\rangle$. It is apparent that motion along the X_i-axis with the average velocity $\boldsymbol{u}(\boldsymbol{X})$ will not affect the particle's displacement in the directions of X_2- and X_3-axes. Therefore the values $d_{22}(\tau)$, $d_{33}(\tau)$, and $d_{23}(\tau)$, which depend on Lagrangian velocities $V_2(\tau)$ and $V_3(\tau)$, have the form (5.38), just as in the case of $\boldsymbol{u}=const$. However, the components $d_{11}(\tau)$ and $d_{13}(\tau)$ will have a different form, because they depend on the form of the distribution $\langle \boldsymbol{u}(\boldsymbol{X})\rangle$, namely,

$$d_{11}(\tau) = \langle X_1^2(\tau)\rangle = \int_0^\tau\int_0^\tau \Big[(\dot{\tau})^2\langle X_3(t_1)X_3(t_2)\rangle + \gamma\langle X_3(t_1)V'(t_2)\rangle$$

$$+\dot{\gamma}\langle X_3(t_2)V_1'(t_1)\rangle + b_{11}^{(L)}(t_1-t_2)\Big]dt_1dt_2,$$

$$d_{13}(\tau) = \langle X_1(\tau)X_3(\tau)\rangle = \int_0^\tau\int_0^\tau [\dot{\gamma}\langle X_3(t_1)V_3'(t_2)\rangle + \langle V_1'(t_1)V_3'(t_2)\rangle]dt_1dt_2.$$

Substituting the expressions for $X_i(t)$ into these relations, using the formulas (5.38) for $d_{ii}(\tau)$, we obtain after repeated integration by parts

$$d_{11}(\tau) = \frac{(\dot{\gamma})^2}{2}\int_0^\tau (2\tau^3 - 3\tau^2 s + s^2) b_{33}^{(L)}(s)ds + \dot{\gamma}\int_0^\tau (\tau - s)^2 b_{31}^{(L)}(s)ds$$

$$+\dot{\gamma}\int_0^\tau (\tau^2 - s^2) b_{13}^{(L)}(s)ds + 2\int_0^\tau (\tau - s) b_{11}^{(L)}(s)ds,$$

$$d_{13}(\tau) = \dot{\gamma}\tau\int_0^\tau (\tau - s) b_{33}^{(L)}(s)ds + \int_0^\tau (\tau - s)\left[b_{13}^{(L)}(s) + b_{31}^{(L)}(s)\right]ds.$$

We are primarily interested in the asymptotic expressions for components of the dispersion tensor at large values of time, $\tau \gg T^{(L)}$, where, just as before, we take $T^{(L)} = \max(T_1^{(L)}, T_2^{(L)}, T_3^{(L)})$. As follows from Eq. (5.45), the components $d_{22}()$, $d_{33}()$, and $d_{23}()$ are proportional to , whereas $d_{11}()$ and $d_{13}()$ are given by

$$d_{11}(\tau) \approx \frac{2}{3}(\dot{\gamma})^2 \langle u_3' 2\rangle T_3^{(L)}\tau^3, \quad d_{13}(\tau) \approx \dot{\gamma}^2 \langle u_3' 2\rangle T_3^{(L)}\tau^2. \tag{5.51}$$

We conclude from these asymptotic expressions that the shear flow gives rise to anisotropy of particle displacement. The dispersion of particle displacement in the direction of the X_1-axis grows with time much faster ($\sim\tau^3$) than in the transverse directions ($\sim\tau$ along the X_3- and X_1-axes), and the correlation of displacements along the X_1- and X_3-axes is different from zero. There exists for the limiting case of $\tau \gg T^{(L)}$ the following asymptotic expression for the correlation coefficient:

$$\Psi_{13} = \frac{d_{13}(\tau)}{d_{11}(\tau)d_{33}(\tau)^{1/2}} \approx \frac{\sqrt{3}}{2}.$$

It should be noted that similar dependences exist for Brownian motion of particles as well (see Section 3.7).

The above-described analysis of random motion of an elementary volume (fluid particle) can be applied to the study of behavior of a macrovolume (say, one mole) when the latter is visualized as a set of elementary volumes. Such a macrovolume gets deformed during its turbulent motion, because the distance between any two particles grows with time. Therefore, in order to find the extent of macrovolume deformation, one has to consider the relative motion of two fluid particles.

If the two particles are initially spaced far apart, that is, the interparticle distance exceeds the integral scale of turbulence, the interaction between the particles can be neglected, and the statistical characteristics of each particle's motion can be determined in the same manner as we did earlier for one fluid particle. If the initial distance between the fluid particles is small in comparison with the integral scale of turbulence, their motion will be mutually interrelated until they get farther apart.

Therefore it should be anticipated that as time goes on, statistical characteristics of motion of two fluid particles that were initially close together will get closer to those of isolated particles.

We see that the most interesting situation arises when the particles are initially close enough to each other, at the distance, say, of the order of the Kolmogorov scale of turbulence. The relative motion of particles under consideration occurs under the action of fluctuations, which have various scales. Smallness of particle sizes and smallness of the initial interparticle distance means that large-scale fluctuations will transport the fluid volume containing both particles as a whole without changing the interparticle distance. Small-scale fluctuations whose scale is of the order of the interparticle distance will change that distance. As the particles get farther apart with time, the larger-scale fluctuations will be able to affect the relative motion.

It was shown in Section 4.9 that small-scale motion is isotropic. Therefore, at the initial stage the relative motion of two fluid particles takes place in an isotropic turbulent field, although the distance between the particles and the difference of their velocities are both functions of time.

Consider two fluid particles a and b at the moment τ, when they are located at their respective spatial points $\boldsymbol{X}_a$ and $\boldsymbol{X}_b$. The radius vector connecting the two points is $\boldsymbol{r} = \boldsymbol{X}_a - \boldsymbol{X}_b$. The relative motion of particles is characterized by the tensor of relative dispersion, whose components are

$$d_{ij}^{(r)} = \langle r_i r_j \rangle = \langle (X_i X_j)_b \rangle + \langle (X_i X_j)_a \rangle - \langle X_{bi} X_{aj} \rangle - \langle X_{bj} X_{ai} \rangle .$$

Turbulence is assumed to be homogeneous, so

$$\langle (X_i X_j)_b \rangle = \langle (X_i X_j)_a \rangle = \langle X_i X_j \rangle$$

and

$$d_{ij}^{(r)} = 2\langle X_i X_j \rangle - \langle X_{bi} X_{aj} \rangle - \langle X_{bj} X_{ai} \rangle . \tag{5.52}$$

Introduce a new coordinate system $\boldsymbol{X} = \boldsymbol{X} - \boldsymbol{X}_a$, whose origin coincides with particle a. Then, in view of Eq. (5.28), we find that $\langle \boldsymbol{X}_i \boldsymbol{X}_j \rangle = \langle \boldsymbol{Y}_i \boldsymbol{Y}_j \rangle$. At large values of τ ($\tau \gg T^{(L)}$) the last two terms in Eq. (5.52) are small, and

$$d_{ij}^{(r)} \approx 2\langle X_i X_j \rangle = 2\langle Y_i Y_j \rangle = 2\langle Y_i' Y_j' \rangle = 2 d_{ij} = \left(\langle u_i' 2 \rangle \langle u_j' 2 \rangle \right)^{1/2} T_{ij}^{(L)} \tau, \tag{5.53}$$

where d_{ij} is the dispersion tensor given by the formula (5.45).

If $\tau \ll T^{(L)}$, the asymptotic relation (5.35) is valid:

$$d_{ij}^{(r)} = \{ 2 b_{ij}^{(0)} - b_{ij,ba}^{(0)} + b_{ij,ba}^{(0)} \} \tau^2, \tag{5.54}$$

where $b_{ij,ab}^{(0)} = \langle u_{i,b}' (\boldsymbol{\xi}^{(a)}, 0) u_{j,a}' (\boldsymbol{\xi}^{(b)}, 0) \rangle$ are components of the correlation tensor of fluctuations at the initial moment.

Hence, the initial stage of relative motion of two fluid particles is characterized by the Eulerian correlation tensor of fluctuations $b_{ij,ba}^{(0)}$ that corresponds to the initial positions of these particles. For homogeneous isotropic turbulence, components of the tensor of relative dispersion can be determined from Eq. (4.88) and Eq. (4.90):

$$d_{ij}^{(r)} = \tau^2 \left\{ \left(\frac{r_i r_j}{r^2} - \delta_{ij} \right) r \frac{\partial b_{LL}}{\partial r} + 2[1 - B_{LL}(r)] \right\} + r_i r_j, \tag{5.55}$$

where b_{LL} is the component of the tensor b_{ij} in the direction of $\boldsymbol{r}$.

A further simplification of Eq. (5.55) is possible if we make use of the relation (4.125), which is valid for locally isotropic developed turbulence:

$$d_{ij}^{(r)} \approx \frac{2}{3} A \tau^2 \left(4\delta_{ij} - \frac{r_i r_j}{r^2} \right) (\bar{\varepsilon} r)^{2/3} + r_i r_j,$$

where A is a universal constant.

Putting $i = j$ into the last formula, one obtains the root-mean-square displacement of one particle relative to the other at the initial stage of relative motion:

$$(d_{ii}^{(r)} - r^2)^{1/2} \approx \left(\frac{22}{3} A \right)^{1/2} (\bar{\varepsilon} r)^{1/3} \tau. \tag{5.56}$$

A comparison of the relations (5.53) and (5.56) shows that the speed with which the particles recede from each other is different in the two cases. At the initial stage of the process, this speed is smaller than at $\tau \to \infty$.

5.3
Turbulent Diffusion

In Sections 5.1 and 5.2 we used the concept of a fluid particle – a volume that is small enough to be thought of as a material point moving together with the carrier fluid. One can use the Lagrangian approach to keep track of a given particle and to determine statistical Lagrangian characteristics of its random motion. In practice, in order for us to be able to observe the motion of fluid particles, these particles should somehow differ from the surrounding medium. However, the density of particles is usually the same or, at any rate, almost the same as that of the surrounding medium. Therefore a particle under consideration becomes distinguishable from the surrounding medium if it has a different color, chemical properties, or temperature. Such particles are known as impurities; sometimes they are also called "passive" because they do not exert any effect on the fluid motion.

The motion of a passive particle has some important features. The first one is that its Lagrangian velocity at any instant of time coincides with Eulerian velocity of the surrounding fluid at the point where the particle is located at this very instant. The second feature is the so-called turbulent diffusion – a rapid spreading of the impurity injected into the flow at some spatial point. This spreading is caused mainly by the impurity transport via turbulent fluctuations, hence the origin of the term. Since molecules of the impurity differ from molecules of the surrounding medium, their spreading may be caused not only by turbulent diffusion, but also by molecular diffusion and Brownian motion. The characteristic time of the diffusion process can be estimated as $t_D \sim L^2/D$, where D is the diffusion coefficient and L is the characteristic linear size of the region under consideration. Since the coefficient of turbulent diffusion D_t is much greater than the coefficient of molecular diffusion D_m ($D_t/D_m \sim 10^2 - 10^6$), molecular diffusion in a region with length scale of the same order as the integral scale of turbulence requires much longer time than turbulent diffusion. Therefore when considering the process of impurity propagation in a macroscopic region, molecular diffusion can be neglected in comparison with turbulent diffusion. It is just this turbulent diffusion that is responsible for the rapid propagation of impurities in the atmosphere and in fluid streams. However, if one considers diffusion in a microscopic region whose size is of the same order as the microscale of turbulence, molecular diffusion can become noticeable. One example of such processes is the mixing of substances entering a chemical reaction, which will be discussed in Section 5.8.

Impurity is usually injected into the flow as a liquid or gaseous admixture or as small solid particles. It can be thought of as a substance continuously distributed in space and characterized by a Eulerian concentration field $C(\boldsymbol{X}, t)$. For every separate instance of turbulent flow, the concentration field (in the absence of impurity sources) is described by the equation of convective diffusion

$$\frac{\partial C}{\partial t} + \frac{\partial (u_i C)}{\partial X_i} = D_m \Delta C, \tag{5.57}$$

where D_m is the coefficient of molecular diffusion, which is assumed to be constant, and u_i are the velocity components.

The assumption that the impurity is passive means that the velocity field $\boldsymbol{u}$ does not depend on C. The velocity $\boldsymbol{u}$ should then be known from the solution of the corresponding hydrodynamic problem, and one sees that Eq. (5.57) is linear. If any sources or sinks are present in the flow (this could happen, for example, due to homogeneous chemical reactions taking place in the given volume), the right-hand side of Eq. (5.57) will contain the corresponding term, which could be nonlinear with respect to C for certain kinds of reactions.

In order to solve Eq. (5.57), one must provide the initial and boundary conditions,

$$C(\boldsymbol{X}, t_0) = C_0(\boldsymbol{X}) \tag{5.58}$$

and

$$D_m \frac{\partial C}{\partial n} + \beta C = 0 \tag{5.59}$$

The parameter β determines the conditions for the heterogeneous reaction at the surface S – the boundary surface of the flow region V. If $\beta = \infty$, then the surface is completely absorbing, and Eq. (5.59) reduces to the condition $C = 0$. In the other limiting case, when $\beta = 0$, the surface is completely impermeable for the impurity, and $D_m \partial C/\partial n = 0/$. In the general case, β can depend on C. This dependence follows from the kinetics of the chemical reaction on the surface.

Istantaneous sources of impurity at given spatial points are described by initial conditions that have the form of delta functions of time and spatial coordinates. Sources acting continuously are described by delta functions of spatial coordinates only. These delta functions appear in the boundary conditions if the sources are located at the boundary; otherwise, if the sources are located within the continuum volume, they appear as source–sink terms on the right-hand side of the diffusion equation (5.57).

To summarize, we say that under the given homogeneous boundary conditions and in the absence of homogeneous chemical reactions, the concentration field of passive impurity $C(\boldsymbol{X}, t)$ is shaped exclusively by the turbulent transport, whose velocity is $\boldsymbol{u}(\boldsymbol{X}, t)$, and by the molecular diffusion; the velocity field is determined independently (regardless of the concentration), by solving the corresponding hydrodynamic problem. The diffusion problem (5.57)–(5.59) thus reduces to a linear boundary value problem.

When deriving the velocity field $\boldsymbol{u}(\boldsymbol{X}, t)$ of the turbulent flow, one must supplement the hydrodynamic equations with the initial $\boldsymbol{u}(\boldsymbol{X}, 0) = \boldsymbol{u}_0(t)$ and boundary conditions. When studying a turbulent flow, the initial velocity field $\boldsymbol{u}_0(\boldsymbol{x}, t)$ is assumed to be random, in other words, there should exist a corresponding probability density distribution in the functional space of solenoidal vector fields. Then the passive impurity concentration $C(\boldsymbol{X}, t)$ that depends on the velocity distribution $\boldsymbol{u}(\boldsymbol{X}, t)$ will also be a random quantity characterized by some probability density distribution.

If the molecular diffusion is negligibly small compared to the turbulent one, Eq. (5.57) takes the form

$$\frac{\partial C}{\partial t} + \frac{\partial (u_i C)}{\partial X_i} = 0. \qquad (5.60)$$

Consider some general properties of this equation. First of all, it is linear, therefore, given the initial condition $C(\boldsymbol{X}, t) = C_0(\boldsymbol{X})$ and the linear boundary conditions (5.58), (5.59) with $\beta = const$, its solution can be represented in the operator form as

$$C(\boldsymbol{X}, t) = \mathbf{L}[u_0(\boldsymbol{X}, t), t] C_0(\boldsymbol{X}), \qquad (5.61)$$

where $\mathbf{L}$ is a linear operator that depends on the velocity $\boldsymbol{u}_0(\boldsymbol{X}, t)$, on the time t, and on the boundary conditions.

Since $\boldsymbol{u}_0(\boldsymbol{X}, t)$ is a random function, $\mathbf{L}$ is also a random operator in the space of linear operators. Then, averaging Eq. (5.61) over all possible realizations of the initial vector field $\boldsymbol{u}_0(\boldsymbol{X}, t)$ assuming a fixed initial concentration distribution $C_0(\boldsymbol{X})$, one obtains

$$\langle C(\boldsymbol{X}, t)\rangle = \langle \mathrm{L}[\boldsymbol{u}_0(\boldsymbol{X}, t), t]\rangle C_0(\boldsymbol{X}). \qquad (5.62)$$

Since **L** is linear, the operator $\langle \mathbf{L} \rangle$ is linear as well, and, given a fixed $C_0(\boldsymbol{X})$, the average concentration $\langle C(\boldsymbol{X}, t) \rangle$ obeys some linear equation.

Equations for correlation functions and higher-order moments of concentration $C(\boldsymbol{X}, t)$ can be derived from Eq. (5.57), but these equations will not be linear. In most cases, when considering turbulent diffusion of a non-reacting impurity, it is sufficient to look at equation (5.62) for the average concentration. If one decides to bring chemical reactions into the picture, it becomes necessary to look at the equations for moments as well. If an instantaneous impurity source of intensity Q was positioned at a point $\boldsymbol{\xi}$ at the initial moment, operator $\langle \mathbf{L} \rangle$ can be found explicitly. Then

$$C_0(\boldsymbol{X}, t) = Q\delta(\boldsymbol{X} - \boldsymbol{\xi}).$$

In the absence of molecular diffusion, the total mass of impurity concentrated initially within a fluid particle will remain inside that particle even as the particle undergoes random motion. After time t the particle will reach the point $\boldsymbol{X} = \boldsymbol{X}(\xi, t)$ where the impurity concentration is equal to

$$C(\boldsymbol{X}, t) = Q\delta(\boldsymbol{X} - \boldsymbol{X}(\boldsymbol{\xi}, t)). \tag{5.63}$$

Let us bring into consideration the quantity $p(\boldsymbol{X}|\boldsymbol{\xi}, t)$ – probability density distribution of finding a fluid particle at the point $\boldsymbol{X}$ at the instant of time t under the condition that the particle was initially located at the point $\boldsymbol{\xi}$. Than, using the averaging rule, we obtain

$$Q\langle \delta(\boldsymbol{X} - \boldsymbol{X}(\boldsymbol{\xi}, t)) \rangle = p(\boldsymbol{X}|\boldsymbol{\xi}, t). \tag{5.64}$$

Consider now an arbitrary initial concentration distribution $C_0(\boldsymbol{X})$. If we model the continuum as a set of fluid particles, each particle obeying condition (5.64), then, recalling that $\langle \mathbf{L} \rangle$ is a linear operator, applying the superposition principle and replacing summation by integration, one gets

$$\langle C(\boldsymbol{X}, t) \rangle = \langle \mathrm{L}[\boldsymbol{u}_0(\boldsymbol{X}, t), t] \rangle C_0(\boldsymbol{X}) = \int p(\boldsymbol{X}|\boldsymbol{\xi}, t) C_0(\boldsymbol{X}) d\boldsymbol{X}. \tag{5.65}$$

This relation tells us that determination of the average concentration boils down to the problem of finding the probability density distribution $p(\boldsymbol{X}|\boldsymbol{\xi}, t)$ for one fluid particle. It should be noted that, by virtue of Eq. (5.63) and Eq. (5.64), the distribution $p(\boldsymbol{X}|\boldsymbol{\xi}, t)$ itself can be interpreted as a concentration field $\langle C(\boldsymbol{X}, t) \rangle$ from an instantaneous source of unit intensity at the point $\boldsymbol{X} = \boldsymbol{\xi}$.

So, if the distribution $p(\boldsymbol{X}|\boldsymbol{\xi}, t)$ is known, the formula (5.64) allows to determine the concentration field $\langle C(\boldsymbol{X}, t) \rangle$ from different types of sources: an instantaneously acting source distributed in space; a continuously acting point source; or a continuously acting source distributed in space – in other words, from virtually any type of sources occurring in practice.

As we noted earlier in Section 5.2, in the case of stationary and homogeneous turbulence, the probability density distribution $p(\boldsymbol{X}|\boldsymbol{\xi}, t)$ of fluid particle coordinates at any time $\tau = t = t_0$ is close to a normal one. Let us direct the coordinate axes along the principle axes of the dispersion tensor d_{ij} (see Eq. (5.34)). Then

$$p(\boldsymbol{X}|\boldsymbol{\xi}, t_0+\tau) = \frac{1}{(2\pi)^{3/2}[d_{11}(\tau)d_{22}(\tau)d_{33}(\tau)]^{1/2}} \times \exp\left\{-\frac{(X_1-\xi_1)^2}{2d_{11}(\tau)} - \frac{(X_2-\xi_2)^2}{2d_{22}(\tau)} - \frac{(X_3-\xi_3)^2}{2d_{33}(\tau)}\right\}, \quad (5.66)$$

where the values of $d_{ij}(\tau)$ are given by the relations (5.39).

Substituting Eq. (5.66) into Eq. (5.65), one obtains the average impurity concentration:

$$\langle C(\boldsymbol{X}, t)\rangle = \int_{-\infty}^{\infty}\int_{-\infty}^{\infty}\int_{-\infty}^{\infty} C_0(X_1, X_2, X_3) \times \exp\left\{-\frac{(X_1-\xi_1)^2}{2d_{11}(\tau)} - \frac{(X_2-\xi_2)^2}{2d_{22}(\tau)} - \frac{(X_3-\xi_3)^2}{2d_{33}(\tau)}\right\} dX_1 dX_2 dX_3. \quad (5.67)$$

To further simplify the expression (5.67), one can use asymptotic relations for principal values of the dispersion tensor for large values of time $\tau = (t - t_0) \gg T^{(L)}$; here $T^{(L)}$ is the maximum value of three Lagrangian time scales $T_1^{(L)}$, $T_2^{(L)}$, and $T_3^{(L)}$ (see Eq. (5.40)), for which asymptotical expressions (5.43) are valid:

$$d_{11}(\tau) = 2\langle (u_1')^2\rangle t_1^{(L)}, \quad d_{22}(\tau) = 2\langle (u_2')^2\rangle t_2^{(L)}, \quad d_{33}(\tau) = 2\langle (u_3')^2\rangle t_3^{(L)},$$

where u_1', u_2', u_3' are velocity fluctuation components.

Let us see now how turbulent diffusion relates to molecular diffusion. At the beginning of this section, we mentioned that turbulent diffusion of an impurity is happening much faster than molecular diffusion. Does it imply that molecular diffusion can always be ignored? The special feature of turbulent diffusion that distinguishes it from molecular one is that the region initially occupied by the impurity can subsequently, in the course of its motion, be deformed in a most fanciful manner, but will still hold the total original volume of the impurity. Therefore, when the impurity is contained exclusively in the given fluid volume and is absent outside of the volume, the distribution of impurity concentration at a fixed point in space has the form of a step function: as long as pure fluid passes through the point, the concentration is equal to zero, but as soon as the observed fluid volume reaches this point, the concentration jumps to a constant value.

In reality, of course, the distribution of concentration has no discontinuities. In a very short time after being injected into the fluid volume, the impurity diffuses into the surrounding fluid, which leads to an increase of the fluid volume, with smoothing down of concentration differences in the adjacent layers. This process is caused by molecular diffusion; the greater the coefficient of molecular diffusion, the faster it happens. Hence, when describing small-scale statistical structure of the field of concentrations, it is impermissible to neglect molecular diffusion: otherwise after a certain amount of time we would get an absolutely unrealistic picture of concentration distribution inside the flow.

The role of molecular diffusion becomes evident if we consider the diffusion of a passive impurity from an instantaneous point source of unit intensity in the field of homogeneous turbulence with zero average velocity. Suppose the source is initially ($t = t_0$) placed the point $\boldsymbol{X} = 0$. The distribution of concentration $C(\boldsymbol{X}, t)$ is described by Eq. (5.57) with the initial condition $C(\boldsymbol{X}, t_0) = \delta(\boldsymbol{X})$. In a quiescent fluid with no turbulence, molecular diffusion causes the impurity to gradually spread out in a spherical cloud, where the concentration distribution exhibits spherical symmetry with the variance $2D_m(t = t_0)$. Let us examine the consequences of including turbulence into the picture.

Consider the diffusion along the X_i-axis. The distinguishing characteristics of cloud spreading are as follows:

- variance of the average concentration distribution $\langle C(\boldsymbol{X}, t)\rangle$ relative to the initial position of the source,

$$d_0^2(t) = \int_{-\infty}^{\infty} X_i^2 \langle C(\boldsymbol{X}, t)\rangle d\boldsymbol{X}; \tag{5.68}$$

- variance of the center of gravity of concentration distribution relative to the source,

$$\langle X_0^2(t)\rangle = \left\langle \left\{ \int_{-\infty}^{\infty} X_i C(\boldsymbol{X}, t) d\boldsymbol{X} \right\}^2 \right\rangle; \tag{5.69}$$

- variance of concentration distribution relative to the center of gravity,

$$d_0^2(t) = \left\langle \int_{-\infty}^{\infty} [X_i - X_0(t)]^2 C(\boldsymbol{X}, t) d\boldsymbol{X} \right\rangle. \tag{5.70}$$

To ensure conservation of the total mass of the impurity, we write the normalization condition,

$$\int_{-\infty}^{\infty} C(\boldsymbol{X}, t) d\boldsymbol{X} = 1,$$

after which the equations above result in

$$d_c^2(t) = d_0^2(t) - X_c^2(t). \tag{5.71}$$

In order to determine the parameters $d_0^2(t)$, $\langle X_c^2(t)\rangle$ and $d_c^2(t)$, one should switch from the fixed frame of reference $\boldsymbol{X}$ to a moving frame $\boldsymbol{Y} = \boldsymbol{X} - \boldsymbol{X}(0, t)$, whose origin at each instant of time coincides with the fluid particle's position ($\boldsymbol{X} = 0$ being the initial position of the particle). Then, using the transformed diffusion equation (5.57) plus the relations (5.68)–(5.70), and expanding the functions $d_0^2(t)$, $\langle X_c^2(t)\rangle$, and $d_c^2(t)$ in Taylor series over the powers of $t - t_0$, one gets

$$d_c^2(t) = 2D_m(t - t_0) + \frac{2}{3}\langle(\nabla u_i)^2\rangle(t - t_0)^3 \ldots, \tag{5.72}$$

$$d_0^2(t) = \langle X_i^2(0, t)\rangle + 2D_m(t - t_0) - \frac{1}{3}\langle(\nabla u_i)^2\rangle(t - t_0)^3 \ldots, \tag{5.73}$$

$$\langle X_c^2(t)\rangle = \langle X_i^2(0, t)\rangle - D_m\langle(\nabla u_i)^2\rangle(t - t_0)^3 \ldots. \tag{5.74}$$

From Eq. (5.72), it follows that at the initial stage of the process, when $t - t_0 \ll 1$, there should hold

$$d_c^2(t) \approx 2D_m(t - t_0),$$

that is, the impurity transfer relative to the center of mass is driven by molecular diffusion. As time goes on, the influence of the second term in Eq. (5.73) grows. This term characterizes turbulence through the average value of the square of the velocity gradient, $\langle(\nabla u_i)^2\rangle$. As a result, turbulence causes an acceleration of molecular diffusion, and thereby a faster spreading of the impurity. By and large, this effect is explained by significant deformation of the moving fluid volume in the turbulent flow. But in order to be noticeably deformed, the fluid volume should first undergo a significant expansion, which is possible only through the molecular diffusion mechanism. That is why molecular diffusion is vital at the initial stage, where it acts as a necessary prerequisite for the further spreading of the fluid volume.

The formula (5.74) for X_c^2 shows that the variance of the center of gravity is smaller than the variance of the fluid particle's coordinate $X_1^2(0, t)$. This time lag is due to the fact that, because of molecular diffusion, the impurity particle falls behind the fluid particle (on the average). As a result, molecular diffusion slows down the turbulent diffusion. The same conclusion can be made with respect to the formula (5.73) for d_0^2.

We conclude from the foregoing discussion that molecular diffusion exerts a major influence on the average concentration for a limited time, just after the beginning of the process. As time goes on, this influence becomes less and less important. So if we are interested in the spreading of the impurity in macroscopic volumes at high Reynolds numbers, it is safe to neglect molecular diffusion when determining the average concentration.

Up to this point, we were concerned with turbulent diffusion of a single fluid particle. In practice, one is primarily interested in relative diffusion, defined as spreading of a macroscopic volume (say, one mole) of the impurity consisting of a large number of fluid particles. To solve such a problem, one must bring into consideration statistical characteristics, namely, multidimensional (multiparticle) probability density distributions of coordinates $\boldsymbol{X}$ and velocities $\boldsymbol{V}$ of n given particles at arbitrary instants of time $t_1, t_2, \ldots, t_m$ under the condition that initial coordinates of particles are given: $\boldsymbol{\xi}_1 = \boldsymbol{X}_1(t_{01}), \boldsymbol{\xi}_2 = \boldsymbol{X}_2(t_{02}), \ldots, \boldsymbol{\xi}_n = \boldsymbol{X}_n(t_{0n})$, though the initial moments $t_{01}, \ldots, t_{0n}$ are not necessarily the same. Up till now, we have considered only probability density distributions of a single particle's coordinates $\boldsymbol{X}(t)$. Consideration of multidimensional distributions is a very complicated problem. Let us therefore consider pair distributions only, that is, we shall be interested in the probability density distribution of coordinates $\boldsymbol{X}_1, (t)$ and $\boldsymbol{X}_2, (t)$ of two particles positioned different spatial points $\boldsymbol{\xi}_1$ and $\boldsymbol{\xi}_2$ at one and the same initial instant t_0.

Instead of the vectors $\boldsymbol{X}_1$ and $\boldsymbol{X}_2$, it is more convenient to use two other vectors: the displacement vector of the first particle $\boldsymbol{Y}(\tau) = \boldsymbol{X}_1(t_0 + \tau) - \boldsymbol{\xi}_1$, and the vector $\boldsymbol{r}(\tau) = \boldsymbol{X}_2(t_0 + \tau) - \boldsymbol{X}_1(t_0 + \tau)$ connecting the particles at the moment $t = t_0 + \tau$) and characterizing their mutual arrangement. The pair probability density distribution can then be written as $p(\boldsymbol{Y}, \boldsymbol{r}|\boldsymbol{\xi}_1, \boldsymbol{r}_0, \tau, t_0)$.

This notation means that we are looking at the joint distribution of $\boldsymbol{Y}$ and $\boldsymbol{r}$ at the moment τ after the beginning of the process, given that at the initial moment t_0, the first particle was at $\boldsymbol{\xi}_1$ and the position of the second particle relative the first is described by the vector $\boldsymbol{r}$. In the particular case of stationary, homogeneous turbulence, the probability density distribution does not depend on t_0 and $\boldsymbol{\xi}_1$. Of primary interest is the distribution $p(\boldsymbol{r}|\boldsymbol{\xi}_1, \boldsymbol{r}_0, \tau, t_0)$, equal to

$$p(\boldsymbol{r}|\xi_1, \boldsymbol{r}_0, \tau, t_0) = \int p(\boldsymbol{Y}, \boldsymbol{r}|\boldsymbol{\xi}_1, \boldsymbol{r}_0, \tau, t_0) d\boldsymbol{Y}. \tag{5.75}$$

Following Richardson [3], who first introduced such a distribution, it is called the distance function between neighbors. We shall list the main properties of this distribution.

1. After a sufficiently long time $\tau \gg T^{(L)}$, the two particles will get so far apart that their velocities $\boldsymbol{V}_1(t_0 + \tau) = \boldsymbol{u}(\boldsymbol{X}_1(t_0 + \tau), t_0 + \tau)$ and $\boldsymbol{V}_2(t_0 + \tau) = \boldsymbol{u}(\boldsymbol{X}_2(t_0 + \tau), t_0 + \tau)$ will become practically independent from each other. Then the distance function is derived from the distribution of displacements for a single particle:

$$p(\boldsymbol{Y}, \boldsymbol{r}|\boldsymbol{\xi}_1, \boldsymbol{r}_0, \tau, t_0) = p(\boldsymbol{Y}|\boldsymbol{\xi}_1, \tau, t_0)\, p(\boldsymbol{Y} + \boldsymbol{r} - \boldsymbol{r}_0|\boldsymbol{\xi}_1 + \boldsymbol{r}_0, \tau, t_0),$$

where $p(\mathbf{Y}|\boldsymbol{\xi}_1, \tau, t_0)$ is the displacement distribution for a single fluid particle, and Eq. (5.75) can be written as

$$p(\mathbf{r}|\boldsymbol{\xi}_1, \mathbf{r}_0, \tau, t_0) = \int p(\mathbf{Y}|\boldsymbol{\xi}_1, \tau, t_0)\, p(\mathbf{Y} + \mathbf{r} - \mathbf{r}_0|\boldsymbol{\xi}_1 + \mathbf{r}_0, \tau, t_0) d\mathbf{Y}. \quad (5.76)$$

If $p(\mathbf{Y}|\boldsymbol{\xi}_1, \tau, t_0)$ satisfies the diffusion equation with constant diffusion coefficients D_{ij}, then $p(\mathbf{r}|\boldsymbol{\xi}_1, \mathbf{r}_0, \tau, t_0)$ obeys the same equation, except that the diffusion coefficients are now equal to $2D_{ij}$ (see Eq. (5.95)). If we have $r_0 \gg L$ at the initial moment, where L is the integral scale of turbulence, then the formulas (5.75) and (5.76) will be valid for any instant of time.

Owing to the results obtained in Section 5.2, for large values of τ we have the following asymptotic relations for the relative dispersion tensor $d_{ij}^{(r)}(\tau)$ and the mean square deviation $\langle r^2(\tau)\rangle$:

$$d_{ij}^{(r)}(\tau) \approx 2\langle(u_i')^2\rangle T_i^{(L)} \delta_{ij}\tau, \quad \langle r^2(\tau)\rangle \approx 2\sum_i \langle(u_i')^2\rangle T_i^{(L)} \tau. \quad (5.77)$$

2. If at the initial moment the particles were sufficiently close to one another, that is, $|\mathbf{r}_0| \ll L$, then large-scale fluctuations with $\lambda \sim L$ will only move the two particles as a whole, without changing their mutual arrangement. Therefore relative motion of particles is possible only under the action of small-scale fluctuations. To describe the behavior of the distance function for such a case, one should use the theory of local isotropic turbulence (see Section 4.9).

Let us remind ourselves that when studying the local structure of developed turbulence, we classified small-scale motions with $\lambda \ll L$ as belonging to one of the two intervals: inertial interval with $\lambda_0 \ll \lambda \ll L$, where λ_0 is the inner (Kolmogorov) scale defined by the formula (4.114), and dissipative interval with $\lambda \leq \lambda_0$. Since the initial relative position of the two particles under consideration is characterized by the vector $\mathbf{r}_0$, the character of subsequent dependence of the vector $\mathbf{r}$ on time τ is given by the relation between $\mathbf{r}_0$, λ_0, and L. As far as $\mathbf{r}(\tau)$ varies continuously, there always exists a time τ_1 such that the condition $|\mathbf{r}(\tau)| \ll L$ is valid at $\tau < \tau_1$. When $\tau < \tau_1$, fluctuations with $\lambda \sim L$ do not affect the mutual arrangement of particles, that is, the vector $\mathbf{r}$. At $\tau_3 < \tau < \tau_1$, the interparticle distance is comparable with fluctuation scales in the inertial interval. In this time interval, it is possible to imagine the particles as moving in a straight line due to their inertia during the length of time τ_2. Finally, at $\tau \leq \tau_3$, the interparticle distance is comparable with the inner scale of turbulence.

This division of time into separate intervals allows one to investigate the general properties of the distance function between neighbors $p(\mathbf{r}|\xi_1, \mathbf{r}_0, \tau, t_0)$ on the basis of the Kolmogorov hypotheses within the framework of the theory of local isotropic turbulence. Rather than focusing on minute details pertaining to the distance function, we refer the reader to [2]. We shall list here only two important formulas,

$$d_{ij}^{(r)} = \frac{1}{3} g\bar{\varepsilon}\tau^3 \delta_{ij}, \quad \langle \mathbf{r}^2 \rangle = g\bar{\varepsilon}\tau^3, \tag{5.78}$$

which are valid in the interval $\tau_3 < \tau < \tau_1$, that is, for $|\mathbf{r}_0| \sim \lambda_0$, $\tau \gg \tau_{\lambda 0}$ or for $|\mathbf{r}_0| \ll \lambda_0$, $\tau > \tau(\mathbf{r}_0)$, where g is a universal constant and $\bar{\varepsilon}$ is the specific dissipation of energy.

As an example, consider the relative diffusion accompanying the spreading of an impurity cloud consisting of a large number of particles. Let us select some particle from the cloud and call it a test particle. Then the distribution of concentration in the cloud relative to the test particle at the instant of time $t = t_0 + \tau$ coincides with the distribution of values $\mathbf{r}(\tau)$ for all possible pairs of particles that include the test particle. The spreading of a cloud of N particles is described by the Richardson function $q(\mathbf{r}, \tau)$ such that $Nq(\mathbf{r}, \tau)d\mathbf{r}$ is equal to the number of particles for which the small volume surrounding the point $X_i(\tau) + \mathbf{r}$ at the instant $t_0 + \tau$ contains at least one cloud particle. This function obeys the normalization condition

$$\int q(\mathbf{r}, \tau) d\mathbf{r} = 1 \tag{5.79}$$

and defines the relative number of pairs of cloud particles whose coordinates differ by $\mathbf{r}$. The average value of the Richardson function $q(\mathbf{r}, \tau)$ is

$$\langle q(\mathbf{r}, \tau) \rangle = \int p(\mathbf{r}|\mathbf{r}_0, \tau) q_0(\mathbf{r}_0) d\mathbf{r}_0,$$

where $p(\mathbf{r}|\mathbf{r}_0, \tau)$ is the probability density of the vector $\mathbf{r}(\tau)$ and $q_0(\mathbf{r}) = q(\mathbf{r}, 0)$ is the initial value of $q(\mathbf{r})$. For example, when the impurity initially fills a sphere of radius R, we have $4\pi R^3 q_0(\mathbf{r}) = 1 - 3r/4R - r^3/16R^3$where $r = |\mathbf{r}|$.

Let us introduce the tensor of relative dispersion of the cloud, having the components

$$\Lambda_{ij}(\tau) = \overline{\langle r_i r_j \rangle} = \int r_i r_j \langle q(r, \tau) \rangle dr, \tag{5.80}$$

which relates the quantities $\Lambda_{ij}(\tau)$ to components of the relative dispersion tensor $d_{ij}^{(r)}(\tau|\mathbf{r}_0|) = \langle r_i r_j \rangle$ for a particle pair. The horizontal bar at the top denotes averaging over all particles pairs. Then the effective diameter of the cloud can be defined as

$$\Delta(\tau) = [\Lambda_{ij}(\tau)]^{1/2}. \tag{5.81}$$

The quantity

$$\Xi = \frac{1}{6}\frac{d\Delta^2(\tau)}{d\tau} \tag{5.82}$$

is called the virtual turbulent diffusion coefficient of the cloud, and

$$\Xi_i = \frac{1}{2}\frac{d\Lambda_{ii}(\tau)}{d\tau} = \frac{1}{2}\frac{\langle dr_i^2(\tau)\rangle}{d\tau} \tag{5.83}$$

is called the turbulent diffusion coefficient along the X_i-axis.

If the initial distance between cloud particles satisfies the condition $|\boldsymbol{r}_0| \ll L$, then for $\tau \gg \tau_{\lambda 0}$ and $\tau_3 < \tau < \tau_1$, where τ_1 and τ_3 are related to the initial cloud diameter Δ_0, Eq. (5.78) gives us $\Delta^2(\tau) \sim \tau^2$, and the virtual turbulent diffusion coefficient of the cloud is equal to

$$\Xi = \alpha\Delta^{4/3}. \tag{5.84}$$

This law is called the law of four thirds or the Richardson law. Since this law is a consequence of general ideas about the structure of small-scale turbulence, we could also have derived it from dimensionality considerations in the same manner as we derived the relations that appear in Section 4.9. Indeed, the transport coefficient Ξ in the inertial interval of fluctuations, that is, at $\lambda_0 \ll \lambda \ll L$, should be defined in terms of dimensional quantities $\bar{\varepsilon}$ and Δ. Since the dimensionality of Ξ is m^2/s, it is easy to obtain

$$\Xi = \alpha\bar{\varepsilon}^{1/3}\Delta^{4/3}.$$

Examination of relative motion of two fluid particles shows that the interparticle distance grows with time. So, turbulent motion causes the cloud (a mole, say) to change its shape in such a way that in the end, any small spherical volume stretches into a long thin band whose length, width, and thickness change exponentially with time, though with different values of coefficients in the exponent. Needless to say, this initial volume should be sufficiently small for all our original assumptions to be satisfied.

5.4 A Semiempirical Model of Turbulent Diffusion

Consider the diffusion equation (5.57). If we neglect molecular diffusion, it takes the form

$$\frac{\partial C}{\partial t} + \frac{\partial\langle u_i C\rangle}{\partial X_i} = 0. \tag{5.85}$$

The velocity $\boldsymbol{u}$ is equal to the sum of the average $\langle \boldsymbol{u} \rangle$ and fluctuational $\boldsymbol{u}'$ velocities: $\boldsymbol{u} = \langle \boldsymbol{u} \rangle + \boldsymbol{u}'$. In a similar fashion, we shall represent concentration as a sum of its average value $\langle C \rangle$ and fluctuation C':

$$C = \langle C \rangle + C'. \tag{5.86}$$

Averaging Eq. (5.85), we obtain

$$\frac{\partial \langle C \rangle}{\partial t} + \frac{\partial (\langle u_i \rangle \langle C \rangle)}{\partial X_i} = -\frac{\partial (\langle u_i' C' \rangle)}{\partial X_i}. \tag{5.87}$$

The expression $J_i = \langle u_i' C' \rangle$ on the right can be interpreted as the flux density of the impurity in the direction of the X_i-axis.

The structure of Eq. (5.87) is the same as that of Reynolds equations (see Section 4.6). It is unclosed, because it contains a new unknown. Therefore the closure problem (this time with respect to Eq. (5.87)) is just as imperative in the theory of turbulent diffusion as it is in the theory of turbulence.

The simplest closure relation can be constructed by assuming that diffusion is isotropic and making use of the Taylor hypothesis [4]

$$J_i = \langle u_i' C' \rangle = -D_t \frac{\partial \langle C \rangle}{\partial X_i}. \tag{5.88}$$

The coefficient D_t entering this equation is called the coefficient of turbulent diffusion. It should be noted that relation (5.88) is similar to the Boussinesq hypothesis (4.140). The Taylor hypothesis implies that impurity transport via random turbulent fluctuations is similar to the transport via molecular diffusion, since the impurity flux is proportional to the concentration gradient, just as in Fick's law.

The coefficient of turbulent diffusion D_t has nothing to do with the coefficient of molecular diffusion D_m. Indeed, D_t characterizes the impurity transport via chaotic turbulent motion, while D_m characterizes transport via chaotic molecular motion. The only common feature is the randomness of the relevant processes and the resulting need to employ statistical methods.

Let us estimate D_t by the order of magnitude. As we noted in the previous section, turbulent diffusion is characterized by the scale of large-scale turbulence, namely, by a linear size L such as the pipe diameter; by the change of the average flow velocity ΔU; and by fluid density ρ_e. From these quantities one can construct a combination ΔUL, which has the dimensionality of the diffusion coefficient. Therefore, just as in the kinetic theory of gases, we can write

$$D_t \sim \Delta UL. \tag{5.89}$$

Note that D_t is equal by the order of magnitude to the coefficient of turbulent kinematic viscosity (see Eq. (4.104)). Therefore the coefficient of turbulent diffusion is akin to the coefficient of turbulent viscosity and has the same order of magnitude.

Let us compare the coefficients of turbulent and molecular diffusion:

$$\frac{D_t}{D_m} \sim \frac{\Delta U L}{D_m} = \frac{\mathrm{Re}\nu_e}{D_m}.$$

Since for fluids, $\nu_e \sim 10^{-7}\ \mathrm{m^2/s}$ and $D_m \sim 10^{-9}\ \mathrm{m^2/s}$, we have $D_t/D_m \sim 10^{-2}\ \mathrm{Re} \gg 1$. Such a large value of D_t ensures a rapid mixing of the fluid. So if a chemical reaction occurs at the surface, turbulent diffusion will quickly equalize concentration in the flow even at small distances from the wall. As a result, the concentration becomes practically constant in the whole region except for a thin layer adjacent to the wall.

Near the wall, the characteristic linear scale is equal to the distance from the wall. If one coordinate axis, for example, Z, is directed perpendicularly to the wall, Z will be the characteristic linear scale in this region, and

$$D_t \sim Z^2 \frac{\partial U}{\partial Z}. \tag{5.90}$$

So, as we approach the wall, the coefficient of turbulent diffusion decays as Z^2. A more detailed analysis of the flow structure that takes into account the existence of a viscous sublayer shows that near the wall, D_t decays as Z^4.

The flux of matter toward the wall is by the order of magnitude equal to

$$J_w \sim Z^2 \frac{\partial U}{\partial Z}\frac{\partial\langle C\rangle}{\partial Z}. \tag{5.91}$$

It is well known that distribution of the average velocity near the wall obeys a logarithmic law (see Section 3.10). Then one can see from Eq. (5.91) that $\langle C\rangle$ is also distributed logarithmically. But in the immediate vicinity of the wall, the logarithmic law needs to be corrected in order to account for the viscous sublayer.

In the anisotropic case, the diffusion coefficient is a second-rank tensor D_{ij}. It is introduced through a relation that is similar to Eq. (5.88):

$$J_i \sim \langle u_i' C'\rangle = -D_{ij}\frac{\partial\langle C\rangle}{\partial X_j}. \tag{5.92}$$

As usual, summation over repeated indices is implied by default.

When coefficients D_{ij} depend on the coordinates X_i and time t, Eq. (5.87) takes the form

$$\frac{\partial\langle C\rangle}{\partial t} + \frac{\partial(\langle u_i\rangle\langle C\rangle)}{\partial X_i} = \frac{\partial}{\partial X_i} D_{ij}\frac{\partial\langle C\rangle}{\partial X_j}. \tag{5.93}$$

Consider the case of stationary homogeneous turbulence. Then the tensor of fluctuation dispersions d_{ij} is defined by the expression (5.38). The average value of fluid particle's displacement along the X_i-axis during the length of time τ is $\langle Y_i(\tau)\rangle = \langle u_i\rangle\tau$ (see Eq. (5.36)), where the constants $\langle u_i\rangle$ are components of the average velocity. As was mentioned in Section 2.2, in the case under consideration the joint probability density distribution of particle displacements $\langle Y_i(\tau)\rangle$ for all τ can be considered as a three-dimensional Gaussian distribution (see Eq. (1.77)). Then the distribution $p(\boldsymbol{X}|\boldsymbol{\xi}, t)$ of fluid particles initially located at $\boldsymbol{X} = \boldsymbol{\xi}$ over spatial coordinates $\boldsymbol{X}$ will also be Gaussian, with the mean value $a_j = \xi_j + \langle u_j\rangle(t - t_0)$ and the matrix $\boldsymbol{b} = ||d_{ij}||$. The distribution $p(\boldsymbol{X}|, \boldsymbol{\xi}\ t)$ obeys Eq. (5.94), where D_{ij} should be expressed through components of the dispersion tensor in accordance with the relation

$$D_{ij} = \frac{1}{2}\frac{d}{dt}\left[d_{ij}(\tau)\right] = \frac{1}{2}\int_0^{\tau}\left[b_{ij}^{(L)}(\tau) + b_{ji}^{(L)}(\tau)\right]d\tau, \tag{5.94}$$

where $\tau = t - t_0$. Since average concentration $\langle C(\boldsymbol{X}, t)\rangle$ of the impurity obeys the same equation, the condition (5.94) means that fluid particle displacements in a stationary homogeneous turbulent field are distributed according to the normal (Gaussian) law.

At $\tau \gg T^{(L)}$, the asymptotic representation (5.45) is valid:

$$d_{ij} \approx \left(\langle u_i'^2\rangle\langle u_j'^2\rangle\right)^{1/2} T_{ij}^{(L)}\tau$$

so we can write the following approximate equality:

$$D_{ij}(\tau) \approx \frac{1}{2}\left(\langle u_i'^2\rangle\langle u_j'^2\rangle\right)^{1/2} T_{ij}^{(L)}, \tag{5.95}$$

where $T_{ij}^{(L)} = \int_0^{\infty}\left[\Psi_{ij}^{(L)}(\tau) + \Psi_{ji}^{(L)}(\tau)\right]d\tau.$

The discussion of Brownian motion in Chapter 5 was centered around the Fokker–Planck equation that describes the variation of probability density distribution $p(\boldsymbol{X}, t|\boldsymbol{\xi}, t_0)$, where p refers to the probability for the particle to arrive at the point $\boldsymbol{X}$ at the moment t given that it was at the point $\boldsymbol{\xi}$ at the moment t_0. The derivation of the Fokker–Planck equation hinges on the assumption that the process is Markowian (see Section 1.11). Recall that our model of Brownian motion involved the division of particle trajectory into discrete intervals by the points $\boldsymbol{X}_1, \boldsymbol{X}_2, \ldots$ corresponding to the instants of time $t_1, t_2, \ldots$. We required the time intervals $t_2 - t_1, \ldots$ to be much longer than the time between successive collisions of the particle with molecules of the surrounding fluid, but at the same time much shorter than the characteristic time of the process under consideration. The probability density distribution $p(\boldsymbol{X}, t|\boldsymbol{\xi}, t_0)$ at two successive instants of time t_0 and t is then described by the Fokker–Planck equation (1.124), whose solution is the Gaussian distribution (1.126).

In the turbulent flow, we have a similar situation. Consider the random motion of a fluid particle that is described using the Lagrangian approach. Let us identify the time between two successive collisions with Lagrangian correlation time $T^{(L)}$ and the characteristic time of Brownian motion – with the characteristic time of turbulence t_t. We now divide the trajectory of the particle's random motion in a turbulent flow into discrete intervals by spatial points $\boldsymbol{X}_1$, $\boldsymbol{X}_2$, ... corresponding to the particle's positions at the instants $t_1, t_2, \ldots$, and assume that time intervals between successive particle positions are much longer than $T^{(L)}$ but much shorter than t_t. Then, as was mentioned above, probability density distribution $p(\boldsymbol{X}, t|\boldsymbol{\xi}, t_0)$ obeys Eq. (5.93), whose solution (under the condition (5.95)) is a Gaussian distribution. We conclude that $\boldsymbol{X}(\boldsymbol{\xi}, t)$ is a Markowian function and that $p(\boldsymbol{X}|\boldsymbol{\xi}, t)$ satisfies the Fokker–Planck equation

$$\frac{\partial p}{\partial t} + \frac{\partial}{\partial X_i}(\langle V_i(\boldsymbol{X}, t)\rangle\, p) = \frac{\partial^2}{\partial X_i \partial X_j}(D_{ij}(\boldsymbol{X}, t)\, p), \tag{5.96}$$

where $\langle V_i \rangle$ is the average velocity of a (fluid) impurity particle. This velocity is equal to the sum of the average flow velocity $\langle u_i \rangle$ and the additional velocity $\partial D_{ij}/\partial X_j$ induced by the inhomogeneity of turbulent diffusion coefficients:

$$\langle V_i \rangle = \langle u_i \rangle \frac{\partial D_{ij}}{\partial X_j}. \tag{5.97}$$

To summarize, under the conditions we imposed above, the sequence of particle positions $\boldsymbol{X}(\boldsymbol{\xi}, t_k)$, as well as the process of turbulent diffusion at $\tau \gg T^{(L)}$, wil be Markowian.

Of course, representations (5.88) and (5.92) are but idealizations of a real process. Indeed, at $\tau \gg T^{(L)}$ the dispersion is $b_{ij} \sim \tau^2$ (instead of being proportional to τ), the condition (5.95) is no longer fulfilled, and the process is not Markowian. In this case it becomes impossible to use the semiempirical model, and in order to close the diffusion equation (5.87), one should employ other methods such as the method of moments.

Hence, the semiempirical equation of turbulent diffusion (5.93) with effective turbulent diffusion coefficient (5.95) is useful when solving problems that involve diffusion of a passive impurity in a stationary homogeneous turbulent flow whose characteristic time exceeds the Lagrangian scale of turbulence (5.40). We should remark that in the problems that involve spreading of an impurity in the air, the Lagrangian time scale has the order of seconds.

The turbulent diffusion equation (5.93) has helped to solve many practical problems on diffusion in homogeneous turbulent fields and in shear flow fields: impurities spreading away from sources; longitudinal spreading of impurities in pipes and channels; diffusion in free jets, in the atmosphere, and so on.

5.5
Models of Two-phase Disperse Turbulent Flows

Two-phase disperse media are the media consisting of a continuous (carrier) phase (fluid, gas) and a disperse phase (solid particles of some mechanical impurity, droplets, bubbles) whose properties differ from those of the surrounding medium. Later on we shall call them simply "particles", providing a clarification as to their physical nature whenever necessary.

An important parameter of such medium is the volume concentration of the disperse phase φ, equal to the total volume of particles contained in a unit volume of the medium. A medium with $\varphi \ll 1$ is called dilute, or rarified. If, in addition, the particles are small and their properties are not much different from those of the continuous phase, the disperse phase has a weak influence on the hydrodynamic characteristics of the continuum, and the particles (as well the disperse phase as a whole) can be regarded as passive. To describe the behavior of a passive disperse phase, one can use the theory of turbulent diffusion, which essentially states that particle motion is defined by the turbulent motion of the continuous phase, whereas the inverse influence of the disperse phase on the continuous phase flow is absent. An increase of φ (or, for small φ, an increase of particle size) or a noticeable difference of particle properties (such as density or viscosity) from those of the continuous phase would make this reverse influence significant. In this case the problem becomes more complicated, as one is forced to take into account the motion and mutual interaction of both phases.

Impurity particles are drawn into chaotic motion as a result of fluctuations of the viscous drag force, which depends on the particle's velocity relative to the carrier phase. For Stokesian particles, this force is proportional to the difference of velocities $\boldsymbol{V}$ and $\boldsymbol{u}$ of the particle and the fluid. The difference $\boldsymbol{V}-\boldsymbol{u}$ by itself may be due to inertia and (or) to the action of an external force such as gravity. The influence of the external force is especially noticeable for large particles.

In applications, we are mostly concerned with microparticles, that is, particles whose size does not exceed the microscale of turbulence. Relative motion of such particles obeys the Stokes resistance law, hence the term "Stokesian particles". Hydrodynamic aspects of the motion of such particles in a laminar flow have been discussed in Chapter 2.

Another important parameter characterizing the particles' inertia and affecting their motion is the viscous (dynamic) relaxation time. For spherical Stokesian particles, it is equal to

$$t_v = \frac{d^2 \rho_p}{18 \nu_e \rho_e}, \qquad (5.98)$$

where ρ_p and ρ_e are the densities of the particle and the surrounding fluid, ν_e is the kinematic viscosity of the surrounding fluid, d is the diameter of the particle.

Dynamic relaxation time has the meaning of the time it takes a particle moving by inertia to be stopped by the viscous drag force. Therefore t_ν characterizes the influence of particle inertia on its motion relative to the fluid. For large particles and (or) for the relative motion of a particle under the action of large-scale fluctuations, dynamic relaxation time depends on the particle's Reynolds number:

$$t_\nu = \frac{4}{3}\frac{d^2\rho_p}{\nu_e\rho_e}\frac{1}{\mathrm{Re}_p C_p}. \tag{5.99}$$

where Re_p is the Reynolds number determined by the particle's parameters and C_p is the coefficient of resistance for the particle, equal to

$$C_p = \frac{24}{Re_p}(1 + 0.179 Re_p^{0.5} + 0.013 Re_p).$$

In the absence of external forces, the relative motion of large particles is caused only by turbulent fluctuations of the carrier fluid, while the relative motion of small particles may also be caused by Brownian motion.

If particle temperature differs from the temperature of the carrier phase, one can introduce the characteristic time of thermal relaxation t_T, defined as the time it takes for the particle temperature to reach the temperature of the surrounding medium.

Particle inertia is characterized by a dimensionless parameter called the Stokes number St, equal to the ratio between the particle's dynamic relaxation time t_ν and the Lagrangian correlation time $T^{(L)}$:

$$\mathrm{St} = \frac{t_\nu}{T^{(L)}}. \tag{5.100}$$

Depending on how strongly the particles are involved into turbulent motion, they can be divided into two categories based on their inertia. The first category includes the particles with $\mathrm{St} > 1$. These particles are called inertial. They exhibit only a weak response to small-scale fluctuations. Particles with $\mathrm{St} < 1$ are called inertialess. Their motion is defined by the microstructure of turbulence. In the limit $\mathrm{St} \to 0$ they become passive particles, whose motion is described by the turbulent diffusion model.

At the present moment, the turbulent flow theory is far from complete, even for a single-phase medium, to say nothing of two-phase media. In Section 4.10, we discussed the existing turbulent flow models for single-phase media (fluids). Following the review [5], we now list some theoretical models of two-phase disperse turbulent flows.

When simulating two-phase turbulent flows, we focus our attention on the following problems: interaction of particles with the turbulent flow of the continuous phase; the inverse effect – the action of the particles on the turbulent flow; interactions between particles; accounting for polydispersity (difference in particle sizes) and for the time evolution of the size spectrum of particles; the effect of turbulence on the rate of phase transitions, chemical reactions and combustion.

Currently, there exist two simulation methods. The first one is based on the use of phenomenological or semiempirical description of the continuous phase, with introduction of algebraic and (or) differential equations as closure relations (see Section 4.10). The second method, known as the direct numerical simulation [32], relies on numerical solution of nonstationary equations of motion without the closure relations. This method requires the employment of supercomputers; it is still in the process of development and is far from perfection. Therefore we shall confine ourselves to the first method.

Within this method, we can, in turn, identify two subcategories. The first one comprises techniques based on the combined Euler–Lagrange approach to the description of the medium's motion. In this paradigm the momentum and energy equations for the continuous phase (for an isothermic flow of an incompressible fluid, it is possible to take the momentum equations only) are represented and solved in Eulerian variables, whereas equations for the disperse phase are written and solved in Lagrangian variables. This means that the motion of each particle is described by the Langevin stochastic equation in the relaxation approximation (see Section 1.13),

$$\frac{d\boldsymbol{V}_p}{dt}=\frac{\boldsymbol{u}(\boldsymbol{X}_p(t),t)-\boldsymbol{V}_p(t)}{t_v}+\boldsymbol{F}(\boldsymbol{X}_p(t),t)+\boldsymbol{f}(\boldsymbol{X}_p(t),t),\quad \boldsymbol{V}=\frac{d\boldsymbol{X}_p}{dt} \tag{5.101}$$

whose right-hand side is a sum of the random viscous drag force, the external force, and the random Brownian force (in that order).

The heat exchange equation for the particle in the relaxation approximation also has the form of the Langevin stochastic equation,

$$\frac{d\vartheta_p}{dt}=\frac{\vartheta(\boldsymbol{X}_p(t),t)-\vartheta_p(t)}{t_T}. \tag{5.102}$$

where $\vartheta_p(t)$ and $\vartheta(\boldsymbol{X}_p(t),\ t)$ are the temperatures of the particle and the carrier medium, t_T is the characteristic time of thermal relaxation for the particle. The Langevin equations are integrated for different values of initial particle position $\boldsymbol{X}_p(0)$, velocity $\boldsymbol{V}_p(0)$, and temperature $\vartheta_p(0)$, which corresponds to different particles. The obtained solutions are then averaged over the ensemble of initial data. This method is known as the method of stochastic simulation. A similar method for Brownian motion has been considered in Section 3.12.

To have confidence in the obtained results, it is necessary to have a representative ensemble of the initial data, that is, the number of particles must be sufficiently large, though it comes at the cost of increased volume of computations. Besides, as we reduce the particle size, smaller-scale fluctuations also begin to interact with particles, which increases the number of possible variants of the initial data. Thus application of the method of stochastic simulation to discrete particles is sensible only for inertial particles satisfying the condition $\mathrm{St}>1$. It should be noted that in the limiting case $\mathrm{St} \gg 1$ (highly inertial particles) it becomes possible to use the deterministic Lagrangian description based on equations for the average values only, disregarding interactions of particles with random fluctuations of the velocity and temperature fields.

Another difficulty associated with stochastic simulation is that as we increase the volume concentration of particles φ, the average distance between particles gets smaller, which means higher probability of particle collisions that may result in aggregation and/or breakup of particles. The difficulties associated with Lagrangian simulation become even more formidable when we take into account the possibility of nucleation of the disperse phase as a result of fluid boiling or condensation of supersaturated vapor.

Another simulation method is based on the Eulerian continuous representation of momentum and energy equations for both the continuous and the disperse phase. Such models are called two-fluid models. The essence of this method is to model the disperse phase as yet another continuous phase with its own properties (density, viscosity, and so on) and phenomenological relation between the stress tensor with the rate-of-strain tensor. This brings us to the question of whether or not it is possible to describe the motion of a large group of particles by methods of continuum mechanics. To answer this question, one should turn to the so-called continuity hypothesis, which is commonly used in continuum mechanics, and states that the continuous approximation is applicable to scales that are sufficiently small compared to the characteristic scale of macroscopic flow parameter changes, yet large enough to be able to contain many particles. The appropriate scale can be estimated in the following way. Assuming statistically independent behavior of individual particles, the relative fluctuation of distributed density is of the order $\varepsilon \sim N^{-1/2}$, where N is the number of particles in the given volume. Consider a cubic volume with N spherical particles of diameter d lined up along the edge L. For a given volume concentration of particles φ, we can make the approximate estimation $L/d \sim (2\varepsilon^2\varphi)^{-1/3}$. At $\varphi = 10^{-3}$, $\varepsilon \sim 10^{-2}$, and $d \sim 100\,\mu\mathrm{m}$, this estimation yields $L \sim 1\,\mathrm{cm}$. Thus, a disperse phase with such parameters can be modeled as continuous only on scales far larger than 1 cm.

In the Eulerian approach, the main task is to determine the force and energy interactions between phases. The advantage of Eulerian two-fluid approach as compared to the Euler–Lagrange trajectory modeling is that equations of the same type (Reynolds equations plus the closure relations), as well as the same algorithm for solving the resulting system of equations, are used for both phases. Within the framework of the two-fluid approach, it is possible to go to the $\mathrm{St} \to 0$ limit (the case of very small particles of passive impurity).

Two approaches have emerged as the most popular methods of solving the problem of turbulent motion of a disperse medium. Namely, one can study turbulent motion based on semiempirical models of Prandtl mixing length, or one can use second-order one-point moments of velocity and temperature, by analogy with single-phase flows. In the framework of the first method, Prandtl mixing length models for a single-phase medium are generalized for the case of disperse medium.

At of today, the second method, which involves equations for second moments of velocity and temperature fluctuations, is used more often.

When describing motion and heat exchange of the disperse phase, it is necessary to determine turbulent stresses, as well as diffusion and heat fluxes resulting from the particle's involvement into fluctuational motion of the carrier phase. Among the various methods of finding disperse phase characteristics, we should single out local equilibrium (aka algebraic) models. In one of these models, which relies on the locally homogeneous approximation, turbulent stresses in the disperse phase $\langle V_i' V_j' \rangle$ are connected with Reynolds stresses of the carrier phase $\langle u_i' u' \rangle$ through the relations

$$\langle V_i' V_j' \rangle = f_v \langle u_i' u_j' \rangle, \quad f_v = \int_0^\infty {}_v(\tau) \exp\left(-\frac{\tau}{t_v}\right) d\tau, \tag{5.103}$$

where $\boldsymbol{u}'$ and $\boldsymbol{V}'$ are velocity fluctuations of the continuous and disperse phases, Ψ_v is the two-time autocorrelation function (see Eq. (1.25)) of velocity fluctuations of the continuous phase along the particle trajectory.

The relation between turbulent heat flux in the disperse phase $\langle V_i' \vartheta'_p \rangle$ and turbulent heat flux in the continuous phase $\langle u_i' \vartheta' \rangle$ has a similar form:

$$\langle V_i' \vartheta'_p \rangle = \frac{t_v f_{vt} + t_T f_{Tv}}{t_v + t_T} \langle u_i' \vartheta' \rangle,$$
$$f_{vt} = \int_0^\infty \Psi_{vT}(\tau) \exp\left(-\frac{\tau}{t_v}\right) d\tau, \quad f_{Tv} = \int_0^\infty \Psi_{Tv}(\tau) \exp\left(-\frac{\tau}{t_T}\right) d\tau, \tag{5.104}$$

where ϑ'_p and ϑ' are temperature fluctuations in the disperse and continuous phases, t_T is the heat relaxation time for a particle, and Ψ_{vT} and Ψ_{Tv} are two-time autocorrelation functions of velocity and temperature fluctuations.

The relations (5.103) and (5.104) will follow from differential equations for second moments of velocity and temperature fluctuations if we drop the terms in these equations that describe convective transport, diffusion, and production from the averaged motion; this is justified only for relatively small particles.

Another way to determine turbulent characteristics of the disperse phase is to use expression of the gradient type according to the rules of the semiempirical theory. In this paradigm, turbulent stresses and heat flux of the disperse phase are represented in the form that is similar to the relations for a single-phase medium, namely,

$$\langle V_i' V_j' \rangle = \frac{2}{3} e_{kp} \delta_{ij} - \nu_p \left(\frac{\partial \langle V_i \rangle}{\partial X_j} + \frac{\partial \langle V_j \rangle}{\partial X_i} - \frac{2}{3} \frac{\partial \langle V_k \rangle}{\partial X_k} \right), \tag{5.105}$$

$$\langle V_i' \vartheta'_p \rangle = -\frac{\nu_p}{Pr_p} \frac{\partial \langle \vartheta'_p \rangle}{\partial X_i}, \tag{5.106}$$

where $e_{kp} = \langle V_i' V_i' \rangle /2$ is the turbulent energy of the disperse phase, ν_p and χ_p are coefficients of turbulent viscosity and thermal diffusivity, respectively, and $\mathrm{Pr}_p = \nu_p/\chi_p$ is the Prandtl number of the disperse phase.

Models of the gradient type are usually constructed on a purely phenomenological basis, therefore they contain additional empirical constants. Models of this type are valid at $\mathrm{St} < 1$, that is, for particles with small inertia.

In addition to local equilibrium algebraic models describing turbulent transfer of momentum and heat in the disperse phase, non-local differential models based on the energy balance equations and equations for the second moments of velocity and temperature are gaining broad recognition. Application of differential models (you can think of them as transport models by analogy with single-phase media, see Section 4.10) makes it possible to describe non-local effects of transport of velocity and temperature fluctuations by inertial particles, that is, turbulent transport of momentum and heat via convection and diffusion. These effects are especially important in the flow regions adjacent to the wall.

We showed earlier in Section 4.11 that a complete statistical description of turbulence can be obtained from the probability density function (PDF) of particle positions and velocities. Knowledge of the PDF enables us to find the average values of the velocity and temperature fields as well as their correlations. So the main problem is to determine the PDF. As shown in the same Section 4.11, in the general case the PDF is described by the Hopf equation – equation in functional derivatives, whose solution still presents difficulties at the time of writing this book.

In the theory of turbulence of disperse media, one comes up against the same difficulties. If the equation for the PDF is known, one can derive from it the systems of equations (moments equations) describing the motion and heat-and-mass-exchange of the disperse phase in context of the Eulerian approach in the same manner as one derives the Euler and Navier–Stokes equations from the Boltzmann equation in the kinetic theory of rarefied gases.

As was shown earlier, in the Lagrangian approach particle motion is described by the Langevin equation, which enables us to keep track of each particle's movements in a random force field. By examining the PDF we can obtain statistical description of the behavior of a particle ensemble, rather than dynamical description of individual particles. From the mathematical viewpoint, it means that the problem of integration of stochastic ordinary differential equations in a physical space is replaced by the problem of solving a deterministic partial differential equation in the phase space of coordinates, velocities, and temperature.

Interaction of particles with turbulent fluctuations of the carrier medium is described by the same diffusion operator in the velocity space as the one we encountered when studying Brownian diffusion. For delta-correlated random fields (for a disperse phase, it means that the particles are inertial, that is, $\mathrm{St} \gg 1$), the equation for the PDF reduces to the Fokker–Planck equation. A more general form of the PDF equation is derived in [6–11].

Comparing the two approaches, we may conclude that some information about the behavior of individual particles is lost in the transition from the Lagrangian to the

Eulerian approach, but, to compensate for that, we are gaining information about statistical regularities of particle motion. Besides (and one can argue that this is, in fact, the main advantage of the Eulerian approach), the equation for the PDF can be used to build a system of continual equations for the averaged values of hydrodynamic and heat-and-mass-transport characteristics of the disperse phase. The obtained system for the moments is similar to an infinite system of equations in the theory of single-phase turbulent flows (the Friedmann–Keller chain, see Sections 4.6 and 4.7). Any finite subsystem of these equations is unclosed because the equation for n-th moment already contains $(n+1)$-th moment.

In order to close a finite subsystem, we must introduce closure relations similar to those introduced in Section 4.19 for single-phase media. In particular, we can use the equation for energy of the disperse phase e_p; as for the turbulent stress and the heat flux, they can be obtained from the relations (5.105) and (5.106), in which we set

$$\nu_p = G_\nu \nu_t + t_\nu e_{kp}/3, \qquad Pr_p = \frac{(t_\nu + t_T)(f_\nu \nu_t + t_\nu e_{kp}/3)}{(t_\nu f_{\nu t} + t_T f_{T\nu})\nu_t/Pr_t + 2t_\nu t_T e_{kp}/3}, \tag{5.107}$$

where ν_t and Pr_t are, respectively, the turbulent viscosity coefficient and the turbulent Prandtl number for the carrier phase.

We should mention one more method called the inertial diffusion model. It is used in hydrodynamical calculations of two-phase flows that contain a finely dispersed impurity consisting of particles whose inertia is small ($\mathrm{St} \ll 1$) but whose density is much higher than the density of the carrier phase. The essence of this method is to solve the diffusion equation for the impurity, taking into proper account some inertial transport mechanisms: turbulent migration of particles from regions of high turbulence to regions of low turbulence; the action of mass forces; and the deviation of particle trajectories from stream lines of the carrier medium due to their curvature and the nonstationary character of the flow.

Fig. 5.1 shows how the range of applicability of the above-considered models of rarefied disperse media depends on the value of the Stokes number St. The analysis of these calculation models convinces us that for rarefied disperse media ($\varphi \ll 1$), it makes sense to use the Eulerian approach at $\mathrm{St} < 1$, whereas the Lagrangian approach is more practical at $\mathrm{St} > 1$. However, depending on the effectiveness and

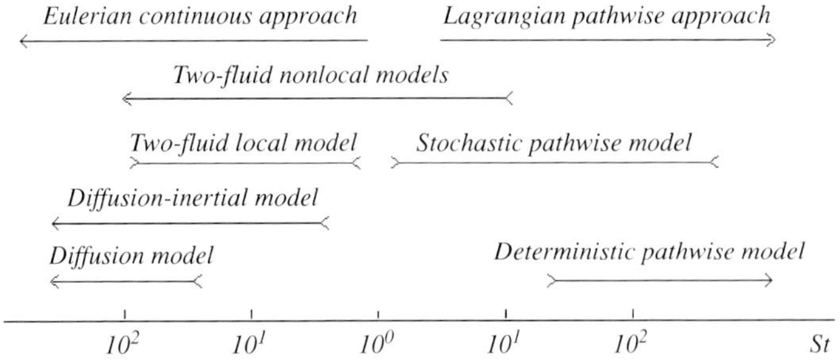

Fig. 5.1 Regions of applicability of the models for rarefied disperse media.

complexity of a chosen model, its region of applicability could be widened or narrowed.

The character of particle influence on the turbulent flow of the carrier phase is not definitely known. Depending on their inertia and concentration, the presence of particles could make the flow more laminar or more turbulent. The "feedback" effect of particles on turbulence is proportional to the mass concentration of particles γ and becomes noticeable at $\gamma > 0.1$. The mass concentration γ is related to the volume concentration φ as $\gamma = \varphi \rho_p/\rho_e$, where ρ_p is the particle density and ρ_e is the density of the carrier medium. Therefore if condition $\rho_p/\rho_e \gg 1$, smallness of φ does not guarantee that mass concentration γ will be small. Therefore the often-used assumption that $\varphi \ll 1$ does not mean that the feedback effect of particles on the turbulence can be ignored.

Relatively small particles with $\mathrm{St} \leq 1$ are not completely involved in fluctuational motion, and their velocity differs from that of fluctuations. As a result, this effect, called fluctuational phase slippage, produces additional energy dissipation and decreases the intensity of turbulent fluctuations. For very small particles with $\mathrm{St} \ll 1$, whose relaxation time is comparable to the inner (Kolmogorov) scale of turbulence, energy dissipation increases as a result of their interaction with high-frequency small-scale fluctuations of the carrier phase. As was shown in [12], this effect lowers the hydrodynamic resistance when a small admixture of impurity is injected into a turbulent flow. As the Stokes number St gets larger, additional dissipation caused by the fluctuational phase slippage attenuates and at $\mathrm{St} \gg 1$ becomes inessential.

Some possible mechanisms of turbulence generation caused by the feedback influence of the disperse phase are: additional gradient production of turbulent energy by the averaged motion; formation of a nonstationary vortex wake behind large particles as they are bypassed by the flow; diffusional transport of particles as a result of non-uniform distribution of the disperse phase in space; generation of disturbances due to particle collisions. As we increase particle inertia, the laminarization effect of particles gets replaced by turbulization effect.

To model a two-phase turbulent flow while taking into account the feedback effect of particles on the flow, we start from the equations for turbulent energy and its rate of dissipation, and bring in additional terms that describe the effect of the disperse phase; these terms have the form of integrals in the phase space of the PDF.

The character of particle interactions with the surface depends on properties of the disperse phase (solid particles or droplets) and the surface (rigid wall or liquid film). Interactions between solid particles and the wall are described by heat and momentum recovery (or retention) coefficients and the rebound angle; these quantities characterize elasticity or inelasticity of the impact. They depend on velocity, angle of impact, roughness of the surface, and relative hardnesses of the touching surfaces.

When droplets interact with a dry surface, we may observe droplet spreading or quasi-elastic droplet rebound, depending on the droplet velocity, angle of incidence, and wetting conditions. When a solid particle or a droplet collides with a liquid film, we may observe a splash accompanied by formation of secondary droplets. Besides, small droplets can get stripped away from the wave crest (dynamical ablation) due to the interfacial friction at the film surface.

Deposition of particles on surfaces can be induced by different mechanisms: particle inertia; Brownian and turbulent diffusion; turbulent migration resulting from particle interactions with turbulent eddies of the carrier fluid; thermo- and electrophoresis; external mass forces; transverse force caused by the shear of the carrier fluid's average velocity (the Saffman force); particle rotation (the Magnus force), and so forth. Besides, the presence of a liquid film on the wall might lead to evaporation or condensation at the film surface, initiating transverse flow of gas (the so-called Stefan flux), which, in its turn, will affect the deposition of particles on the wall.

When considering particle interactions with the surface boundary of a two-phase flow, we formulate boundary conditions for the equations of motion and heat transfer for the disperse phase. As we use Eulerian continuous simulation to determine disperse phase's parameters in the region adjacent to the wall, we lose information about the details of particles' interaction with the surface because of summation of momenta, energies, and other parameters of incident and reflected particle fluxes. In the Eulerian paradigm, we determine the boundary conditions by finding the PDF of velocity and temperature near the wall from the corresponding kinetic equation (an analogy with the rarefied gas theory comes to mind). A more sophisticated treatment of this problem may be carried out on the basis of Lagrangian trajectory approach.

The boundary conditions thus obtained describe interaction of particles with the surface in terms of the reflection coefficient (which is equal to the probability of rebound in a flow of particles colliding with the wall) and the velocity and temperature recovery coefficients. It follows from the boundary conditions that, due to the dynamic and thermal inertia, velocity and temperature "slipping" may take place at the wall surface (again, this phenomenon is akin to similar phenomena in the rarefied gas theory).

When simulating particle motion in rarefied disperse gaseous media, that is, in media with low volume concentration of the disperse phase, the main attention is focused on particle interactions with turbulent eddies of the carrier phase, as the role of interparticle interactions is of little importance. As we increase concentration and particle size, interparticle interactions begin to make larger and larger contribution to the transport of momentum and energy in the disperse phase. Chaotic motion of particles caused by their interactions has come to be known as pseudoturbulent motion; the purpose of this term is to distinguish this motion from simple "turbulent motion" caused by the particles' involvement into the turbulent flow of the carrier medium. Pseudoturbulent motion may arise from hydrodynamic interaction between particles (that is, through momentum and energy exchange with random velocity and pressure fields of the surrounding medium) as well as from direct interactions via collisions.

Pseudoturbulent motion resulting from hydrodynamic (collisionless) interactions is anisotropic. Collisional interactions, on the other hand, result in an isotropic distribution similar to the Maxwellian distribution of velocity fluctuations. The role of momentum and energy exchange between particles via collisions becomes more important as we increase the size and concentration of particles. In concentrated disperse media, collisions become a major factor defining statistical properties of the system. A theoretical investigation of this problem may be carried out in the same manner as in the kinetic theory of dense gases.

The effective relaxation time t_{eff} describing particle interactions with the surrounding medium and with each other is found as

$$t_{eff} = \frac{t_v t_c}{t_v + t_c},$$

where t_v is the dynamic relaxation time of particles and t_c is the effective time between particle collisions, estimated as

$$t_c = \frac{d_p (1-\varphi/\varphi_*)^{1/3}}{e_p^{1/2} \varphi},$$

where d_p is the particle diameter, φ_* is the limiting volume concentration, and e_p is the turbulent energy of the disperse phase.

In a rarified disperse medium, $t_c \gg t_v$ and $t_{eff} \sim t_v$, while in a concentrated disperse medium, $t_c \ll t_v$ and $t_{eff} \sim t_c$.

With the growth of volume concentration of particles, the Lagrangian approach runs into difficulties as one is forced to keep track of ever-larger number of particle trajectories, and the Eulerian approach becomes preferable. The boundary separating the regions of applicability of these two approaches can be estimated from the equality $t_{eff}/T^{(L)} \sim 1$. For $t_{eff}/T^{(L)} < 1$, it is makes more sense to use the Eulerian approach, whereas for $t_{eff}/T^{(L)} > 1$, the Lagrangian approach is preferable. Stokes number St (it tells us which of the two approaches is preferable) is shown qualitatively in Fig. 5.2 as a function of volume concentration of the disperse phase.

Up till now we were dealing with monodisperse media, that is, media containing particles of the same size. The need to consider a polydisperse medium arises when we study combustion, phase transitions, breakup, coagulation, and other processes leading to variation of particle size spectrum. This spectrum is characterized by particle distribution over sizes $n(V, t, \mathbf{X})$ such that $n(V, t, \mathbf{X})dV$ is the number of particles whose volumes belong to the interval (V, V, dV) (here the size of a particle is

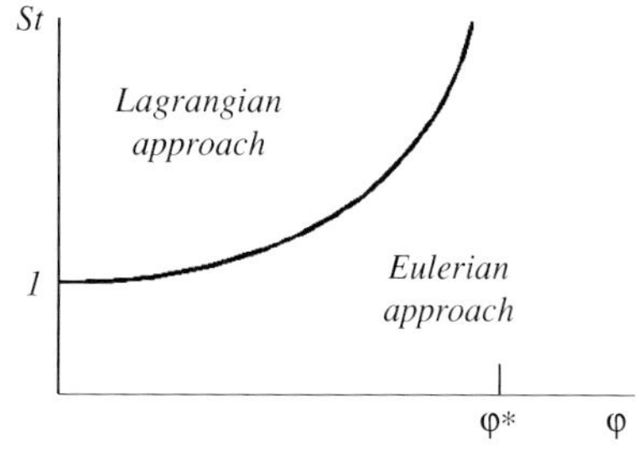

Fig. 5.2 Regions of applicability of the Eulerian approach and the Lagrangian approach.

represented by its volume v). All currently existing methods fall into one of the two groups. Methods belonging to the first group aim to determine the distribution function of particles over sizes by calculating the time evolution of the entire system of particles. The second group consists of methods that involve partitioning of the whole particle spectrum into discrete fractions, with further consideration of the dynamics of each fraction. In the first group, we should single out the methods based on direct solution of the kinetic equations for the size distribution of particles and for the PDF moments. The applicability of such methods is limited to the case of small particles whose velocities and temperatures only slightly differ from those of the carrier medium. The methods from the second group are more universal and allow for a multi-speed and multi-temperature description of the disperse phase, in other words, they are capable of taking into account the differences in velocity and temperature between different fractions of the spectrum.

Physical processes involving a change in the size spectrum of particles can be either continuous (combustion, condensation, evaporation) or discrete (breakup, coagulation). These processes are described, respectively, by differential and integral operators working on the mass coordinate in the phase space of the kinetic equation for the PDF. The birth of new particles in the flow volume may take place due to nucleation in the course of spontaneous homogeneous condensation of vapor in supercooled flows. A variation of the particle spectrum may also occur as a result of interaction of particles (or the flow itself) with boundary surfaces. The examples include deposition, breakup, and secondary ablation. These processes should be taken into account by writing the appropriate boundary conditions or by including source-type terms in the balance-of-mass equation.

If we want to control the coagulation process in a turbulent flow, it is necessary to know beforehand the frequency of particle collisions that are caused by their chaotic motion. To calculate this frequency (i.e., the rate of coagulation), we can employ one of the two approaches. The first one focuses on finding the correlation coefficients of relative velocity and its derivatives for a pair of particles in a homogeneous turbulent flow by solving the equation of motion for the particles. The other approach is based on the diffusion model that is similar to the Smoluchowski model for Brownian diffusion. The most effective method, however, is to use the equation for the PDF of relative velocity of a pair of particles.

5.6 Deposition of Particles from a Turbulent Flow

When considering the problem of particle deposition onto a surface, we can model the disperse phase motion by using either Eulerian or Lagrangian approach. In the Eulerian continual approach, the form of equations depends on Stokes number *St*. In the most interesting case of $\mathrm{St} \ll 1$, the task of finding the concentration of particles boils down to solving some diffusion equation given by a diffusion or inertial diffusion model. The question about the appropriate boundary conditions at the surface remains open, because the Eulerian continual approach does not provide information about interactions between the particles and the surface. So, in order to determine the boundary conditions, one has to find the PDF of velocity and temperature in the region adjacent to the wall, having solved the kinetic equation for the PDF.

Lagrangian trajectory approach enables us to formulate the boundary conditions for the equations of motion and heat-and-mass-transfer equations written for the disperse phase. As an example, let us try to derive the kinetic equation for the PDF of particles in a turbulent flow, the diffusion equation and the boundary conditions that would allow us to handle the problem of particle deposition on the wall for the isothermal case. Consider the equation of motion for an individual solid spherical particle:

$$\frac{d\boldsymbol{V}_p}{dt}=\frac{\boldsymbol{u}(\boldsymbol{X}_p(t),t)-\boldsymbol{V}_p(t)}{t_v}+\boldsymbol{F}(\boldsymbol{X}_p(t),t)+\boldsymbol{f}(\boldsymbol{X}_p(t),t),$$

$$\frac{d\boldsymbol{X}_p}{dt}=\boldsymbol{V}_p,\quad t_v=\frac{d_p^2\rho_p}{18\nu_e\rho_e},\tag{5.108}$$

where $\boldsymbol{X}_p(t)$ and $\boldsymbol{V}_p(t)$ are the coordinate and velocity of the particle at the instant t, $\boldsymbol{u}(\boldsymbol{X}t)$ – Eulerian velocity of the flow, d_p – particle diameter, ρ_p and ρ_e – densities of the particle and the carrier fluid.

The first term on the right-hand side describes the force of viscous interaction between phases in Stoksian approximation. The second term is the external force, for example, gravity. The third term is the random force acting on a unit particle mass in the course of Brownian motion. This force must obey the condition (1.150), in other words, the random Brownian force $\boldsymbol{f}$ must be delta-correlated.

The equation of motion (5.108) is written in the approximation of large difference between particle and fluid densities and smallness of the particle size so that additional forces acting on the particle could be ignored, specifically, the force caused by the pressure gradient, the force due to the virtual mass, and the Basset force, which arises when a nonstationary flow bypasses the particle. The Saffman and Magnus forces are also neglected. Additional information about forces acting on the particle can be found in [13–16]. Later on we shall need the solution of Eq. (5.108), which, when taken together with the initial conditions $\boldsymbol{X}(0)$ and $\boldsymbol{V}(0)$, can be represented in the integral form

$$\boldsymbol{V}_p(t)=\boldsymbol{V}_p(0)\exp\left(-\frac{t}{t_v}\right)$$

$$+\int_0^t\left(\frac{\boldsymbol{u}(\boldsymbol{X}_p(t_1),t_1)}{t_v}+\boldsymbol{F}(\boldsymbol{X}_p(t_1),t_1)+\boldsymbol{f}(\boldsymbol{X}_p(t_1),t_1)\right)\exp\left(-\frac{t-t_1}{t_v}\right)dt_1,$$

$$\times\boldsymbol{X}_p(t)=\boldsymbol{X}_p(0)+\int_0^t\boldsymbol{X}_p(t_1)d(t_1).\tag{5.109}$$

Eq. (5.108) is the Langevin equation that depends on two random uncorrelated quantities: velocity $\boldsymbol{u}$ and force $\boldsymbol{f}$. For simplicity's sake, assume the mass concentration of particles to be small ($\gamma \ll 1$). Then the influence of particles on the characteristics of the carrier flow, as well as interparticle collisions can be neglected.

Introduce the probability density distribution of particles $p(\boldsymbol{X}, \boldsymbol{V}, t)$ over coordinates$\boldsymbol{X}$ and velocities $\boldsymbol{V}$ in the $(\boldsymbol{X}, \boldsymbol{V})$ phase space (see Eq. (1.145)):

$$p(\boldsymbol{X}, \boldsymbol{V}, t) = \langle \delta(\boldsymbol{X}-\boldsymbol{X}_p(t))\delta(\boldsymbol{V}-\boldsymbol{V}_p(t))\rangle. \tag{5.110}$$

Here averaging is performed over all possible instances of the turbulent flow and the random force $\boldsymbol{f}$.

From Eq. (5.110), there follows

$$\langle \delta(\boldsymbol{X}-\boldsymbol{X}_p(t))\delta(\boldsymbol{V}-\boldsymbol{V}_p(t))\boldsymbol{V}_p(t)\rangle = \boldsymbol{V}_p(\boldsymbol{X}, \boldsymbol{V}, t). \tag{5.111}$$

Differentiating $p(\boldsymbol{X}, \boldsymbol{V}, t)$ with respect to t and making use of Eq. (5.108), we write:

$$\begin{aligned}\frac{\partial p}{\partial t} &= -\left\langle \frac{\partial}{\partial X_k}\left(\delta(\boldsymbol{X}-\boldsymbol{X}_p(t))\delta(\boldsymbol{V}-\boldsymbol{V}_p(t))\frac{dX_{pk}}{dt}\right)\right\rangle \\ &\quad -\frac{\partial}{\partial V_k}\left\langle \delta(\boldsymbol{X}-\boldsymbol{X}_p(t))\delta(\boldsymbol{V}-\boldsymbol{V}_p(t))\frac{dV_{pk}}{dt}\right\rangle \\ &= \left\langle \frac{\partial}{\partial X_k}\delta(\boldsymbol{X}-\boldsymbol{X}_p(t))\delta(\boldsymbol{V}-\boldsymbol{V}_p(t))V_{pk}\right\rangle \\ &\quad -\frac{\partial}{\partial V_k}\left\langle\left(\delta(\boldsymbol{X}-\boldsymbol{X}_p(t))\delta(\boldsymbol{V}-\boldsymbol{V}_p(t))\frac{dV_{pk}}{dt}\right)\left(\frac{u_k-V_{pk}}{t_v}+F_k+f_k\right)\right\rangle.\end{aligned} \tag{5.112}$$

Velocity of the carrier phase is a sum of the average value $U_k(\boldsymbol{X}, t)$ and the fluctuational component $u'_k(\boldsymbol{X}, t)$:

$$u_k(\boldsymbol{X}, t) = U_k(\boldsymbol{X}, t) + u'_k(\boldsymbol{X}, t), \tag{5.113}$$

where, as usual, we take $\langle u_k(\boldsymbol{X},t)\rangle = U_k(\boldsymbol{X},t)$ and $\langle u'_k(\boldsymbol{X}, t)\rangle = 0$. Then from the relation (5.112) plus Eq. (5.111) and Eq. (5.113), one gets an equation for the probability density distribution (the Liouville equation):

$$\frac{\partial p}{\partial t} + V_k\frac{\partial p}{\partial X_k} + \frac{\partial}{\partial V_k}\left(\frac{U_k-V_{pk}}{t_v} + F_k\right)p + \frac{\partial}{\partial V_k}\left(\frac{\langle p u'_k\rangle}{t_v} + \langle p f_k\rangle\right) = 0. \tag{5.114}$$

Here

$$\langle pu'_k\rangle = \langle\delta(\boldsymbol{X}-\boldsymbol{X}_p(t))\delta(\boldsymbol{V}-\boldsymbol{V}_p(t))u'_k\rangle,$$

$$\langle p f_k\rangle = \langle\delta(\boldsymbol{X}-\boldsymbol{X}_p(t))\delta(\boldsymbol{V}-\boldsymbol{V}_p(t)) f_k\rangle.$$

Eq. (5.114) is unclosed because it contains unknown correlations $\langle pu'_k\rangle$ and $\langle pf_k\rangle$. To derive these correlations, we assume the random fields u'_k and f_k to be Gaussian and use the relations (5.110) for $p(\boldsymbol{X}, V, t)$ and the Furutsu–Donsker–Novikov formula (see Eq. (1.201)):

$$\langle Z(\boldsymbol{X})R(Z)\rangle = \int\langle Z(\boldsymbol{X})Z(\boldsymbol{X}_1)\rangle\left\langle\frac{\delta R(Z(\boldsymbol{X}))}{\delta Z(\boldsymbol{X}_1)}\right\rangle d\boldsymbol{X}_1, \tag{5.115}$$

where $Z(\boldsymbol{X})$ is a random process in $\boldsymbol{X}$-space, $R(Z)$ – a functional dependent on the random process, and $\delta R/\delta Z$ – the functional derivative.

Let us substitute into Eq. (5.115) successively $Z(\boldsymbol{X}) = u'_j(\boldsymbol{X}, t)$, $Z(\boldsymbol{X}) = f'_j(\boldsymbol{X}, t)$, and $R(Z) = p(\boldsymbol{X}, V, t)$. Then

$$\langle pu'_i\rangle = \int\int\langle u'_i(\boldsymbol{X}, t)u'_k(\boldsymbol{X}_1, t_1)\rangle\left(\frac{\delta p(\boldsymbol{X}, \boldsymbol{V}, t)}{\delta u'_k(\boldsymbol{X}_1, t_1)}\right)d\boldsymbol{X}_1 dt_1, \tag{5.116}$$

$$\langle p f_i\rangle = \int\int\langle f_i(\boldsymbol{X}, t) f_k(\boldsymbol{X}_1, t_1)\rangle\left(\frac{\delta p(\boldsymbol{X}, \boldsymbol{V}, t)}{\delta f_k(\boldsymbol{X}_1, t_1)}\right)d\boldsymbol{X}_1 dt_1, \tag{5.117}$$

where because of Eq. (5.109), the functional derivative in the integrand is equal to

$$\frac{\delta p(\boldsymbol{X}, \boldsymbol{V}, t)}{\delta u'_k(\boldsymbol{X}_1, t_1)} = -\frac{\partial}{\partial X_j}\left\langle\delta(\boldsymbol{X}-\boldsymbol{X}_p(t))\delta(\boldsymbol{V}-\boldsymbol{V}_p(t))\frac{\delta X_{pj}(t)}{\delta u'_k(\boldsymbol{X}_1, t_1)}\right\rangle$$
$$-\frac{\partial}{\partial V_j}\left\langle\delta(\boldsymbol{X}-\boldsymbol{X}_p(t))\delta(\boldsymbol{V}-\boldsymbol{V}_p(t))\frac{\delta V_{pj}(t)}{\delta u'_k(\boldsymbol{X}_1, t_1)}\right\rangle.$$

Let us apply to Eq. (5.109) the operator of functional differentiation, using the equality $\delta u_i(\boldsymbol{X}, t)/\delta u_j(\boldsymbol{X}_1, t_1) = \delta_{ij}\delta(\boldsymbol{X}-\boldsymbol{X}_1)$, $\delta(t-t_1)$ (see Eq. (1.186)); the condition of causality, which states that the solution of the stochastic equation (5.108) at the moment t is defined by the behavior of the random field $f(\tau)$ on the time interval $(0, t)$ and does not depend on the values of $f(\tau)$ at $\tau > t$; and the absence of dependence of $V_{pi}(0)$ and $X_{pi}(0)$ on u'_i. The result is

$$\frac{\delta V_{pi}(t)}{\delta u'_j(\boldsymbol{X}_1, t_1)} = \frac{1}{t_v}e^{-b}\delta_{ij}\delta(\boldsymbol{X}_1-\boldsymbol{X}_p(t_1)), \quad b = \frac{t-t_1}{t_v},$$

$$\frac{\delta X_{pi}(t)}{\delta u'_j(\boldsymbol{X}_1, t_1)} = \int_{t_1}^{t}\frac{\delta V_{pi}(t_2)}{\delta u'_j(\boldsymbol{X}_1, t_1)}dt_2 = (1-e^{-b})\delta_{ij}\delta(\boldsymbol{X}_1-\boldsymbol{X}_p(t_1)).$$

Now, putting the derived relations into Eq. (5.116), we get

$$\langle pu_i'\rangle = t_v g\langle u_i'u_k'\rangle\frac{\partial p}{\partial X_k} - f\langle u_i'u_k'\rangle\frac{\partial p}{\partial V_k}, \tag{5.118}$$

where

$$\begin{aligned} g\langle u_i'u_k'\rangle &= \frac{1}{t_v}\int_0^\infty \langle u_i'(\boldsymbol{X},t)u_k'(\boldsymbol{X}_p(t_1),t_1)\rangle(1-e^{-b})dt_1, \\ f\langle u_i'u_k'\rangle &= \frac{1}{t_v}\int_0^\infty \langle u_i'(\boldsymbol{X},t)u_k'(\boldsymbol{X}_p(t_1),t_1)\rangle e^{-b}dt_1, \end{aligned} \tag{5.119}$$

$\langle u_i'u_j'\rangle$ are second single-point simultaneous moments of velocity fluctuation of the carrier flow, and g and f are the coefficients of the particles' involvement into fluctuational motion of the carrier phase.

Entrainment of particles by the turbulent flow can be characterized by the time $T^{(L)}$ (Lagrangian correlation time) of particle interaction with a macroscopic volume (a "turbulent mole"). It is equal by the order of magnitude to the integral time scale of turbulence. Physically, $T^{(L)}$ stands for the length of time between the two given instants that is required for the single-point velocity correlation to disappear. In other words, the velocities measured at t_1 and t do not correlate if $(|t_1 - t| > T^{(L)})$. Therefore correlation functions at $\boldsymbol{X}=\boldsymbol{X}_p$ can be approximated by the step function

$$\langle u_i'(\boldsymbol{X},t)u_j'(\boldsymbol{X}_p(t_1),t_1)\rangle = \begin{cases} \langle u_i'u_j'\rangle & \text{at } |t_1-t| \le T^{(L)}, \\ 0 & \text{at } |t_1-t| > T^{(L)}. \end{cases}$$

Substituting this relation into Eq. (5.119), one finds the particle involvement coefficients:

$$g = \frac{T^{(L)}}{t_v} - 1 + \exp\left(-\frac{T^{(L)}}{t_v}\right), \quad f = 1 - \exp\left(-\frac{T^{(L)}}{t_v}\right). \tag{5.120}$$

Let us now turn to determination of $\langle pf_i\rangle$. The random force $\boldsymbol{f}$ is a Brownian force, therefore its components are delta-correlated in time and, according to Eq. (3.50), obey the condition

$$\langle f_i(\boldsymbol{X},t)\, f_j(\boldsymbol{X}_p(t_1),t_1)\rangle = \frac{1}{t_v} D_{br}^0 \delta_{ij}\delta(t-t_1)$$

where $D_{br}^0 = \mathrm{k}\vartheta/6\pi\mu_e a$ is the coefficient of unhindered Brownian diffusion, ϑ – the absolute temperature of the carrier phase, k – the Boltzmann constant, a – the particle radius, μ_e – the dynamical viscosity coefficient for the carrier phase.

From Eq. (5.117), there follows

$$\langle p f_i \rangle = \int\int \langle f_i(\boldsymbol{X}, t)\, f_k(\boldsymbol{X}_1, t_1)\rangle \left(\frac{\delta p(\boldsymbol{X}, \boldsymbol{V}, t)}{\delta f_k(\boldsymbol{X}_1, t_1)}\right) d\boldsymbol{X}_1 dt_1 = -\frac{D^0_{br}}{t_v^2} \delta_{ik} \frac{\partial p}{\partial V_k}. \tag{5.121}$$

Substituting the relations (5.118) and (5.121) into Eq. (5.114), we obtain a closed equation for probability density distribution of particles over coordinates and velocities in a turbulent flow:

$$\begin{aligned} &\frac{\partial p}{\partial t} + V_k \frac{\partial p}{\partial X_k} + \frac{\partial}{\partial V_k}\left(\frac{U_k - V_k}{t_v} + F_k\right) p \\ &= g\langle\langle u'_i u'_k\rangle\rangle \frac{\partial^2 p}{\partial X_i \partial X_k} + \frac{f}{t_v}\langle\langle u'_i u'_k\rangle\rangle \frac{\partial^2 p}{\partial V_i \partial V_k} + \frac{D^0_{br}}{t_v^2} \frac{\partial^2 p}{\partial V_k \partial V_k}. \end{aligned} \tag{5.122}$$

Plugging $\langle u'_i u'_k\rangle = 0$ into Eq. (5.122), we get the Fokker–Planck equation for Brownian particles in a laminar flow:

$$\frac{\partial p}{\partial t} + V_k \frac{\partial p}{\partial X_k} + \frac{\partial}{\partial V_k}\left(\frac{U_k - V_k}{t_v} + F_k\right) p = \frac{D^0_{br}}{t_v^2} \frac{\partial^2 p}{\partial V_k \partial V_k}. \tag{5.123}$$

In the other limiting case – that of highly inertial particles with Stokes number $\mathrm{St} = t_v/T^{(L)} \gg 1$ – one can neglect Brownian diffusion ($D^0_{br} = 0$) and take $f \approx 1/\mathrm{St} \sim 0$ and $g \approx 1/(2\mathrm{St}^2) \sim 0$, thereby obtaining the Liouville equation for deterministic particle motion:

$$\frac{\partial p}{\partial t} + V_k \frac{\partial p}{\partial X_k} + \frac{\partial}{\partial V_k}\left(\frac{U_k - V_k}{t_v} + F_k\right) p = 0. \tag{5.124}$$

Eq. (5.122) is a multidimensional partial differential equation whose direct solution presents difficulties. Consider the widely used method of moments that yields equations for statistical characteristics of the disperse phase.

Let us write disperse phase's velocity as a sum of the average value $\langle \boldsymbol{V} \rangle$ and the fluctuation $\boldsymbol{V}'$,

$$\boldsymbol{V} = \langle \boldsymbol{V} \rangle + \boldsymbol{V}'$$

and introduce the moments

$$C = \int p(\boldsymbol{X}, \boldsymbol{V}, t) d\boldsymbol{V}, \quad \langle V_k \rangle = \int V_k\, p(\boldsymbol{X}, \boldsymbol{V}, t) d\boldsymbol{V}, \quad \langle V'_i V'_j \rangle = \frac{1}{C}\int V'_i V'_j\, p d\boldsymbol{V},$$

that have the meaning of concentration, average velocity of particles, and stress tensor components in the disperse phase, respectively. Integration of Eq. (5.122) over the entire velocity space $\boldsymbol{V}$ yields a balance-of-mass equation for the disperse phase:

$$\frac{\partial C}{\partial t}+\frac{\partial}{\partial X_k}(C\langle V_k\rangle)=0. \tag{5.125}$$

Multiplying (5.122) by V_k and integrating with respect to $\boldsymbol{V}$, we then get an equation for the average velocity of the disperse phase:

$$\frac{\partial\langle V_i\rangle}{\partial t}+\langle V_k\rangle\frac{\partial\langle V_i\rangle}{\partial X_k}=-\frac{\partial\langle V_i'V_k'\rangle}{\partial X_k}+\frac{U_i-\langle V_i\rangle}{t_v}+F_i-\frac{\tilde{D}_{ik}-\langle V_i\rangle}{t_v}\frac{\partial(\ln C)}{\partial X_k}, \tag{5.126}$$

where $\tilde{D}_{ik}$ is the diffusion tensor equal to

$$\tilde{D}_{ik}=t_v\langle V_i'V_k'\rangle+g\langle V_i'V_k'\rangle. \tag{5.127}$$

Terms on the right-hand side of Eq. (5.126) have the following meaning. The first term describes Brownian motion and generation of stresses in the disperse phase as a result of particles' involvement into the turbulent motion of the carrier phase. The second term characterizes the action of the friction force exerted on the particles by the continuous phase. The third term stands for the external force, and the last term – for the thermodynamic diffusional force.

Disregarding convective terms in Eq. (5.126), we get the following relation for the average velocity of the disperse phase:

$$\langle V_i\rangle=U_i+t_vF_i-t_v\frac{\partial\langle V_i'V_k'\rangle}{\partial X_k}-\tilde{D}_{ik}\frac{\partial(\ln C)}{\partial X_k}. \tag{5.128}$$

We get the equation for second moments $\langle V_i'V_j'\rangle$ by multiplying both sides of Eq. (5.122) by $V_i'V_j'$ and integrating it with respect to $\boldsymbol{V}$. The result is

$$\frac{\partial\langle V_i'V_j'\rangle}{\partial t}+\langle V_k\rangle\frac{\partial\langle V_i'V_j'\rangle}{\partial X_k}+(\langle V_i'V_k'\rangle+g\langle u_i'u_k'\rangle)\frac{\partial\langle V_j'\rangle}{\partial X_k}+(\langle V_i'V_k'\rangle+g\langle u_i'u_k'\rangle)$$
$$\times\frac{\partial\langle V_i\rangle}{\partial X_k}+\frac{1}{C}\frac{\partial}{\partial X_k}C\langle V_i'V_j'V_k'\rangle=\frac{2}{t_v}\left(f\langle u_i'u_j'\rangle+\frac{D_{br}}{t_v}\delta_{ij}-\langle V_i'V_j'\rangle\right). \tag{5.129}$$

The system of equations (5.125), (5.126), and (5.129) is unclosed because Eq. (5.129) contains a new unknown – the third moment $\langle V_i'V_j'V_k'\rangle$. The process of deriving higher-order moments can be continued, but new equations will also contain new unknowns. As a result, we get an infinite system of equations similar to the Friedman–Keller chain in the turbulence theory for a single-phase medium. Although the derived equations cannot be solved in the ordinary sense of the word, they still provide important information about statistical characteristics of the disperse phase.

In the special case when the random process is stationary and the turbulent flow is homogeneous, it follows from Eq. (5.129) that

$$\langle V_i' V_j' \rangle = f \langle u_i' u_j' \rangle + \frac{D_{br}}{t_v} \delta_{ij}. \tag{5.130}$$

We see that very small particles with $\mathrm{St} = t_v / T^{(L)} \to 0$ (and consequently, $f \to 1$ according to Eq. (5.120) are completely involved into the turbulent flow of the carrier phase, whereas relatively large particles with $\mathrm{St} \to \infty$ and $f \to 0$ do not get involved in fluctuational motion.

After we substitute the relations (5.128) and (5.130) into the balance-of-mass equation (5.125), the latter turns into a diffusion equation:

$$\frac{\partial C}{\partial t} + \frac{\partial}{\partial X_k}((U_k + t_v F_k)C) = \frac{\partial}{\partial X_k}\left(\tilde{D}_{ik}\frac{\partial C}{\partial X_i} + C\frac{\partial}{\partial X_i}(q D_{ik} + D_{br}\delta_{ik})\right), \tag{5.131}$$

where $\tilde{D}_{ik} = D_{ik} + D_{br}\delta_{ik}$, D_{ik} is the coefficient of turbulent diffusion, and $q = \mathrm{St}\cdot f$ is the migration coefficient.

Compare the diffusion equation (5.131) for particles with the turbulent diffusion equation (5.93) for a passive impurity. The first difference is the presence of an external force (gravity, for instance) that gives rise to an additional convective transport. Secondly, in addition to diffusional transport, there appears migrational transport (second term in parenthesis on the right-hand side) caused mainly by the inhomogeneity of the turbulent fluctuation field of the carrier flow. Migration coefficient q is proportional to Stokes number and increases as the particle size goes from the size of passive particles to the size of inertial ones. So, the deposition of small particles with $t_v/T^{(L)} \ll 1$ is occurs mostly via turbulent and Brownian diffusion, while the deposition of relatively large particles with $t_v/T^{(L)} \geq 1$ occurs (to a greater extent) via turbulent migration.

To solve the diffusion equation (5.131), one has to formulate the boundary conditions at the surface that is partially or completely absorbing. To this end, let us examine the stationary solution of Eq. (5.122) in a thin kinetic layer near the wall. Suppose that in this layer, the only important terms are those associated with projections onto the normal direction to the wall (the Y-axis). Let the asymptotic relation (5.43), $D_t = T^{(L)}\langle (u_y')^2 \rangle$, stand for the coefficient of turbulent diffusion. Then Eq. (5.122) reduces to

$$\begin{aligned} &\frac{(qD_t + D_{br})}{t_v}\frac{\partial^2 p}{\partial V_y^2} + \frac{\partial (V_y p)}{\partial V_y} \\ &= t_v V_y \frac{\partial p}{\partial Y} + (U_y + t_v F_y)\frac{\partial p}{\partial V_y} - (1-q) D_t \frac{\partial^2 p}{\partial Y \partial V_y}. \end{aligned} \tag{5.132}$$

The solution will be obtained for the case of small deviations of p from the equilibrium value p_{eq},

$$p_{eq} = C\left(\frac{t_v}{2\pi(qD_t + D_{br})}\right)^{1/2} \exp\left(-\frac{t_v V_y^2}{2(qD_t + D_{br})}\right),$$

taking the right-hand side of this equation as the perturbing factor.

Let us take $p = p_{eq} + p'$, where p' is a small perturbation of the distribution. Substituting this representation into Eq. (5.132) and ignoring small quantities of higher orders, one obtains an equation for p':

$$\frac{(qD_t + D_{br})}{t_v}\frac{\partial^2 p'}{\partial V_y^2} + \frac{\partial(V_y p')}{\partial V_y}$$
$$= t_v V_y \frac{\partial p_{eq}}{\partial Y} + (U_y + t_v F_y)\frac{\partial p_{eq}}{\partial V_y} - (1-q)D_t \frac{\partial^2 p_{eq}}{\partial Y \partial V_y}.$$

$$p = C\left(\frac{t_v}{2\pi(qD_t + D_{br})}\right)^{1/2} \exp\left(-\frac{t_v V_y^2}{2(qD_t + D_{br})}\right)\left\{1 - \frac{t_v V_y}{qD_t + D_{br}}\right.$$
$$\times\left[(D_t + D_{br})\frac{d(\ln C)}{dY} + \frac{1}{2}(2qD_t - D_t + D_{br}) + \frac{(D_t + D_{br})t_v V_y^2}{3(qD_t + D_{br})}\right]$$
$$\left.\times\frac{d\ln(qD_t + D_{br})V_y^2}{dY} - (U_y + t_v F_y)\right\}.$$

The distribution above enables us to find the incident J_f and reflected J_r particle fluxes:

$$J_f = -\int_{-\infty}^{0} V_y p dV_y = C\left(\frac{qD_t + D_{br}}{2\pi t_v}\right)^{1/2} - (U_y + t_v F_y)\frac{C}{2}$$
$$+ \frac{(D_t + D_{br})}{2}\frac{dC}{dY} + \frac{C}{2}\frac{d(qD_t + D_{br})}{dY}, \tag{5.133}$$

$$J_r = -\int_{0}^{\infty} V_y p dV_y = C\left(\frac{qD_t + D_{br}}{2\pi t_u}\right)^{1/2} + (U_y + t_u F_y)\frac{C}{2}$$
$$- \frac{(D_t + D_{br})}{2}\frac{dC}{dY} - \frac{C}{2}\frac{d(qD_t + D_{br})}{dY}. \tag{5.134}$$

The quantities on the right-hand side of relations (5.133) and (5.134) correspond to their values at the wall. A surface's ability to reflect and absorb particles is characterized by the reflection factor χ equal to the probability for the particle to get detached from the wall, or by the absorption factor $1-\chi$ equal to the probability for the particle to adhere to the wall. The reflection coefficient is equal to the ratio between the reflected and incident particle fluxes,

$$\chi = \frac{J_r}{J_f}. \tag{5.135}$$

Substitution of relations (5.133) and (5.134) for fluxes into Eq. (5.135) gives the following boundary condition connecting particle concentration at the wall C_w with the flux of particles deposited at the wall $J_w = J_f - J_r$:

$$C_w = \frac{(1+\chi)}{(1-\chi)}\left(\frac{\pi t_v}{2(qD_t + D_{br})}\right)^{1/2} J_w, \tag{5.136}$$

where

$$J_w = (D_t + D_{br})\frac{dC}{dY} - C\frac{d(qD_t + D_{br})}{dY} - C(U_y + t_u F_y).$$

It follows that for a completely reflecting surface, $\chi = 1$ and $J_w = 0$ whereas for a completely absorbing surface, $\chi = 0$ and $C_w \neq 0$, contrary to a widespread opinion. It should be noted that when solving concrete problems, the boundary condition (5.136) is specified at some distance from the wall, outside the viscous sublayer, rather than at the wall itself. Sometimes this distance is taken to be equal to the particle radius. When considering the final stage of particle's approach to the wall and the chance that it will be captured by the wall, it is necessary to take into account the interaction force between the particle and the surface at small values of clearance between the particle and the wall.

Paper [17] performs numerical solution of the problem of plane-parallel turbulent flow of a two-phase disperse medium in the near-wall region and suggests an approximate expression for particle flux at the completely absorbing wall, which takes into account Brownian and turbulent diffusion, turbulent migration, convection, and the external force

$$J_w = \varphi\left(\frac{(0.115/\mathrm{Sc}_{br}^{3/4} + 2.5\cdot 10^{-4}\tau_+^{2,5})u^*}{(1+10^{-3}\tau_+^{2,5})\max[0.61;\ \min(1.32 - 0.27\ln\tau_+;\ 1)]} - U_y - t_u F_y\right),$$

where $\tau_+ = t_v u^{*2}/\nu_e$; u^* is the dynamic velocity; $\mathrm{Sc}_{br} = \nu_e/D_{br}$ is the Schmidt number for Brownian diffusion; U_y and F_y are the normal (i.e., perpendicular to the wall) components of velocity and external force-driven acceleration of the carrier phase (gas); φ is the volume concentration of particles. The formula is valid at $\tau_+ < 100$ for relatively small particles inside the logarithmic layer $30 < y_+ < 100$, where $y_+ = yu^*/\nu_e$.

As an example, consider particle deposition from a stationary hydrodynamic developed turbulent flow in a plane-parallel or cylindrical channel. In the boundary layer approximation and in the absence of external forces, the diffusion equation (5.131) takes the form

$$\frac{d}{dX}(r^a U_y C) = \frac{d}{dr}\left\{r^a\left[(D_t + D_{br})\frac{dC}{dr} + C\frac{d}{dr}(qD_t + D_{br})\right]\right\}, \tag{5.137}$$

where X and $r = 1 - Y$ are the longitudinal and transverse (again, with respect to the wall) coordinates; $a = 0$ and 1 indicate plane-parallel and cylindrical channels, respectively.

Integration of Eq. (5.137) over the channel cross section leads to the equation

$$\frac{d(U_m C_m)}{dX} = -\frac{2^a}{r_w} J_w,$$

$$U_m = 2^a \int_0^{r_w} r^a U_X \frac{dr}{r_w^{1+a}}, \quad C_m = 2^a \int_0^{r_w} r^a C U_X \frac{dr}{U_m r_w^{1+a}}, \tag{5.138}$$

where U_m and C_m are the average mass velocity and particle concentration, r_w is the radius of the channel.

In the region of hydrodynamic stabilized flow one can take

$$\frac{\partial(U_X C)}{dX} = \frac{d(U_m C_m)}{dX}.$$

Then in view of Eq. (5.138), Eq. (5.137) reduces to

$$(D_t + D_{br})\frac{dC}{dr} + C\frac{d}{dr}(qD_t + D_{br}) = \frac{r}{r_w} J_w. \tag{5.139}$$

Suppose the coefficient of turbulent diffusion is equal to the coefficient of turbulent viscosity of the carrier fluid $D_t = \nu_t$. The latter is taken in the form

$$\frac{\nu_t}{\nu_e} = \frac{\kappa}{3} y_+ (2 - y_0)\left[\frac{1}{2} + (1 - y_0)^2\right]\left[1 - \exp\left(-\frac{\kappa y_+}{A^2}\right)\right],$$

which tends to the Reichardt formula far away from the wall and to the van Driest–Deissler formula near the wall. Here $y_0 = Y/r_w$, $y_+ = Yu_*/\nu_e$, $\kappa = 0.4$ and $A = 26$.

When solving Eq. (5.139), the boundary condition (5.136) is taken at the distance from the wall equal to the particle radius a, and $T^{(L)} = 200\nu_e/u_*^2$ is taken for the Lagrangian correlation time [18].

Look at Fig. 5.3a and fig. 5.3b to compare theoretical rates of deposition of suspended particles $V_w = J_w/C_w$ or $J_+ = V_m/u_*$ with experimental data. Curves 1 and 2 in Fig. 5.3a correspond to the experimental data [19] for the Reynolds numbers $Re = 2r_w U_m/\nu_e = 2.9 \cdot 10^5$ and $5 \cdot 10^5$ and curves 1–3 in Fig. 5.3b – to the experimental data [20] for the flow velocities $U_m = 7.6$; 17.6 and 26.6 m/s.

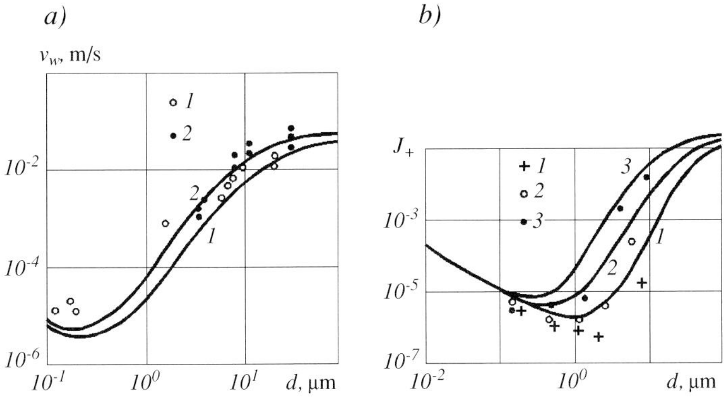

Fig. 5.3 Dependence of particle deposition rate on particle diameter.

The dependence of the rate of deposition on particle diameter has a minimum. The initial drop of the deposition rate is associated with the decrease of Brownian diffusion coefficient that accompanies the increase of the size of particles. In the region where particle size is smaller then 1 μm, Brownian diffusion becomes the predominant mechanism of particle deposition. Increasing the size of particles, we observe a significant increase of the velocity of turbulent migration due to the non-uniform distribution of turbulent fluctuation intensity over the channel cross section, which leads to further growth of the deposition rate. For particle sizes on the order of 100 μm, the dimensional deposition rate reaches its maximum value: $J_+ \approx 0.2$. The rate of deposition V_w increases with the flow velocity U_m as a consequence of more intense particle deposition under the action of turbulent diffusion as well as turbulent migration.

Fig. 5.4 demonstrates the dependence of the dimensional deposition rate V_w on the Reynolds number in a pipe of radius 2.5 mm for particles of diameters 0.01; 0.27; 2; and 8 μm (curves 1–5, respectively). The same figure also shows experimental

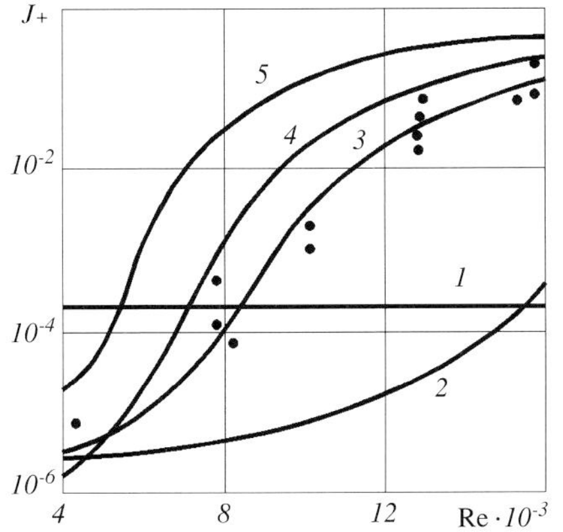

Fig. 5.4 Deposition velocity vs. Re.

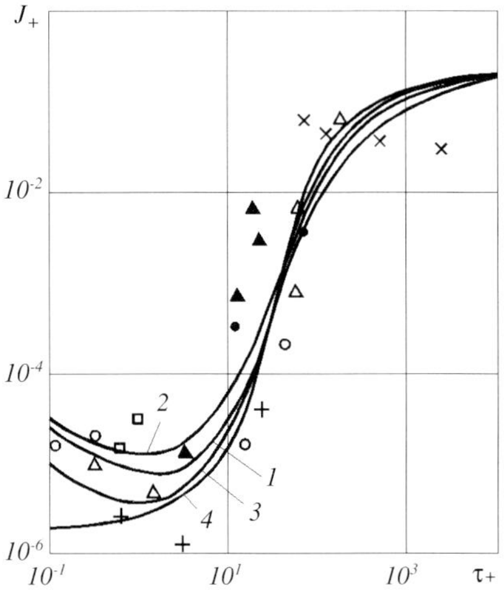

Fig. 5.5 Comparison with experiments.

points for particles of $d = 0.18\,\mu$m [21]. One can see that the rate of deposition of submicron particles practically does not depend on Re because the rate of deposition V_w of Brownian particles is proportional to the dynamic velocity u_*.

The correlation between numerical results and experimental data is shown in Fig. 5.5 as a dependence of J_+ on the dimensional relaxation time of the particle $\tau_+ = t_+ u_*^2/\nu_e$. Curves 1, 2 correspond to the pipe radius $r_w = 0.015$ m and the flow velocities $U_m = 10$ and 30 m/s; curve 3 – to $r_w = 0.3$ m, $U_m = 30$ m/s; curve 4 – to $r_w = 0.025$ m, $U_m = 30$ m/s; the experimental data is taken from [16]. One can see that experimental results are in good agreement with theoretical ones only for the inertial particles ($_+ > 10$), whereas for small (inertialess) particles, the particle size becomes essential. This is easy to understand once we realize that the intensity of particle deposition is determined by processes of turbulent nature, that is, turbulent diffusion and turbulent migration. For small (Brownian) particles with $_+ < 1$, the rate of deposition is determined not only by the intensity of turbulent transport, but also by Brownian diffusion, and thus essentially depends on the particle size.

5.7 Interaction of Particles in a Turbulent Flow

Particle interactions in a turbulent flow, as well as in a laminar flow, may be of two kinds: implicit interaction, that is, the influence exerted on the motion of the particle under consideration (test particle) by its neighbors through perturbation of hydrodynamic and concentration parameters, and explicit interactions, that is, direct particle collisions. The type of interaction depends on two parameters: volume concentration of suspended particles φ and Stokes number St responsible for the inertia of particles.

When we model particle motion in a rarefied disperse medium ($\varphi \ll 1$) containing particles with low inertia (St < 1), interparticle interactions are insignificant and result in a very small perturbation of hydrodynamic parameters of the carrier medium, which in its turn, exerts but a minuscule effect on the motion of particles. Moreover, the probability of particle collisions is low. We conclude that interparticle interactions can be neglected. In this statement of the problem, the main emphasis is on particle interactions with turbulent fluctuations of the carrier medium. When we increase volume concentration φ and Stokes number St (i.e., particle size), both types of interparticle interactions begin to make a larger contribution to the transport of momentum and energy in the disperse phase. Chaotic motion of particles that arises from interparticle interactions is called pseudoturbulent and should be distinguished from particle motion arising from turbulent fluctuations of the carrier phase.

The motion of particles in a turbulent flow of a concentrated disperse medium is affected by the medium itself and by interparticle interactions; the mechanism of these interactions can be collisionless hydrodynamic (implicit interaction) or collisional (explicit interaction). These mechanisms affect the motion of particles in different ways: particle motion caused by collisionless interactions is anisotropic, whereas collisional interactions result in isotropic motion whose distribution of velocity fluctuations in the disperse phase is close to Maxwellian.

With an increase of size and concentration of particles, the role of momentum and energy exchange between particles is growing in importance as compared to that of hydrodynamic interactions. In concentrated disperse media, interparticle interactions assume a leading role in the formation of statistical properties. Theoretical investigation of this problem is performed in the same way as in the kinetic theory of dense gases: one writes the Boltzmann equation and solves it with the Enskog method.

Various kinetic models whose aim is to calculate momentum transport in highly concentrated disperse media have been suggested in a number of works. Some of them that take into account particle interactions with turbulent fluctuations as well as with other particles.

The motion of a heavy spherical particle in a turbulent flow is described by the Langevin equation (5.101), into which one should insert additional terms in order to account for both types of particle interactions:

$$\frac{d\boldsymbol{V}_p}{dt}=\frac{\boldsymbol{u}-\boldsymbol{V}_p}{t_v}+\boldsymbol{F}+\boldsymbol{w}+\boldsymbol{W},\quad \frac{d\boldsymbol{X}_p}{dt}=\boldsymbol{V}_p, \tag{5.140}$$

where $\boldsymbol{u}$ and V_p are, respectively, the velocities of the carrier medium and the particle; $\boldsymbol{X}_p$ – particle coordinates; $\boldsymbol{F}$ – acceleration caused by the external force; t_v – the particle's dynamic relaxation time, which takes into account the inertia forces (they manifest themselves in deviations from the Stokes law) and the hindered character of particle motion, that is, dependence on φ; $\boldsymbol{w}$ and $\boldsymbol{W}$ – terms that account for the hydrodynamic collisionless and collisional interactions between particles (the former term represents a continuous, and the latter – a discrete random process). In contrast to Eq. (5.101), Brownian motion is not taken into consideration in Eq. (5.140).

Lagrangian approach can be used to model the behavior of the disperse phase while properly accounting for interparticle interactions. We start from Eq. (5.140) for a single particle and then proceed to the ensemble of a finite number of particles by varying the initial position $\boldsymbol{X}(0)$ and initial velocity $\boldsymbol{V}(0)$ of the particle. We can switch from Lagrangian trajectory description of particle motion (which is based on the Langevin equation (5.140)) to Eulerian description of the particle ensemble in the same manner as we did in Section 5.6. The probability density distribution of particles over coordinates and velocities $p(\boldsymbol{X}, \boldsymbol{V}, t) = \langle \delta(\boldsymbol{X} - \boldsymbol{X}_p)\delta(\boldsymbol{V} - \boldsymbol{V}_p) \rangle$ will then be described by the equation similar to Eq. (5.114): [9]

$$\frac{\partial p}{\partial t} + V_k \frac{\partial p}{\partial X_k} + \frac{\partial}{\partial V_k}\left[\left(\frac{U_k - V_{pk}}{t_v} + F_k\right) p\right] = -\frac{1}{t_v}\frac{\partial \langle u'_k p \rangle}{\partial V_k} - \frac{\partial \langle w_k p \rangle}{\partial V_k} + \boldsymbol{I}(p), \tag{5.141}$$

where U_k and u'_k are the average and fluctuational components of velocity of the carrier phase, and $\boldsymbol{I}(p)$ is the Boltzmann collision operator written in the Enskog form.

Suppose that the random fields of velocity fluctuations of the carrier phase u' and of hydrodynamic interactions w_k are Gaussian. Then, making the additional assumption that the field w_k is delta-correlated and using the method of functional differentiation, we can obtain differential equations for the moments (average concentration and average velocity, second moments, and higher-order moments) of velocity fluctuations in the carrier phase. The interested reader will find more details and further verification of the model in [8,9,22].

5.8 Chemical Reactions in a Turbulent Flow

The problem of modeling the turbulent flow of a chemically reacting mixture can be divided into three parts [23–29]:

– macromixing that results from turbulent diffusion on the distances on the order of the characteristic linear size of the flow region;
– micromixing, or, to use another term, mixing to the molecular level, which results in the formation of "genuine" local concentrations and in the possibility of reactions between formely unmixed reagents;
– calculation of average rates of chemical reactions with proper account taken of local fluctuations of reagent concentrations and temperature.

Of greatest practical interest is the calculation of average rates of chemical reactions in a turbulent flow, because the composition of reaction products in chemical engineering devices (chemical reactors) corresponds to the value obtained by averaging over the volume in which the reaction is taking place.

Depending on the relation between the characteristic times of turbulent mixing and those of chemical reactions, reactions are classified as slow, fast, or very fast.

If only a small fraction of reagents gets consumed in the reaction during the characteristic time of turbulent mixing (i.e., the time it takes for a uniform concentration to be established due to the action of turbulent mixing and molecular diffusion), we call such reaction slow. In other words, for slow reactions, the characteristic time of turbulent mixing is far shorter than the characteristic time of the reaction. This allows us to consider the two processes – turbulent mixing and chemical reaction – independently from one another. First, we solve the problem on turbulent mixing of the given substances; then, once the hydrodynamic and concentration fields have been obtained, we proceed to determine reaction rates.

For fast reactions, the characteristic times of turbulent mixing are of the same order as the characteristic times of reactions. Then the local rate of chemical reaction strongly depends on temperature fluctuations and fluctuations of reagent concentration. Therefore the processes of turbulent mixing and chemical transformations should be considered jointly. The modeling of chemical reactions is complicated by the fact that turbulent mixing and chemical reactions affect each other. Therefore turbulent mixing occurs differently than it would in the absence of chemical reactions, because local concentration gradients depend on the peculiarities of chemical reactions.

In the limiting case of very fast chemical reactions, the characteristic times of reactions are negligibly small compared to the characteristic time of turbulent mixing, and the reaction can be regarded as instantaneous. Then at each space point, we can expect to find one of the reagents involved in the chemical reaction, or the reaction products, because there is no chance to find reacting components simultaneously at the same point. In this case, the reaction zone changes into a surface of rather complicated topology.

To illustrate the points made above, let us perform dimensional analysis of the convective diffusion equation (5.57), adding a source-type term describing the chemical reaction to the right-hand side:

$$\frac{\partial C_i}{\partial t} + \boldsymbol{u}\cdot\nabla C_i = D_m \,\Delta C_i + \boldsymbol{W}_i. \tag{5.142}$$

The second term on the left and the two terms on the right correspond, respectively, to three different processes: convective transport via turbulent motion; transfer via molecular diffusion; production or consumption of i-th component in the course of chemical reaction. Each of them is described by its own characteristic time:

$$\tau_t = L_0/u_0, \quad \tau_D = L_0^2/D_m, \quad \tau_w = C_0/|W_0|,$$

where L_0, u_0, C_0, and $|W_0|$ are the characteristic values of the turbulent linear scale, velocity, passive impurity concentration, and reaction rate. Let τ be the characteristic time of the process under consideration. If we replace t with dimensionless time t/τ and divide all other dimensional quantities in Eq. (5.142) by their characteristic values, this will bring Eq. (5.142) to the dimensionless form. Conserving the old symbols, we can write

$$\frac{\partial C_i}{\partial t} + N_t \boldsymbol{u} \cdot \nabla C_i = N_D \Delta C_i + N_W W_i. \tag{5.143}$$

Here we have introduced dimensionless parameters $N_t = \tau/\tau_t$, $N_D = \tau/\tau_D$, and $N_W = \tau/\tau_W$.

If any of the above-listed processes dominates, the time corresponding to this process should be taken as τ. Let us consider these three cases successively.

1. *Turbulent transport (macromixing) dominates.*
   Then $\tau = \tau_t$ and the dimensionless parameters in Eq. (5.143) are equal to

   $$N_t = 1;\ N_D = \tau_t/\tau_D = D_m/(L_0 u_0) = \mathrm{Pe}^{-1};\ N_W = \mathrm{Da}_1$$
   $$= \tau_t/\tau_W = |W_0| L_0/u_0 C_0,$$

   where Pe is the Peclet number and Da is the Damko?hler number.
2. *Molecular diffusion (micromixing) dominates.*
   Then $\tau = \tau_D$ and the dimensionless parameters in Eq. (5.143) are equal to

   $$N_t = \tau_D/\tau_t = (L_0 u_0)/D_m = \mathrm{Pe};\ N_D = 1;$$
   $$N_W = \mathrm{Da}_2 = \tau_D/\tau_W = Da_1 \mathrm{Pe},$$

3. *Chemical reaction dominates.*
   Then $\tau = \tau_W$ and the dimensionless parameters in Eq. (5.143) are equal to

   $$N_t = \tau_W/\tau_t = \mathrm{Da}_1^{-1};\ N_D = \tau_W/\tau_t = \mathrm{Da}_2^{-1};\ N_W = 1.$$

For slow chemical reactions, the conditions $\tau_W \gg \tau_D$ and $\tau_W \gg \tau_t$ should hold. Therefore in cases 1 and 2 we have $N_W \ll 1$ and the source term in Eq. (5.143) can be ignored.

For very fast chemical reactions, the conditions $\tau_W \gg \tau_D$ and $\tau_W \gg \tau_t$ should be satisfied, and we also have $N_t \gg 1$ and $N_D \ll 1$. Then the convective transport and molecular diffusion terms in Eq. (5.143) can be neglected. Since N_D is a small parameter by high-order derivative, diffusion can be ignored everywhere except narrow regions adjacent to reaction surfaces.

In case of fast chemical reactions we have $\tau_W \sim \tau_D$, $\tau_W \sim \tau_t$. Therefore no process is given a preference and all terms in Eq. (5.143) should be taken into account, which seriously complicates the solution of the problem.

Mixing of components entering into a chemical reaction is a necessary prerequisite for this reaction. Insufficient mixing of components leads to concentration inhomogeneities (local concentration gradients) and thereby to inhomogeneous concentrations of reaction products.

The main characteristic of the degree (level) of mixing of *i*-th reacting component is the variance of concentration distribution

$$\langle (C_i')^2 \rangle = \langle (C_i - \langle C_i \rangle)^2 \rangle.$$

The degree of mixing depends on size of the volume over which the averaging is carried out. This size may vary from the characteristic macroscale of the volume filled with reacting mixture to the microscale that can be considered as a material point (or fluid particle) within the framework of the hydrodynamic approach. The quantity $\langle (C')^2 \rangle$ that depends on time and spatial coordinates will appear later on as a characteristic of micromixing.

The principal methods for theoretical description of turbulent mixing accompanied by fast chemical reactions are the method of moments and the probabilistic method that employs the equation for probability density distribution (the PDF-method). When using the method of moments, one usually considers only the equations for the first two moments of a scalar random quantity – concentration (and of temperature as well if the process is not isothermal). These two moments are $\langle C_i \rangle$ and $\langle (C_i)^2 \rangle$. The corresponding equations are obtained by the averaging of the convective diffusion equation in the same manner as we did earlier for hydrodynamic equations. Just like in the turbulence problem, the equations thus obtained turn out to be unclosed. Therefore in order to close the system of equations for the moments, one needs additional relations, which, as a rule, have a semiempirical character. The attempt to account for chemical reactions compels us to insert into the diffusion equation an additional source-type term (reaction term), which, generally speaking, has a nonlinear dependence on the concentrations of reagents. Therefore in addition to the problem of correct representation of the terms describing turbulent transport and mixing, one encounters the closure problem for the reaction terms.

The PDF method [30–32] is based on the equation for probability density distribution of fluid particles over coordinates, velocities, and concentrations. The knowledge of this distribution enables us to determine all statistical moments, that is, to solve the problem completely. Thus the PDF-method is more logical and informative in comparison with the method of moments. But solution of the problem is seriously complicated by multidimensionality of the distribution (it depends not only on time and spatial coordinates, but also on functions – velocities and concentrations).

Before we turn our attention to the method of moments and the PDF method, let us dwell briefly on some basic concepts of chemical kinetics.

5.8.1 Concepts of Chemical Kinetics

The rate of chemical reaction is defined by the amount of substance z produced or consumed in a unit volume per unit time. In the general case, the volume of the system under consideration can vary in the course of the reaction. When the volume is constant, the reaction rate is given by the quantity $\pm dC_z/dt$, where C_z is the concentration of substance z measured in mol/m^3. The plus sign corresponds to production and minus – to consumption of the matter. Therefore in chemical kinetics, the reaction rate is always positive. But when conservation equations are written in the general form (see Eq. (5.142)), the source term usually appears as

$$W = \frac{dC_z}{dt}, \tag{5.144}$$

so that production or consumption of matter in the course of the chemical reaction is indicated by the sign of W.

If a chemical reaction with n reagents $A, B, \ldots, E, F, \ldots$ is taking place in a single-phase system, and $A, B, \ldots$ are the reacting components, while $E, F, \ldots$ are the products of reaction, the kinetic scheme of such a reaction is written as

$$\nu_A A + \nu_B B + \cdots \rightarrow \nu_E E + \nu_F F \cdots, \tag{5.145}$$

where $\nu_A, \ldots, \nu_F$ are the stoichiometric coefficients; they are negative for "reactants" (components that are being consumed) and positive for "products" of reaction (it is worth noting, however, that all coefficients must change sign if the reaction is viewed as going "from the right to the left"). For example, the stoichiometric coefficients for the reaction

$$2NH_3 \rightleftarrows N_2 + 3H_2$$

are $\nu_{NH_3} = -2, \nu_{N_2} = 1, \nu_{H_2} = 3$.

The sum of stoichiometric coefficients on the left-hand side of Eq. (5.145) defines the so-called molecularity, which is defined as the number of particles involved in a single reaction step.

It follows from the law of definite proportions that the increase (decrease) of mass m_i of i-th component produced (consumed) in the reaction is proportional to its molecular weight M_i and stoichiometric coefficient ν_i for the given reaction. Therefore, denoting through N_i^0 the number of moles of i-th component at the initial moment and through and N_i – at the current moment, and keeping in mind that $m_i = N_i M_i$, we get

$$N_i - N_i^0 = \nu_i \xi, \quad (i = A, B, \cdots, E, F, \cdots). \tag{5.146}$$

where $\xi(t)$ is the degree of reaction completion. At the initial moment, $\xi = 0$. The value $\xi = 1$ corresponds to the moment when $\nu_A, \nu_B, \ldots$ moles of components $A, B, \ldots$ are transformed into $\nu_E, \nu_B, \ldots$ moles of components $E, F, \ldots$. If the system has made a transition from the state $\xi = 0$ to the state $\xi = 1$, we say that one reaction equivalent has occurred.

From the equation of mass conservation there follows the stoichiometric equation

$$\sum_i \nu_i M_i = 0. \tag{5.147}$$

Differentiating Eq. (5.146) with respect to t, we get

$$\frac{dN_A}{\nu_A} = \frac{dN_B}{\nu_A} = \cdots = \frac{dN_F}{\nu_A} = d\xi. \tag{5.148}$$

The reaction rate is equal to

$$W = \frac{d\xi}{dt}, \tag{5.149}$$

and we see from Eq. (5.148) that this rate can be written as

$$W = \frac{1}{\nu_i}\frac{dN_i}{dt}$$

or, in terms of i-th component concentration $C_i = N_i/V$, where V is the volume occupied by the reacting mixture, as

$$W = \frac{d(C_i V)}{\nu_i dt}.$$

If the reaction is not accompanied by a change of volume, the last relation gives us

$$\frac{1}{\nu_A}\frac{dC_A}{dt} = \frac{1}{\nu_B}\frac{dC_B}{dt} = \cdots = \frac{1}{\nu_E}\frac{dC_E}{dt} = \frac{1}{\nu_F}\frac{dC_F}{dt} = \cdots. \tag{5.150}$$

The dependence of reaction rate on reagent concentration is called the kinetic equation of reaction. For one-stage reactions of type (5.145) taking place in a homogeneous system, this dependence is given by the law of mass action:

$$\frac{dC_A}{dt} = k(C_A)^{n_A}(C_B)^{n_B}\ldots. \tag{5.151}$$

Constant k is known as reaction rate constant and has the meaning of specific reaction rate, that is, reaction rate for a unit concentration of reagents. Its dependence on temperature is given by the Arrhenius equation:

$$k = A\exp\left\{-\frac{E_A}{R\vartheta}\right\}, \tag{5.152}$$

where E_A is the activation energy, that is, the minimum energy a molecule must possess in order to enter into the reaction, A is a reaction constant, and R is the gas constant.

Exponent n_i in the formula (5.151) is called the order of reaction, and $n = \sum n_i$ – the total, or kinetic order of reaction. Kinetic order for single-stage reaction coincides with its molecularity. Among single-stage reactions, first- and second-order reactions are the most common. For multi-stage reactions, the order of reaction could be fractional.

The simplest example of reaction requiring a prior mixing of components is a second order reaction of the form

$$A + B \rightarrow \text{products}.$$

Among such reactions, one may single out (based on the reaction rate constant k) slow reactions with $k < 10^3$ 1/mole · s, fast reactions with 10^3 1/mole · s $< k < 10^6$ 1/mole · s, and very fast reactions with $k > 10^6$ 1/mole · s.

It should be noted that dimensionality of the reaction rate constant is defined by the kinetic order of reaction in the formula (5.151).

5.8.2
Method of Moments

Let us consider the convective diffusion equation (5.57), adding a term of the form (5.144) to the right-hand side:

$$\frac{\partial C_i}{\partial t} + \boldsymbol{u}\cdot\nabla C_i = D_m \Delta C_i + W_i, \tag{5.153}$$

where C_i is i-th component's concentration, $\boldsymbol{u}$ is the flow velocity, and D_m is a constant molecular diffusion coefficient of i-th component.

Suppose the turbulent velocity field $\boldsymbol{u}$ is known from the solution of the corresponding hydrodynamic problem under the assumption that i-th component is a passive impurity. Since the velocity field is random, concentration C_i will also be a random variable. Let us write C_i as a sum of the average value and the fluctuation:

$$C_i = \langle C_i \rangle + C_i' \tag{5.154}$$

and perform the averaging of Eq. (5.153) in the conventional way (see Section 4.6). As a result, we get an equation for the first moment – the average concentration of i-th component (see Eq. (4.45)):

$$\frac{\partial \langle C_i \rangle}{\partial t} + \langle \boldsymbol{u} \rangle \cdot \nabla \langle C_i \rangle = D_m \Delta \langle C_i \rangle - \nabla \cdot \langle u' C_i' \rangle + \langle W_i \rangle, \tag{5.155}$$

where $\langle \boldsymbol{u} \rangle$ and $\boldsymbol{u}'$ are the average and fluctuational components of velocity.

The average concentration $\langle C_i \rangle$ is of interest to us in the macromixing problem, when molecular diffusion can be ignored as compared to the turbulent diffusion (see Section 5.4).

Equation (5.155) is not closed, since it contains the unknown function $\langle \boldsymbol{u}' C_i' \rangle$. To close the problem, we can use the semiempirical gradient hypothesis (5.88), according to which $\langle \boldsymbol{u}' C_i' \rangle = -D_t \nabla \langle C_i \rangle$, where D_t is the coefficient of turbulent diffusion (see Section 5.4). Then Eq. (5.155) reduces to

$$\frac{\partial \langle C_i \rangle}{\partial t} + \langle \boldsymbol{u} \rangle \cdot \nabla \langle C_i \rangle = \nabla \cdot (D_t \Delta \langle C_i \rangle) + \langle W_i \rangle. \tag{5.156}$$

The problem of determination of $\langle W_i \rangle$ will be considered later.

Let us now look at the equation for the second-order moment $\langle (C_i')^2 \rangle$, which functions as the primary characteristic of micromixing. As was shown in Section 5.4, when studying micromixing, it is essential to include molecular diffusion into the picture, along with small-scale turbulent fluctuations. The equation for $\langle (C_i')^2 \rangle$ is derived by successive multiplication of the equations (5.156) and (5.155) by $2C_i'$, followed by term-by-term subtraction and subsequent averaging. The result is

$$\frac{\partial \langle (C_i')^2 \rangle}{\partial t} + \langle \boldsymbol{u} \rangle \cdot \nabla \langle (C_i')^2 \rangle = -\nabla \cdot \langle \boldsymbol{u}' (C_i')^2 \rangle - 2 \langle \boldsymbol{u}' C_i' \rangle \cdot \nabla \langle C_i \rangle \\ + 2D_m \langle C_i' \Delta C_i' \rangle + 2 \langle C_i W_i \rangle. \tag{5.157}$$

This equation, in its turn, is also unclosed because it contains unknown moments $\langle \boldsymbol{u}' C_i' \rangle$ and $\langle \boldsymbol{u}' (C_i')^2 \rangle$ associated with turbulent diffusion, the moment $\langle C_i' \Delta C_i' \rangle$ characterizing micromixing, and the moments $\langle W_i \rangle$ and $\langle C_i W_i \rangle$ containing reaction terms. For the moments $\langle \boldsymbol{u}' C_i' \rangle$ and $\langle \boldsymbol{u}' (C_i')^2 \rangle$, one can use the semiempirical gradient hypothesis

$$\langle \boldsymbol{u}' (C_i') \rangle = -D_t \nabla \langle C_i \rangle, \quad \langle \boldsymbol{u}' (C_i')^2 \rangle = -D_t' \langle (\nabla C_i')^2 \rangle. \tag{5.158}$$

Here we have introduced an additional diffusion coefficient D_t', which, generally speaking, should differ from D_t, but is often taken equal to D_t.

We should note that the second fluctuation moment $\langle (C_i')^2 \rangle$ is an important characteristic of mixing even in the absence of chemical reactions. As will be shown later, in a turbulent flow with chemical reactions taking place in the mixture, one can use the first and second moments to determine the average rates of chemical reactions $\langle W_i \rangle$ on the basis of the corresponding empirical hypotheses.

Following Corssin, we represent the term that describes micromixing in the form

$$2D_m \langle C_i' \Delta C_i' \rangle = -2D_m \langle (\Delta C_i')^2 \rangle. \tag{5.159}$$

The derivation of this relation can be demonstrated for the case of a one-dimensional process, assuming that C_i' depends on only one coordinate X. Since micromixing is considering in a microscopic region on small scale l, the random field C_i can be taken as homogeneous, that is, $\partial \langle C_i \rangle / \partial X = 0$. Then

$$\left\langle C_i' \frac{\partial^2 C_i'}{\partial X^2} \right\rangle = \frac{1}{l} \int_0^l C_i' \frac{\partial^2 C_i'}{\partial X^2} dX = \frac{1}{l} C_i' \frac{\partial C_i'}{\partial X} \bigg|_0^l - \frac{1}{l} \int_0^l C_i' \left(\frac{\partial C_i'}{\partial X} \right)^2 dX.$$

Neglecting the first term, we obtain

$$\left\langle C_i' \frac{\partial^2 C_i'}{\partial X^2} \right\rangle \approx -\left\langle \left(\frac{\partial C_i'}{\partial X} \right)^2 \right\rangle.$$

The expression (5.159) immediately follows from this relation. In Section 4.9, we introduced the parameter $\overline{N} = D_m(\nabla C')^2$ (see Eq. (4.127)) called the dissipation of concentration inhomogeneities. One can see now that this parameter is the main characteristic of micromixing.

Hence, the rate of micromixing is determined by local gradients of the random concentration field. Therefore in order to determine the rate of micromixing, we must know the correlation functions $B_{ab} = \langle(\nabla C_i')_a\rangle\langle(\nabla C_i')_b\rangle$ relating the values of concentration fluctuations $\langle C_a'\rangle$ and $\langle C_b'\rangle$ at two neighboring points a and b with coordinates $\boldsymbol{X}_a$ and $\boldsymbol{X}_b = \boldsymbol{X}_a + r$ (see Section 4.8).

In the micromixing problem, the scalar field can be assumed locally isotropic. For such a field, the intensity of concentration fluctuations $(C_i')^2$ obeys Eq. (4.93) that follows from the Corssin equation (4.92):

$$\frac{d\langle(C_i')^2\rangle}{dt} = -12\frac{D_m}{(\lambda_c^2)_i}\langle(C_i')^2\rangle, \tag{5.160}$$

where D_m is the coefficient of molecular diffusion, $(\lambda_c)_i$ – microscale of scalar concentration field C_i (see Eq. (4.91)). Eq. (5.160) could in fact be considered as the definition of microscale – the characteristic distance on which the intensity of concentration fluctuation decays. Further on, we shall omit the index i.

The microscale of a scalar field λ_c depends on the structure of random field inhomogeneities and for nonstationary process it, generally speaking, changes in the course of mixing.

It follows from Eq. (5.160) that micromixing leads to dissipation of the scalar field, that is, to a decrease of $\langle(C')^2\rangle$. It is worth mentioning that this dissipation is caused solely by molecular diffusion. This can be illustrated by a simple example. Consider the Corssin equation (4.92) in the absence of molecular diffusion ($D_m = 0$). In view of the gradient hypothesis (5.88) for the correlation $B_{La,b} = (u_L)_a C_a C_b$, the Corssin equation reduces to

$$\frac{\partial B_{ab}}{\partial t} = \frac{1}{r^2}\frac{\partial}{\partial r}\left(r^2 D_t \frac{\partial B_{ab}}{\partial r}\right). \tag{5.161}$$

Since $\langle(C')^2\rangle = \lim_{r\to 0} B_{ab}$, the change of $\langle(C')^2\rangle$ with time at $D_m = 0$ is obtained from Eq. (5.161) by going to the limit in both sides of the equation. Suppose the mixture was completely segregated at the initial moment so that impurity concentration C assumes the value C_0 in the region where the impurity is present and 0 in the region where it is absent. Since we are interested in the behavior of B_{ab} at $r \to 0$, it makes sense to consider the points that lie in the vicinity of the interface. At small values of r, the quantity B_{ab} can be represented to the first approximation as

$$B_{ab}(r, 0) \approx \langle(C')^2\rangle - \Sigma r, \tag{5.162}$$

where Σ is the specific area of the initial interface between regions with different values of C. Transition to the limit $r \to 0$ takes us to the scale interval that belongs to the dissipation region where, as we mentioned in Section 4.9, turbulent fluctuations are by their nature viscous. Thus the dependence of D_t on r is defined by Eq. (5.90), namely, $D_t \sim r^2$. It is easy to verify by substituting this dependence into Eq. (5.161) that for the initial distribution (5.162), the equality $\partial B_{ab}/\partial t = 0$ is valid, and $\langle (C')^2 \rangle$ remains constant under the condition that $D_m = 0$.

Another important characteristic of local structure of the scalar field is integral scale L_c (macroscale), equal by the order of magnitude to the average size of inhomogeneities and defined by the relation (see Eq. (4.91))

$$L_c = \int_0^\infty \Psi_{cc} dr, \tag{5.163}$$

where the integrand is the correlation coefficient

$$\Psi_{cc} = \frac{\langle C'_a C'_b \rangle}{(C'_i)^2}.$$

In contrast to the microscale λ_c, the macroscale L_c varies even in the absence of molecular diffusion as the inhomogeneities get deformed by turbulent fluctuations.

Thus the process of micromixing can be pictured as follows. The first stage is deformation of sufficiently large inhomogeneities by turbulent fluctuations. The integral scale of the scalar field L decreases, whereas $\langle (C')^2 \rangle$ remains practically unchanged because the specific surface area available for the transport of matter via molecular diffusion is small. With decrease of the integral scale, molecular diffusion begins to play a more important role. As a result, the quantity $\langle (C')^2 \rangle$ decreases, which indicates dissipation of the scalar field.

Micromixing is takes place in small regions where turbulence is characterized by local isotropy. It was shown in Section 4.9 that, depending on the predominance of one or another physical mechanism, the spectrum of fluctuations can be divided into several intervals. The same approach is justified for the scalar field (see Section 4.9). Depending on the character of deformation of inhomogeneities by turbulent fluctuations and on the role of molecular diffusion in the mixing process, all scales λ_c of concentration inhomogeneities are divided into three subranges: inertial-convective with $\lambda_c \gg \lambda_0$; viscous-convective with $\lambda_0 \gg \lambda_c \gg \lambda_b$, and viscous-diffusional with $\lambda_b \gg \lambda_c$, where $\lambda_0 = (\nu_e^3/\bar{\varepsilon})^{1/4}$ is the inner (Kolmogorov) scale of turbulence, λ_b – the Batchelor scale (see Eq. (4.133))

$$\lambda_b = \left(\frac{\nu_t D_m^2}{\bar{\varepsilon}} \right)^{1/4}; \tag{5.164}$$

$\bar{\varepsilon}$ – specific dissipation of turbulent energy; ν_e – kinematic viscosity coefficient of the carrier phase. The Batchelor scale is found from the condition that the characteristic time of diffusion is equal to that of velocity fluctuations, assuming that the equalization of concentration on this scale happens as a result of the molecular diffusion.

The relation between viscous and diffusional effects is characterized by the Schmidt number $Sc = \nu_e / D_m$. For liquids, the Schmidt number is on the order of $10–10^3$, and thus there exists a sufficiently large interval of scales $\lambda_c (\lambda_0 > \lambda_c > \lambda_b)$ where the flow can be regarded as viscous while the influence of diffusion is insignificant. In this region, convective transport of impurities happens mostly due to the deformation of fluid particles, and the quantity $\langle (C')^2 \rangle$ can be written as a sum of three components,

$$\langle (C')^2 \rangle = \sum_{i=1}^{3} \langle (C'_i)^2 \rangle \tag{5.165}$$

corresponding to different scale subranges: inertial-convective, viscous-convective, and viscous-diffusional.

Scalar field evolution is described by the Corrsin dynamic equation (4.92), which is unclosed because of the unknown third order moment $B_{La,b}$. One usually employs the spectral representation of Eq. (4.92) rather than Eq. (4.92) itself. This results in a transition from scales of concentration inhomogeneities λ_c to the corresponding wave numbers $k_c \sim \lambda_c^{-1}$. Then micromixing time gets reinterpreted as the time of motion of perturbations along the wave number axis. Spectral representation has proved to be more convenient for the purpose of formulating closure hypotheses for the dynamic equations.

In the study of random hydrodynamic or scalar fields, there arises the problem of finding the spectral density of a random field given some assumptions about the term that is responsible for the transport of perturbations along the wave number axis. If turbulent mixing is accompanied by chemical reactions, we have to address the additional problem of estimating the mixing time to compare it with the characteristic time of the chemical reaction, so that we could judge whether the reaction is slow, fast, or very fast. While doing so, we should use the expected form of quasi-stationary spectral density of the scalar field for different wave number ranges. The relevant forms of the spectral density are as follows:

– for the inertial-convective region with $k_c \ll k_0$,

$$E_c(k_c) = A_1 \bar{\varepsilon}_c (\bar{\varepsilon})^{-1/3} k_c^{-5/3}; \tag{5.166}$$

– for the viscous-convective region with $k_0 < k_c < k_b$,

$$E_c(k_c) = A_2 \left(\frac{\nu_e}{\bar{\varepsilon}} \right)^{1/2} \frac{\bar{\varepsilon}_c}{k_c} \exp\left\{ -A_2 \frac{k_c^2}{k_b} \right\}, \tag{5.167}$$

– for the region with $k_c \ll k_b$,

$$E_c(k_c) \approx A_2 \left(\frac{\nu_e}{\bar{\varepsilon}} \right)^{1/2} \frac{\bar{\varepsilon}_c}{k_c}. \tag{5.168}$$

In the viscous-diffusional region ($k_c \gg k_b$), molecular diffusion plays an important role, and the spectral function $E_c(k_c)$ decreases more and more rapidly as k_c grows.

In the relations above, A_1 and A_2 are constants approximately equal to 1, $\bar{\varepsilon}$ is the specific dissipation of turbulent energy, $\bar{\varepsilon}_c$ – specific dissipation rate of the scalar field, $k_c \ll k_b$ and k_b – wave numbers corresponding to the Kolmogorov and Batchelor scales.

Consider the velocity of disturbance propagation V_k along the wave number axis. Suppose it does not depend on the form of spectral distribution but instead is a function of a point on the wave number axis. This hypothesis is similar to the hypothesis about the spectral transport of turbulent energy [33]. Then the length of time t_{12} during which perturbations pass through the wave number range (k_1, k_2) equals

$$t_{12} = \int_{k_1}^{k_2} \frac{dk}{V_k}. \tag{5.169}$$

For a quasi-stationary field, the velocity V_k can be expressed through the specific dissipation rate of the scalar field $\bar{\varepsilon}_c$ (it is connected with the dissipation of scalar inhomogeneities $\bar{N}$ introduced in Section 4.9 by the relation $\bar{\varepsilon}_c = 2\bar{N}$) and the spectral distribution $E_c(k)$:

$$V_k = \frac{\bar{\varepsilon}_c(k)}{E_c(k)}, \tag{5.170}$$

where $\bar{\varepsilon}_c(k)$ and $E_c(k)$ are related to one another by

$$\bar{\varepsilon}_c(k) = 2D_m \int_k^{\infty} k^2 E_c(k) dk \tag{5.171}$$

and spectral distribution $E_c(k)$ obeys the normalization condition

$$\int_0^{\infty} E_c(k) dk = \langle (C')^2 \rangle. \tag{5.172}$$

Substituting Eq. (5.170) into Eq. (5.169) and using the relations (5.166) and (5.167), we obtain expressions for the micromixing time τ_c for different wave number ranges. Thus, for the inertial-convective region we have

$$(\tau_c)_1 = \frac{3}{2} A_1 \bar{\varepsilon}_c (\bar{\varepsilon})^{-1/3} k_m^{-2/3}, \tag{5.173}$$

where k_m is the wave number corresponding to the maximum size of inhomogeneities (the right end of the inertial-convective region where micromixing begins). In the viscous-convective region, the micromixing time equals

$$(\tau_c)_2 = \frac{A_2}{2}\left(\frac{\nu_e}{\bar{\varepsilon}}\right)^{1/2} \ln\left(\frac{\nu_e}{D_m}\right). \tag{5.174}$$

The assumption that local velocity of perturbation propagation along the wave number axis must be independent from the general form of the spectrum is mostly applicable in the viscous flow region. In this region the equations of motion become linear, and interactions between fluctuations having different scales can be neglected.

The above-mentioned relations make it possible, at least in principle, to estimate micromixing time, but in doing so we are facing the challenge of obtaining spectral distributions for concrete types of turbulent inhomogeneities. This prompts us to use various empirical equations for the dissipation rate of the scalar field $\bar{\varepsilon}_c$; it is related to the micromixing time $_c$ by

$$\bar{\varepsilon}_c = \frac{\langle (C')^2 \rangle}{\tau_c}. \tag{5.175}$$

As an example, we shall adduce two frequently-used relations,

$$\frac{d\bar{\varepsilon}_c}{dt} = -\left(\frac{B_1}{\tau_t} + \frac{B_2}{\tau_c}\right)\bar{\varepsilon}_c, \tag{5.176}$$

$$\frac{d\bar{\varepsilon}_c}{dt} = -C_1\frac{\bar{\varepsilon}_c}{\tau_c} + C_2\left(\frac{\bar{\varepsilon}_e}{\nu}\right)^{1/2}\bar{\varepsilon}_c, \tag{5.177}$$

where τ_t is the relaxation time of the hydrodynamic field, and B_1, B_2, C_1, C_2, are empirical constants.

The form of equations depends on the procedure used to mix the reagents. For example, the mixing can take place in a turbulent flow behind a grid; this process is characterized by simultaneously dissipation of hydrodynamic and scalar fields. Another possible procedure is the mixing of two streams moving in a pipe, where the hydrodynamic field is assumed to be stationary, and the main focus is on the time evolution of the scalar field.

Summarizing our efforts to describe turbulent transport and mixing on the basis of the method of moments, we can make the following conclusions. The equations (5.156) and (5.157) for first two moments of the scalar field are unclosed due to the presence of mixed moments in these equations. In most problems that involve turbulent transport and macromixing, one adopts the gradient hypotheses that is expressed by the relation (5.158). The disadvantages of this hypothesis are most evident at the initial stages of fast chemical reactions, when micromixing can produce a noticeable change of mixture composition in the course of fluid particle's motion. To describe micromixing, one should go beyond the scope of the gradient hypothesis, which necessitates the use of one additional equation for the mixed moments $\langle \boldsymbol{u}'C' \rangle$. One way to obtain such an equation is to multiply the convective diffusion equation for C' by $\boldsymbol{u}'$ and the equation of motion for $\boldsymbol{u}'$ by C', add up the two equations and average the result. But the equation derived in this way turns out to be unclosed as well. The situation here is the same as for hydrodynamic equations. Therefore in order to close the obtained system of equations, one resorts to various semi-empirical approximations for the third-order mixed moments $\langle u_i' u_j' C' \rangle$ and for the moment $\langle C' \Delta p' / \rho_e \rangle$ that contains pressure.

5.8.3
Approximations for Chemical Reaction Rates

In the beginning of this section we said that inclusion of chemical reactions into the picture results in the appearance of an additional source-type term (reaction term) in the diffusion equation. This term defines the intensity of production or consumption of matter in course of the chemical reaction. In the general case, this term depends nonlinearly on reagent concentrations, and so in addition to the above-considered problem of modeling the terms describing turbulent transport and mixing (in the two-moment approximation), we are now facing the closure problem for the reaction terms, in particular, for the moments $\langle W_i \rangle$ and $\langle C_i W_i \rangle$. In some cases it is convenient to switch from component concentrations to conserved scalar variables, so that our equations do not contain source-type terms that are due to the chemical reactions. Thus, for the reaction

$$A + nB \rightarrow (n+1)P,$$

where mass quantities (rather than molar quantities conventionally used in chemical kinetics) serve as concentrations (one should read this reaction as follows: 1 kg of matter A reacts with n kg of matter B, produce $(n+1)$ kg of matter P), we have the following conserved variables:

$$Z_1 = C_A - \frac{1}{n} C_B; \quad Z_1 = C_A + \frac{1}{n+1} C_P; \quad Z_3 = C_B + \frac{1}{n+1} C_P, \qquad (5.178)$$

where C_A, C_B, and C_P stand for the corresponding matter concentrations.

According to the rule (5.145), the stoichiometric coefficients of this reaction are $\nu_A = -1$, $\nu_B = -n$, $\nu_P = -n+1$ and the equality (5.150) provides the following connection between the reaction rates of different components:

$$\frac{dC_A}{dt} = \frac{1}{n} \frac{dC_B}{dt} = -\frac{1}{n+1} \frac{dC_P}{dt}. \qquad (5.179)$$

It is easy to convince ourselves in the validity of the relations

$$\frac{dZ_i}{dt} = 0, \quad (i = 1, 2, 3). \tag{5.180}$$

Thus, as we switch from the variables C_A, C_B, C_P to the variables Z_1, Z_2, Z_3, the source-type terms in the corresponding equations for Z_1, Z_2, Z_3 will be absent in view of the expression (5.180). The variables (5.178) are referred to as Schwab–Zeldovich variables. They are widely used in the combustion theory.

Suppose that diffusion coefficients of different components are identical. Then only one of these variables will be independent. If the reaction is also very fast, then substances A and B cannot be found simultaneously at one and the same point and their concentrations can be expressed through conserved variables. So, if Z_1 is chosen as our independent variable, then

$$Z_1 = \begin{cases} C_A \text{ at } Z_1 > 0, \\ -\dfrac{1}{n} C_B \text{ at } Z_1 < 0. \end{cases} \tag{5.181}$$

However, the use of conserved variables has one negative aspect – the difficulty of reverse transition from variables Z_i to the concentrations. This difficulty arises because of the nonlinear or, to be more precise, piecewise linear, dependence (5.181). Therefore our knowledge of the moments $\langle Z_i \rangle$ and $\langle (Z_i')^2 \rangle$ gives no way to determine the corresponding moments of concentrations C_A and C_B. Further on we are going to show that in order to obtain the moments of concentrations, it is necessary to know the probability density distribution of the variables Z_i. For a single-step reaction of the second order

$$\mathrm{A} + \mathrm{B} \rightarrow \text{products}$$

the averaged values of reaction terms in the equations for the average concentrations $\langle C_A \rangle$ and $\langle C_B \rangle$ are equal to

$$\langle W_A \rangle = \langle W_B \rangle = -k(\langle C_A \rangle \langle C_B \rangle + \langle C_A' C_B' \rangle), \tag{5.182}$$

where k is the reaction rate constant.

For isothermal reactions, the average reaction rate is calculated at constant temperature. If the reaction is accompanied by heat release, then local temperature is also a random scalar variable and the average reaction rate should be calculated with temperature fluctuations properly taken into account. One good example is combustion. Chemical reactions in this process are extremely exothermal, have very high activation energy and are therefore highly sensitive to local temperature changes [34]. If the reaction is taking place in a gas, it is necessary to take into account density fluctuations within the gas, which means we have to employ the equations of gas dynamics. Another useful procedure that helps account for density variations in a turbulent gas flow is to average all parameters except pressure over the mass. Such averaging is called Favre averaging [35].

We see from Eq. (5.182) that if the kinetic order of reaction is higher than one, the average reaction rate cannot be expressed through the average concentrations. The second term inside the brackets on the right-hand side of Eq. (5.182) (this term characterizes the degree of local correlation between reagent distributions) is zero only in the case of a totally homogeneous mixture. Depending on the character of the correlation between concentrations of reagents A and B, the average reaction rate could be higher or lower than the reaction rate calculated on the basis of average concentration values. Additional equations for $\langle C'_A C'_B \rangle$ are therefore necessary. To this end, we construct semi-empirical relations that express this moment through the moments $\langle C'_A \rangle$, $\langle C'_B \rangle$, $\langle (C'_A)^2 \rangle$, and $\langle (C'_B)^2 \rangle$. But the equations for second moments $\langle (C'_i)^2 \rangle$ will contain new unknown reaction terms, which in their turn will require new closure relations. The most common strategy is to calculate the second order fluctuation moments, taking hint from Toor's hypothesis [25] that it might be safe to neglect the third-order moment containing the reaction term.

To take into account the mutual correlation of reagents A and B, the average reaction rate in Eq. (5.182) is represented in the form

$$W_A = W_B = -k(1-U)\langle C_A \rangle \langle C_B \rangle, \tag{5.183}$$

in agreement with the immiscibility model. Here U is the degree of immiscibility for the whole reaction, which is expressed through the degrees of immiscibility for each reagent and through the correlation between average concentration gradients of these reagents. The degree of immiscibility U_i of reagent i is defined as the fraction of time during which this reagent is absent at the given point:

$$U_i = 1 - \frac{\langle (C_i)^2 \rangle}{\langle (C_i)^2 \rangle + \langle (C_i)^2 \rangle}. \tag{5.184}$$

The sign of the moment $\langle C'_A C'_B \rangle$ depends on the direction of average concentration gradients of the reagents $\nabla \langle C_A \rangle$ and $\nabla \langle C_B \rangle$. Suppose these gradients have opposite directions at the given point, as is the case, for instance, when we are mixing two previously segregated reagents A and B. Then we will still have partial segregation even as the reagents are being mixed together. Consequently, $\langle C'_A C'_B \rangle < 0$. If, on the other hand, the previously mixed reagents $A + B$ are now being mixed with some inert solvent, then the two local gradients $\nabla \langle C_A \rangle$ and $\nabla \langle C_B \rangle$ are parallel and $\langle C'_A C'_B \rangle > 0$.

The first case is of greater interest to us, especially for fast reactions. As a rule, we consider this case when choosing the closure model. For example, according to Patterson's model [27] for equimolar ratio of reagents A and B, the moment $\langle C'_A C'_B \rangle$ is represented as

$$\langle C' C' \rangle = \frac{\langle (C'_A)^2 \rangle \langle (C'_B)^2 \rangle}{\langle C_A \rangle \langle C_B \rangle}. \tag{5.185}$$

For more complicated two-step reactions of the form

$$\mathrm{A + B \rightarrow R, \quad R + B \rightarrow P} \tag{5.186}$$

we also have consider the selectivity of the reaction. Selectivity is defined as the parameter $R/(R+P)$ characterizing the relative product yield (note that selectivity depends on the intensity of mixing).

The reaction terms for the two stages of reaction (5.186) are

$$\langle W_1 \rangle = -k_1(\langle C_A \rangle \langle C_B \rangle + \langle C'_A C'_B \rangle), \quad \langle W_2 \rangle = -k_2(\langle C_R \rangle \langle C_B \rangle + \langle C'_R C'_B \rangle). \tag{5.187}$$

Note that one cannot use the same closure model for the moment $\langle C'_R C'_B \rangle$ as for the moment $\langle C'_A C'_B \rangle$ because local correlations between R and B differ from the corresponding correlations between A and B. Many possible closures have been suggested for the moment $\langle C'_R C'_B \rangle$. One of them is

$$\langle C'_R C'_B \rangle = -K \langle C_R C_B \rangle$$

with the empirical coefficient K varying from –1 to 1.

The eventual choice of the closure model for reaction terms can be made by running a comparison with experimental data or with results obtained from the PDF-method; remember that the PDF method makes it possible to calculate the rate of an arbitrary nonlinear reaction without any additional assumptions once the probability density distribution of concentrations is known.

5.9 The PDF Method

The PDF method is an abbreviation for the method that employs the equation for the probability density function (distribution) to describe the behavior of random hydrodynamic and scalar fields. The central idea is to represent the fluid as a large set of fluid particles, each particle characterized by a set of time-varying parameters: particle's position in space, velocity, temperature, and composition (component concentrations).

The currently existing methods of describing turbulence can be divided into two groups:

- the method of moments, based on the use of transport equations and equations for the moments (which, in their turn, are obtained from transport equations) complemented by semi-empirical closure relations;
- the PDF method, based on the equation for the probability density function (distribution) of random hydrodynamic and scalar fields.

The main weakness of the first method is the need to use additional equations (closures) for higher-order moments, since any system of equations for the first N moments turns out to be unclosed due to the presence of moments of even higher order. Additional equations typically contain a large number of empirical constants and functions. To obtain receive these constants and functions, one has to carry out a multiple experiments, and even then the values obtained in this way are not universal, because they may be different for different flows. The shortcomings of semiempirical models manifest themselves particularly strongly when one needs to describe turbulent flows of multi-phase, multi-component media accompanied by chemical reactions.

The positive feature of semi-empirical models is that they reduce to a system of partial differential equations for velocities, concentrations, and temperature as functions of coordinates and time, whose solution for many types of flows such as flows inside channels, pipes, boundary layers, jet flows, flows bypassing a body, and so forth, can be obtained by well-developed numerical methods.

The idea behind the PDF method is to introduce one single function – joint PDF p of several random functions (velocity vector $\boldsymbol{u}$, concentrations C_i, enthalpy H or temperature ϑ, etc.) and solve the Fokker–Planck equation for p. If the function p is known, we can find all statistical characteristics of the random functions (moments), which saves us the effort of constructing closure relations. Therefore the PDF method is more logical and informative then the method of moments. Its main disadvantage is the large number of dimensions in the problem (time t, coordinates $\boldsymbol{X}$, functions $\boldsymbol{u}$, C, ϑ, ...), which makes it difficult to obtain even a numerical solution, to say nothing of the analytical one. As of today, there are several different methods based on the models that use simplified equations for the PDF together with semi-empirical models of turbulence such as the $k-\varepsilon$ model. A comprehensive treatment of the PDF method with appropriate references can be found in [30–32].

The state of an incompressible fluid in an isothermal process is completely described by three components of the velocity vector $\boldsymbol{u}$ and by N scalar concentrations of passive impurities $C_1, C_2, \ldots, C_N$. Let us introduce a joint single-point PDF $p(\boldsymbol{u}, C_1, C_2, \ldots, C_N; \boldsymbol{X}, t)$ at the point $\boldsymbol{X}$ at the instant of time t. By definition (see Section 1.2), the quantity $p(\boldsymbol{u}, C_1, C_2, \ldots, C_N; \boldsymbol{X}, t)\ d\boldsymbol{u}\ dC_1dC_2,\ldots,dC_N$ is the probability that at the moment t, at the point $\boldsymbol{X}$, we are going to find the velocity $\boldsymbol{u}$ in the interval $(\boldsymbol{u}, \boldsymbol{u}+d\boldsymbol{u})$ and each of the concentration values C_i in its respective interval (C_i, C_i+dC_i). This function is non-negative and obeys the normalization condition

$$\int p(\boldsymbol{u}, C_1, C_2, \ldots, C_N; \boldsymbol{X}, t)\ d\boldsymbol{u}\ dC_1 dC_2 \ldots dC_N = 1,$$

where the integral is taken over the entire phase space of velocities and concentrations.

Having written the distribution $p(\boldsymbol{u}, C_1, C_2, \ldots, C_N; \boldsymbol{X}, t)$, we can always derive distributions with fewer dimensions (marginal distributions) by integrating p over one variable or several variables, for example,

$$p(C_1, C_2, \ldots, C_N; \boldsymbol{X}, t) = \int p(\boldsymbol{u}, C_1, C_2, \ldots, C_N; \boldsymbol{X}, t) d\boldsymbol{u}. \tag{5.188}$$

More importantly, we can find the average value of a random function F,

$$\langle F\rangle = \int F\, p(\boldsymbol{u}, C_1, C_2, \ldots, C_N; \boldsymbol{X}, t) d\boldsymbol{u}\, dC_1 dC_2 \ldots dC_N,$$

and all moments of this function.

Before we proceed to write the equation for p, let us write the system of conservation equations describing the isothermal flow of an incompressible multi-component fluid. It consists of the continuity equation

$$\frac{\partial u_i}{\partial X_i} = 0, \tag{5.189}$$

the momentum equation

$$\rho\frac{Du_j}{Dt} = -\frac{\partial p}{\partial X_j} + \frac{\partial \tau_{ij}}{\partial X_i} + \rho f_j \equiv A_j, \tag{5.190}$$

and the mass conservation equation for k-th substance

$$\rho\frac{Dm_k}{Dt} = -\frac{\partial J_i^k}{\partial X_i} + \rho S_k \equiv B_k, \tag{5.191}$$

where $D/Dt = \partial/\partial t + u_i \partial/\partial X_i$ is the substantial derivative; f_j – j-th component of the body force (for example, gravity); τ_{ij} – components of the viscous stress tensor; J_i^k – i-th component of the diffusion flux of k-th substance; S_k – specific rate of production or consumption of k-th substance in the course of chemical reaction.

One of the ways to derive the equation for $p(\boldsymbol{u}, C_1, C_2, \ldots, C_N; \boldsymbol{X}, t)$ is to represent p as a product of delta functions (see Eq. (1.152)),

$$p(\boldsymbol{u}, C_1, C_2, \ldots, C_N; \boldsymbol{X}, t) = \delta(\boldsymbol{u} - \boldsymbol{u}(\boldsymbol{X}, t)) \prod_{i=1}^{N} \delta(C_i - C_i(\boldsymbol{X}, t)).$$

Using the Eulerian approach, we write the transport equation for the PDF p in the form

$$\frac{\partial p}{\partial t} = -\frac{\partial}{\partial X_j}(u_j p) - \frac{\partial}{\partial u_j}(\langle A_j(\boldsymbol{u}, \mathbf{C})\rangle p) = -\frac{\partial}{\partial C_k}(\langle B_k(\boldsymbol{u}, \mathbf{C})\rangle p), \tag{5.192}$$

where we have introduced the vector $\mathbf{C} = (C_1, C_2, \ldots, C_N)$.

We see from Eq. (5.192) that the variation of p is caused by convective transport in the physical space $\boldsymbol{X}$ with the velocity $\boldsymbol{u}$ and in the $\boldsymbol{u} - \mathbf{C}$ phase space with the velocities $\langle A_j\rangle$ along the u_j-axis and $\langle B_k\rangle$ along the C_k-axis. An analytic solution of Eq. (5.192) can be obtained only for the simplest cases of no practical interest. Most situations that present any interest for applications allow for a numerical solution only. But the employment of numerical methods such as the finite difference method presents difficulties because of the excessively large number of dimensions in the problem. Even in the case of a single scalar quantity we have 8 dimensions (three velocity components, concentration, three coordinates, and time). To apply this numerical method, it is necessary to divide the spatial volume into N finite elements $\{\boldsymbol{X}_k, \boldsymbol{u}_k, C_k\}$ and to present p in the discrete form

$$p_N = \frac{1}{N}\sum_{k=1}^{N}\delta(\boldsymbol{u}-\boldsymbol{u}_k)\delta(C-C_k)\delta(\boldsymbol{X}-\boldsymbol{X}_k).$$

The transition from the continuous to the discrete distribution is achieved with a statistical error proportional to $N^{-1/2}$, and the average value of any arbitrary quantity is calculated with the root-mean-square deviation $\sim N^{-1/2}$ independently from the number of dimensions. So as N gets larger, the statistical error changes very slowly. For instance, to achieve a 10% error, one needs only 100 elements, but to decrease the error to 1%, one needs as much as 10000 elements. The discrete distribution for given values of t and N is defined by specifying N vectors and a set of $6N$ numbers in a 6-dimensional space. While in a 3-dimensional space one needs 1000 elements, in a 6-dimensional space 6000 elements are needed in order to achieve the same calculation accuracy. That is why the Monte Carlo method is the method of choice for a multidimensional case.

As opposed to the Eulerian approach with its reliance upon Eq. (5.192), the Lagrangian approach uses the system of stochastic Langevin equations for fluid particles. As a rule, these are simply transport equations in the relaxation approximation (see Eq. (5.101)). In the case of a single scalar variable they have the form

$$\frac{D\boldsymbol{X}(\xi,t)}{Dt} = \boldsymbol{u}(\boldsymbol{X}(\xi,t),t),$$
$$\frac{D\boldsymbol{u}(\boldsymbol{X}(\xi,t),t)}{Dt} = -\alpha_L(\boldsymbol{u}(\boldsymbol{X}(\xi,t),t)-\langle \boldsymbol{u}(\boldsymbol{X}(\xi,t),t)\rangle)+\boldsymbol{f}, \tag{5.193}$$
$$\frac{DC(\boldsymbol{X}(\xi,t),t)}{Dt} = -\beta_L(C(\boldsymbol{X}(\xi,t))-\langle C(\boldsymbol{X}(\xi,t))\rangle) + W(C),$$

where it is assumed that initially $\boldsymbol{X}=\xi$, $\boldsymbol{u}=\boldsymbol{v}$, $C=S$. Here $\mathbf{X}(\xi, t)$ are the fluid particle's coordinates at the moment t (ξ being the initial coordinate); $\boldsymbol{u}(\mathbf{X}(\xi, t), t)$ is that particle's Lagrangian velocity; α_L and β_L are dissipation rate constants (i.e., inverse relaxation times) of the hydrodynamic and scalar fields respectively; $\boldsymbol{f}$ is the random force exerted on the fluid particle by the flow.

Using Eq. (5.193) and switching to Eulerian variables, we get the following equation for the PDF $p(\boldsymbol{u}, C; \boldsymbol{X}, t)$:

$$\frac{\partial p}{\partial t} = -u_i\frac{\partial p}{\partial X_i}+\frac{\partial}{\partial u_i}(\alpha_L(u_i-\langle u_i\rangle p)+\langle f_i^2\rangle\frac{\partial^2 p}{\partial u_i^2}+\frac{\partial}{\partial C}(\beta_L(C-\langle C\rangle p)-\frac{\partial}{\partial C}(W(C)p). \tag{5.194}$$

As you may guess from the formula (5.188), the equation for the PDF of concentration only, that is, for $p(C; \boldsymbol{X}, t)$ is derived by integrating both sides of Eq. (5.194) with respect to velocity from $-\infty$ to $+\infty$. Exploiting the condition $\partial p/\partial u_i \to 0$ at $|u_i| \to \infty$ and introducing the coefficient of turbulent diffusion through the gradient hypothesis

$$\int (u_i - \langle u_i \rangle)\, p(\boldsymbol{u}, C; \boldsymbol{X}, t) d\boldsymbol{u} = -D_t \frac{\partial p(C; \boldsymbol{X}, t)}{\partial X_i}, \tag{5.195}$$

we reduce Eq. (5.194) to

$$\frac{\partial p}{\partial t} = -\langle u_i \rangle \frac{\partial p}{\partial X_i} + \frac{\partial}{\partial X_i}\left(D_t \frac{\partial p}{\partial X_i} \right) + \frac{\partial}{\partial C}(\beta_L (C - \langle C \rangle\, p) - \frac{\partial}{\partial C}(W(C)\, p). \tag{5.196}$$

The third term on the right-hand side of Eq. (5.196)

$$\beta K = \frac{\partial}{\partial C}(\beta_L (C - \langle C \rangle\, p) \tag{5.197}$$

describes the process of micromixing [24].

It turns out that the above-described Langevin model of micromixing does not always give satisfactory results. The need to introduce other models of micromixing, including nonlinear ones, is caused by the fact that, according to the Langevin model, all concentration values change simultaneously. Thus if the PDF $p(C; \boldsymbol{X}, t)$ has the form

$$p(C; \boldsymbol{X}, \boldsymbol{t}) = \gamma \delta (C - C_1) + (1 - \gamma) \delta (C - C_2)$$

the Langevin model predicts that both peaks will eventually approach the average value $\langle C \rangle = \gamma C_1 + (1 - \gamma) C_2$, while in a real-life situation, partial mixing of two fluid elements (moles) having respective concentrations C_1 and C_2 will produce mixing layers containing all of the intermediate concentration values.

The arbitrariness in the choice of phenomenological model for micromixing is explained by the fact that we are considering a one-point PDF, whereas micromixing is characterized by local gradients of the scalar field and thus requires the knowledge of spatial correlations that can be determined only from a two-point PDF.

Another method is based on the idea of including local gradients into the PDF as independent variables. But such a procedure complicates the problem and, furthermore, it requires the corresponding new models that become necessary if we are to write an equation for the PDF that would properly account for the new variables.

As the model of micromixing for the operator **K**, one typically chooses some model of the coalescence-dispersion type. Such models describe the process of micromixing as the outcome of random contact (coalescence) of two microvolumes with concentrations C_1 and C_2, hence the name used for these models. The probability of such interaction is taken to be proportional to the product $p(C_1;\ \boldsymbol{X},\ t)\,p(C_2;\ \boldsymbol{X},\ t)$ in accordance with the so-called average field approximation. The contact and mixing of the microvolumes results in the formation of intermediate concentrations c ($C_1 < C < C_2$), whose distribution, in its turn, should be modeled beforehand.

In a coalescence-dispersion model, the operator **K** reduces to the following expression [36]:

$$K = 2\int_0^c dC_1 \int_c^1 dC_2\, p(C_1;\boldsymbol{X},t)\, p(C_2;\boldsymbol{X},t) G(C, C_1, C_2). \tag{5.198}$$

The kernel of this equation $G(C,\ C_1,\ C_2)$ is the distribution density of intermediate concentrations that are formed during the elementary mixing event. The choice of G is what distinguishes various coalescence-dispersion models from each other. For instance, Curl's model proposed in [24] assumes that only one intermediate concentration $C = (C_1 + C_2)/2$ is formed during the contact of the microvolumes and that

$$G(C, C_1, C_2) = \delta\left(C - \frac{C_1 + C_2}{2}\right). \tag{5.199}$$

In the Nedorub model considered in [37], the intermediate concentration distribution is taken to be uniform and the kernel is expressed as

$$G(C, C_1, C_2) = |C_2 - C_1|^{-1}. \tag{5.200}$$

We should note that the value of the constant β in Eq. (5.197) depends on the choice of the micromixing model and can be determined from the Corrsin equation (5.160) for the intensity of concentration fluctuations,

$$\frac{d\langle (C')^2\rangle}{dt} = -\gamma\langle (C')^2\rangle \tag{5.201}$$

with a given value of γ. In the Langevin model, $\beta = \gamma/2$; in Curl's model (5.199), $\beta = 2\gamma$; and in the Nedorub model (5.200) with a uniform distribution of intermediate concentrations, $\beta = 3\gamma$. So, models of the type (5.198) predict an exponential dependence of $\langle (C')^2\rangle$ on time.

In the general case, the dependence of contact frequency of fluid microvolumes with concentrations C_1 and C_2 on these concentrations is more complex. For example, in Frost's model [29] the contact frequency is taken in the form

$$\beta|C_2 - C_1|\, p(C_1;\boldsymbol{X},t)\, p(C_2;\boldsymbol{X},t). \tag{5.202}$$

For this model, the change of $\langle (C')^2\rangle$ with time cannot be described by Eq. (5.201) with a constant value of γ.

In order to have criteria that a micromixing model should satisfy, we usually demand that the PDF of concentration at the early stage of mixing should be represented by a sum of two delta functions. We may also demand that for long mixing times the distribution should asymptotically approaches a Gaussian distribution. The proximity of an asymptotic distribution to a Gaussian one is usually estimated by the value of the excess β_2,

$$\beta_2 = \frac{\langle (C')^4 \rangle}{\langle (\langle (C')^2 \rangle)^2 \rangle}. \tag{5.203}$$

For coalescence-dispersion models of the type (5.198), the excess is equal to

$$\beta_2 = 1 + 4(\exp(at) - 1), \tag{5.204}$$

where the parameter a is expressed through the moments of the distribution $G(C, C_1, C_2)$. Consequently, for any coalescence-dispersion model, the excess tends to infinity when $t \to \infty$, whereas for the Gaussian distribution, $\beta_2 = 3$. Notice that in the model (5.202), the asymptotic value of the excess is also close to 3, even though this is hard to explain from a physical point of view, because the contact frequency in this model increases with $|C_2 - C_1|$.

The form of the PDF at the initial stage of mixing can be estimated by considering the formation of a mixing layer caused by molecular diffusion between two fluid volumes with concentrations C_1 and C_2 as shown in Fig. 5.6.

The diffusion equation gives a distribution with two maximum values that are close to C_1 and C_2 (Fig. 5.6 a). If we assume (as an approximation) that in a mixing layer of thickness δ, concentration is distributed linearly, then the distribution of intermediate concentrations will also be linear, which corresponds to a uniform PDF (Fig. 5.6 b).

Let us say a few words about the influence of chemical reactions on the process of mixing. The fact that such an influence exists becomes evident if we look at the

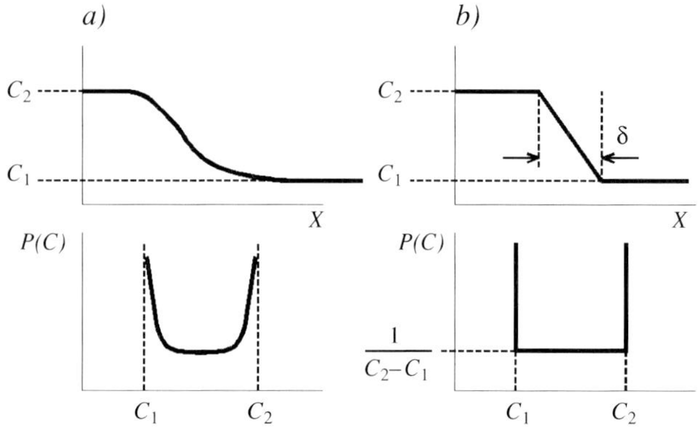

Fig. 5.6 Mixing layer between two fluid volumes.

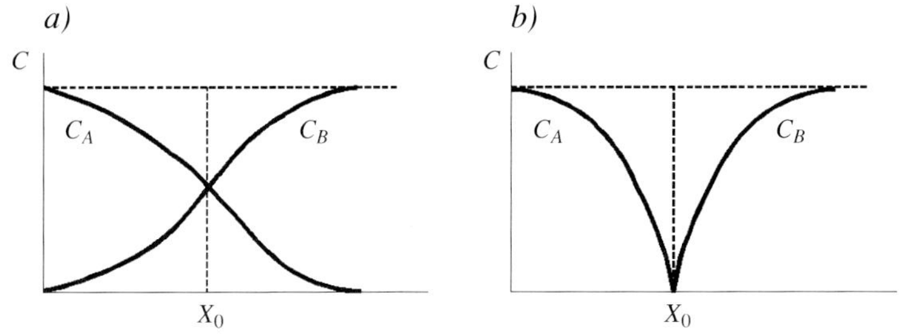

Fig. 5.7 Concentration profiles: (a) without chemical reaction; (b) with chemical reaction.

mutual diffusion of two substances A and B in a quiescent fluid. Concentration profiles for two limiting cases of mixing – without chemical reactions and with infinitely fast, irreversible chemical reactions of the type $\boldsymbol{A} + \boldsymbol{B} \rightarrow$ product – are shown schematically in Fig. 5.7.

The effect of chemical reactions on the rate of micromixing is difficult to estimate, because turbulent fluctuations cause significant deformation of the surface where the mass transport is taking place. Of particular interest in this context is the structure of surfaces at which mass transport is observed in the course of turbulent mixing. In the case under consideration, these surfaces are isoconcentration surfaces. Theoretical studies suggest that isoconcentration surfaces formed in the course of mutual diffusion of substances A and B in a motionless fluid layer have a fractal structure, and that structure self-similarity exists in some scale interval. Fractal structure is characterized by the parameter d known as fractal dimensionality. Determination of d boils down to estimating the surface area by calculating number $N(r)$ of cubes of linear size r that are needed to cover the whole surface. Then surface area is $N(r)r^2$. As the cube size decreases, the area of surfaces one usually encounters in practical situations quickly reaches its limiting value no longer depends on r. For a fractal surface of dimensionality d, the number of cubes needed to cover the surface and the surface area increase as $N{\sim}r^{-d}$ and $S{\sim}r^{2-d}$. For each system, there are constraints from above as well as from below, which must be satisfied by such power laws; note that the exponent may have different values in different intervals. Determination of surface structure and surface area is an important practical problem because the rate of mass exchange through any surface is given by the product of the molecular diffusion coefficient, the local concentration gradient, and the surface area (whose deformation due to the turbulent fluctuations should, of course, be taken into account). In particular, for scales that exceed the inner scale of turbulence λ_0 and thus belong to the inertial-convective region, fractal dimensionality of isoconcentration surfaces is equal to 7/3, which is confirmed by experiments that involve visualization of isoconcentration surfaces $C_A = C_B = 0$ during a very fast neutralization reaction between acid and alkali solutions. Estimates of fractal dimensionality of these surfaces in the viscous-convective scale interval give the value 3, in other words, such a surface would fill the entire volume under consideration. It is obvious that this scale subrange has to be bounded from below by the Batchelor scale λ_b, starting from which molecular diffusion leads to a completely uniform distribution of reagents.

The above-considered models of micromixing have been developed for the PDF distributions of one variable – concentration. But there are many problems where several variables (velocity, concentration, temperature, etc.) are involved. Until recently, such problems were solved by considering Curl's model [23] and using the Monte Carlo method to solve the corresponding stochastic equations. It should be noted that if we have than one variable, especially if these variables have different physical nature (e.g., concentration and velocity, concentration and temperature, and so on), we are facing two additional problems. Namely, we have to determine the regions where intermediate values of the variables arise during one elementary step of micromixing, and to find out how micromixing intensity for different variables depends on molecular transport coefficients, specifically, on diffusion, viscosity, and heat conductivity.

When deriving Eq. (5.196) for the PDF, we started from the Langevin equations (5.193). Derivation of this equation directly from the convective diffusion equation has been attempted in [34] for the limiting case of strongly developed turbulence (**Re** $\to \infty$). The expression $-\,\partial^2(\langle \bar{N}_c\rangle\, p)/\partial C^2$ was obtained for the term describing micromixing; here $\langle \bar{N}_c\rangle$ is the conditionally averaged rate of dissipation of the scalar field for a given value of concentration C, so that the total dissipation rate $\bar{\varepsilon}_c$ is equal to

$$\bar{\varepsilon}_c = \int dC\langle \bar{N}_c\rangle\, p(C;\boldsymbol{X},\boldsymbol{t}).$$

However, this results in an inverse parabolic equation for $p(C;\, \boldsymbol{X},\, t)$, whose solution presents difficulties.

When considering the Schwab–Zeldovich conserved variables Z_i in Section 5.8, we mentioned that use of these variables makes it possible to exclude source-type reaction terms from the convective diffusion equations. Consequently, a transition to the Schwab–Zeldovich variables leads to the exclusion these terms from the equation for the PDF as well. We must know the PDF of conserved variables if we want to switch back to ordinary concentrations. Sometimes the form of conserved variable distributions is given a priori and one only needs to find the parameters of these distributions. Let us look at the examples of conserved variables for some reactions.

For infinitely fast reactions of the type $A + B \to P$ the conserved variable is $Z = C_A - C_B$. Because reagents A and B cannot exist simultaneously at one and the same point, the reverse transition to concentrations C_A and C_B is accomplished by means of the relations $Z = C_A$ for $Z > 0$ and $Z = -C_B$ for $Z < 0$. The model of micromixing (which is accompanied by an infinitely fast reaction) at the (C_A, C_B) plane is replaced by a one-dimensional model for the variable Z.

For a two-step reaction of the type

$$A + B \xrightarrow{k_1} P_1, \quad A + C \xrightarrow{k_2} P_2,$$

in which the first step is considered as an infinitely fast reaction ($k_1 \to \infty$), $Z = C_A - C_B$ and C_C can be chosen as the independent variables. If both steps of the reaction are thought to be infinitely fast, that is, if $k_1 \to \infty$, $k_2 \to \infty$ and $k_1/k_2 = const$, then it is convenient to switch to new variables $Z = C_B + C_C - C_A$ and C_B and to describe the macrokinetics of this reaction with turbulent mixing models for these variables. The intermediate concentration boundary can be found from equations of chemical kinetics; one must keep in mind that each elementary step of micromixing is accompanied by an infinitely fast chemical reaction.

For a two-step sequential-parallel reaction of the type

$$A + B \to R, \quad B + R \to P,$$

when the first step is infinitely fast, the conserved variables are $Z = C_A - C_B$ and C_R. If both steps are infinitely fast, then the process is described with turbulent mixing models for the variables $Z = 2C_A - C_B + C_R$ and C_R.

To conclude, the use of conserved variables allows to reduce the number of independent variables and thereby decrease the dimensionality of the problem, especially in cases when some reaction stages can be considered as infinitely fast.

Finally, let us make a remark about the modeling of macromixing and turbulent transport in equations for the PDF. Turbulent transport in Eq. (5.196) is described by the second term on the right-hand side. This term is written in the gradient hypothesis approximation. When discussing the method of moments in Section 5.8, we already commented on the disadvantages of this approximation and on the need to go beyond the scope of the gradient hypothesis by using additional equations for the mixed moments, in particular, for $\langle \boldsymbol{u} C \rangle$. Use of the joint PDF gives an exact expression for the moment $\langle \boldsymbol{u} C \rangle$, namely,

$$\langle \boldsymbol{u} C \rangle = \int d\boldsymbol{u}\, \boldsymbol{u} \int dC\, C\, p(\boldsymbol{u}, C; \boldsymbol{X}, t). \quad (5.205)$$

Transition to the joint PDF $p(\boldsymbol{u}, C, \boldsymbol{X}, t)$ would certainly complicate the problem by increasing its dimensionality. For the sake of simplicity, one can use instead of $p(\boldsymbol{u}, C; \boldsymbol{X}, t)$ the distribution $p(C; \boldsymbol{X}, t)$ for concentration only, which should be considered in combination with the equation for conditionally averaged velocity $\langle \boldsymbol{u}_c \rangle$,

$$\langle \boldsymbol{u}_c \rangle\, p(C, t) = \int d\boldsymbol{u}\, \boldsymbol{u}\, p(\boldsymbol{u}, C, t). \quad (5.206)$$

References

1 Sedov, L.I. (1970) *Mechanics of Continuum*, Nauka, Moscow, **1**, **2** (in Russian).

2 Monin, A.S. and Yaglom, A.M. (1971) *Statistical Fluid Mechanics: Mechanics of Turbulence*, **1**, MIT Press, Cambridge,

MA; Monin, A.S. and Yaglom, A.M. (1975) *Statistical Fluid Mechanics: Mechanics of Turbulence*, **2**, MIT Press, Cambridge, MA.
3 Richardson, L.E. (1926) Atmospheric Diffusion shown on a Distance–Neighbour Graph. *Proc. Roy. Soc. A*, **110** (756), 709–737.
4 Taylor, G.I. (1921) Diffusion by Continuous Movements. *Proc. London Math. Soc.*, **20** (2), 196–211.
5 Zaichik, L.I. and Pershukov, V.A. (1996) Problems of Modeling Gas–Particle Turbulent Flows with Combustion and Phase Transitions. Review. *Fluid Dynamics*, **31** (5), 635–646.
6 Derevich, I.V. and Zaichik, L.I. (1988) Particle Deposition from a Turbulent Flow. *Fluid Dynamics*, **23** (5), 722–729.
7 Derevich, I.V. and Zaichik, L.I. (1990) An Equation for the Probability Density Velocity and Temperature of Particles in a Turbulent Flow Modeled by a Random Gaussian Field. *J. Appl. Mathematics and Mechanics*, **54** (5), 631–636.
8 Zaichik, L.I. (1992) A Kinetic Model of Particle Transport in Turbulent Flows with Allowance for Collisions. *Eng.-Phys. J.*, **63** (1), 44–50 (in Russian).
9 Zaichik, L.I. and Pershukov, V.A. (1995) Modeling of Particle Motion in a Turbulent Flow with Allowance for Collisions. *Fluid Dynamics*, **30** (1), 49–63.
10 Alipchenkoy, V.M. and Zaichik, L.I. (2000) Modeling of the Motion of Particles of Arbitrary Density in a Turbulent Flow on the Basis of a Kinetic Equation for the Probability Density Function. *Fluid Dynamics*, **35** (6), 883–900.
11 Alipchenkov, V.M. and Zaichik, L.I. (2001) Particle Collision Rate in Turbulent Flow. *Fluid Dynamics*, **36** (4), 608–618.
12 Buevich, Yu.A. (1970) Drag Reduction Model for Particle Injection into a Turbulent Viscous Fluid Stream. *Fluid Dyn.*, **5** (2), 271–276.
13 Tchen, C.M. (1947) *Mean Value and Correlation Problems Connected with the Motion of Small Particles Suspended in a Turbulent Fluid.* Martinus Nijhoff, Haague.
14 Fuks, N.A. (1955) *Mechanics of Aerosols*, Acad. Nauk SSSR, Moscow (in Russian).
15 Hinze, J.O. (1975) *Turbulence*, 2nd ed.,McGraw–Hill, New York.
16 Mednikov, E.P. (1981) *Turbulent Transport and Deposition of Aerosols*, Nauka, Moscow (in Russian).
17 Gusev, I.N. and Zaichik, L.I. (1992) Numerical Modeling of Two-Phase Turbulent Flows in a Furnace. *J. Appl. Math. Techn. Phys.*, **2**, 116–122 (in Russian).
18 Kirillov, P.L. (1986) About the Effect of Thermophysical Properties of a Surface on Heat Transfer by Turbulent Motion. *Eng.-Phys. J.*, **50** (3), 501–512 (in Russian).
19 Sehmel, G.A. (1973) Particle Eddy Diffusivities and Deposition Velocities for Isothermal Flow and Smooth Surfaces. *J. Aerosol. Sci.*, **4**, 125–138.
20 Friedlander, S.K. and Johnstone, H.F. (1957) Deposition of Suspended Particles from Turbulent Gas Stream. *Ind. and Eng. Chem.*, **49** (7), 1151–1156.
21 Sehmel, G.A. (1970) Particle Deposition from Turbulent Air Flow. *J. Geophys. Res.*, **75** (9), 1766–1781.
22 Alipchenkov, V.M. and Zaichik, L.I. (1998) Modeling the Dynamics of Colliding Particle in a Turbulent Shear Flow. *Fluid Dynamics*, **33** (4), 552–558.
23 Curl, R.L. (1963) Disperse Phase Mixing: I. Theory and Effects in Simple Reactors. *AIChE J.*, **9** (2), 175.
24 Chung, P.M. (1969) A Simplified Statistical Model of Turbulent Chemical Reacting Shear Flows. *AIAA J.*, **7**, (1982).
25 Hill, J.C. (1976) Homogeneous Turbulent Mixing with Chemical Reaction. *Ann. Rev. Fluid Mech.*, **8**, 135.

26 Kompaneez, V.Z., Ovsyannikov, A.A. and and Polak, L.S. (1979) *Chemical Reactions in Turbulent Flows of Gas and Plasma*, Nauka, Moscow (in Russian).

27 Patterson, G.K. (1981) Application of Turbulence Fundamentals to Reactor Modelling and Scaleup. *Chem. Engng. Commun.*, **8**, 24.

28 Libby P.A. and Williams F.A. (eds.) (1980) *Turbulent Reacting Flows*, Springer, Berlin/New York.

29 Kaminski, V.A., Fedorov, A.J. and Frost, V.A. (1994) Calculation Methods for Turbulent Flows with Fast Chemical Reactions. *Theor. Fund. Chem. Techn.*, **28** (6), 591–599 (in Russian).

30 Pope, S.B. (1985) PDF Methods for Turbulent Flows. *Prog. Energy Combust. Sci.*, **11**, 119–192.

31 Pope, S.B. (1994) Lagrangian PDF Methods for Turbulent Flows. *Ann. Rev. Fluid Vtch.*, **26**, 23–63.

32 Pope, S.B. (2000) *Turbulent Flows*, Cambridge Univ. Press, Cambridge.

33 Pao, Y-H. (1965) Structure of Turbulent Velocity and Scalar Fields at Large Wave Numbers. *Phys. Fluids*, **86**, 1063–1075.

34 Kuznetsov, V.R. and Sabel'nikov, V.A. (1990) *Turbulence and Combustion*, Hemisphere, New York.

35 Favre, A. (1969) *Statistical Equations of Turbulent Gases In Problems of Hydrodynamics and Continuum Mechanics*, SIAM, Philadelphia p. 231.

36 Janicka, J. , Kolbe, W. and Kollman, W. (1979) Closure of the Transport Equation for the Probability Density Function of Turbulent Scalar Fields. *J. Non-Equilib. Thermodyn.*, **4**, 47–66.

37 Nedorub, S.A. (1979) Investigation of Models for Calculation of Probability Density Functions of Impurity Concentrations in Turbulent Flows, Dis. Thesis Cand. of Phys.-Math. Sci, MFTI, Dolgoprudni (in Russian).

6
Coagulation and Breakup of Inertialess Particles in a Turbulent Flow

6.1
Kinetic Equations of Coagulation

When modeling particle motion in low-concentrated disperse media ($\varphi \ll 1$), we notice that the average distance between particles is large as compared to the particle size, and analysis of particle interactions can be restricted to pair interactions. If, in addition, the disperse phase consists of particles with low inertia ($St < 1$), the primary focus should be on particle interactions with turbulent fluctuations of the carrier phase.

Examples of such a system include water–oil emulsions and gas–condensate mixtures, both of which are disperse media whose disperse phase consists of droplets sized from 0.1 μm to 100 μm.

When particle coagulation is caused solely by pair collisions, and assuming a spatially homogeneous polydisperse system, the dynamics of the process is described by the following kinetic equation [1]:

$$\frac{\partial n(V,t)}{\partial t} = I_k, \tag{6.1}$$

$$I_k = \frac{1}{2}\int_0^V K(V-\omega,\omega)n(V-\omega,t)n(\omega,t)d\omega - n(V,\omega)\int_0^V K(V,\omega)n(\omega,t)d\omega,$$

where $n\,(V, t)$ is the distribution of particles over volumes V at the instant of time t, and $K(V, \omega)$ is the coagulation kernel having the meaning of collision frequency of particles of volumes V and ω per unit volume of the disperse phase at unit concentration of these particles.

The first term on the right-hand side of Eq. (6.1) corresponds to the rate of production of particles of volume V due to collisions between particles of volumes $V - \omega$ and ω, and the second term – to the rate of decrease in the number of particles of volume V due to their coagulation with other particles.

Statistical Microhydrodynamics. Emmanuil G. Sinaiski and Leonid Zaichik

ISBN: 978-3-527-40656-2

By solving Eq. (6.1) for a given initial distribution n_0 (V), we can observe how the volume distribution of particles changes with time and find the main parameters of this distribution: number concentration of particles, that is, the number of particles per unit volume of the medium

$$N(t) = \int_0^\infty n(v, t)dv, \tag{6.2}$$

volume concentration, that is, the volume of particles per unit volume of the medium

$$\varphi(t) = \int_0^\infty vn(v, t)dv, \tag{6.3}$$

and the average volume of particles

$$V_{av}(t) = \varphi(t)/N(t). \tag{6.4}$$

The kernel $K(V, \omega)$ defines the mechanism of particle interaction. The reader should keep in mind that study of general properties of the kernel and determination its specific form for various processes should be treated as two separate problems.

An important feature of the coagulation kernel is its symmetry with respect to the sizes of colliding particles, in other words, $K(V, \omega) = K(\omega, V)$. Multiplying both sides of Eq. (6.1) by V, integrating the result over V from 0 to ∞ and taking into account the relation (6.3), we obtain

$$\frac{d\varphi}{dt} = \frac{1}{2}\int_0^\infty\int_0^V VK(V-\omega, \omega)n(V-\omega, t)m(\omega, t)dVd\omega$$
$$-\int_0^\infty\int_0^\infty VK(V, \omega)n(V, t)n(\omega, t)dVd\omega. \tag{6.5}$$

A change of variables, $z = V - \omega$, $\omega = \omega$ in the first integral (6.5) makes it possible to transform the region of integration $0 \leq V < \infty$, $0 \leq \omega < V$ into $0 \leq z < \infty$, $0 \leq \omega < \infty$. Then Eq. (6.5) takes the form

$$\frac{d\varphi}{dt} = \frac{1}{2}\int_0^\infty\int_0^\infty (z+\omega)K(z, \omega)n(z, t)m(\omega, t)dzd\omega$$
$$-\int_0^\infty\int_0^\infty VK(V, \omega)n(V, t)n(\omega, t)dVd\omega. \tag{6.6}$$

Let us change the variable of integration by replacing z with V. Then Eq. (6.6) will be rewritten as

$$\frac{d\varphi}{dt} = \frac{1}{2}\int_0^\infty\int_0^\infty (\omega K(V,\omega) - VK(v,\omega))n(V,t)m(\omega,t)dVd\omega. \qquad (6.7)$$

By interchanging the positions of V and ω in the first term of the integrand in Eq. (6.7), which obviously does not affect the value of the integral, we get

$$\frac{d\varphi}{dt} = \frac{1}{2}\int_0^\infty\int_0^\infty V(K(\omega,V) - K(V,\omega))n(V,t)m(\omega,t)dVd\omega. \qquad (6.8)$$

Since we are considering only the coagulation of particles, the total volume of particles, that is, the volume concentration φ, remains constant, and

$$\int_0^\infty\int_0^\infty V(K(\omega,V) - K(V,\omega))n(V,t)m(\omega,t)dVd\omega = 0. \qquad (6.9)$$

One can readily see from Eq. (6.9) that symmetry of the coagulation kernel $K\,(V,\ \omega) = K\,(\omega,\ V)$ is a sufficient condition for the constancy of the volume concentration of particles. Let us show that it is also a necessary condition. As a corollary of the kinetic equation (6.1), the relation (6.8) must be valid for any physically permissible particle distribution over volumes n (V, t), including bi-disperse systems, that is, systems composed of particles of only two volumes V_1 and V_2. The particle distribution for such a system is

$$n(V,t) = n_0(\xi\delta(V - V_1) + (1 - \xi)\delta(V - V_2)). \qquad (6.10)$$

where ξn_0 is the number of particles of volume V_1, $(1-\xi)\ n_0$ – the number of particles of volume V_2 and $\delta(x)$ – the delta function.

Substituting Eq. (6.10) into Eq. (6.9) and recalling the property (1.11.b) of the delta function, we can write

$$n_0^2(V_1 - V_2)\xi(1 - \xi)(K(V_2,V_1) - K(V_1,V_2)) = 0. \qquad (6.11)$$

For a bi-disperse system, $n_0 \neq 0$, $V_1 \neq V_2$, $\xi \neq 0$, and $\xi \neq 1$, and therefore

$$K(V_2,V_1) = K(V_1,v_2). \qquad (6.12)$$

Since volumes V_1 and V_2 have been chosen arbitrarily, the equality (6.12) should be satisfied for all particle volumes present in the system.

If the symmetry condition for the coagulation kernel is violated, it means that the volume concentration of particles does not remain constant, which is equivalent to saying that the system contains sources and sinks whose intensities depend on the degree and specific form of kernel asymmetry. Another consequence Eq. (6.12) is that when the sum of particle volumes is fixed, in other words, when $V+\omega=z=const$, the function $K(V,\omega)$ has a local extremum at $V=\omega$.

Coagulation kernel characterizes particle collision frequency and is usually determined by solving the particle interaction problem numerically and approximating the obtained numerical results. Being a necessary condition, the condition of symmetry thus imposes restrictions on the choice of approximating relations. An efficient method of achieving symmetrization of the kernel by using available numerical data is outlined in [2].

Kinetic Eq. (6.1) is nonlinear integro-differential equation, which solution presents greatly difficulties. Currently available exact solutions are based on use of operational method as applied to linear dependence of $K(V,\omega)$ on particle volumes [3]. In order to solve Eq. (6.1) with kernels of more general form are used method of moments and numerical methods.

The method of moments makes it possible to reduce Eq. (6.1) to a system of differential equations for the moments the distribution of particles over volumes. But the resulting system of equations is usually unclosed because, in addition to moments of integer orders, it contains fractional-order moments, which appear due to the power law dependence of the kernel $K(V,\omega)$ on particle volumes. To close the system of equations, one needs additional relations or additional constraints on the form of the volume distribution. For example, the parametric method is based on the assumption that the volume distribution of particles belongs to some definite class of distributions (logarithmically normal distributions, gamma distributions, etc.) with time-varying parameters that need to be determined. Another method, known as the method of interpolation of fractional moments, is independent of the form of particle distribution but requires additional relations that express fractional moments in terms of integer ones.

In the majority of practical cases one is interested not so much in the distribution $n(V,t)$ as in its first several moments or their combinations, which have definite physical meaning and can be found experimentally with relative ease. Experimental data should tell us whether the chosen model of particle interaction is realistic, because the choice of particle coagulation mechanism is usually based on certain assumptions.

Integer moments of particle distribution over volumes are defined by the following expressions:

$$m_k=\int_0^\infty V^k n(V,t)dV;\quad (k=0,1,2,\ldots). \tag{6.13}$$

The value $k=0$ corresponds to the zero-order moment m_0 having the meaning of the number of particles per unit volume of the medium (i.e., number concentration).

The value $k=1$ corresponds to the first order moment m_1 equal to the volume of particles per unit volume of the medium (i.e., volume concentration, aka volume content φ). Several combinations of moments are worth noting: $(3/4\pi)^{1/3}(m_{1/3}/m_0)$ – mean particle radius; m_1/m_0 – mean particle volume; $(36\pi)^{1/3}m_{2/3}$ – total particle surface area per unit volume of the medium (i.e., specific area of the interface).

Furthermore, information about the type of the given distribution n (V, t) can be garnered from the first five moments $m_i(i=\overline{0,4})$, by using a Pearson diagram [4] (see Fig. 6.1). The quantities β_1 and β_2 plotted along the coordinate axes of the diagram are respectively called the asymmetry square and the excess. These distribution parameters are expressed through the moments as follows:

$$\beta_1=\left(\frac{\mu_3}{\mu_2^{3/2}}\right)^2,\ \beta_2=\frac{\mu_4}{\mu_2^2},$$

$$\mu_2=\frac{m_2}{m_0}-\left(\frac{m_1}{m_0}\right)^2,$$

$$\mu_3=\frac{m_3}{m_0}-3\frac{m_2m_1}{m_0^2}+2\left(\frac{m_1}{m_0}\right)^3,\quad \mu_4=\frac{m_4}{m_0}-4\frac{m_3m_1}{m_0^2}+6\frac{m_2m_1^2}{m_0^3}-3\left(\frac{m_1}{m_0}\right)^4.$$

To make a transition from the kinetic equation (6.1) to moment equations, we multiply both parts by V^k and integrate the result with respect to V from 0 to ∞:

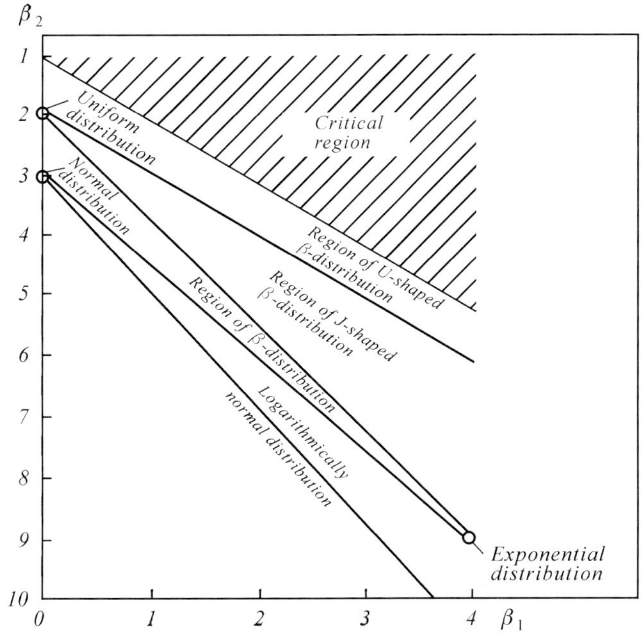

Fig. 6.1 The Pearson diagram.

$$\frac{d}{dt}\left(\int_0^\infty V^k n(V,t)dV\right) = \frac{1}{2}\int_0^\infty\int_0^V V^k K(V-\omega,\omega)n(V-\omega,t)d\omega dV$$
$$-\int_0^\infty\int_0^\infty V^k K(V,\omega)n(V,t)d\omega dV. \quad (6.14)$$

Making a change of variables in Eq. (6.14) in the same way as in the derivation of Eq. (6.6), we rewrite this equation as

$$\frac{dm_k}{dt} = \frac{1}{2}\int_0^\infty\int_0^\infty (z+\omega)^k K(z,\omega)n(z,t)d\omega dz - \int_0^\infty\int_0^\infty V^k K(V,\omega)n(V,t)d\omega dV.$$

Now, replace z in the first integral with V and rewrite this equation in the form that is regularly used in the method of moments:

$$\frac{dm_k}{dt} = \int_0^\infty\int_0^\infty \left(\frac{1}{2}(V+\omega)^k - V^k\right) K(V,\omega)n(V,t)d\omega dV. \quad (6.15)$$

The majority of known coagulation kernels have the form of power functions of particle volume. Since the kernel $K(V,\omega)$ must be symmetrical with respect to V and ω, it can be represented in the general case as

$$K(V,\omega) = \sum_{j=0}^{l} G_j(V^{\alpha_j}\omega^{\beta_j} + \omega^{\alpha_j}V^{\beta_j}). \quad (6.16)$$

Let us substitute the relation (6.16) into Eq. (6.15) and expand the binomial in the integral on the right-hand side:

$$\frac{dm_k}{dt} = \sum_{j=0}^{l} G_j\left(\frac{1}{2}\sum_{i=0}^{k}\binom{i}{k}\left(m_{i+a_j}m_{k-i+\beta_j} + m_{i+\beta_j}m_{k-i+\alpha_j}\right)\right.$$
$$\left. -(m_{\beta_j}m_{k+\beta_j} + m_{a_j}m_{k+\alpha_j})\right); \quad (k = 0, 1, 2, \ldots). \quad (6.17)$$

Expression (6.17) is an infinite system of ordinary differential equations for the moments m_k. If α_j and β_j are nonzero, the number of equations in any finite subsystem is less than the number of unknowns, so all finite subsets of the system of equations are unclosed.

Sometimes the exponent in the power function of volume appearing in the coagulation kernel will be fractional. Then fractional moments will appear in the right-hand side of Eq. (6.17). In such a case regularization of the system (6.17) can be achieved by interpolating the fractional moments through the integer ones [2] or by using the parametric method. Let us consider both methods successively.

Suppose we know the moments m_k (0) at the initial instant $t=0$. Introduce dimensionless moments $\widehat{m}_k(t) = m_k(t)/m_k(0)$, where k can be either integer or fractional. The left-hand side of Eq. (6.17) contains only integer moments, whereas the right-hand side may contain fractional moments as well. Our goal is to express fractional moments through integer ones. Consider a fractional moment of the order $j+$, where j is the integral part and $0< <1$, and integer moments m_s, m_{s+1}, ..., m_{s+r} where $s<j+ <s+r$. The logarithm of the fractional moment is sought in the form of a Lagrange interpolation polynomial

$$\ln(\widehat{m}_{j+\xi}) = \sum_{q=s}^{s+r} L_q^{(r)}(j+\xi)\ln(\widehat{m}_q(t)); \quad (j=1,2,\ldots), \tag{6.18}$$

where $L_q^{(r)}(j+\xi)$ are the coefficients of the interpolation polynomial when interpolating over $r+1$ nodes

$$L_q^{(r)}(j+\xi) = \prod_{\substack{p=s \\ p\neq q}}^{s+r} \frac{j+\xi-r}{q-p}. \tag{6.19}$$

With the help of Eq. (6.19), fractional moments can be represented as

$$\frac{m_{j+\xi}(t)}{m_{j+\xi}(0)} = \prod_{\substack{p=s \\ p\neq q}}^{s+r} \left(\frac{m_q(t)}{m_q(0)}\right)^{L_q^{(r)}(j+\xi)} ; s\leq j\leq s+r, 0<\xi<1 \tag{6.20}$$

and thus the system (6.17) becomes regularized.

Let us apply this method to a kinetic equation with the kernel

$$K(V,\omega) = GV^{\alpha}\omega^{\alpha}; \quad 0\leq\alpha\leq 1. \tag{6.21}$$

We shall restrict ourselves to the first four moments in the system of equations (6.17); for the fractional moment entering the right-hand side, we shall make use of two-point interpolation, expressing the moments $m_{j+\alpha}$ through the integer moments m_j and m_{j+1}. This will give us a closed system of equations for the first four moments:

$$\frac{dm_0}{dt} = -\frac{1}{2}Gm_{\alpha}^2(0)\left(\frac{m_0(t)}{m_0(0)}\right)^{2-2\alpha}\left(\frac{m_1(t)}{m_1(0)}\right)^{2\alpha},$$

$$\frac{dm_1}{dt} = 0,$$

$$\frac{dm_2}{dt} = Gm_{1+\alpha}^2(0)\left(\frac{m_1(t)}{m_1(0)}\right)^{2-2\alpha}\left(\frac{m_2(t)}{m_2(0)}\right)^{2\alpha}, \tag{6.22}$$

$$\frac{dm_3}{dt} = 3Gm_{2+\alpha}^2(0)m_{1+\alpha}^2(0)\left(\frac{m_1(t)}{m_1(0)}\right)^{1-\alpha}\left(\frac{m_2(t)}{m_2(0)}\right)^{2\alpha},$$

whose solution is as follows:

– for $\alpha \neq 0.5$:

$$m_0(\tau) = m_0(0)\left(1+\frac{B_0\tau}{\delta}\right)^{-\delta}, \quad m_1(\tau) = m_1(0),$$

$$m_2(\tau) = m_2(0)\left(1+\frac{B_0\tau}{\delta}\right)^{-\delta}, \tag{6.23}$$

$$m_3(\tau) = m_3(0)\left(1+\frac{B_3}{2B_2}\left(1+\frac{B_2\tau}{\delta}\right)^{1+\delta}\right)^{\frac{1}{1-\alpha}};$$

– for $\alpha = 0.5$:

$$m_0(\tau) = m_0(0)\exp(B_0\tau), \quad m_1(\tau) = m_1(0), \tag{6.24}$$

$$m_2(\tau) = m_2(0)\exp(B_2\tau), \quad m_3(\tau) = m_3(0)\left(\frac{B_3}{B_2}\exp(B_2\tau)+1\right)^2,$$

where the following dimensionless parameters have been introduced:

$$\tau = Gm_1^{2\alpha}t; \quad B_0 = \frac{m_\alpha^2(0)}{2m_0(0)m_1^{2\alpha}}; \quad \delta = \frac{1}{1-2\alpha};$$

$$B_2 = \frac{m_{1+\alpha}^2(0)}{m_2(0)m_1^{2\alpha}}; \quad B_3 = \frac{m_{1+\alpha}^2(0)m_{2+\alpha}(0)}{m_3(0)m_1^{2\alpha}}.$$

Now, let us discuss the parametric method. It is based on the assumption that the sought-for distribution belongs to some definite class. The choice of this class is usually based on general physical reasonings about the possible shape of the distribution produced in a specific process. For example, when studying the developed turbulent flow of emulsion in a pipe, the distribution of droplets is assumed to be a logarithmically normal distribution or a gamma distribution. Consider these two distributions in sequence.

Let the particle distribution belong to the class of logarithmically normal distributions [1]

$$n(\mathsf{V},t)=\frac{N}{3\sqrt{2\pi}\sigma \mathsf{V}}\exp\left(-\frac{\ln^2(\mathsf{V}/\mathsf{V}_0)}{18\,\ln^2\sigma}\right),\tag{6.25}$$

where N is the number concentration of particles, σ^2 – the variance, and V_0 – a parameter related to the average particle volume V_{av} through the expression $\mathrm{v}_{av}=\mathrm{v}_0\,\exp\,(1.5\ln^2\sigma)$. The form of the distribution (6.25) does not change with time, although its parameters $N(t)$, $\mathsf{V}_0(t)$, and $\sigma(t)$ are time-dependent.

We then substitute Eq. (6.25) into the moment equations (6.15) and take successively $k=0$, 1, 2. In the case of power dependence of K (V, ω) on V and ω the right-hand side may contain other moments (fractional or integer) different from the first three. Then the obtained system of equations can be regularized by writing additional relations that follow from the property of the distribution (6.25)

$$m_k=m_1\mathsf{V}_0^{k-1}\exp\left(-\frac{9}{2}(k^2-1)\ln^2\sigma\right).\tag{6.26}$$

As a result we obtain a closed system of equations for m_1, V_0, and σ. In fact, there are only two equations, because during the process of coagulation, $m_1=const.$

Suppose now that the particle distribution belongs to the class of gamma distributions [1]

$$n(\mathsf{V},t)=\frac{\varphi\;(i+1)}{\mathsf{V}_0^2\quad i!}\left(\frac{\mathsf{V}}{\mathsf{V}_0}\right)^i\exp\left(-\frac{\mathsf{V}}{\mathsf{V}_0}\right),\tag{6.27}$$

where φ is the volume concentrations of particles, which remains constant during coagulation, and V_0 and t are distribution parameters connected with the average volume and the variance through the relations $\mathsf{V}_0=\mathsf{V}/(i+1)$, $=\mathsf{V}/\sqrt{i+1}$. The form of the distribution does not change with time; only N and V_0 change. The distribution (6.27) allows us to seek the solution of equations (6.15) as an expansion of $n(\mathsf{V},\,t)$ in terms of associated Laguerre polynomials $L_{\mathrm{k}i}$ [5]:

$$n(\mathsf{V},t)=\left(\frac{\mathsf{V}}{\mathsf{V}_0}\right)^i\exp\left(-\frac{\mathsf{V}}{v_0}\right)\sum_{\mathrm{k}=0}^{\infty}a_{\mathrm{k}}L_{\mathrm{k}i}\left(\frac{\mathsf{V}}{\mathsf{V}_0}\right).\tag{6.28}$$

The first two polynomials are

$$L_{0i}=1;\quad L_{1i}=i+1-\mathsf{V}/\mathsf{V}_0,$$

and other are defined by the recurrent relation

$$(n+1)L_{n+1,i}-(\mathsf{V}/\mathsf{V}_0-2n+i+1)L_{ni}+(n+i)L_{n-1,i}=0.$$

The orthogonally conditions for polynomials give us expansion coefficients:

$$a_k = \frac{k!}{\Gamma(k+i+1)} \int_0^\infty L_{ki}\left(\frac{V}{V_0}\right) n(V,t) d\left(\frac{V}{V_0}\right), \tag{6.29}$$

where $\Gamma(x)$ is the gamma function. In particular, the first two coefficients are equal to

$$a_0 = \frac{k!}{\Gamma(k+1)} \int_0^\infty n(V,t) d\left(\frac{V}{V_0}\right); \quad a_1 = 0.$$

Using the two-moment approximation, we can write

$$n(V,t) = \frac{N(t)}{V_0(t) i!} \left(\frac{V}{V_0}\right)^i \exp\left(-\frac{V}{V_0}\right),$$

$$\frac{d(N/V_0)}{dt} = -\frac{1}{2}\left(\frac{N}{V_0}\right)^2 \int_0^\infty \int_0^\infty \exp\left(-\frac{\omega+V}{V_0}\right) \tag{6.30}$$

$$\times K(V,\omega)\left(\frac{V}{V_0}\right)^i \left(\frac{\omega}{V_0}\right)^i d\omega dV, \quad (i+1)NV_0 = \varphi.$$

The system of equations (6.30) enables us to find the volume distribution of particles as a function of time, and to determine the parameters of this distribution.

Practice has shown that the method of moments describes the kinetics of coagulation in a satisfactory manner only at the initial stage. When certain restrictions are imposed on the form of the coagulation kernel and on the initial distribution, there exists self-similar solution at $t \to \infty$ [6]. In the general case, however, one has to rely on numerical methods to obtain the solution of the kinetic equation.

6.2 Fundamental Features of the Coagulation of Particles

We may distinguish two stages in the process of particle coagulation. During the first stage (the transport stage) particles approach each other and come into contact; during the second (kinetic stage) they coagulate if they are solid) or coalesce if they are liquid or gaseous.

Experimental studies of droplet coalescence on a plane interface between two fluids [7] give grounds to suggest that the time of particle coalescence is much shorter than the time of their mutual approach. Thus it can be assumed that the duration of the coalescence process is limited by the transport stage. Then the coagulation kernel can be found from detailed analysis of the process of mutual approach of particles – all the way to their eventual collision, assuming that each collision of particles results in their coagulation or coalescence.

The character of particles' mutual approach (and thereby of interparticle collisions) depends on the hydrodynamic regime of the flow. In a laminar flow, or in the process of particle sedimentation in a quiescent fluid, collision frequency can be determined by studying trajectories of particles' relative motion right up to the collision. In a turbulent flow, particle motion is caused by random turbulent fluctuations of the carrier fluid. The mutual approach and collision of particles can thus be considered as a random process as well. Let us examine both cases.

Under the action of gravity, particles of different sizes settle with different velocities. As a result large particles overtake small ones, and collisions become possible. By studying particles' trajectories relative to other particles, we can determine the collision frequency of particles. The volume concentration of particles is low, so the analysis can be restricted to the relative motion of two particles and to pair collisions.

Examination of relative motion of particles is conducted in a spherical coordinate system (r, θ, Φ) with the origin in center of the largest particle (Fig. 6.2). In this frame of reference the carrier fluid moves relative to the largest particle with a constant velocity U; on a large distance from the particle this velocity is equal to the

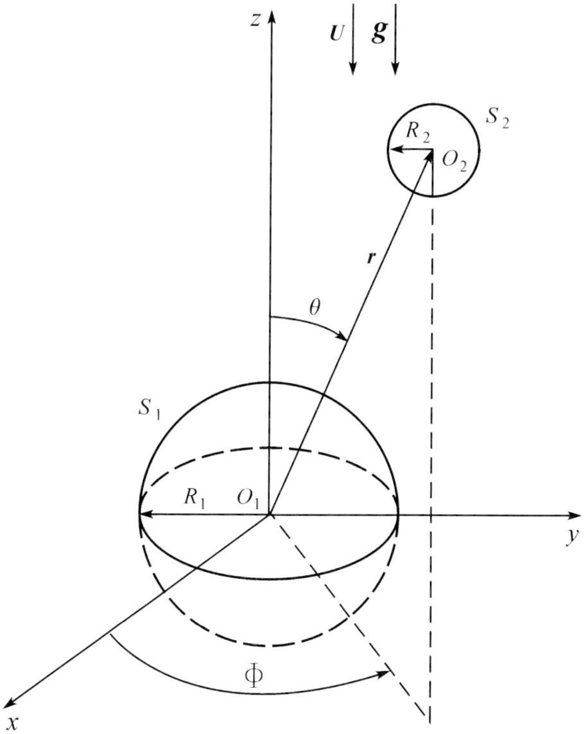

Fig. 6.2 Spherical coordinate system.

sedimentation velocity for this particle. Another particle of smaller size moving together with the carrier flow goes around the large particle, taken as the frame of reference, either touching it or passing by. Due to the small size of particles, their motion should be regarded as inertialess. So when the small particle is sufficiently far from the large one, its trajectory coincides with the streamline of the carrier fluid. In the vicinity of the large particle, one can observe a deviation of the trajectory from the streamline due to hydrodynamic interactions with the carrier fluid and interactions between the particles.

Interaction forces can be hydrodynamic, molecular, and electrostatic. Hydrodynamic forces are resistance forces acting on a particle that increase unboundedly as the gap between particle surfaces gets smaller. Molecular forces are attractive van der Waals forces acting only at relatively small distances between particles. Electrostatic forces are repulsive forces created by double electrical layers at particle surfaces and (or) interaction forces between electrically conductive particles (charged or uncharged) placed in an external electric field. To determine the relative trajectories of particles, it is necessary to know all of the above-mentioned forces; the evaluation of each force presents an independent hydrodynamical or physical problem.

The whole set of trajectories of the smaller particle can be divided into two groups: trajectories that passing by the larger particle, and trajectories that lead to particle collision. These two trajectory groups are separated by the so-called limiting trajectories (Fig. 6.3).

In the remote regions of the carrier flow (i.e., far away from the large particle) the limiting trajectories form a flow tube whose cross-sectional area $G(V, \omega)$ is called the collision cross section of particles having volumes V and ω $(V > \omega)$. Then the collision frequency of particles with volumes V and ω may be written as

$$K(V, \omega) = G(V, \omega)|U_V - U_\omega|, \tag{6.31}$$

where U_V and U_ω are velocities of unhindered particle sedimentation, that is, sedimentation velocities that would be observed in the absence of hydrodynamic interactions. These velocities can be assumed equal to the corresponding Stokesian velocities. Trajectories of a particle i are described by the equations of motion

$$m_i \frac{d\boldsymbol{u}_i}{dt} = \sum_k \boldsymbol{F}_k^{(i)}; \quad J_i \frac{d\boldsymbol{\Omega}_i}{dt} = \sum_k \boldsymbol{L}_k^{(i)}; \quad \boldsymbol{u}_i = \frac{d\boldsymbol{r}_i}{dt}, \tag{6.32}$$

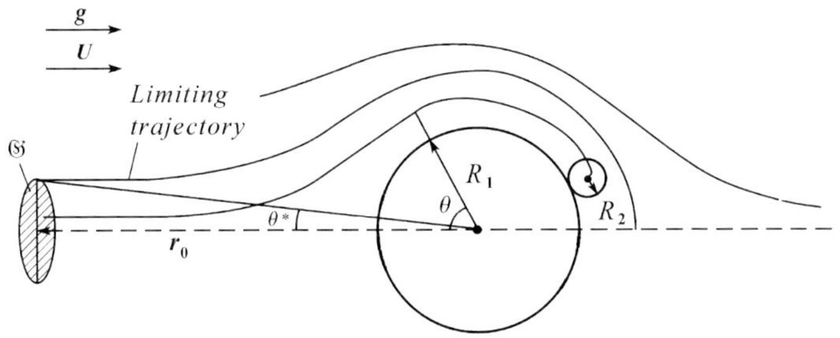

Fig. 6.3 Trajectories of particle 2 relative to particle 1.

where m_i is the particle mass; $F_k^{(i)}$, $L_k^{(i)}$ are, respectively, forces and torques acting on the particle; J_i is the particle's moment of inertia; and $i = 1, 2$.

If particle sizes are small enough and the density ratio of the particle and the external fluid is practically equal to unity, fluid flow can be taken as slow, that is, Stokesian, and particle motion – as inertialess. Then equations (6.32) reduce to the equations of inertialess motion in a quasistationary approximation, namely,

$$\sum_k \boldsymbol{F}_k^{(i)} = 0; \quad \sum_k \boldsymbol{L}_k^{(i)} = 0; \quad \boldsymbol{u}_i = \frac{d\boldsymbol{r}_i}{dt}. \tag{6.33}$$

Assigning various initial positions to particle 2 and integrating equations (6.33), one obtains family of trajectories of particle 2 relative to particle 1. In a spherical coordinate system, the position of particle 2 is given by the initial coordinates of its center, r_0, θ_0, Φ_0. Consider a flow tube whose surface contains limiting trajectories, and let θ^* (Φ) be the angle between the vector r_0 and the second vector that connects the particle center with a point on the surface of a fixed cross section of the flow tube (whose radius is therefore $z = r_0 \cos\theta^*$). At large distances from particle 1 (that is, at $r_0 \to \infty$ and $\theta^* \to 0$), the flow tube's cross-sectional area can be found from simple geometrical reasonings. Thus, in the general case when the cross section is not circular, its area equals

$$G = \lim_{\substack{r_0 \to \infty \\ \theta^* \to 0}} \left(\frac{1}{2} r_0^2 \int_0^{2\pi} \sin^2(\theta^*(\Phi)) d\Phi \right). \tag{6.34}$$

If the fluid flow and the relative motion of particles are both axisymmetric, then θ^* does not depend on Φ, the cross section is circular, and

$$G = \lim_{\substack{r_0 \to \infty \\ \theta^* \to 0}} (\pi r_0^2 \sin^2(\theta^*)). \tag{6.35}$$

Collision frequency is higher in a turbulent flow than in a laminar flow or during sedimentation in a quiescent fluid. Particles suspended in the fluid are carried along by turbulent fluctuations and move chaotically inside the flow region. Since their fluctuational motion is alike to Brownian motion, it can be characterized by the effective diffusion coefficient D_t and the problem of finding particles' collision frequency can be reduced to a diffusion problem in the same manner as it was first done by Smoluchowski for Brownian motion [8]. Such an approach was attempted by Levich [9] in application to coagulation of hydrodynamically non-interacting particles in a turbulent flow. The decision to ignore hydrodynamic interactions ahs resulted in exaggerated values of collision frequency as compared to experimental data on turbulent motion of emulsions in pipes and agitators [10,11].

The influence of hydrodynamic interactions between particles on the value of particles' interdiffusion coefficient has been investigated in [12–16]. It was shown that correct accounting for hydrodynamic interactions between particles assures good agreement with the experiment.

When diffusion is the main mechanism responsible for the mutual approach of particles, the number of collisions between particles having radii R_1 and R_2 per unit time is equal to the flux of particles of radii R_2 toward the particle of radius R_1. If diffusion equilibrium is established much faster than concentration equilibrium, the problem reduces to that of solving the stationary diffusion equation in a force field [17,18]

$$\nabla \cdot \left(D_t \nabla n - \frac{\mathbf{F}}{h} n \right) = 0, \tag{6.36}$$

where n is the number concentration of particles 2; $\mathbf{F}$ is the force of interaction between particles 1 and 2 (resulting from electrodynamic gravitational, molecular, or electrostatic fields); h is the coefficient of hydrodynamic resistance for particles approaching each other along their line of centers. Let u be the relative velocity of such particles and $\mathbf{F}$ – the force of hydrodynamic resistance experienced by particle 2. Then

$$\mathbf{F}_h = h\mathbf{u}. \tag{6.37}$$

Instead of the hydrodynamic resistance coefficient one sometimes works with mobility $\upsilon = h^{-1}$, which helps express velocity in terms of force:

$$\mathbf{u} = \upsilon \mathbf{F}_h. \tag{6.38}$$

If every collision of fluid particles leads to their coalescence, the boundary conditions for Eq. (6.36) are

$$n = 0 \quad \text{at} \quad r = R_1 + R_2; \qquad n = n_0 \quad \text{at} \quad r \to \infty. \tag{6.39}$$

where n_0 is the concentration of particles 2 far away from the test particle 1.

By solving Eq. (6.36) under the conditions (6.39) we can find n and thereby the flux of particles 2 toward the test particle 1:

$$J = \iint\limits_S \left(D_t \nabla n - \frac{\mathbf{F}}{h} n \right) \cdot \mathbf{n}\, ds. \tag{6.40}$$

where $\mathbf{n}$ is the outward normal to the surface element ds. Hence in order to determine the collision frequency of particles, it is necessary to obtain the interaction forces acting between the particles and then to find particles' trajectories or the diffusion flux of particles. In the latter case one needs to know the effective coefficient of turbulent diffusion. This procedure gives us the coagulation kernel. Sometimes the obtained kernel turns out to be asymmetric with respect to particles' volumes, and then symmetrization of the kernel is in order. Only after that one can proceed to study the kinetics of coagulation, that is, to solve the kinetic equation of coagulation.

6.3
A Model of Turbulent Diffusion

Transport of small particles in a turbulent flow can be considered as diffusion and characterized by an effective diffusion coefficient. Before we proceed to determine this coefficient, it is worthwhile to recall the basic features of fluid particle motion in turbulent flows. For developed turbulence, these features have been studied extensively in [9,19–21]. Let us therefore restrict ourselves to those features that pertain to particle collisions.

A turbulent flow means that a random fluctuational motion (characterized by a set of fluctuational velocities $\boldsymbol{u}_\lambda$) is superimposed onto the average flow that has a definite speed U and a definite direction. Turbulent fluctuations are characterized not just by velocities, but also by the distances at which these velocities experience a noticeable change. These distances are known as fluctuation scales and denoted by λ. The full set of λ's represents a spectrum of turbulent fluctuations, where λ can vary from 0 to the maximum value, which is equal by the order of magnitude to the characteristic linear scale L of the flow region. Every fluctuational motion is characterized by its Reynolds number $\mathrm{Re}_\lambda = \lambda u_\lambda/\nu_e$, where ν_e is the coefficient of kinematic viscosity of the carrier flow. Fluctuations with $\lambda \sim L$ are called large-scale fluctuations. For such fluctuations, $\mathrm{Re}_\lambda \gg 1$, so the fluid flow induced by these fluctuations is inviscid. The Reynolds number decreases as the fluctuation scale gets smaller. At $\lambda = \lambda_0$, which is called the inner (or Kolmogorov) turbulence scale, the Reynolds number is approximately equal to

$$\mathrm{Re} = \mathrm{Re}_{\lambda_0} = \lambda u_{\lambda_0}/\nu_e \sim 1. \tag{6.41}$$

It means that fluctuations with $\lambda \leq \lambda_0$ are viscous, and fluctuational motion on such a scale is accompanied by energy dissipation. Fluctuations with $\lambda \ll L$ are referred to as small-scale. They are induced by larger fluctuations. The energy is transmitted from large-scale fluctuations to small-scale ones and then dissipates into heat. Thus turbulent motion is accompanied by considerable dissipation of energy. Energy loss per unit mass per unit time $\bar{\varepsilon}$ is called the specific dissipation of energy and is one of the most fundamental parameters of turbulence. Since energy is drawn from large-scale fluctuations, $\bar{\varepsilon}$ depends on U and L. Thus $\bar{\varepsilon}$ can be estimated from dimensionality considerations:

$$\bar{\varepsilon} \sim U^3/L. \tag{6.42}$$

The velocity of small-scale fluctuations with $\lambda_0 \ll \lambda \ll L$ can depend only on $\bar{\varepsilon}$ and λ. Therefore

$$u_\lambda \sim (\bar{\varepsilon}\lambda)^{1/3}. \tag{6.43}$$

Since at $\lambda \sim \lambda_0$, the condition $\mathrm{Re}_{\lambda_0} \sim 1$ must hold, the inner scale of turbulence follows from Eq. (6.41) and Eq. (6.43):

$$\lambda_0 \sim \nu_e/u_{\lambda_0} \sim (\nu_e^3/\bar{\varepsilon})^{1/4}. \tag{6.44}$$

According to the hypothesis put forward by Landau and Levich, viscous fluctuations with $\lambda < \lambda_0$ attenuate rather gradually instead of disappearing abruptly. The motion of the fluid reduces to a set of mutually independent periodic motions whose periods T are constant for all $\lambda < \lambda_0$. The value of T can be estimated as $T \sim \sqrt{\nu_e/\bar{\varepsilon}}$. The velocities of fluctuations with $\lambda < \lambda_0$ are given by

$$u_\lambda \sim \lambda/T \sim \lambda\sqrt{\bar{\varepsilon}/\nu_e}. \tag{6.45}$$

Combining the relations (6.43) and (6.45), one gets a general expression for the velocity of turbulent fluctuations:

$$u_\lambda \sim \begin{cases} (\bar{\varepsilon}\lambda)^{1/3} & \text{at} \quad \lambda > \lambda_0, \\ \lambda(\bar{\varepsilon}/\nu_e)^{1/2} & \text{at} \quad \lambda < \lambda_0 \end{cases} \tag{6.46}$$

Consider now the motion of small particles of radius R in a turbulent fluid flow. Volume concentration of particles is assumed to be small so that their influence on the carrier fluid can be ignored. Large-scale fluctuations with $\lambda \gg R$ transport particles together with the adjacent fluid layers, whereas small-scale fluctuations with $\lambda \ll R$ are not capable of drawing particles into their motion. In the latter case, particles behave as immovable bodies relative to the fluid. Fluctuations having intermediate scales can draw particles into their motion only to a moderate extent. Consider a situation that is typical for particles with low inertia, when densities of the particle ρ_i and the surrounding fluid ρ_i are roughly the same, and the particle radius is much smaller than the inner scale of turbulence, that is, $R \ll \lambda_0$. For example, for water–in–oil emulsions, $\rho_i/\rho_e \sim 1.1$–1.5. Denote through $\boldsymbol{u}_0$ the fluid velocity at the point where the particle is located, and through $\boldsymbol{u}_1$ – the particle's velocity relative to the fluid. If the particle were completely entrained by the surrounding fluid, the force acting on the particle would be the same as if the particle were composed of this fluid, that is, $\frac{4}{3}\pi R^3 \rho_e \frac{d\boldsymbol{u}_0}{dt}$. But because the entrainment is only partial, the particle experiences a resistance force F, which, given the conditions $R \ll \lambda_0$ and $u_1 < u_{\lambda_0}$, is determined by the Stokes formula $\boldsymbol{F} = -6\pi\rho \boldsymbol{v}_e R \boldsymbol{u}_1$. The equation of motion for the particle takes the form

$$m\frac{d\boldsymbol{u}_1}{dt} + \frac{4}{3}\pi R^3 \rho_e \left(\frac{d\boldsymbol{u}_0}{dt} + \frac{d\boldsymbol{u}_1}{dt}\right) = \frac{4}{3}\pi R^3 \rho_e \frac{d\boldsymbol{u}_0}{dt} - 6\pi\rho_i \nu_e R \boldsymbol{u}_1, \tag{6.47}$$

where $m = 2\pi R^3 \rho_e/3$ is the particle's virtual mass.

Let us estimate particle's acceleration that appears in Eq. (6.47). The period T is constant for motions with scales $\lambda < \lambda_0$, so

$$\frac{d\boldsymbol{u}_1}{dt} \sim \frac{\boldsymbol{u}_1}{T} \sim \boldsymbol{u}_1\sqrt{\frac{\varepsilon}{\nu_e}}. \tag{6.48}$$

To estimate the acceleration of the fluid at the point occupied by the particle, let us take the maximum acceleration of fluctuational motion in the interval of scales under consideration, that is, at $\lambda = \lambda_0$:

$$\frac{du_0}{dt} \sim \frac{u_{\lambda_0}}{T} \sim \varepsilon_0\frac{\lambda_0}{\lambda} \sim \left(\frac{\varepsilon_0^3}{\nu_e}\right)^{1/4}. \tag{6.49}$$

Substitution of Eq. (6.48) into Eq. (6.47) and comparison with Eq. (6.49) shows that the vectors $\boldsymbol{u}_1$ and $\mathrm{d}\boldsymbol{u}_0/dt$ are approximately collinear. Then, in view of Eq. (6.49), the approximate solution of Eq. (6.47) is given by

$$\boldsymbol{u}_1 \sim \frac{2R^2|\rho_e - \rho_i|\lambda_0\varepsilon_0\nu^{-1}}{2R^2(\rho_i + 0.5\rho_e)\varepsilon_0^{1/2} + 9\rho_e\nu_e}. \tag{6.50}$$

The ratio u_1/u_{λ_0} tells us to which extent the particle is entrained by the fluctuation λ. When $u_1/u_{\lambda_0} \ll 1$, the entrainment is complete, whereas at $u_1/u_{\lambda_0} \gg 1$, entrainment is absent. For fluctuations with scales $\lambda < \lambda_0$, Eq. (6.46) and Eq. (6.50) give us

$$\frac{u_1}{u_\lambda} \sim \frac{2R^2|\rho_e - \rho_i|\varepsilon_0^{1/2}\nu_e^{-1/2}}{R^2(\rho_e + 2\rho_i)\varepsilon_0^{1/2}\nu_e^{-1/2} + 9\rho_e\nu_e}. \tag{6.51}$$

For water-in-oil emulsions, where $\rho_e \sim 800\,\mathrm{kg/m^3}$, $\rho_i \sim 1200\,\mathrm{kg/m^3}$, $\nu_e \sim 10^{-5}\,\mathrm{m^2/s}$, $\lambda_0 \sim 10^{-4}\,\mathrm{m}$, $\varepsilon_0 \sim 10\,\mathrm{J/kg\,s}$, Eq. (6.51) gives

$$\frac{u_1}{u_\lambda} \sim \frac{8R^2}{28R^2 + 72\cdot 10^{-3}} \ll 1.$$

So, water droplets of radius $R < \lambda < \lambda_0$ are, for all practical purposes, fully entrained by fluctuations whose scale is λ.

Transport of particles of radii $R \ll \lambda_0$ is defined by two parameters: velocity u_λ and fluctuation scale λ. From these parameters, one can form a combination having the dimensionality of diffusion coefficient:

$$D_t^{(0)} \sim \lambda u_\lambda. \tag{6.52}$$

Substitution of fluctuational velocity (6.46) into Eq. (6.52) gives the diffusion coefficient for particles suspended in a turbulent flow:

$$D_t^{(0)} \sim \begin{cases} (\bar{\varepsilon}\lambda^4)^{1/3} & \text{at} \quad \lambda > \lambda_0, \\ \lambda^2(\bar{\varepsilon}/\nu_e)^{1/2} & \text{at} \quad \lambda < \lambda_0. \end{cases} \tag{6.53}$$

Let us see how the obtained formulas relate to diffusional interaction of particles. Let r be the distance between particle centers. In the course of diffusion, resistance forces that obstruct the mutual approach of two particles are mostly confined to the region where the gap between the particles is very small: $\delta < \lambda_0 - (R_1 + R_2)$. In this region, Eq. (6.53) gives the following diffusion coefficient:

$$D_t^{(0)} \sim \bar{\varepsilon}^{1/2} \lambda^2 / \nu_e^{1/2} \sim \nu \lambda^2 / \lambda_0^2. \tag{6.54}$$

This still leaves open the question of the scale of fluctuations that might cause the particles to come closer together, because large fluctuations will transport the particle pair as a whole without changing the interparticle distance. It is evident that particles of the same size ($R_1 = R_2$) can be brought closer together only by fluctuations with $\lambda \sim r$, and particles whose sizes are largely different ($R_1 \gg R_2$) – only by fluctuations with $\lambda \sim r - R_1$, because the larger particle (when put next to the smaller one) can be considered as a plane wall, in whose vicinity small-scale fluctuations get attenuated. If the relative size of the two particles is in the intermediate zone, we can determine the relevant scale of fluctuations from the approximation that obeys both of the above-considered limiting cases together with the condition of symmetry of the diffusion coefficient with respect to particle radii, namely, $D_t^{(0)}(R_1, R_2) = D_t^{(0)}(R_2, R_1)$. As a result we obtain

$$\lambda \sim (R_1 + R_2)\left(\frac{r - R_1 - R_2}{R_1 + R_2} + \frac{R_1 R_2}{R_1^2 + R_2^2 - R_1 R_2}\right). \tag{6.55}$$

Going back to Eq. (6.54), we get the following expression for the interdiffusion coefficient for particles of radii R_1 and R_2:

$$D_t^{(0)} = \beta \frac{\nu_e}{\lambda_0^2} (R_1 + R_2)^2 \left(\frac{r - R_1 - R_2}{R_1 + R_2} + \frac{R_1 R_2}{R_1^2 + R_2^2 - R_1 R_2}\right)^2, \tag{6.56}$$

where we have introduced a correction factor β having the order of unity. This factor is required because the relations (6.54) and (6.55) are merely approximations accurate to the order of magnitude.

The formula (6.56) was obtained under the assumption that particles are fully involved into relative motion by fluctuations having the scale λ. So, this formula can be used only if the particles are spaced relatively far from each other. When they get closer together so that the gap δ between particle surfaces is equal by the order of magnitude to the smaller particle's radius, hydrodynamic resistance force begins to influence the particles' relative velocity; the force goes to the infinity at $\delta \to 0$. To take this force into account, we employ the approach that is based on the Langevin equation [22,23]; this approach is also used in statistical physics when considering Brownian motion of a particle subject to an external random force.

Driven by random turbulent fluctuations, the particle changes its direction many times during a short time interval. Thus it is next to impossible to trace its trajectory visually. Besides, particle displacements cannot serve as useful characteristics of particle motion, because the average displacement during a finite time interval is equal to zero. Instead, the mean-square displacement emerges as the main characteristic of particle motion for finite time intervals. One should keep in mind that "mean" here implies ensemble averaging rather than time averaging.

Consider two instants of time, t_1 and $t > t_1$. We make the assumption that during the relevant time intervals the particle will change its direction quite often, and, moreover, any two displacements taking place during non-overlapping time intervals are fully independent. For simplicity's sake, we restrict ourselves to the case of one-dimensional motion. Let s denote the path traveled by the particle during the time interval $(0, t)$. Similarly, the path s_1 will correspond to the interval $(0, t_1)$ and the path s_2 – to the interval (t_1, t). From the condition of statistical independence of s_1 and s_2 and keeping in mind that positive and negative displacements should have equal probabilities, we derive the mean-square displacement:

$$\langle s^2 \rangle = \left\langle (s_1 + s_2)^2 \right\rangle = \langle s_1^2 \rangle + \langle s_2^2 \rangle. \tag{6.57}$$

Denoting $\langle s^2 \rangle = \psi(t)$, we rewrite the last equation as

$$\psi(t) = \psi(t_1) + \psi(t - t_1). \tag{6.58}$$

The solution of this equation has the form

$$\psi(t) = \langle s^2 \rangle = 2 D_t t, \tag{6.59}$$

where D_t is the coefficient of turbulent diffusion that still needs to be determined.

Consider now the equation of motion of a particle subject to a force exerted by the surrounding medium. This force consists of a systematical part – the friction force, and a random part – the force F whose average value is zero. The systematic force may also include the force of gravity; however, it is negligible for sufficiently small particles. Let us also ignore any other external forces. Then in the inertialess approximation the equation of motion of such a particle (i.e., the Langevin equation) along the X-axis has the form

$$h \frac{dx}{dt} + F = 0, \tag{6.60}$$

where h is the coefficient of hydrodynamic resistance.

From Eq. (6.60), it follows that the particle's displacement from its initial position X_0 during the time t is equal to

$$X - X_0 = \frac{1}{h} \int_0^t F(\tau)\, d\tau. \tag{6.61}$$

The integral on the right-hand side of Eq. (6.61) stands for the impulse of the random force F during the time t. For any instant t_1 in the range $0 < t_1 < t$, this integral can be written as

$$\int_0^t F(\tau)d\tau = \int_0^{t_1} F(\tau)d\tau + \int_{t_1}^t F(\tau)d\tau. \tag{6.62}$$

Taking the square of both sides of Eq. (6.62), averaging the result and remembering that random impulses for any two non-overlapping time intervals should be independent, we obtain:

$$\left\langle \left(\int_0^t F(\tau)d\tau \right)^2 \right\rangle = \left\langle \left(\int_0^{t_1} F(\tau)d\tau \right)^2 \right\rangle + \left\langle \left(\int_{t_1}^t F(\tau)d\tau \right)^2 \right\rangle. \tag{6.63}$$

This equation is similar to Eq. (6.57), so its solution is

$$\left\langle \left(\int_0^t F(\tau)d\tau \right)^2 \right\rangle = 2Bt. \tag{6.64}$$

The particle's mean-square displacement is given by Eq. (6.61):

$$\langle (X - X_0)^2 \rangle = \frac{1}{h^2} \left\langle \left(\int_0^t F(\tau)d\tau \right)^2 \right\rangle. \tag{6.65}$$

Substituting the relations (6.59) and (6.64) into this expression, we get

$$D_t = \frac{B}{h^2}; \quad B = \frac{1}{2t} \left\langle \left(\int_0^t F(\tau)d\tau \right)^2 \right\rangle. \tag{6.66}$$

In view of the fact that a similar approach is applied to Brownian diffusion, it is worthwhile to point out the main distinction of turbulent diffusion from Brownian one. In the course of Brownian diffusion, particles perform random thermal motion due to collisions with molecules of the surrounding medium. In [22], the corresponding force acting on the test particle is modeled as a quasi-elastic force $\boldsymbol{F} = -\alpha \boldsymbol{X}$ proportional to the displacement X. As a result the form of Eq. (6.60) changes: a term proportional to X appears in the equation, and the condition of thermodynamic equilibrium of the system leads us to

$$B = h\mathrm{k}T, \quad D_{br} = \frac{\mathrm{k}T}{h}. \tag{6.67}$$

Consequently, the coefficient of Brownian diffusion is inversely proportional to the first degree of the particle's coefficient of hydrodynamic resistance h.

In the case of turbulent diffusion, the situation is different. Particle motion driven by turbulent fluctuations is quite independent from any thermal fluctuations. Therefore $B = const$ and the coefficient of turbulent diffusion is inversely proportional to the second degree of the coefficient of hydrodynamic resistance.

If the distance at which h can vary significantly is greater than the fluctuation-driven displacement of a particle, then h does not depend on displacement X. This is the case, for instance, when the particle is moving in a region adjacent to a wall or approaches another particle.

Consider a pair of spherical particles with radii R_1 and R_2 ($R_1 \geq R_2$) approaching each other along their line of centers. In the inertialess approximation, the equation of motion of one particle relative to the other (which is taken as motionless) is

$$\boldsymbol{F}_{1r} + \boldsymbol{F}_r = 0, \quad \boldsymbol{F}_{1r} = h_r^0 \boldsymbol{u}_r^0, \quad \boldsymbol{F}_r = -h_r \boldsymbol{u}_r. \tag{6.68}$$

where $\boldsymbol{F}_{1r}$ is the force that would be exerted on the particle by the surrounding fluid if the particle were at rest; $\boldsymbol{F}_r$ is the drag force caused by the particle's own motion; u_r^0 is the fluid's velocity at the point where the particle is located; $\boldsymbol{u}_r$ is the particle's velocity.

When the interparticle distance is much larger than the size of each particle, both particles will be completely entrained by the fluid, so that $h_r^0 = h_r, \boldsymbol{u}_r^0 = \boldsymbol{u}_r$. A decrease of the interparticle distance leads to a change of the resistance coefficients h_r^0 and h_r. While the first coefficient does not change much, the second one increases rapidly and becomes infinite when the particles touch.

One can see from Eq. (6.68) that the motion of one particle in the vicinity of another is constrained and its velocity is equal to

$$\boldsymbol{u}_r = \frac{h_r^0}{h_r} \boldsymbol{u}_r^0. \tag{6.69}$$

Equation (6.68) for small displacements describes a motion that is similar to the motion of an unconstrained particle with the similarity coefficient h_r^0/h_r. This allows us to write the coefficient of turbulent diffusion (6.66) in a similar form, namely,

$$D_t = D_t^{(0)}(r) \left(\frac{h_r^0}{h_r}\right)^2. \tag{6.70}$$

where $D_t^{(0)}(r)$ is the coefficient of turbulent diffusion for particles performing unconstrained motion (see Eq. 6.56).

The expression (6.70) does not take into account the influence of the second particle. To see how the influence of both particles will affect the velocity with which they approach each other, we shall proceed as follows. Let u be the velocity of one particle relative to the other, and $\boldsymbol{u}_1$ and $\boldsymbol{u}_2$ – particle velocities relative to a frame of reference whose origin is placed into a point on the line of centers of the two particles. Then $\boldsymbol{u} = \boldsymbol{u}_1 - \boldsymbol{u}_2$. The forces $\boldsymbol{F}_i$ acting on the particles are equal in magnitude but opposite in sign. Then the mutual approach of these two particles will be characterized by the hydrodynamic resistance coefficient equal to

$$h_r = \frac{F}{u_1 - u_2} = \frac{F}{F/h_1 + F/h_2} = \frac{h_1 h_2}{h_1 + h_2}, \tag{6.71}$$

where h_i is the hydrodynamic resistance coefficient of i-th particle ($i = 1,2$).

When the particles are still far from each other, their mutual influence is insignificant and $h_i = 6\pi\rho\nu_e R_i$. Then Eq. (6.71) yields

$$h_r = h_r^0 = 6\pi\rho\nu \frac{R_1 R_2}{(R_1 + R_2)}; \quad r \gg R_1, R_2. \tag{6.72}$$

At relatively small distances $\delta = r - R_1 - R_2$ between particle surfaces, the resistance coefficient behaves as $1/\delta$ for solid particles [24] and as $1/\sqrt{\delta}$ for droplets with mobile surfaces [25]. We shall restrict ourselves to the case of solid particles. Then we have for small clearances between the particles

$$h_r = h_r^0 \frac{R_1 R_2}{(R_1 + R_2)^2} \frac{(R_1 + R_2)}{(r - R_1 - R_2)}. \tag{6.73}$$

Combining the far and near asymptotics (6.72) and (6.73), we get

$$h_r = 6\pi\rho\nu_e \frac{R_1 R_2}{(R_1 + R_2)} \left(1 + \frac{R_1 R_2}{(R_1 + R_2)^2} \frac{(R_1 + R_2)}{(r - R_1 - R_2)} \right). \tag{6.74}$$

Let us now substitute the relations (6.56), (6.72), (6.74) into (6.70). As a result we obtain the coefficient of mutual turbulent diffusion for spherical particles that takes hydrodynamic interactions into account:

$$D_t = \beta \frac{\nu_e}{\lambda_0^2} \frac{(R_1 + R_2)^2 (s + \gamma)^2 s^2}{(s + \kappa)^2}. \tag{6.75}$$

We have introduced the following dimensionless parameters:

$$\Delta = (r - R_1 - R_2)/(R_1 + R_2), \quad \gamma = R_1 R_2/(R_1^2 + R_2^2 - R_1 R_2),$$

$$\kappa = R_1 R_2/(R_1 + R_2)^2.$$

6.4
Hydrodynamic, Molecular, and Electrostatic Forces

Hydrodynamic forces acting on particles reveal themselves when the particles move relative to each other and to the surrounding fluid. Generally speaking, these forces can deform surfaces of particles (droplets, bubbles), especially when the gap between the particles becomes smaller than the particle size. Bur when the particles are sufficiently small and their surfaces are covered by adsorbed impurities which stabilize the surface, this deformation can be ignored. Henceforth, we shall assume the particles to be nondeformable and spherical.

Among the factors influencing the value of the hydrodynamic force are the presence of neighboring particles and the velocity field of the carrier medium. For particles with radius up to 100 μm, it is safe to suppose that their motion is taking place at small Reynolds numbers. Thus, if both translational and angular velocities of a particle at each instant of time are given, the hydrodynamic force acting on the particle can be found from the Stokes equations. In this type of problems, one usually assumes that the disperse phase has a low volume concentration, so that it becomes permissible to consider pair interactions only. In the frame of reference placed at the center of the larger particle S_1, the Stokes equations with the corresponding boundary conditions are

$$\nabla \cdot \boldsymbol{u} = 0; \quad \mu_e \nabla \boldsymbol{u} = \Delta p, \tag{6.76}$$

$$\boldsymbol{u}|_{S_i} = v_i + \boldsymbol{\Omega}_i \times (\boldsymbol{r} - \boldsymbol{r}_i), \quad |\boldsymbol{r} - \boldsymbol{r}_i| = R_i; \quad (i = 1, 2), \tag{6.77}$$

$$\boldsymbol{u} \to \boldsymbol{U}^{\infty} = (-U^{\infty} \cos\theta, \quad U^{\infty} \sin\theta) \quad \text{at} \quad r \to \infty,$$

where $\mathbf{v}_i$ is translational velocity of the center of the particle S_i, $\boldsymbol{U}^{\infty}$ – velocity of the carrier fluid far from the particle (further on, we shall simply take $U = const$), $\boldsymbol{\Omega}_i$ – angular (rotational) velocity of the particle S_i, $\boldsymbol{r}_i$ – radius vector of the particle center, and $\boldsymbol{R}_i$ – particle radius.

The force and torque exerted on the particle by the surrounding fluid are, respectively [24],

$$\boldsymbol{F}_{ih} = \iint_{S_i} d\boldsymbol{s} \cdot (p\boldsymbol{I} + 2\mu_e \boldsymbol{E}); \quad \boldsymbol{L}_{ih} = \iint_{S_i} (r - r_i) \times (-p\boldsymbol{I} + 2\mu_e \boldsymbol{E}) \cdot d\boldsymbol{s}, \tag{6.78}$$

where E is the rate-of-strain tensor, I is the unit tensor, and $d\boldsymbol{s} = \boldsymbol{n}ds$ is the surface element pointing outward (i.e., toward the fluid).

Hence the solution of the boundary problem (6.76)–(6.77) gives $\boldsymbol{F}^0_{ih}$ and $\boldsymbol{L}^0_{ih}$ as functions of particle sizes and velocities, interparticle distance and viscosity μ_e of the external fluid. If the particles are droplets with mobile surfaces, then in addition to the listed parameters there appears one extra parameter, namely, the internal fluid viscosity μ_i:

$$\begin{aligned} \boldsymbol{F}_{ih} &= \boldsymbol{F}_i(R_1, R_2, \boldsymbol{v}_1, \boldsymbol{v}_2, \boldsymbol{\Omega}_1, \boldsymbol{\Omega}_2, \boldsymbol{r}_2 - \boldsymbol{r}_1, \mu_e, \mu_i), \\ \boldsymbol{L}_{ih} &= \boldsymbol{L}_i(R_1, R_2, \boldsymbol{v}_1, \boldsymbol{v}_2, \boldsymbol{\Omega}_1, \boldsymbol{\Omega}_2, \boldsymbol{r}_2 - \boldsymbol{r}_1, \mu_e, \mu_i). \end{aligned} \tag{6.79}$$

Linearity of the Stokes equations implies that the sought-for expressions for $\boldsymbol{F}_{ih}$ and $\boldsymbol{L}_{ih}$ can be found by superposition and approximation of particular solutions of two well-known problems: motion of particle S_2 relative to particle S_1 in a quiescent fluid; and bypassing of two immovable particles S_2 and S_1 spaced at a certain distance from each other by a flow whose velocity U at the infinity is given. We then conclude that the force $\boldsymbol{F}_h$ and torque $\boldsymbol{L}_h$ can be written as

$$\boldsymbol{F}_h = \boldsymbol{F}_s + \boldsymbol{F}_e; \quad \boldsymbol{L}_h = \boldsymbol{L}_s + \boldsymbol{L}_e. \tag{6.80}$$

The terms $\boldsymbol{F}_s$ and $\boldsymbol{L}_s$ are Stokesian components which correspond to a fixed spherical particle S_1 under the conditions $\boldsymbol{v}_1 = \Omega_1 = 0$ and $\boldsymbol{U} \neq 0$. The terms $\boldsymbol{F}_e$ and $\boldsymbol{L}_e$ are the respective contributions to the hydrodynamic force and torque from the motion of particle S_2 (needless to say, in this case $\boldsymbol{v}_2 \neq 0$ and $\Omega_2 \neq 0$) in a fluid whose velocity at the infinity is zero: $U = 0$. The terms $\boldsymbol{F}_s$ and $\boldsymbol{L}_s$ can be determined as follows. Denote through $\boldsymbol{u}_0$ the velocity of Stokesian flow that bypasses an isolated particle $\boldsymbol{S}_1$. Velocity components in the meridional section ($\Phi = const$) are [9]

$$u_{0r} = U^{\infty} \cos\theta \left(\frac{3R_1}{2r} - \frac{R_1^3}{2r^3} - 1 \right); \quad u_{0\theta} = U^{\infty} \sin\theta \left(\frac{3R_1}{4r} - \frac{R_1^3}{4r^3} + 1 \right). \tag{6.81}$$

The characteristic distance at which the Stokesian velocity field (6.81) experiences a noticeable change is equal to R. Thus as long as the region under consideration is much smaller than R, the flow in the meridional section can be considered as an approximately quasi-planar flow resulting from the superposition of a uniform flow and a simple shear flow. The uniform flow induces the force $\boldsymbol{F}_s$ and the shear flow induces the torque $\boldsymbol{L}_s$ acting on the particle S_2. On a sufficiently large distance from S_2 they are equal to

$$\boldsymbol{F}_s = 6\pi\mu_e R_2 \boldsymbol{u}_0; \quad \boldsymbol{L}_s = 4\pi\mu_e R_2^3 \nabla \times \boldsymbol{u}_0. \tag{6.82}$$

A decrease of the gap δ between the particles S_1 and S_1 would cause a violation of the dependences (6.82). For narrow gaps between the particles, distortion of the velocity field can be taken into account by introducing the corresponding resistance coefficients into Eq. (6.82). While doing so, it is necessary to keep in mind that the form of resistance coefficients is different for the motion of the particle S_2 along and perpendicular to the surface S_1. Thus

$$F_{sr} = 6\pi\mu_e R_2 f_{sr} u_{0r}; \quad F_{s\theta} = 6\pi\mu_e R_2 f_{s\theta} u_{0\theta}; \quad L_{s\varphi} = 8\pi\mu_e R_2^3 t_{s\varphi} G, \tag{6.83}$$

where f_{sr} and $f_{s\theta}$ are translational resistance coefficients that correspond to the motion perpendicular and parallel to the surface S_1, $t_{s\varphi}$ is a rotational resistance coefficient, and $\mathbf{G} = 0.5|\times \mathbf{u}_0$. Resistance coefficients in Eq. (6.83) depend on the relative gap between the particles, $s = (r - R_1 - R_2)/(R_1 + R_2)$, and on the ratio of particle radii, $k = R_2/R_1$.

As evidenced by the solutions of the corresponding hydrodynamic problems [24–36], the coefficients f_{sr}, $f_{s\theta}$, and $t_{s\varphi}$ are not much different from unity when the distance from S_1 is on the order of several radii of S_2 or smaller, and remain finite as $\delta \to 0$. Since the analogous coefficients in the expressions for $\boldsymbol{F}_e$ and $\boldsymbol{L}_e$ increase unboundedly as $\delta \to 0$, we can claim without sacrifice of precision that f_{sr}, $f_{s\theta}$, and $t_{s\varphi} \approx 1$.

Consider now the hydrodynamic force and torque resulting from particle S_2's own motion. This problem has been extensively studied, and the currently available solutions embrace practically all types of particle's motion: along the line of centers; perpendicular to the line of centers in the meridional plane; rotation about an axis that is perpendicular to the meridional plane. In the case when one spherical particle moves in the vicinity of another particle or a plane, that is, when the gap is small compared to the smaller particle's radius, the obtained numerical results are supplemented by near asymptotic and far asymptotic relations. The three types of particle motion we just mentioned give the following expressions for the components of force $\boldsymbol{F}_e$ in the meridional plane and for the torque $\boldsymbol{L}_e$ acting on the particle:

$$F_{er} = -6\pi\mu_e R_2 f_{er} v_r, \quad F_{e\theta} = -6\pi\mu_e R_2 f_{e\theta} v_\theta + 6\pi\mu_e R_2^2 f_{e\theta_1} \Omega,$$
$$L_{e\varphi} = 8\pi\mu_e R_2^2 t_{e\varphi} v_\varphi - 8\pi\mu_e R_2^3 t_{e\varphi_1} \Omega. \tag{6.84}$$

Listed below are asymptotical expressions for the resistance coefficients of solid spherical particles written in dimensionless variables – the distance between the particle centers $x_1 = (r - R_2)/R_1$ and the ratio of particle radii $k = R_2/R_1$:

– the near asymptotic $1 < x_1 < 1.5$ for $k \ll 1$:

$$f_{e\theta} \approx -\frac{2(2 + k + 2k^2)}{15(1 + k)^3} \ln(x_1 - 1) + 0.959,$$

$$f_{e\theta_1} \approx -\frac{2(1 + 4k)}{15(1 + k)^2} \ln(x_1 - 1) - 0.2526,$$

$$f_{e\varphi} \approx -\frac{1 + 4k}{10(1 + k)^2} \ln(x_1 - 1) + 0.29, \tag{6.85}$$

$$f_{e\varphi_1} \approx -\frac{2}{5(1 + k)^2} \ln(x_1 - 1) + 0.3817,$$

$$f_{er} \approx -\frac{1}{(1 + k)^2 (x_1 - 1)} - \frac{1 + 7k + k^2}{5(1 + k)} \ln(x_1 - 1) + 0.97.$$

– the far asymptotic $x_1 \gg 1/k$ for $k \ll 1$:

$$f_{e\theta} \approx \left(1 - \frac{9}{15x_1} + \frac{1}{8x_1^3}\right)^{-1}, \quad f_{e\theta_1} \approx \frac{1}{8x_1^4}, \quad f_{e\varphi} \approx \frac{1}{32x_1^4},$$
$$f_{e\varphi_1} \approx 1 + \frac{5}{16x_1}, \quad f_{er} \approx 1 + \frac{1.125}{15x_1} + \frac{1.266}{x_1^2} \tag{6.86}$$

– finally, we have for $k < 1$:

$$f_{er} \approx \frac{1}{1 - 2.25(\boldsymbol{k}\boldsymbol{x}_1 + 1)^{-2}}, \quad f_{e\theta} \approx \frac{1}{1 - 0.56\boldsymbol{k}(\boldsymbol{k}\boldsymbol{x}_1 + 1)^{-2}},$$
$$\boldsymbol{t}_{e\varphi} \approx \frac{0.56\boldsymbol{k}^2(\boldsymbol{k}\boldsymbol{x}_1 + 1)^{-3}}{1 - 0.56\boldsymbol{k}(\boldsymbol{k}\boldsymbol{x}_1 + 1)^{-2}}. \tag{6.87}$$

In the intermediate region, the resistance coefficients are presented in the form of infinite series, and their numerical values can be found from tables.

An analysis of relative motion of particles along their line of centers at low Reynolds numbers has been carried out in [35]. The approximate expression for the resistance coefficient of the fluid particle S_2 is

$$h = 6\pi\mu_e R_2 f_{er} \approx 6\pi\mu_e \left(\frac{R_1 R_2}{R_1 + R_2}\right)^2 \frac{f(m)}{\delta}, \tag{6.88}$$

where $\delta = r - R_1 - R_2$ is the gap between the particles and

$$f(\boldsymbol{m}) = \frac{1 + 0.402\boldsymbol{m}}{1 + 1.711\boldsymbol{m} + 0.461\boldsymbol{m}^2}; \quad \boldsymbol{m} = \frac{1}{\bar{\mu}}\left(\frac{\boldsymbol{R}_1\boldsymbol{R}_2}{\boldsymbol{R}_1 + \boldsymbol{R}_2}\right)^{1/2}; \quad \bar{\mu} = \frac{\mu_i}{\mu_e}.$$

The parameter m characterizes mobility of the particle surface. For $m \ll 1$, the surface is, for all practical purposes, fully retarded, and the particle's hydrodynamic behavior is that of a solid particle with resistance coefficient

$$h \approx 6\pi\mu_e \left(\frac{R_1 R_2}{R_1 + R_2}\right)^2 \frac{1}{\delta}. \tag{6.89}$$

This expression coincides with the first term of the asymptotic expansion (6.85) for f_{er}. One can see from the form of the resistance coefficient that it has a non-integrable singularity at $\delta \to 0$. Therefore particles with fully retarded surfaces cannot come into contact under the action of a finite force during any finite time interval.

The other limiting case, $m \gg 1$, corresponds to a completely mobile surface. The particle behaves like a gas bubble, and its resistance coefficient is

$$h \approx 6\pi\mu_e \left(\frac{R_1 R_2}{R_1 + R_2}\right)^2 \frac{0.872\bar{\mu}}{\delta^{1/2}} \left(\frac{R_1 + R_2}{R_1 R_2}\right)^{1/2}. \qquad (6.90)$$

At $\delta \to 0$, the expression for h for a particle with a mobile surface has an integrable singularity. Therefore in a laminar flow (in particular, during gravitational sedimentation of particles), particles can come into contact even in the absence of the force of molecular attraction. In a turbulent flow, the coefficient of turbulent diffusion is $D_t \sim 1/h^2$, so in the absence of molecular attraction, any contact of particles with either immobile or fully mobile surfaces is impossible.

Paper [25] derives an asymptotic expression for the resistance coefficient for the case of relative motion of two spherical particles ($\delta \to 0$) with different inner viscosities $\mu_i^{(1)}$ and $\mu_i^{(2)}$ suspended in a fluid whose viscosity is μ_e:

$$h \approx \frac{\pi^2\, k^{1/2}(\mu_i^{(1)} + \mu_i^{(2)})}{16\,(1+k)^2(s-2)^{1/2}}, \qquad (6.91)$$

where $s = 2r/(R_1 + R_2), \bar{\mu}_i^{(1,2)} = \mu_i^{(1,2)}/\mu_e$.

If one of the particles has a fully retarded surface, which corresponds to the case of relative motion of a droplet and a solid particle, the asymptotic expression for the particle's resistance coefficient at $\delta \to 0$ is

$$h \approx \frac{k}{2(1+k)^3(s-2)} + \frac{9\pi^2 k\bar{\mu}_i}{64(1+k)^2(s-2)^{1/2}}. \qquad (6.92)$$

Consider now the forces of molecular and electrostatic interactions between particles. In the absence of external forces (gravitational, centrifugal, electrical), uncharged particles and droplets dispersed in a quiescent fluid should be distributed homogeneously in space. But interaction always exists between particles, even in a quiescent fluid: molecular attraction (van der Waals–London forces) and electrostatic repulsion (for charged particles or for particles enveloped by electric double layers). Forces of electrostatic repulsion between particles with like charges help ensure a homogeneous distribution of particles. The ability of a system to maintain a homogeneous distribution of particles in the fluid for a long time characterizes its stability.

In practice, in the majority of two-phase disperse systems, the number of particles decreases with time, while the particle size gets larger. Collision of two particles in an emulsion results in their coalescence with the formation of a single large particle. Colliding solid particles can form aggregates. The phenomena of coagulation, coalescence, and aggregation of particles occur due to the presence of van der Waals–London attractive forces. System in which aggregation, coagulation, or coalescence is taking place are called unstable. On the other hand, systems devoid of these processes are called stable.

In the absence of external and hydrodynamic forces, stability of a disperse system depends on particle interactions that are due to surface forces: molecular attraction and electrostatic repulsion [37].

The potential of electrostatic interaction between two spheres of radii $\boldsymbol{R}_1$ and $\boldsymbol{R}_2$ ($R_1 > R_2$), whose centers are separated by the distance r is equal to [38]

$$V_R^s = \frac{\varepsilon R_1 \phi_1^2 R_2}{4(R_1+R_2)}\left(-2\frac{\phi_2}{\phi_1}\ln\left(\frac{1-e^{-\chi h_0}}{1+e^{-\kappa h_0}}\right)+\left(1+\frac{\phi_2^2}{\phi_1^2}\right)\ln(1-e^{-2\chi h_0})\right), \tag{6.93}$$

where ϕ_1 and ϕ_2 are the potentials of particle surfaces, χ is the inverse Debye radius, and $\boldsymbol{h}_0 = \boldsymbol{r} - \boldsymbol{R}_1 - \boldsymbol{R}_2$.

The force of electrostatic repulsion between two particles is found from the following relation:

$$F_R^s = -\frac{dV_R}{dr} = \frac{\varepsilon R_1 \phi_1^2 \chi}{2} f_R, \tag{6.94}$$

$$f_R = \frac{\boldsymbol{k}}{\boldsymbol{k}+1}\frac{e^{-a}\varepsilon}{1-e^{-2a}}\left(2\frac{\phi_2}{\phi_1}-\left(1+\frac{\phi_2^2}{\phi_1^2}\right)e^{-a}\right), \quad \boldsymbol{a} = 0.5\boldsymbol{R}_1\chi(1+\boldsymbol{k})\left(\frac{2\boldsymbol{r}}{\boldsymbol{R}_1+\boldsymbol{R}_2}\right).$$

This relation leads us to a frequently-used formula for the force of electrostatic repulsion between two identical particles of radius R with equal surface potentials ϕ_0:

$$F_R^s = \frac{\varepsilon R \phi_0^2 \chi}{2(1+e^a)}. \tag{6.95}$$

The forces of molecular attraction between two parallel plates and between two identical spherical particles were derived in [39]. For parallel plates, the specific interaction potential is

$$V_A^p = -\frac{\Gamma}{12\pi h^2}, \tag{6.96}$$

where h is the distance between the plates and Γ is the Hamaker constant equal to 10^{-20}–10^{-10} J. For two identical spherical particles of radii R under the condition $h_0 \ll R$, where h_0 is the minimum distance between their surfaces, this potential is equal to

$$V_A^s = -\frac{R\Gamma}{12h_0}. \tag{6.97}$$

The formula (6.97) can be generalized for the case of an arbitrary distance r between the particle centers:

$$V_A^s = -\frac{\Gamma}{6}\left(\frac{2R^2}{r^2-4R^2}+\frac{2R^2}{r^2}+\ln\left(\frac{r^2-4R^2}{r^2}\right)\right) \tag{6.98}$$

and for the case of different radii R_1 and R_2:

$$V_A^s = -\frac{\Gamma}{6}\left(\frac{8k}{(s^2-4)(1+k)^2}+\frac{8k}{s^2(k+1)^2-4(1-k)^2}+\ln\left(\frac{(s^2-4)(1+k)^2}{s^2(k+1)^2-4(1-k)^2}\right)\right), \tag{6.99}$$

where $s = 2r/(R_1 + R_2)$; $k = R_2/R_1$.

The force of molecular interaction between two arbitrary spherical particles is equal to

$$F_A^s = -\frac{dV_A^s}{dr} = -\frac{2\Gamma}{3R_1} f_A, \tag{6.100}$$

$$f_A = -\frac{1}{(1+k)}\left(\frac{s}{(s^2-4)(1+k)^2}(8k-(1+k)^2(s^2-4)) + \frac{s(1+k)^2}{(s^2(1+k)^2-4(1-k)^2)^2}(8k+s^2(1+k)^2-4(1-k)^2\right).$$

An approximate expression for the force of molecular interaction that takes electromagnetic retardation into account (leading to a decrease of the van der Waals force) is presented in [40] and has the form

$$F_A^s = \frac{2\Gamma}{3R_1} f_A \Phi(p), \tag{6.101}$$

$$\Phi(p) = \begin{cases} \dfrac{1}{1+1.77p} & \text{at} \quad p \le 0.57, \\ -\dfrac{2.45}{60p}+\dfrac{2.17}{120p^2}-\dfrac{0.59}{420p^3} & \text{at} \quad p > 0.57, \end{cases}$$

where $p = 2\pi\ (r - R_1 - R_2)\lambda_L$, and $\lambda_L \sim 10^3$ A? is the London wavelength.

The total potential energy of interaction between two spherical particles is equal to the sum of electrostatic and molecular potentials:

$$V = V_A + V_R. \tag{6.102}$$

The energy of repulsion decreases exponentially with increase of h_0 with the characteristic linear scale of energy variation λ_D. The energy of attraction falls off as $1/h_0$. Attraction therefore prevails on relatively small and relatively large distances between the particles, whereas repulsion dominates in the intermediate distance range (Fig. 6.4).

Coagulation can be fast or slow. We speak of fast coagulation when only molecular interaction is considered. If, in addition, electrostatic repulsion is taken into account, we call such coagulation slow. It is a known fact that the electrolyte concentration C_0 at which fast coagulation is initiated depends on the charge of counterions, that is, ions carrying the charge opposite in sign to the charge of particles. On the other hand, stability of the system is practically independent of the charge of ions or concentration of particles. This fact is in agreement with the Schulze–Hardy rule, according to which the valence of counterions is the main factor influencing the system's stability. The values of molecular and electrostatic potentials of interparticle interaction for the boundary state of the system that separates stable states from unstable ones are determined from the following system of equations:

$$V = V_A + V_R = 0, \tag{6.103}$$

$$\frac{dV}{dr} = \frac{dV_A}{dr} + \frac{dV_R}{dr} = 0. \tag{6.104}$$

If the system is completely instable, in other words, if repulsion forces are not taken into account, each collision of two particles results in their coagulation. The presence of a stabilizer (electrolyte) in the system leads to the emergence of

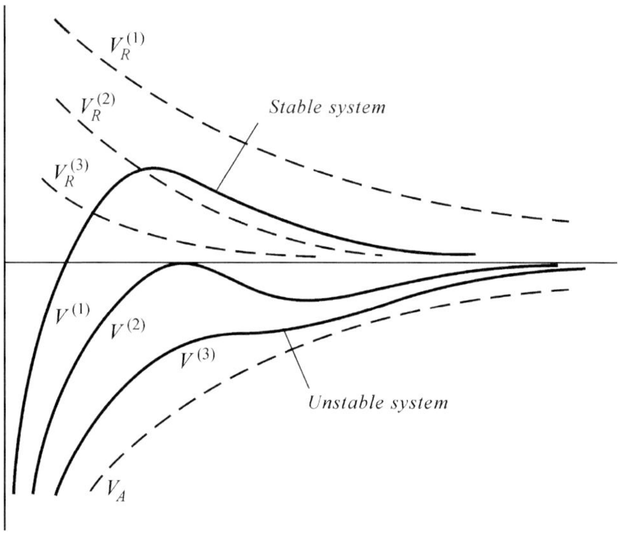

Fig. 6.4 Potential energy of particle interaction.

repulsion forces created by double electric layers on particle surfaces. Consequently, the rate of coagulation decreases, hence the name "slow coagulation".

The rate of coagulation is characterized by the stability factor W equal to the ratio between the number of particle collisions in the presence and in the absence of electrostatic repulsion [41]:

$$W = \frac{I}{I_R}. \tag{6.105}$$

We then have $I_R = I$ and $W = 1$ for fast coagulation, and $I_R < I$ and $W > 1$ for slow coagulation.

6.5
Conducting Particles in an Electric Field

Consider the behavior of conducting particles in an external electric field. When the particles are relatively far from each other, the influence of neighboring particles is small and our analysis can be restricted to a single particle in an unbounded fluid. Smaller interparticle distances would cause a distortion of the external electric field near the particle surface, which could noticeably affect the shape of a deformable particle (droplet).

If the internal and external fluids are both ideal dielectrics and there are no free charges at the interface, or if the internal fluid is highly conductive and the external fluid is an insulator, the external electric field gives rise to a force distributed over the particle surface; this force emerges because the electric field is discontinuous at the interface [42]. This force is perpendicular to the interface and is directed away from the fluid with higher dielectric permittivity (conducting fluid) toward the fluid with lower permittivity (insulator). For the equilibrium shape of a motionless droplet in a quiescent fluid to be stable, it is necessary that the electrostatic surface force should be equal to the surface tension force. As a result, at static conditions, the shape of a droplet is that of a body of revolution – a prolate ellipsoid stretched along the direction of the external electric field.

The theory of static equilibrium of droplets in an electric field (electrohydrostatics) is most thoroughly developed for ideal media (dielectrics and conductors) [43–49]. But the real fluids are media with finite conductivity and finite dielectric permittivity. The only exception is superconductivity, which takes place at very low temperatures for such fluids as, for example, liquid helium. When we take conductivity to be finite, the problem becomes very complicated both mathematically and physically, because the possible shapes of droplets are different from those of ideal conducting droplets. Thus a droplet may assume the shape of a prolate ellipsoid stretched in the direction parallel or perpendicular to the direction of the external electric field, or even be spherical [50]. A theoretical explanation of these phenomena is given in [51]. It is shown that a droplet with a finite conductivity accumulates electric charge in its surface layer, thereby giving rise to a non-uniform surface tangential electrical stress. This stress, in its turn, induces tangential hydrodynamic stresses in both the internal and external fluids, which affects droplet deformation. The magnitude of these stresses depends on the fluids' properties and on the strength of the external electric field. Hence, depending on the relation between the electrical and hydrodynamic surface stresses, the droplet may take one of above-mentioned shapes. The solution of this problem carried out in [51] takes into account circulation of the internal fluid. In this work the flow is assumed to be slow and Stokesian, and deformations of the droplet surface are assumed to be small; with these assumptions, it becomes possible to get an approximate asymptotic solution.

Let us restrict ourselves to simple estimates. Suppose an ideally conductive droplet is suspended in an ideal dielectric. The motion of internal as well as external fluid will be neglected. The external electric field $\boldsymbol{E}_0$ polarizes and deforms the droplet, which then takes the shape of an ellipsoid whose major axis is parallel to the field. Consider the behavior of an isolated conductive uncharged spherical droplet of radius R freely buoyant in a quiescent dielectric fluid with constant dielectric permittivity ε in the presence of a uniform external electric field with constant strength $\boldsymbol{E}_0$. Since outside the sphere the electric charge is absent, electrostatic potential ϕ obeys the Laplace equation:

$$\Delta\phi = 0. \tag{6.106}$$

The surface of the conductive droplet S_R is equipotential, therefore

$$\phi = 0 \quad \text{at} \quad S_R. \tag{6.107}$$

On large distances from the droplet, the total electric field is equal to the external field:

$$\nabla\phi \to -\boldsymbol{E}_0 \quad \text{at} \quad \boldsymbol{r} \to \infty. \tag{6.108}$$

There is no electric field inside the droplet, and the electric charge induced by the external field is distributed over the surface so that the surface density σ is equal to

$$\sigma = -\frac{1}{2\pi}\frac{\partial\phi}{\partial n}, \tag{6.109}$$

where n is the outward normal to S_R.

The total charge of the droplet is zero, therefore

$$\int\limits_{S_R} \frac{\partial\phi}{\partial n} ds = 0. \tag{6.110}$$

Suppose the droplet is slightly deformed. In the first approximation, its shape can be considered spherical. In spherical coordinates, the Laplace equation (6.106) for a spherically symmetrical problem is written as

$$\sin\theta\frac{\partial}{\partial r}\left(r^2\frac{\partial\phi}{\partial r}\right)+\frac{\partial}{\partial\theta}\left(\sin\theta\frac{\partial\phi}{\partial\theta}\right)=0. \tag{6.111}$$

The boundary conditions are

$$\phi=0 \quad \text{at} \quad r=R; \quad -\frac{\partial\phi}{\partial r}\rightarrow E_0\cos\theta \quad \text{at} \quad r\rightarrow\infty. \tag{6.112}$$

The solution of this boundary value problem is

$$\phi=E_0\left(r-\frac{R^3}{r^2}\right)\cos\theta. \tag{6.113}$$

From Eq. (6.113) we obtain the strength of electric field at the droplet surface:

$$E_R=\left(-\frac{\partial\phi}{\partial r}\right)_{r=R}=3E_0\cos\theta. \tag{6.114}$$

Let us now find the force acting on the conductive sphere. Momentum flux density in an electric field is defined by the Maxwell stress tensor

$$\boldsymbol{t}_{ik}=-\frac{1}{4\pi}\left(\frac{\boldsymbol{E}^2}{2}\delta_{ik}-\boldsymbol{E}_i\boldsymbol{E}_k\right), \tag{6.115}$$

and the force acting on an oriented surface element ds is

$$t_{ik}ds_k=t_{ik}n_k ds.$$

where $t_{ik}n_k$ is the force acting on a unit surface area. At the surface of the droplet, the electrical stress is directed along the normal to the surface, so $E_iE_k=E_R^2\delta_{ik}$. Consequently, the force acting from the inside on a unit surface area of the droplet is equivalent to the pressure

$$\Delta p=\frac{E_R^2}{8\pi}-\frac{9E_0^2}{8\pi}\cos^2\theta. \tag{6.116}$$

This pressure achieves its maximum value at $\theta=0$ and $\theta=\pi$, that is, at the droplet poles lying along a straight line parallel to $\boldsymbol{E}_0$. As a result, the maximum deformation is observed in the vicinity of these points, and the droplet assumes the shape of a prolate ellipsoid stretched in the direction of the external electric field (Fig. 6.5).

The equilibrium condition for the droplet boils down to the equality of two forces: the force that arises from electrical pressure (6.116) and the surface tension force. As long as electrical pressure dominates, the droplet keeps changing its shape until at

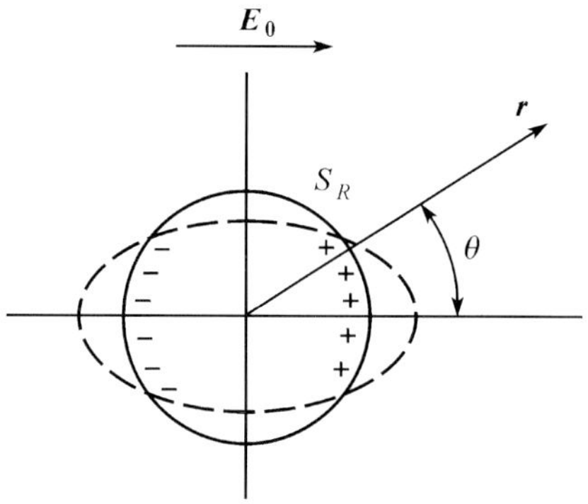

Fig. 6.5 Deformation of a conductive droplet in an electric field.

some point the decrease of the principal curvature radii results in a stronger surface tension that can counterbalance the internal pressure. Note, however, that a considerable deformation of the droplet might cause a loss of stability and a breakup of the droplet. The critical strength E_{cr} of the external electric field can be estimated from the approximate equality

$$\frac{9E_{cr}^2}{8\pi} \sim \frac{2\Sigma}{R}, \tag{6.117}$$

from which it follows that droplet stability is characterized by the dimensionless parameter

$$\kappa^2 = \frac{\text{electric force}}{\text{force of surface tension}} = E_0^2 \frac{R}{\Sigma}. \tag{6.118}$$

The simple estimation (6.117) gives $\kappa_{cr} \sim 1.77$ for the critical value of the parameter κ. A more accurate calculation that takes into account the deformation of the droplet [45,52] gives $\kappa_{cr} = 1.625$. For example, for a water-in-oil emulsion, we have $\sum = 3\cdot 10^{-2}$ N/m, and for a water droplet of 1 cm radius, the critical electric field strength is $\boldsymbol{E}_{cr} = 2.67$ kV/cm. Thus electric fields whose strength is $\boldsymbol{E}_0 < 3$ kV/cm will fail to produce any noticeable deformation of spatially separated droplets of radius $R \ll 1$ cm.

A different situation arises when droplets are located next to each other. Smaller distance between droplets leads to a considerable increase of the local electric field strength. It was shown in [53] that stability of a system of two conductive droplets is defined by the same parameter κ, but in contrast with the case of an isolated droplet, it depends not only on E_0, R, and Σ, but also on the relative clearance δ/R between the particle surfaces. Table 6.1 lists the values of κ_{cr} for different values of δ/R.

From the adduced values of κ_{cr}, one can see that even very small droplets can lose their stability in relatively weak external electric fields as long as they are sufficiently close to each other. However, the relevant distances are so small that forces of

Table 6.1

δ/R	∞	10	1	0.1	0.01	0.001
κ_{cr}	1.625	1.555	0.9889	0.0789	$3.91 \cdot 10^{-3}$	$1.9 \cdot 10^{-4}$

molecular interaction come into play, encouraging capture and coalescence of small droplets formed after breakup of the larger droplets.

Consider now two motionless conducting spherical particles of radii R_1 and R_2 carrying the charges q_1 and q_2 in an external uniform electric field of strength $\boldsymbol{E}_0$ (Fig. 6.6) Denote through θ the angle between the particles' line of centers and the vector $\boldsymbol{E}_0$. The space between the particles is filled with a quiescent isotropic uniform dielectric medium having the permittivity ε. Since there are no free charges outside the spheres, electrostatic potential ϕ in the region outside the spheres obeys the Laplace equation

$$\Delta\phi = 0. \tag{6.119}$$

Far away from the particles, strength of the electric field tends to that of the external field:

$$\boldsymbol{E} = -\nabla\phi \to \boldsymbol{E}_0 = -\nabla\phi_0. \tag{6.120}$$

Since the frame of reference is chosen in such a way that the vector $\boldsymbol{E}_0$ lies in the XOZ-plane, ϕ_0 does not depend on y and is equal to

$$\phi_0 = -E_0(Z\cos\theta + X\sin\theta). \tag{6.121}$$

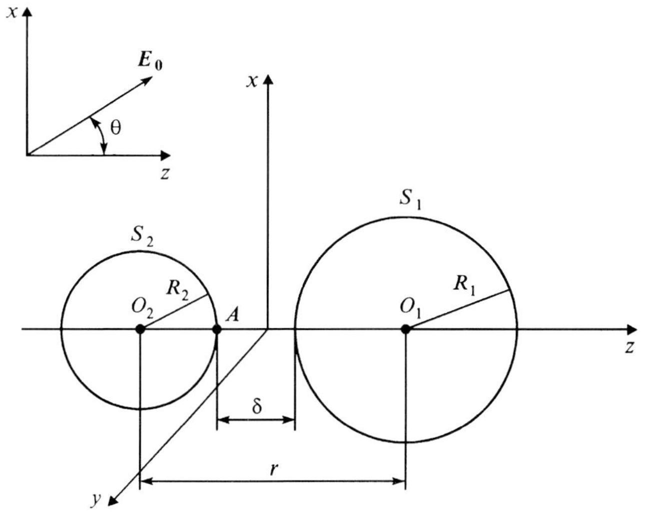

Fig. 6.6 Two conductive particles in an electric field.

The condition of equipotentiality of particle surfaces gives us

$$\phi_{S_1} = V_1, \quad \phi_{S_2} = V_2, \tag{6.122}$$

where V_1 and V_2 are constant surface potentials. In order to determine them, we must use the condition that the charge of a conducting particle is distributed over the particle surface:

$$-\frac{\varepsilon}{4\pi}\int_{S_i}\frac{\partial\phi}{\partial n}ds = q_i \quad (i = 1, 2). \tag{6.123}$$

where n is the outward normal to the particle surface.

The projection of the force acting on particle 2 onto the unit vector $\boldsymbol{p}$ is equal to

$$\boldsymbol{F}\cdot\boldsymbol{p} = \frac{\varepsilon}{8\pi}\int_{S_2}\left(\frac{\partial\phi}{\partial n}\right)^2 \boldsymbol{n}\cdot\boldsymbol{p}ds. \tag{6.124}$$

Without going into details (see [1] for a comprehensive treatment of the problem), we adduce the expression for the electric field strength near the point A in the interparticle region (Fig. 6.6):

$$E_A = \frac{1}{\varepsilon R_2^2}(E_1 q_1 + E_2 q_2) + E_3 E_0 \cos\theta, \tag{6.125}$$

and the longitudinal and transverse (with respect to the line of centers) components of the force acting on particle 2:

$$\begin{aligned} E_{2z} &= \varepsilon R_2^2 E_0^2 (f_1 \cos^2\theta + f_2 \cos^2\theta) + E_0 \cos\theta (f_3 q_1 + f_4 q_2) \\ &\quad + \frac{1}{\varepsilon R_2^2}(f_5 q_1^2 + f_6 q_1 q_2 + f_7 q_2^2) + E_0 q_2 \cos\theta, \end{aligned} \tag{6.126}$$

$$E_{2x} = \varepsilon R_2^2 E_0^2 f_8 \sin 2\theta + E_0 \sin\theta (f_9 q_1 + f_{10} q_2) + E_0 q_2 \sin\theta.$$

The coefficients E_i and f_i are dimensionless quantities that depend on the relative clearance between the particle surfaces $\Delta = (r - R_1 - R_2)/R_1$ and on the ratio of particle radii $k = R_2/R_1$. The dependences of E_3 and f_1 on Δ and k are shown in Fig. 6.7 and Fig. 6.8.

6.6
Coagulation of Particles in a Turbulent Flow

Consider the coagulation of particles in a developed turbulent flow. It is assumed that particles are spherical, non-deformable, and inertialess, and that their size is

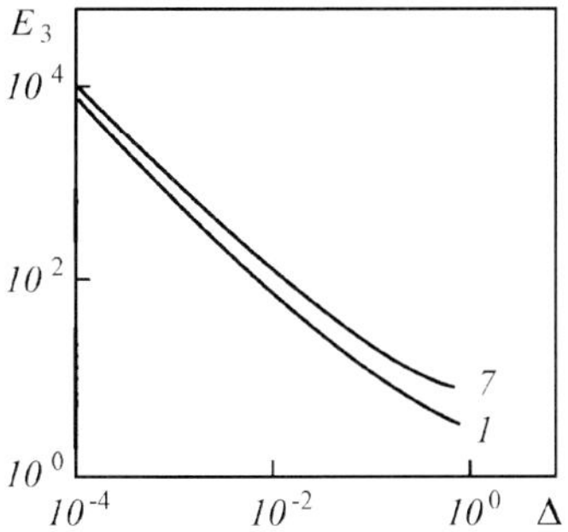

Fig. 6.7 E_3 vs. Δ and k: 1 – $k = 1$; 2 – $k = 0.01$.

much smaller than the inner scale of turbulence, that is, $R \ll \lambda_0$. Under these conditions the particles' interdiffusion coefficient (with hydrodynamic interactions taken into account) is given by the expression (6.70). To determine the frequency of collisions between particles of radii R_1 and R_2 ($R_1 < R_2$), one has to solve the diffusion equation (6.36) with the boundary conditions (6.39). Let us place the origin of a spherical coordinate system (r, θ, Φ) at the center of the larger particle R_1. If the forces of particle interaction possess spherical symmetry, Eq. (6.36) with the conditions (6.39) takes the form

$$\frac{1}{r^2}\frac{d}{dr}\left(r^2\left(D_t\frac{dn_2}{dr} - \frac{F}{h}n_2\right)\right) = 0, \tag{6.127}$$

$$n_2 = 0 \quad \text{at} \quad r = R_1 + R_2; \quad n_2 = n_{20} \quad \text{at} \quad r \to \infty, \tag{6.128}$$

where n_2 is the number concentration of particles of radius R_2, D_t is the interdiffusion coefficient, F is the force of interaction (molecular, electrostatic, hydrodynamic) between the particles, and h is the coefficient of hydrodynamic resistance to particle

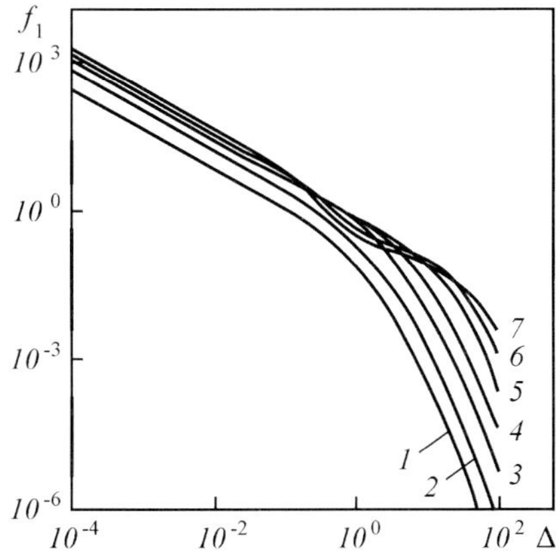

Fig. 6.8 f_1 vs. Δ and k: 1–7 – $k = 1$; 0.5; 0.2; 0.1; 0.05; 0.02; 0.01.

motion. Eq. (6.127) has a general solution

$$n_2(r) = e^{g(r)}\left(C_1 - \int_r^\infty C_2 \frac{e^{-g(\rho)}}{\rho D_t^2(\rho)}\, d\rho\right), \tag{6.129}$$

$$g(\rho) = \int_\rho^\infty C_2 \frac{F(r)}{h(r) D_t(r)}\, dr.$$

The boundary conditions (6.28) enable us to find the constants C_1 and C_2:

$$C_1 = n_{20}; \quad C_2 = n_{20}\left(\int_{R_1+R_2}^\infty \frac{e^{-g(z)}}{z^2 D_t(z)}\, dz\right)^{-1},$$

resulting in the solution

$$n_2(r) = n_{20} e^{g(r)}\left(1 - \int_r^\infty \frac{e^{-g(z)}}{z^2 D_t(z)}\, dz \Bigg/ \int_{R_1+R_2}^\infty \frac{e^{-g(z)}}{z^2 D_t(z)}\, dz\right). \tag{6.130}$$

The diffusion flux of particles 2 toward the test particle 1 is determined from the expression (6.40), which in the spherically symmetric case reduces to

$$j = 4\pi (R_1 + R_2)^2 \left(D_t \frac{dn_2}{dr} - \frac{F}{h} n_2\right). \tag{6.131}$$

Substitution of Eq. (6.130) into Eq. (6.131) gives

$$j(R_1, R_2) = 4\pi n_{20}\left(\int_{R_1+R_2}^\infty \frac{dr}{r^2 D_t(r)} \exp\left(\int_r^\infty \frac{F(z)}{h(z) D_t(z)}\right)\right)^{-1}. \tag{6.132}$$

Let us now consider the interaction of particles, consecutively introducing hydrodynamic, molecular, electrostatic, and electric forces.

We start with the case when particle surface is fully retarded, in other words, particles can be thought of as rigid and non-deformable; coagulation occurs the under the joint action of turbulent fluctuations and forces of molecular attraction. Molecular attraction between two spherical particles is described by Eq. (6.100), which says that this force is defined by the distance between particle surfaces and does not depend on the orientation of the particle pair, that is to say, the force possesses spherical symmetry relative to the center of particle R_1. Since the molecular force manifests itself only when the gap Δ between the particles is small, it is possible to take its asymptotic expression at $\Delta \to 0$:

$$F_A \approx \frac{\Gamma R_1 R_2}{6(R_1+R_2)^3}\frac{1}{\Delta^2}, \tag{6.133}$$

where $\Delta = (r - R_1 - R_2)/(R_1 + R_2)$ is the dimensionless gap between the particle surfaces and Γ is the Hamaker constant.

For the coefficient of hydrodynamic resistance we take the approximation

$$h = h^{(0)}\left(1+\frac{R_1R_2}{(R_1+R_2)^2\Delta}\right), \quad h^{(0)} = 6\pi\nu_e\rho_e\frac{R_1R_2}{R_1+R_2}, \tag{6.134}$$

which is obtained by combining the far asymptotic and near asymptotic expressions for the force of hydrodynamic interaction, when the particles approach each other along the line of centers.

The coefficient of particle interdiffusion is

$$D_t = D_t^{(0)}\left(\frac{h^{(0)}}{h}\right)^2, \tag{6.135}$$

where

$$D_t^{(0)} = \frac{\nu_e}{\lambda_0^2}(R_1+R_2)^2\left(\frac{r-R_1-R_2}{R_1+R_2}+\frac{R_1R_2}{R_1^2+R_2^2-R_1R_2}\right)^2.$$

Eqs. (6.132)–(6.135) yield the dimensionless flux:

$$J = \frac{\lambda_0^2 j(R_1,R_2)}{4\pi\nu_e R_1^3 n_{20}} = (1+k)^3\varphi_1, \tag{6.136}$$

$$\varphi_1^{-1} = \int_0^\infty \frac{(\Delta+\Delta_1)^2}{\Delta^2(1+\Delta)^2(\Delta+\Delta_2)^2}\exp\left(-S_A\int_A^\infty \frac{(\Delta_1+z)dz}{z^3(z+\Delta_2)^3}\right)d\Delta,$$

where $\Delta_1 = R_1R_2/(R_1+R_2)^2, \Delta_2 = R_1R_2/R_1^2 + R_2^2 - R_1R_2$, and $S_A = \Gamma\lambda_0^2/36\pi\rho_e\nu_e^2 \times (R_1+R_2)^3$ is the parameter of molecular interaction. For water–in–oil emulsions, where $\Gamma \sim 10^{-20}$ J, $\lambda_0 \sim 10^{-3}$ m, $\nu_e \sim 5\cdot 10^{-5}\,\mathrm{m^2/s}$, $\rho_e \sim 10^3\,\mathrm{kg/m^3}$, this parameter equals $S_A \sim 4\cdot 10^{-23}\,(R_1+R_2)^{-3}$. For particles with R_1, $R_2 \sim 10\,\mu$m, we get $S_A \sim 10^{-8}$. Smallness of S_A makes it possible to find an asymptotic expression for φ_1. For particles of comparable sizes, the quantity $\Delta \ll (1, \Delta_1, \Delta_2)$ gives the major contribution to the integral in the expression for φ_1^{-1}. Expanding the integrand as a power series of Δ and keeping the dominant terms, we get

$$\varphi_1^{-1} = \int_0^\infty \frac{\Delta_1^2}{\Delta^2\Delta_2^2}\exp\left(-S_A\int_0^\infty \frac{\Delta_1 dz}{z^3\Delta_2^3}\right)d\Delta. \tag{6.137}$$

We can take this integral by using the method of quadratures:

$$\varphi_1^{-1} = \frac{\Delta_1^{3/2}}{\Delta_2}\left(\frac{\pi}{2S_A}\right)^{1/2}. \tag{6.138}$$

Substitution of Eq. (6.138) into Eq. (6.136) gives an asymptotic expression for the diffusion flux at $S_A \ll 1$:

$$j(R_1, R_2) = \frac{2\sqrt{2}\Gamma^{1/2}(R_1+R_2)^{9/2}}{\lambda_0\sqrt{\rho_e R_1 R_2 (R_1^2+R_2^2-R_1R_2)}} n_{20}. \tag{6.139}$$

The obtained expression for the diffusion flux corresponds to pair interaction of particles freely buoyant in the fluid. It is interesting to compare this flux with the flux caused by pair interactions when one of the particles is fixed:

$$j_1(R_1, R_2) = \frac{2\sqrt{2}\Gamma^{1/2}(R_1+R_2)^{11/2}}{\lambda_0\sqrt{\rho_e} R_1^3 R_1^{1/2} R_2^{1/2}} n_{20}. \tag{6.140}$$

The ratio between the fluxes (6.140) and (6.139) is estimated as $j_1/j \sim (R_1^3+R_2^3)R_1^{-3}$ for $R_1 \geq R_2$, with $j_1/j \sim 1$ for $R_1 \gg R_2$ and $j_1/j \sim 2$ for $R_1 \sim R_2$. 2. So the fixed particle approximation is pretty accurate: the resulting flux and consequently the resulting collision frequency is exaggerated (as compared to the case of free particles) by the factor of 2 at most.

The main flaw of the Levich model of turbulent coagulation is that it overstates the particle collision frequency by 1–2 orders of magnitude. We can demonstrate that accounting for the interaction forces in a proper manner would eliminate this disadvantage.

Let j_0 be the diffusion flux in the case of unconstrained particle motion (i.e., in the absence of interactions). Let us take $h = h_0 = 6\pi\rho_e\nu_e R$, $F_A = 0$, and $S_A = 0$. Then Eq. (6.136) yields

$$j_0 = \frac{4\pi\nu_e(R_1, R_2)^3}{\lambda_0^2}\left(\frac{1}{\Delta_2-1}+\frac{2}{(\Delta_2-1)^2}-\frac{2}{(\Delta_2-1)^3}\ln\Delta_2\right)^{-1} n_{20}. \tag{6.141}$$

At $R_1 = R_2$ the flux (6.141) coincides with the flux obtained by Levich.

To estimate the effect of hydrodynamic forces on the frequency of collisions, we compose the ratio j_0/j, where j_0 obeys Eq. (6.141) and j obeys Eq. (6.139) for identical particles $R_1 = R_2 = R$. Then

$$\frac{j_0}{j} = \frac{3\sqrt{2}\pi\nu_e}{\Gamma^{1/2}\lambda_0} R^{3/2}. \tag{6.142}$$

Taking the parameter values that are typical for water–in–oil emulsions, $\lambda_0 \sim 10^{-3} - 10^{-4}$ m, $R \sim 10^{-5}$ m, $\nu_e \sim 10^{-5}$ m^2/s, $\rho_e \sim 10^{-3}$ kg/m^3, $\Gamma \sim 10^{-20}$ J, we get

$j_0/j \sim 10 - 10^2$. It means that the hydrodynamic resistance of particles reduces the collision frequency by 1–2 orders of magnitude, which is consistent with experimental data; thus the main disadvantage of the Levich model is eliminated.

The effect of hydrodynamic interaction on the diffusion flux decreases as the particle size gets smaller. For $R \sim 10^{-8}$ m, we have $j_0/j \sim 1$, and for $R \leq 10^{-8}$ m, hydrodynamic interaction becomes insignificant. For particles that small, the range of action of molecular forces exceeds the particle radius.

It should be noted that in the case of Brownian coagulation, hydrodynamic interactions also reduce the collision frequency, though but as strongly (the attenuation factor is 1.5–2 for Brownian coagulation vs. $10-10^2$ for turbulent coagulation). There are two factors that explain why the effect of hydrodynamic interactions on the collision frequency is so different in these two cases: the difference in characteristic particle sizes (particles involved in Brownian motion are much smaller than those involved in turbulent motion) and different character of dependence of diffusion coefficients on the hydrodynamic resistance coefficient ($D_{br} \sim h^{-1}$ for Brownian motion and $D_t \sim h^{-2}$ for turbulent motion).

The asymptotic expression (6.139) for diffusion flux is suitable for particles of roughly the same size. For particles whose sizes differ by a lot ($k \to 0$), we have the following expression:

$$\varphi_1^{-1} = \int_0^\infty \frac{1}{\Delta^2} \exp\left(-S_A \int_0^\infty \frac{dz}{z^4}\right) d\Delta = \frac{3^{1/3}\Gamma(4/3)\lambda_0^2}{S_A^{1/3}\nu_e(R_1+R_2)^3}, \tag{6.143}$$

where $\Gamma(x)$ is the gamma function.

The dependence (6.136) of φ_1 on $k = R_2/R_1$ and S_A is shown in Fig. 6.9. The dotted line illustrates the corresponding asymptotic dependence (6.143).

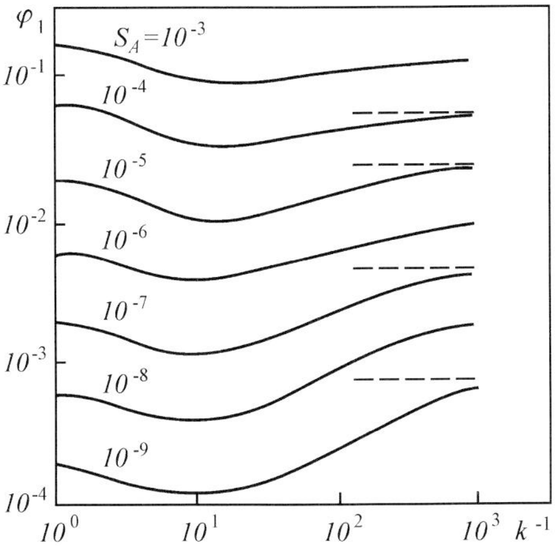

Fig. 6.9 Dependence φ_1 on k and S_A.

Going back to Eq. (6.136), we get the following asymptotic expression for the diffusion flux of small particles toward a larger particle ($R_2 \ll R_1$):

$$j(R_1, R_2) = \frac{2^{4/3}\pi}{3\Gamma(4/3)}(R_1 + R_2)^2 \left(\frac{\Gamma \nu_e}{\pi \rho_e \lambda_0^4}\right)^{1/3} n_{20}. \tag{6.144}$$

Hence, for particles of different sizes the diffusion flux turns out to be proportional to the surface of the larger particle, not to the third power of radius as is the case for particles of comparable sizes.

Let us now find the coagulation kernel $K(V, \omega)$. When particles coagulate via diffusion, the coagulation kernel is equal to the flux of particles of radius R_2 (at a unit concentration) toward the particle of radius R_1. Making use of equations (6.139) and (6.144) for the fluxes of particles having the same size and different sizes, respectively, we get

$$K_1^0(V, \omega) = \frac{\sqrt{2}\Gamma^{1/2}(V^{1/3} + \omega^{1/3})^{9/2}(V\omega)^{-1/6}}{\lambda_0\sqrt{3\rho_e \pi}(V^{2/3} + \omega^{2/3} - V^{1/3}\omega^{1/3})} \quad \text{at} \quad V \sim \omega, \tag{6.145}$$

$$K_2^0(V, \omega) = \frac{1}{3^{1/3}\Gamma(4/3)}(V^{1/3} + \omega^{1/3})^2 \left(\frac{\Gamma \nu_e}{\pi \rho_e \lambda_0^4}\right)^{1/3} \quad \text{at} \quad V \gg \omega. \tag{6.146}$$

We can match the asymptotic expressions (6.145) and (6.146) by introducing a weight function $\alpha = 4V\omega/(V + \omega)$:

$$K^0(V, \omega) = \alpha K_1^0(V, \omega) + (1 - \alpha)K_2^0(V, \omega). \tag{6.147}$$

Because the coagulation kernel (6.147) is inconvenient to use in calculations, we replace it with the approximate expression

$$K^0(V, \omega) = 16\left[\frac{\Gamma}{3\pi\rho_e\lambda_0^4}\right]^{1/2}(V^{1/3}\omega^{1/6} + V^{1/6}\omega^{1/3}). \tag{6.148}$$

that has the form (6.16) and easily renders itself to the method of interpolation of fractional moments, which is the preferred way to solve the kinetic equation of coagulation [1].

Determination of collision frequency for particles with mobile surfaces (droplets, bubbles) having viscosity μ_i that is different from the viscosity of the carrier medium μ_e is carried out in a similar manner [1]. One major difference from the case considered above is the ability of the particle (droplet) surface to deform, which is especially noticeable when particles come close together (to the distances that are small compared to their sizes). The other difference is in the form of the resistance coefficient. If the particles are located far from each other, the coefficient of hydrodynamic resistance $h^{(0)}$ that pertains to the particles' relative motion is given by Eq. (6.71) in which the two coefficients h_1 and h_2 are determined from the Hadamar–Rybczynski formula

$$h^{(0)} = \frac{h_1^{(0)} h_2^{(0)}}{h_1^{(0)} + h_2^{(0)}} = 6\pi\mu_e \frac{R_1 R_2}{R_1 + R_2} \frac{2 + 3\bar{\mu}}{3 + 3\bar{\mu}}. \tag{6.149}$$

When the gap δ between the droplets is small, one can use the following asymptotic expressions (see Eq. (6.188)):

$$h_\delta = 6\pi\mu_e \left(\frac{R_1 R_2}{R_1 + R_2}\right)^2 \frac{f(m)}{\delta}, \quad f(m) = \frac{1 + 0.402m}{1 + 1.711m + 0.461m^2}, \tag{6.150}$$

$$m = \frac{1}{\bar{\mu}} \left(\frac{R_1 R_2}{R_1 + R_2}\right)^{1/2}; \quad \delta = r - R_1 - R_2,$$

where we just introduced a dimensionless parameter $\bar{\mu} = \mu_i/\mu_e$. The case $\bar{\mu} \to \infty$ corresponds to droplets with fully retarded surfaces (rigid particles) and the case $\bar{\mu} = 0$ – to gas bubbles.

The resistance coefficient can be approximated by the expression

$$h = h^{(0)} \left[1 + \frac{R_1 R_2}{R_1 + R_2} \frac{3 + 3\bar{\mu}}{2 + 3\bar{\mu}} \frac{f(m)}{\delta}\right], \tag{6.151}$$

which correctly describes the behavior of h for large and small gaps between the particles.

We can introduce a modified capillary number

$$\mathrm{Ca} = \frac{\mu_e u R_1 R_2}{\Sigma(R_1 + R_2)\delta}, \tag{6.152}$$

which tells us how strongly the particles (droplets) are deformed. Here Σ is the surface tension coefficient of a droplet, and δ – the clearance between droplet surfaces; droplets approach each other with the speed u.

If $\mathrm{Ca} \ll 1$, deformation of the droplet surface is negligible. But at $\delta \to 0$ this condition may no longer be true. However, at such narrow gaps between droplet surfaces, the force of molecular attraction becomes important, which helps to promote coalescence of droplets at the final stage. Therefore surface deformation of interacting droplets can be noticeable only at the final stage of their approach and while it can reduce the rate of coalescence, it does not essentially affect the collision frequency of particles.

In the case under consideration, with molecular and electrostatic interaction forces taken into account, the dimensionless diffusion flux of particles 2 toward particle 1 is given by

$$J = \frac{2\lambda_0^2 j}{\pi \nu_e R_1^3 n_{20}} = (1+k)^3 \varphi_1; \tag{6.153}$$

$$\varphi_1^{-1} = \int_2^\infty \frac{h^2(s)}{s^2(s-\beta)^2} \exp\left[\frac{1+\bar{\mu}}{k(2+3\bar{\mu})} \int_s^\infty \frac{h(x)}{(x-\beta)^2}(-S_A \Phi(p) f_A(x) + S_R \tau f_R(x)) dx\right] ds.$$

The following dimensionless parameters have been introduced: $S_A = 2\Gamma\lambda_0^2/3\pi R_1^3 \nu_e^2 \rho_e$ – the parameter of molecular interaction; $S_R = \varepsilon \phi_1^2 \lambda_0^2 / 2\pi R_1^3 \nu_e^2 \rho_e$ – the parameter of electrostatic interaction; $\tau = R_1 \chi = R_1/\lambda_D$ – the ratio of the droplet radius to the thickness of the double layer on the droplet surface ($\tau \gg 1$); $\bar{\mu} = \mu_i/\mu_e$ – the ratio of viscosity coefficients of the internal end external fluids; $p = \gamma(1+k)(s-2)$; $\gamma = \pi R_1/\lambda_L$; $\beta = 2(1-k)^2/(1+-k^2-k)$.

A detailed analysis of the numerical solution and the appropriate results are presented in [1].

Let us now determine the collision frequency of uncharged conductive spherical particles in a turbulent flow of dielectric fluid in the presence of a uniform external electric field. It is assumed that the turbulent flow is developed and that particle sizes are much smaller than the inner scale of turbulence. The particles are supposed to be non-deformable, which is possible when external electric field strength E_0 does not exceed the critical value E_{cr} and the particle size is sufficiently small. Under these conditions, the coefficient of mutual diffusion of particles of two types 1 and 2 in the presence of hydrodynamic interactions can be taken in the form (6.135), where the drag coefficients h and $h^{(0)}$ are approximated by the expressions (6.134) that correspond to particles with fully retarded surfaces. We shall also take into account molecular and electric interactions of particles.

Consider the relative motion of two particles of radii R_1 and R_2 subject to hydrodynamic, electric, and molecular forces. Introduce a spherical frame of reference (r, θ, Φ) connected to the center of the larger particle 1, angle θ being measured from the direction of the electric field strength vector $\mathbf{E}_0$. The expression (6.125) corresponds to the radial electric force acting on particle 2. If both particles are uncharged, then $q_1 = q_2 = 0$ and

$$F_{2r} = \varepsilon E_0^2 R_1 R_2 (\widehat{f}_1 \cos^2\theta + \widehat{f}_2 \sin^2\theta). \tag{6.154}$$

The force F_{2r} is represented in a form that is symmetric with respect to particle radii. Coefficients $\widehat{f}_i$ are expressed through the coefficients f_i introduced in Eq. (6.125) as follows:

$$\widehat{f}_i(k, \Delta) = k f_i(k, \Delta(1+k^{-1})), \quad (i = 1, 2), \tag{6.155}$$

where $k = R_2/R_1$, $\quad \Delta = (r - R_1 - R_2)/(R_1 + R_2)$.

The total force acting on the particle is equal to the sum of molecular and electric forces:

$$F_2(R_1, R_2, r, \theta) = F_{2A}(R_1, R_2, r) + F_{2r}(E_0, R_1, R_2, r, \theta). \tag{6.156}$$

The force of molecular interaction is taken in the approximate form (6.133). In contrast to the above-considered case of particle coagulation in the absence of electric fields, when the force of particle interaction possessed spherical symmetry, in the present case the electric field causes the force to be dependent on the orientation of the particle pair relative to the external electric field vector $\boldsymbol{E}_0$ and thus possesses axial, rather than spherical, symmetry. As a result, a second-order partial derivative with respect to θ appears in the diffusion equation (6.36), which not only complicates the solution but also causes rotation of the particle pair relative to the direction of the electric field. Let us therefore try to estimate the diffusion flux using the following procedure.

We first determine the diffusion flux $J_0(0)$ per unit solid angle, assuming the interaction force to be purely radial. Then we integrate the resulting expression over the surface of a sphere of radius $R = R_1 + R_2$ (the coagulation radius); while doing so, we must keep in mind the dependence of the interaction force on θ – the orientation angle of the particle pair relative to the direction of electric field strength. The flux determined in this way can be regarded as the first approximation for the total diffusion flux j of particles of radius R_2 toward the particle of radius R_1. Carrying out all these steps, we obtain

$$j_0(0) = \frac{\nu_e}{\lambda_0^2}(R_1 + R_2)\widehat{\varphi}_1(\theta)n_0,$$

$$j = 4\pi(R_1 + R_2)^2 \int_0^{\pi/2} j_0(0)\sin\theta\, d\theta, \tag{6.157}$$

where n_0 is the number concentration of particles 2 and $\widehat{\varphi}_1(\theta)$ is determined by analogy with Eq. (6.136).

Making use of Eq. (6.134) for h, Eq. (6.135) for D_t, and Eq. (6.156) for the interaction force between particles, and introducing the same dimensionless variables as earlier, we get

$$j = 4\pi n_0 \frac{(R_1 + R_2)^3 \nu_e}{\lambda_0^2}\varphi_1, \quad \varphi_1 = \int_0^{\pi/2} \widehat{\varphi}_1(\theta)\sin\theta\, d\theta, \tag{6.158}$$

$$(\widehat{\varphi}_1(\theta))^{-1} = \int_0^{\infty} \frac{(\Delta + \Delta_1)^2}{\Delta^2(1+\Delta)^2(\Delta+\Delta_2)^2}\exp\left[-S_A\int_\Delta^\infty \frac{(\Delta_1 + z)dz}{z^3(z+\Delta_2)^3}\right.$$
$$\left. - S_E\int_\Delta^\infty \frac{(\Delta_1 + z)(\widehat{f}_1\cos^2\theta + \widehat{f}_2\sin^2\theta)}{z(z+\Delta_2)^2}dz\right],$$

$$\Delta_1 = \frac{R_1 R_2}{(R_1 + R_2)^2}, \Delta_2 = \frac{R_1 R_2}{R_1^2 + R_2^2 - R_1 R_2};$$

$$S_A = \frac{\Gamma \lambda_0^2}{36\pi\rho_e \nu_e^2 (R_1 + R_2)^3}, \quad S_E = \frac{\varepsilon E_0^2 \lambda_0^2}{6\pi\rho_e \nu_e^2}.$$

Another method of estimating the diffusion flux j is based on the natural assumption that in the course of mutual approach of the two particles, the orientation of the pair relative to electric field will change a number of times. This will result in the blurring of the concentration profile n_2, which can be determined by averaging over the angle θ. Accordingly, we average the relations (6.156) and (6.154) over the spherical surface, getting the average force acting on particle 2:

$$\begin{aligned} \langle F_2(R_1, R_2, \Delta)\rangle &= \int_0^\pi F_2(R_1, R_2, \Delta, \theta)\sin\theta d\theta \\ &= F_{2A}(R_1, R_2, r) + \langle F_{2r}(E_0, R_1, R_2, r, \theta)\rangle, \end{aligned} \tag{6.159}$$

$$\langle F_{2r}(E_0, R_1, R_2, r, \theta)\rangle = \frac{1}{3}\varepsilon E_0^2 R_1 R_2(\widehat{f}_1 + 2\widehat{f}_2).$$

and substitute these relations into Eq. (6.132). Then the flux j has the form (6.158), but the expression for φ_1 changes to

$$\begin{aligned} \varphi_1^{-1} = \int_0^\infty &\frac{(\Delta + \Delta_1)^2}{\Delta^2(1+\Delta)^2(\Delta+\Delta_2)^2} \exp\left[-S_A \int_\Delta^\infty \frac{(\Delta_1 + z)dz}{z^3(z+\Delta_2)^3}\right. \\ &\left. - S_E \int_\Delta^\infty \frac{(\Delta_1 + z)(\widehat{f}_1 + 2\widehat{f}_2)}{3z(z+\Delta_2)^2} dz\right]. \end{aligned} \tag{6.160}$$

Numerical calculations have shown that while the two methods give different expressions for φ_1, the resulting diffusion fluxes are roughly the same: the maximum difference between the two fluxes is about 5% for a wide range of parameters S_A and S_E. The influence of the external electric field on the collision frequency decreases with growth of S_A and reduction of S_E. Since S_E does not depend on particle size and S_A increases as particles get smaller, we conclude that the smaller the particle size, the less influence electric field has on the collision frequency.

The effect of the electric field on the collision frequency is characterized by the ratio between diffusion fluxes with and without the electric field:

$$\xi = \frac{j(S_A, S_E)}{j(S_A, 0)}. \tag{6.161}$$

Taking the parameter values typical for for water–in–oil emulsions in a turbulent flow, $\varepsilon \sim 2\varepsilon_0$, $E_0 \sim 0.9\,\text{kV/cm}$, $\nu_e \sim 10^{-5}\,\text{m}^2/\text{s}$, $\lambda_0 \sim 10^{-3}\,\text{m}$, $\rho = 900\,\text{kg/m}^3$, $\Gamma \sim 10^{-20}\,\text{J}$, $R_1 + R_2 \sim 10^{-5}\,\text{m} = 10\,\mu\text{m}$ we get $S_E \sim 1$, $S_A \sim 10^{-7}$, and $\xi \sim 30$. Hence, the collision frequency of conductive water droplets in dielectric oil in the presence of an electric field having the strength 0.9 kV/cm is 30 times higher than the same frequency in the absence of the electric field. Since $S_A \sim R^{-3}$, an increased radius of a droplet results in a higher collision frequency (see Fig. 6.10).

That being said, even sufficiently strong electric fields are not able to fully cancel out the effect of hydrodynamic interactions on collision frequency of particles in a turbulent flow. To estimate this effect, consider the ratio of the flux in the absence of hydrodynamic interactions to the flux when such interactions are present:

$$\frac{j_0}{j} = \frac{3}{\varphi_1}. \tag{6.162}$$

At $S_E \sim 5$, this ratio is $j_0/j \sim 10$, and at $S_E \sim 1$ we already have $j_0/j \sim 50$.

Thus even in strong electric fields where S_E is large, hydrodynamic interactions accompanying the mutual approach of particles with fully retarded surfaces decrease the collision frequency by one order of magnitude as compared to the case of non-interacting particles. It should be noted that for particles with mobile surfaces (droplets), the role of hydrodynamic interactions is not as strong, because at small clearances between the particles, hydrodynamic resistance increases slower: $(\sim 1/\Delta^{1/2})$ for droplets as opposed to $(\sim 1/\Delta)$ for rigid particles.

Numerical treatment of the problem leads to the following relation that approximates Eq. (6.158):

$$j \approx 0.24\pi S_E \nu_e \lambda_0^{-2} (R_1^{1/3} R_2^{2/3} + R_2^{1/3} R_1^{2/3}) n_{20}. \tag{6.163}$$

The diffusion flux of particles of radius R_2 (at number concentration $n_{20} = 1$) toward the particle of radius R_1 has the meaning of kernel K (V, ω) of the kinetic equation (6.1) (coagulation kernel). Replacing radii in Eq. (6.163) with volumes, we get

$$K(\text{v}, \omega) = 0.01 \varepsilon E_0^2 \nu_e^{-1} \rho_e^{-1} (V^{1/9}\omega^{2/9} + \omega^{1/9} V^{2/9})^3. \tag{6.164}$$

It is now possible to consider the dynamics of the process of aggregation of particles with fully retarded surfaces. If the electric field is absent and the particles are uncharged, the coagulation kernel is given by Eq. (6.148). We employ the method of moments to solve the kinetic equation (6.1). Eq. (6.15) gives us the following equation for the zero-order moment (i.e., for the number concentration of particles):

$$\frac{dm_0}{dt} = -\frac{1}{2} G[m_{1/3}(t)m_{1/6}(t) + m_{1/6}(t)m_{1/3}(t)], \tag{6.165}$$

where $G = 16(\Gamma/(3\pi\rho_e\lambda_0^2))^{1/2}$.

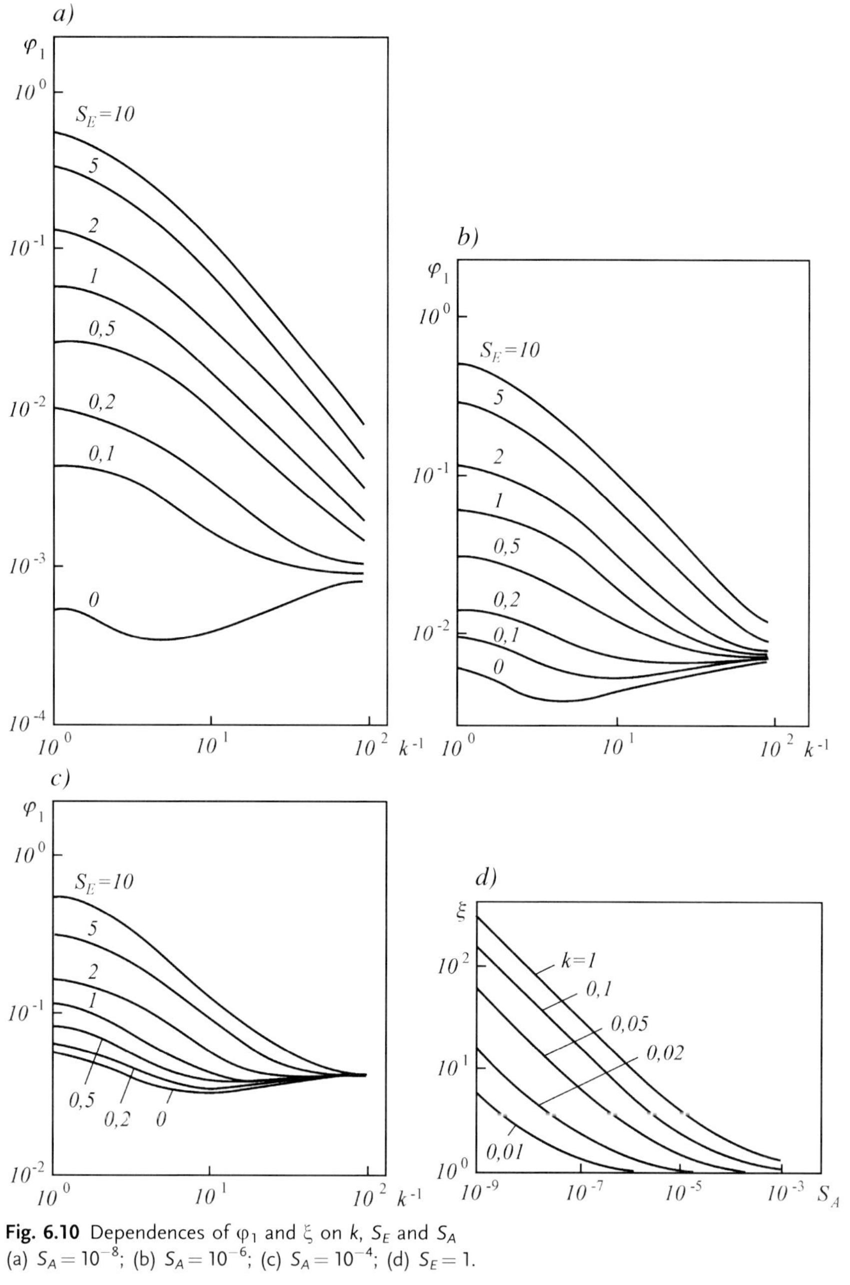

Fig. 6.10 Dependences of φ_1 and ξ on k, S_E and S_A
(a) $S_A = 10^{-8}$; (b) $S_A = 10^{-6}$; (c) $S_A = 10^{-4}$; (d) $S_E = 1$.

The right-hand side or this equation contains fractional moments. They can be determined by the method of interpolation of fractional moments (see Section 6.1). Adopting the two-point interpolation scheme, we express them through the integer moments m_0 and m_1 according to Eq. (6.20). The result is

$$\frac{d\widehat{m}_0}{dt} = G\frac{m_{1/3}(0)m_{1/6}(0)}{m_0(0)}\widehat{m}_0^{3/2}(t)\widehat{m}_1^{1/2}(t), \tag{6.166}$$

where we have introduced the dimensionless moments

$$\widehat{m}_0(t) = \frac{m_0(t)}{m_0(0)}; \quad \widehat{m}_1(t) = \frac{m_1(t)}{m_1(0)}$$

and $m_k(0)$ is the initial value of k-th moment.

In the course of coagulation, the volume concentration does not vary with time and $\widehat{m}_1(t) = 1$. Then the solution of Eq. (6.166) with the initial condition $\widehat{m}_0(t) = 1$ is

$$\widehat{m}_0(t) = \left(1 + 8G\frac{m_{1/3}(0)m_{1/6}(0)}{m_0(0)}t\right)^{-2}. \tag{6.167}$$

Estimating the initial values of fractional moments in the same manner as above, we obtain:

$$m_{1/3}(0)m_{1/6}(0) = (m_0(0))^{3/2}m_1^{1/2} \tag{6.168}$$

Then Eq. (6.167) and Eq. (6.168) give us the number concentration the and average particle volume as functions of time:

$$\widehat{m}_0(t) = (1 + 8G(m_0(0))^{1/2}m_1^{1/2}t)^{-2} \tag{6.169}$$

and

$$\frac{V_{av}(t)}{V_{av}(0)} = \frac{1}{\widehat{m}_0(t)} = (1 + 8G(V_{av}(0))^{-1/2}m_1 t)^{-2}, \tag{6.170}$$

where $V_{av}(0) = m_1/m_0(0)$.

Defining the characteristic time of coagulation t_A as the time it takes for the average particle (droplet) volume to increase by a factor of e, we write

$$t_A \sim \frac{(\sqrt{e}-1)V_{av}^{1/2}(0)}{128m_1}\left(\frac{3\pi\rho_e\lambda_0^2}{\Gamma}\right)^{1/2}. \tag{6.171}$$

Let us take $\boldsymbol{m}_1 = 0.01$; $\lambda_0 = 5\cdot 10^{-4}$ m; $\Gamma = 10^{-20}$ J; $\rho_e = 850\,\text{kg/m}^3$; $V_{\alpha V} = 4\cdot 10^{-15}\,\text{m}^3$ (the average particle radius is $10\,\mu\text{m}$). Then the characteristic time of particle enlargement is $t_A \sim 100$ s. The decision to take hydrodynamic interactions into account decreases the diffusion flux by two orders of magnitude as compared to the case of turbulent coagulation in the absence of hydrodynamic interactions, so the characteristic time of coagulation increases by the same order of magnitude.

Consider now the dynamics of enlargement of conductive particles in an external electric field. The field strength E_0 is supposed to be less than the critical value at which the breakup of approaching particles (droplets) becomes possible. In this case the coagulation kernel has the form (6.164) and the equation for the zero-order moment is written as

$$\frac{d\widehat{m}_0}{dt} = -G_1 \frac{m_{1/3}(0)m_{2/3}(0) + 3m_{4/9}(0)m_{5/9}(0)}{m_0(0)} \widehat{m}_0 \widehat{m}_1. \tag{6.172}$$

where $G_1 = 0.01\varepsilon E_0^2 \nu_e^{-1} \rho_e^{-1}$. The solution of this equation is

$$m_0(\tau) = m_0(0)e^{-B\tau}, \tag{6.173}$$

$$\text{where} \quad \tau = G_1 m_1 t \quad and \quad B = \frac{m_{1/3}(0)m_{2/3}(0) + 3m_{4/9}(0)m_{5/9}(0)}{m_0(0)}.$$

The characteristic time of coagulation follows from Eq. (6.173):

$$t_E \sim \frac{1}{G_1 m_1 B} = \frac{25\mu_e}{\varepsilon E_0^2 m_1}. \tag{6.174}$$

Let us take $m_1 = 0.01$; $\mu_e = 5 \cdot 10^{-3}$ kg/m s; $E_0 = 1$ kV/cm; $\varepsilon \sim 2\varepsilon_0$. Then $t_E \sim 6$ s. If we increase electric field strength to 2 kV/cm with all other conditions being the same, the coagulation time decreases to 1.5 s.

A comparison of the characteristic times of particle enlargement in a turbulent flow t_E^t with the same characteristic time for gravitational sedimentation in an electric field t_E^g shows that $t_E^g / t_E^t \sim 2 \div 3$

6.7
Breakup of Particles

In Section 6.5 we broached the subject of deformation of conductive droplets and the possibility of their breakup. The present section will discuss breakup of non-conducting droplets.

Stokesian motion of small droplets in the carrier flow does not lead to any noticeable deformation of the droplets. Droplet breakup is always preceded by a considerable deformation of its surface, which is possible when in the fluid layers adjacent to both sides of the droplet there exist large gradients of velocity and pressure that can overcome surface tension of the interface. In order to describe droplet deformation, it is necessary to take into account the joint action of inertial and viscous effects and surface tension forces. Data on particle deformation and breakup in viscous fluids at $\mathrm{Re} \leq 1$ are presented in [54], while [55] deals with the case $\mathrm{Re} \gg 1$.

Breakup of small droplets in a turbulent flow of emulsion should be considered as a random process characterized by the following parameters: breakup frequency $f(V)$ of droplets whose volume lies in the interval $(V, V+dV)$; probability $P(V, \omega)$ that a particle whose volume lies in the interval $(V, V+dV)$ will be formed after breakup or a larger droplet whose volume lies in the interval $(\omega, \omega+d\omega)$; minimum droplet radius R_m for which breakup becomes possible. The kinetic equation describing the dynamics of particle distribution over volumes $n(V,\omega)$ during the process of particle coagulation/breakup is given below:

$$\frac{\partial n(V,t)}{\partial t}=\frac{1}{2}\int_0^V K(V-\omega,\omega)n(V-\omega,t)n(\omega,t)d\omega-n(V,\omega)\int_0^\infty K(V,\omega)n(\omega,t)d\omega$$
$$+\int_V^\infty f(\omega)P(V,\omega)n(\omega,t)d\omega-f(V)n(V,t). \tag{6.175}$$

Let us start with the estimation of the minimum particle radius. To this end, we must estimate the forces acting on a particle in a turbulent flow that are capable of deforming the particle.

A droplet suspended in a homogeneous, isotropic turbulence field is subject to the following forces exerted by the carrier fluid: dynamic pressure $Q \sim k_f \rho_e u^2/2$, where $\boldsymbol{k_f} \approx 0.5$ is the resistance coefficient; ρ_e and u – the density and velocity of the surrounding fluid relative to the droplet; $F_v \sim \mu_e \dot{\gamma}_t$ – the force of viscous friction, where μ_e is the viscosity coefficient of the carrier fluid and $\dot{\gamma}_t=(4\bar{\varepsilon}/15\pi\nu_e)^{1/2}$ is the average shear rate of small-scale fluctuations; $\bar{\varepsilon}$ – specific dissipation of energy; $\nu_e=\mu_e/\rho_e$ – the coefficient of kinematic viscosity. In addition to the forces mentioned above, the droplet surface is subject to the force of surface tension $F_{cap}=2\sum/R$, where Σ is the surface tension coefficient and R – the particle radius. There are two possible mechanisms of droplet breakup; which mechanism will actually be employed in the breakup process depends on the dominant force.

Suppose dynamic pressure is the major force [9]. Then particle deformation is caused by the difference between dynamic pressures applied to the opposite sides of the droplet:

$$Q \sim \frac{k_f \rho_e}{2}\left(u_1^2-u_2^2\right), \tag{6.176}$$

where u_1 and u_2 are the velocities at the opposite poles of the droplet separated by the distance $2R$.

First, consider a droplet of size $R>\lambda_0$, where λ_0 is the inner scale of turbulence. Then large-scale fluctuations ($\lambda_0 \ll \lambda \ll \boldsymbol{L}$), which vary slightly over distances of the order of the droplet size, fail to affect a noticeable impact on the droplet. It means that droplet deformation and breakup can be caused only by small-scale fluctuations.

For such fluctuations, Eq. (6.43) gives the change of fluctuational velocity u_λ over a distance of the order of $2R$:

$$u_\lambda \sim (\bar{\varepsilon}\lambda)^{1/3} \sim (\varepsilon_0 2R)^{1/3}. \tag{6.177}$$

Then it follows from Eq. (6.176) and Eq. (6.176) that

$$Q \sim \frac{k_f \rho_e}{2} \bar{\varepsilon}^{2/3} (2R)^{2/3}. \tag{6.178}$$

If the difference of dynamic pressures exceeds the surface tension force, a noticeable droplet deformation takes place and the droplet can be broken. So the breakup condition for the droplet may be written as

$$\frac{k_f \rho_e}{2} \bar{\varepsilon}^{2/3} (2R)^{2/3} \sim \frac{2\Sigma}{R}. \tag{6.179}$$

Substituting the expression (6.42) for $\bar{\varepsilon}$, we obtain a relation for the minimum radius:

$$R_m \sim \sqrt{2} \frac{L^{2/5} \Sigma^{3/5}}{k_f^{3/5} \rho_e^{3/5} U^{6/5}}, \tag{6.180}$$

where L is the characteristic linear scale of the flow region and U is the mean flow velocity.

It should be noted that the formula (6.180) was obtained under the assumption that the inner and outer fluids have roughly the same density, as is the case for water–in–oil emulsions. Otherwise it would be necessary to take into account the dynamic pressure from the inner fluid.

Let us introduce the Weber number

$$\mathrm{We} = \frac{\text{dynamic thrust}}{\text{force of surface tension}} = \frac{2R\rho_e U^2}{\Sigma}. \tag{6.181}$$

Then Eq. (6.180) can be rewritten as

$$\frac{R_m}{L} \sim \frac{2^{11/4} \Sigma^{3/5}}{k_f^{3/2}} We^{-3/2} = C We^{-3/2}. \tag{6.182}$$

Now consider droplets whose sizes is smaller than the inner scale of turbulence $R \ll \lambda_0$. Clearly, breakup of such droplets may cause fluctuations with scales $\lambda < \lambda_0$, that is, fluctuations characterized by considerable viscous friction forces. Therefore the force of viscous friction at the droplet surface emerges as the principal mechanism of droplet deformation, and equality between the forces of viscous friction and surface tension becomes the criterion of strong deformation of the droplet:

$$\mu_e\left(\frac{4\bar{\varepsilon}}{15\pi\nu}\right)^{1/2}\sim\frac{2\Sigma}{R},\tag{6.183}$$

which gives the following minimum droplet radius [56]:

$$R_m\sim\sqrt{15\pi}\frac{\Sigma}{\rho_e(\nu_e\bar{\varepsilon})^{1/2}}\sim C\frac{\Sigma}{\rho_e(\nu_e\bar{\varepsilon})^{1/2}}.\tag{6.184}$$

Introducing another dimensionless parameter – the Ohnesorge number

$$\mathrm{Oh}=\frac{\mu_e}{(2Re\rho_e\Sigma)^{1/2}},\tag{6.185}$$

we rewrite the relation (6.184) as

$$\frac{R_m}{L}=\left(\frac{15\pi}{4}\right)^{1/4}Re^{-3/4}\mathrm{Oh}^{-1},\tag{6.186}$$

where $\mathrm{Re}=UL/\nu_e$ is the Reynolds number.

We now proceed to determine the breakup frequency of droplets $f(\mathrm{V})$, adopting the model of droplet breakup in a locally isotropic developed turbulent flow suggested in [57]. The model is based on the assumption that the possibility of breakup of an isolated droplet is completely predetermined by fluctuations of energy dissipation in the vicinity of the droplet. Droplet breakup takes place when the value of energy dissipation averaged over the fluid volume that is equal by the order of magnitude to the droplet volume V exceeds some critical value $\bar{\varepsilon}_{cr}(\mathrm{V})$. For a given droplet size, the critical value of energy dissipation should be equal to the value of $\bar{\varepsilon}$ in the formula for the minimum radius of breaking droplets. When viscous friction is the dominant force that is responsible for the deformation and breakup of droplets, one can use Eq. (6.136) to derive

$$\bar{\varepsilon}_{cr}(\mathrm{V})\sim\left(\left(\frac{4\pi}{3}\right)^{1/3}C\frac{\Sigma}{\rho}\right)^2\frac{1}{\nu_e\mathrm{V}^{2/3}}.\tag{6.187}$$

Suppose the distribution of energy dissipation in the vicinity of the droplet is uniform and has the average value $\bar{\varepsilon}(t)$. We can then interpret breakup frequency (i.e., the number of breakup events per unit time) as the relative probability of attaining a certain (constant) level $\bar{\varepsilon}(t)$ in a random process $\bar{\varepsilon}(t)$. Consider two instants of time t and $t+\Delta t$. Suppose that at the moment t the dissipation energy is $\bar{\varepsilon}(t)<\bar{\varepsilon}_{cr}(\mathrm{V})$, whereas at the moment $t+\Delta t$ it equals $\varepsilon(t+\Delta t)>\bar{\varepsilon}_{cr}(\mathrm{V})$. Then the breakup frequency may be represented as

$$f(V) = \lim_{\Delta t \to 0}\left(\frac{1}{\Delta t}\frac{P(\bar{\varepsilon}(t) < \bar{\varepsilon}_{cr}(V);\bar{\varepsilon}(t+\Delta t) < \bar{\varepsilon}_{cr}(V))}{P(\varepsilon(t) < \bar{\varepsilon}_{cr}(V))}\right). \tag{6.188}$$

The right-hand side gives the probability for the average dissipation energy to be less than the critical value at the time t while exceeding the critical value at the time $t+\Delta t$.

Expanding $\bar{\varepsilon}(t+\Delta t)$ in series over the powers of Δt and taking random process as stationary, we can transform the relation (6.188) as

$$f(V) = \int_0^{\infty} \dot{\bar{\varepsilon}}\, p(\bar{\varepsilon}_{cr}, \dot{\bar{\varepsilon}})d\dot{\bar{\varepsilon}} \Big/ \int_0^{\bar{\varepsilon}_{cr}} p\bar{\varepsilon}d\bar{\varepsilon}, \tag{6.189}$$

where $p(\bar{\varepsilon}, \dot{\bar{\varepsilon}})$ is the joint distribution density of the random variables $\bar{\varepsilon}(t)$ and $\dot{\bar{\varepsilon}}(t)$ at one and the same instant of time t, and $p(\bar{\varepsilon})$ is a one-dimensional distribution of the random variable $\bar{\varepsilon}(t)$.

According to [58], the joint distribution density $p(\bar{\varepsilon}_{cr}, \dot{\bar{\varepsilon}})$ is expressed through the joint distribution density of one and the same random variable $\bar{\varepsilon}$ taken at different instants of time:

$$p(\bar{\varepsilon}_{cr}, \dot{\bar{\varepsilon}}) = \lim_{\Delta t \to 0} \Delta t p\left(\bar{\varepsilon} + \frac{\Delta t}{2}\dot{\bar{\varepsilon}}, \bar{\varepsilon} - \frac{\Delta t}{2}\dot{\bar{\varepsilon}}\right). \tag{6.190}$$

To determine the one-dimensional distribution $p(\bar{\varepsilon})$, we shall draw on the result obtained in [59], which states that the distribution is well approximated by the logarithmic normal distribution

$$p(\bar{\varepsilon}) = \frac{1}{\sqrt{2\pi}\alpha\bar{\varepsilon}}\exp\left(-\frac{1}{2\alpha^2}(\ln(\kappa\bar{\varepsilon}))^2\right), \tag{6.191}$$

where $\alpha^2 = \ln(\delta^2/\langle\bar{\varepsilon}\rangle^2+1)$, $\kappa = \exp(\alpha^2/2)/\langle\bar{\varepsilon}\rangle^2$, and $\langle\bar{\varepsilon}\rangle$ and δ^2 are the mean value and the variance of the distribution of specific energy dissipation. It has been shown in [60] that the assumption that fourth moments of velocity gradients are connected with second moments in the same fashion as in the case of the normal distribution (the Millionschikov hypothesis) leads us to

$$\delta^2 = 0.4\langle\bar{\varepsilon}\rangle^2 \tag{6.192}$$

and the joint distribution density is equal to

$$p(\bar{\varepsilon}_{cr}, \dot{\bar{\varepsilon}}) = \frac{T_0}{2\pi\alpha c\bar{\varepsilon}^2}\exp\left(-\frac{T_0^2}{2c^2}\left(\frac{\dot{\bar{\varepsilon}}}{\bar{\varepsilon}}\right)^2 - \frac{1}{2\alpha^2}(\ln(\kappa\bar{\varepsilon}))^2\right), \tag{6.193}$$

where $T_0 = \sqrt{\nu_e/\langle\bar{\varepsilon}\rangle}$ and $c = 1 - \exp(-\alpha^2)$.

Substitution of Eq. (6.191) and Eq. (6.193) into Eq. (6.189) gives us the breakup frequency:

$$f(V) = \frac{c}{\sqrt{2\pi}T_0\alpha}\frac{d}{dy}(\ln\Phi(y)), \tag{6.194}$$

where

$$\Phi(y) = \frac{1}{\sqrt{2\pi}}\int_{-\infty}^{y} e^{-y^2}dy, \quad y = \frac{1}{\alpha}\ln(\kappa\bar{\varepsilon}_{cr}(V)).$$

Putting $\bar{\varepsilon}_{cr}$ taken from Eq. (6.187) into this formula and making use of Eq. (6.192), we finally get

$$f(\mathrm{v}) = \frac{1}{2.03\sqrt{2\pi}T_0}\frac{d}{dx}(\ln\Phi(x)), \tag{6.195}$$

where

$$x = -1.1\ln\left(1{,}3\frac{\mathrm{v}}{\mathrm{v}_m}\right), \quad \mathrm{v}_m = \frac{4\pi}{3}R_m^2.$$

It remains for us to find the probability of droplet formation $P(V, \omega)$.

Experiments have been set up to study how the process of breakup of isolated droplets depends on hydrodynamic conditions. Various types of droplet breakup were observed. Most often it resulted in two almost identical droplets (daughter droplets) and several droplets of smaller sizes (satellites). None of the publications has mentioned any correlation between the type of droplet breakup and the size of the splitting droplets. This gives grounds to believe that at fixed parameters of the internal and external fluids, the probability of breakup depends only on the ratio between the volume ω of the splitting droplet and the volume V of the daughter droplets and has the form

$$P(\mathrm{v}, \omega) = k\frac{1}{\omega}g\left(\frac{\mathrm{v}}{\omega}\right). \tag{6.196}$$

Since the total volume of the droplets must be conserved, we may write

$$\int_0^{\omega} VP(V, \omega)dV = \omega \tag{6.197}$$

which together with Eq. (6.196) leads us to conclude that the mean value of the distribution density $g(y)$ equals

$$\langle y\rangle = \int_0^1 y g(y) dy = \frac{1}{k}.$$

Based on the this brief discussion of droplet disintegration, we can assume that $g(y)$ is a bimodal function with two well-defined maxima, one maximum in the region of daughter droplet sizes and the other – in the region of satellite droplet sizes. Then breakup probability can be written as a sum of two weighted single-modal distribution densities $g_1(y)$ and $g_2(y)$ defined in the domain (0, 1) and having the average values $\langle y_1\rangle = V_1/\omega$ and $\langle y_2\rangle = V_2/\omega$:

$$P(V,\omega) = k_1 \frac{1}{\omega} g_1\left(\frac{V}{\omega}\right) + k_2 \frac{1}{\omega} g_2\left(\frac{V}{\omega}\right). \tag{6.198}$$

The condition of conservation of total volume (6.197) requires that

$$k_1 y_1 + k_2 y_2 = 1. \tag{6.199}$$

In the limiting case when the daughter droplets are identical and all satellites have the same size, the variances of the distributions $g_1(y)$ and $g_2(y)$ are zero and Eq. (6.150) takes a simple form

$$P(V,\omega) = k_1 \delta(V - y_1\omega) + k_2 \delta(V - y_1\omega), \tag{6.200}$$

where $\delta(x)$ is the delta function.

In the case of a multimodal breakup probability, the breakup probability has the form

$$P(V,\omega) = \sum_{i=1}^{n} k_i \delta(V - y_i\omega), \quad \sum_{i=1}^{n} k_i y_i = 1. \tag{6.201}$$

In the simplest case when breakup leads to the formation of two identical droplets, the breakup probability is

$$P(v,\omega) = 2\delta(v - 0.5\omega). \tag{6.202}$$

Substituting the breakup probability (6.202) into the kinetic equation (6.175) and neglecting the first two integrals (i.e., droplet breakup is the only process taken into account), we get the following equation:

$$\frac{\partial n(V,t)}{\partial t} = 2 f(2V) n(2V,t) - f(V) n(V,t). \tag{6.203}$$

A solution of this equation is given in [57] as a sum of independent particular solutions with discrete spectra. This paper also discusses two special cases of initial distributions: a monodisperse distribution, and a distribution that is uniform in a certain interval. The obtained solutions make it possible to determine the first four moments of the distribution, and one can show from the Pearson diagram that as time goes on, the solution tends to a logarithmic normal distribution. This conclusion is consistent with the result obtained in [59] for a constant breakup frequency *f* (V) that does not depend on the size of particles.

References

1 Sinaiski, E.G. and and Lapiga, E.J. (2007) *Separation of Multiphase, Multicomponent Systems,* Wiley-VCH, Weinheim.

2 Loginov, V.I. (1979) *Dewatering and Desalting of Oils,* Chemistry, Moscow (in Russian).

3 Voloshuk, V.M. (1984) *Kinetic Theory of Coagulation,* Hydrometeoizdat, Moscow (in Russian).

4 Hahn, G.J. and and Shapiro, S.S. (1967) *Statistical Models in Engineering,* Wiley & Sons, New York.

5 Gradstein, J.S. and Ryszik, I.M. (1971) *Tables of Integrals, Sums, Series, and Products,* Nauka, Moscow.

6 Ruckenstein, E. (1975) Condition for the Size Distribution of Aerosols to be Self-Preserving, *J. Colloid Interface Sci.*, **50** (3), 508–518.

7 Jeffreys, G.V. and Davies, G.A. (1971) Coalescence of Droplets and Dispersions in *Recent Advances in Liquid–Liquid Extraction* (ed. K. Hanson), Pergamon Press, Oxford.

8 Einstein, A. and Smoluchowski, M. (1936) *Brownian Motion,* ONTI, Moscow (in Russian).

9 Levich, V.G. (1962) *Physicochemical Hydrodynamics*. Prentice–Hall, Englewood Cliffs, N.J.

10 Delichatsips, M.A. and Probstein, R.P. (1975) Coagulation in Turbulent Plow: Theory and Experiment, *J. Colloid Interface Sci.*, **51**, 394–405.

11 De Boer, G.B.J., Hoedamakers, G.F.M. and Thones, D. (1989) Coagulation in Turbulent Flow, Part I, *Chem. Eng. Res. and Des.*, **67** (3), 301–307.

12 Entov, V.M., Kaminski, V.A. and Lapiga, E.J. (1976) Calculation of Emulsion Coalescence Rate in a Turbulent Flow. *Izvestia Acad. Sci. USSR, Fluid Dyn.*, **3**, 47–55 (in Russian).

13 Sinaiski, E.G. and Rudkevich, A.M. (1981) Coagulation of Bubbles in a Turbulent Flow of Viscous Fluid. *Colloid J.*, **43** (2), 369–371 (in Russian).

14 Sinaiski, E.G. (1993) Coagulation of Droplets in a Turbulent Flow of Viscous Fluid. *Colloid J.*, **55** (4), 91–103 (in Russian).

15 Sinaiski, E.G. (1994) Brownian and Turbulent Coagulation of Droplets in a Viscous Fluid and Stability of Emulsions. *Colloid J.*, (**1**), 105–112 (in Russian).

16 Sinaiski, E.G. (1994) Coagulation of Droplets with Different Viscosities in a Turbulent Flow of Viscous Fluid. *Colloid J.*, (2),222–225 (in Russian).

17 Tunizki, A.N. , Kaminski, V.A. and Timashov, S.F. (1972) *Methods of Physicochemical Kinetics,* Chemistry, Moscow (in Russian).

18 Kaminski, V.A. (1976) Kinetics of Turbulent Coagulation. *Colloid J.*, **5**, 907–912 (in Russian).

19 Hinze, J.O. (1959) *Turbulence*, McGraw–Hill,New York.

20 Monin, A.S. and Yaglom, A.M. (1971) *Statistical Fluid Mechanics: Mechanics of Turbulence*, Vol **1**, MIT Press, Cambridge, MA Monin, A.S. and Yaglom, A.M. (1975) *Statistical Fluid Mechanics: Mechanics of Turbulence*, Vol **2**, MIT Press, Cambridge, MA.

21 Landau, L.D. and Lifshitz, E.M. (1987) *Fluid Mechanics*, 2nd ed., Pergamon Press, Oxford.

22 Leontovich, M.L. (1983) *Introduction to Thermodynamics. Statistical Physics*, Nauka, Moscow (in Russian).

23 Russel, W.B. (1981) Brownian Motion of Small Particles Suspended in Liquids. *Ann. Rev. Fluid Mech.*, **13**, 425–455.

24 Happel, J. and and Brenner, H. (1965) *Low Reynolds Number Hydrodynamics*, Prentice–Hall, Englewood Cliffs.

25 Zinchenko, A.Z. (1978) Calculation of Hydrodynamic Interactions of a Droplet at Low Reynolds Numbers. *Appl. Math. Mech.*, (5),955–959 (in Russian).

26 Davis, M.H. (1969) The Slow Translation and Rotation of Two Unequal Spheres in a Viscous Fluid. *Chem. Eng. Scl.*, **24**, 1769–1776.

27 O'Neil, M.E. and Majumdar, S.R. (1970) Asymmetrical Slow Viscous Fluid Motion Caused by the Translation or Rotation of Two Spheres, Part I, II. *ZAMP*, **21**, 164.

28 Wacholder, E. and Weihs, D. (1972) Slow Motion of a Fluid Sphere in the Vicinity of Another Sphere or a Plane Boundary. *Chem. Eng. Scl.*, **27**, 1187–1827.

29 Haber, S., Hetsroni, G. and Solan, A. (1973) On the Low Reynolds Number Motion of Two Droplets. *J. Multiphase Flow*, **1**, 57–71.

30 Rushton, E. and Davies, G.A. (1973) The Slow Unsteady Settling of Two Fluid Spheres Along Their Line of Centers. *Appl. Sci. Res.*, **28**, 37–61.

31 Reed, L.D. and Morrison, F.A. (1974) The Slow Motion of Two Touching Fluid Spheres Along Their Line of Centers, *J. Multiphase Flow*, **1**, 573–583.

32 Hetsroni, G. and Haber, S. (1978) Low Reynolds Number Motion of Two Drops Submerged in an Unbounded Arbitrary Velocity Field. *J. Multiphase Flow*, (**4**), 1–17.

33 Zinchenko, A.Z. (1980) Slow Asymmetric Motion of Two Drops in a Viscous Medium. *Appl. Math. Mech.*, **1**, 30–37 (in Russian).

34 Fuentes, Y.O., Kim, S. and Jeffrey, D.J. (1988) Mobility Functions for Two Unequal Viscous Drops in Stokes Flow. Part I. Axisymmertric Motions. *Phys. Fluids. A.*, **31**, 2445–2455. Fuentes, Y.O., Kim, S. and Jeffrey, D.J. (1989) Part II. Nonaxisymmertric Motions. *Phys. Fluids. A.*, **1**, 61–76.

35 Davis, R.H., Schonberg, J.A. and Rallison, J.M. (1989) The Lubrication Force Between Two Viscous Drops. *Phys. Fluids. A.*, **1**, 77–81.

36 Zinchenko, A.Z. (1981) Calculation of Short-Range Interaction of Droplets with Regard to Internal Circulation and Slippage. *Appl. Math. Mech.*, **45**, 759–763 (in Russian).

37 Kruyt, H.R. (ed.) (1952) *Colloid Science. Irreversible Systems*, Elsevier, Amsterdam, Vol **1**.

38 Hogg, R., Healy, T.W. and Fuerstenau, D.W. (1966) Mutual Coagulation of Colloidal Dispersions. *Trans. Faraday Soc.*, **62**, 1638–1651.

39 Hamaker, H.C. (1937) The London–van der Waals Attraction Between Spherical Particles. *Physica.*, **4**, 1058–1078.

40 Shenkel, J.N. and Kitchener, J.A. (1960) A Test of the Derjaguin–Verwey–Overbeek Theory with a

Colloidal Suspension. *Trans. Faraday Soc.*, **56**, 161–173.

41 Fuks, N.A. (1955) *Mechanics of Aerosols*, Acad. Sci,USSR, Moscow, (in Russian).

42 Melcher, J.R. and Taylor, G.I. (1969) Electrohydrodynamics: a Review of the Role of Interfacial Shear Stress. *Ann. Rev. Fluid. Mech.*, **1**, 111–146.

43 O'Konski, C.T. and Thacher, H.C. (1953) The Distortion of Aerosol Droplets by an Electric Field. *J. Phys. Chem.*, **57**, 955–958.

44 Garton, C.G. and Krasucki, Z. (1964) Bubbles in Insulating Liquids: Stability in an Electric Field. *Proc. Roy. Soc. Lond. A.*, **280**, 211–226.

45 Taylor, G.I. (1964) Disintegration of Water Drops in an Electric Field. *Proc. Roy. Soc. Lond. A.*, **280**, 383–397.

46 Rosenklide, C.E. (1969) A Dielectric Fluid Drop in an Electric Field. *Proc. Roy. Soc. Lond. A.*, **312**, 473–494.

47 Miskis, M.J. (1981) Shape of a Drop in an Electric Field. *Phys. Fluids*, **24**, 1967–1972.

48 Adornato, P.M. and Brown, R.A. (1983) Shape and Stability of Electrostatically Levitated Drops. *Proc. Roy. Soc. Lond. A.*, **389**, 101–117.

49 Basaran, O.A. and Scrlven, L.E. (1989) Axisymmetric Shape and Stability of Charged Drops in an Electric Field. *Phys. Fluids A.*, **1**, 799–809.

50 Allan, R.S. and Mason, S.G. (1962) Particle Behaviour in Shear and Electric Fields. I. Deformation and Burst of Fluid Drops. *Proc. Roy. Soc. Lond. A.*, **267**, 45–61.

51 Taylor, G.I. (1966) Studies in Electrohydrodynamics. I. The Circulation Produced in a Drop by an Electric Field. *Proc. Roy. Soc. Lond. A.*, **291**, 159–166.

52 Ausman, E.L. and Brook, M. (1967) Distortion and Disintegration of Water Drops in Strong Electric Fields. *J. Geophys. Res.*, **72**, 6131–6141.

53 Brazier-Smith, P.R. (1971) Stability and Shape of Isolated Pair of Water Drops in an Electric Field. *Phys. Fluids*, **14**, 1–6.

54 Stone, H.A. (1994) Dynamics of Drop Deformation and Breakup in Viscous Fluids. *Ann. Rev. Fluid Mech.*, **26**, 65–102.

55 Nigmatullin, R.I. (1987) *Dynamics of Multiphase Media*, Nauka, Moscow, Vol **1** (in Russian).

56 Sherman, P. (ed.) (1968) *Emulsion Science*, Academic Press,London–New York.

57 Loginov, V.I. (1985) Dynamics of Dropping Liquid in a Turbulent Flow. *Appl. Mech. Techn. Phys.*, (4), 66–73, (in Russian).

58 Tichonov, V.I. (1970) *Overshoots of Random Processes*, Nauka, Moscow (in Russian).

59 Kolmogorov, A.N. (1941) On Logarithmic Normal Low of Particle Distribution During Breakup. *Dokladi Acad. Nauk SSSR*, **31** (2), 99–101 (in Russian).

60 Golizin, G.S. (1962) Fluctuations of Energy Dissipation in a Locally Isotropic Turbulent Flow. *Dokladi Acad. Nauk SSSR*, **144** (3), 520–523 (in Russian).

7
Motion and Collision of Inertial Particles in a Turbulent Flow

Inertness of a particle is characterized by the Stokes number $\mathrm{St} = t_v / T^{(L)}$, where t_v is the characteristic relaxation time for the particle, and $T^{(L)}$ is the Lagrangian time scale (integral scale of turbulence). Particles for which $\mathrm{St} \gg 1$ are called inertial. This class of particles includes sufficiently large particles whose density by far exceeds the density of the carrier phase, for instance, solid particles in a gas. Another important parameter is the volume concentration of particles φ. The region $\varphi \ll 1$ corresponds to a rarefied disperse medium, in which the average distance between the particles is much greater than their average size. Under these conditions, collisions between particles play a minor role, and our attention can be restricted to interactions of particles with turbulent fluctuations (eddies) of the carrier flow. In contrast, when we are interested in the behavior of particles in sufficiently dense disperse media, particle collisions play the key role.

Our goal in both cases is to describe the dynamics of the disperse phase and to determine the collision frequency. The method most suitable for this task is the PDF (Probability Density Function) method, which is based on the use of the kinetic equation for the PDF of particle velocity; this equation takes into account particle interactions with each other as well as with turbulent fluctuations of the carrier phase (gas). An equation for the PDF of particle velocity and temperature has been derived in [1,2] (this equation neglects particle collisions). For relatively large particles in an isotropic turbulent flow this equation reduces to the Fokker–Planck equation for Brownian motion. The effect of particle collisions on the turbulent transport of disperse phase momentum at small values of φ was taken into account in [3.4]. The latter publication deals with the limiting case of inertial particles, when their interactions with turbulent fluctuations can be neglected.

In highly concentrated disperse media, particle collisions play the leading role, and the equation for the PDF reduces to the Boltzmann equation [5]. Two kinetic models of particle transport have been proposed in [6–8]: one based on the solution of the PDF equation by the method of perturbations, and another one representing a modification of the Enskog method of solving the Boltzmann equation for dense gases, specially tailored for the case of colliding particles.

Statistical Microhydrodynamics. Emmanuil G. Sinaiski and Leonid Zaichik

ISBN: 978-3-527-40656-2

Simulation of motion of particles with an arbitrary density based on the kinetic equation for the PDF in inhomogeneous turbulent flows is presented in [8–12]; these papers also attempt to derive the collision frequency and the rate of coagulation.

7.1 Motion of Particles without Mutual Collisions

When particles are moving in sufficiently rarefied, disperse media ($\varphi \ll 1$), the role of particle collisions is insignificant and the main consideration should be given to particle interactions with turbulent fluctuations of the carrier flow. If the characteristic time of dynamic relaxation t_ν for the particles is much shorter than the Lagrangian integral time scale of turbulence $T^{(L)}$ that defines the time of decay of fluctuations carrying large energies, the particles are well involved into the fluctuational motion of the carrier flow.

When the density of the carrier phase is much lower than that of particles, one can neglect the forces arising due to the pressure of a pressure gradient in the fluid, the virtual mass forces, and the Basset forces associated with the particles motion relative to the carrier fluid and arising due to the instability of the flow that bypasses the particle. If, in addition, the particles have a low volume concentration φ, one can also neglect interparticle interactions resulting from the infrequent collisions, and the feedback influence of the particles on the parameters of the carrier flow. Then the equation of motion of an isolated solid spherical particle has the form of the Langevin equation (5.101), which depends on two random uncorrelated fields describing the velocity of the turbulent carrier flow $\boldsymbol{u}$ and the Brownian force $\boldsymbol{f}$.

In general, the motion of an isolated, sufficiently large (i.e., non-Brownian) spherical particle of arbitrary density in a sufficiently rarefied turbulent liquid or gaseous medium is described (neglecting interparticle interactions) by the following equation:

$$\frac{d\boldsymbol{v}_p}{dt} = \frac{3\rho_e}{4\rho_p d_p} C_D|\boldsymbol{u}-\boldsymbol{v}_p|(\boldsymbol{u}-\boldsymbol{v}_p) + \boldsymbol{g} + \frac{\rho_e}{\rho_p}\left(\frac{D\boldsymbol{u}}{Dt}-\boldsymbol{g}\right) + C_A\frac{\rho_e}{\rho_p}\left(\frac{D\boldsymbol{u}}{Dt}-\frac{d\boldsymbol{v}_p}{dt}\right)$$

$$+\frac{9\rho_e}{\rho_p d_p}\sqrt{\frac{\nu_e}{\pi}}\int_0^t \frac{d(\boldsymbol{u}-\boldsymbol{v}_p)}{dt_1}\frac{dt_1}{\sqrt{t-t_1}} + C_L\frac{\rho_e}{\rho_p}(\boldsymbol{u}-\boldsymbol{v}_p)\times(\nabla\times\boldsymbol{u}), \qquad (7.1)$$

$$\frac{d\boldsymbol{R}_p}{dt} = \boldsymbol{v}_p, \quad \frac{D}{Dt} = \frac{\partial}{\partial t} + u_k\frac{\partial}{\partial X_k},$$

where t is time; $\boldsymbol{v}_p$ and $\boldsymbol{R}_p$ are the velocity components and coordinates of a particle; $\boldsymbol{u}$ is the velocity of the carrier flow; ρ_e and ρ_p are the densities of the carrier phase and the particle; d_p is the particles diameter; ν_e is the coefficient of kinematic viscosity.

It should be noted that, unlike Eq. (5.101), Eq. (7.1) is not limited to low Reynolds numbers Re_p of a particle but is valid for larger values of Re_p as well, owing to the dependence of resistance coefficients on the Reynolds number.

The terms on the right-hand side of Eq. (7.1) describe the corresponding specific forces of viscous resistance, gravity, and buoyancy; the effect of the virtual mass (in the form that is commonly used in modeling particle motion in a non-viscous fluid); the Basset force (written in the low Reynolds numbers approximation); and an additional lifting force due to velocity shear in the carrier flow. The resistance coefficient C_D and the coefficient C_L in the lifting force depend on the Reynolds number associated with the particles motion relative to the carrier fluid, on the velocity shear in the carrier flow, and on other parameters.

It is convenient to represent Eq. (7.1) in the relaxation form

$$\frac{d\mathbf{v}_p}{dt} = \frac{\mathbf{u}-\mathbf{v}_p}{t_u} + \mathbf{f}_A + \mathbf{f}_B + \mathbf{f}_L + \mathbf{F}_g, \tag{7.2}$$

where

$$\boldsymbol{f}_A = A\frac{d\boldsymbol{u}}{dt}, \quad \boldsymbol{f}_B = k_B\int_0^t \frac{d(\boldsymbol{u}-\boldsymbol{v}_p)}{dt_1}\frac{dt_1}{\sqrt{t-t_1}}, \quad \boldsymbol{f}_L = L(\boldsymbol{u}-\boldsymbol{v}_p)\times(\nabla\times\boldsymbol{u}),$$

$$\boldsymbol{F}_g = \frac{1-\rho^0}{1+C_A\rho^0}\mathbf{g}, \quad t_u = t_v(1+C_A\rho^0), \quad t_v = \frac{4d_p}{3\rho^0 C_D|\boldsymbol{u}-\boldsymbol{v}_p|},$$

$$A = \frac{(1+C_A\rho^0)\rho^0}{1+C_A\rho^0}, \quad k_B = \frac{9\rho^0}{d_p(1+C_A\rho^0)}\sqrt{\frac{\nu_e}{\pi}}, \quad L = \frac{C_L\rho^0}{1+C_A\rho^0}, \quad \rho^0 = \frac{\rho_e}{\rho_p}.$$

Here t_v is the particles characteristic time of dynamic relaxation, which is a function of the Reynolds number $\mathrm{Re}_p = |\boldsymbol{u}-\boldsymbol{v}_p|d_p/\nu_e$, and t_u is the characteristic relaxation time that takes into account the effect of the virtual mass.

A decision to take into account all of the forces entering Eq. (7.2) would unduly complicate the problem. However, in many practical applications some of the forces can be neglected. Thus, for $\rho_p \gg \rho_e$, the forces associated with fluid acceleration $\boldsymbol{f}_A$, memory $\boldsymbol{f}_B$, and velocity shear $\boldsymbol{f}_L$ are of no importance and can be excluded from consideration.

In order to make a transition from dynamic stochastic description of discrete particles (Eq. (7.2)) to simulation of statistical behavior of a particle ensemble (i.e., of the disperse phase), we introduce the dynamic probability density in the phase space of particle coordinates X and velocities $\boldsymbol{v}$

$$p_p(\boldsymbol{X},\boldsymbol{v},t) = \delta(\boldsymbol{X}-R_p(t))\delta(\boldsymbol{v}-\boldsymbol{v}_p(t)), \tag{7.3}$$

where averaging is performed over all possible instances of the turbulent flow and the random force field $\boldsymbol{f}$.

The procedure used to derive the kinetic equation has been described in Section 5.6. Differentiating Eq. (7.3) with respect to time and taking into account Eq. (7.2), we get the Liouville equation for dynamic probability density in the phase space:

$$\frac{\partial p_p}{\partial t} + v_{pk}\frac{\partial p_p}{\partial X_k}\left[\left(\frac{u_k - v_{pk}}{t_u} + f_{Ai} + f_{Bi} + f_{Li} + F_{gi}\right) p_p\right] = 0. \tag{7.4}$$

Low volume concentration of particles makes it possible to consider them as independent. Then the dynamic probability density of a particle ensemble can be introduced as

$$p = \frac{\mathrm{v}}{V}\sum_p p_p,$$

where **v** is the particles volume and V is the volume of the spatial region under consideration.

Let us now introduce the PDF of velocities for a system of particles $P = \langle p\rangle$ and represent the velocity of the carrier medium as a sum of the average and fluctuational components: $\boldsymbol{u} = \boldsymbol{U} + \boldsymbol{u}'$. Averaging Eq. (7.4) over the ensemble of random realizations of the turbulent velocity field and using the fact that $\langle v_{pi} p\rangle = v_i P$, we obtain:

$$\begin{aligned}&\frac{\partial P}{\partial t} + v_i\frac{\partial P}{\partial X_i} + \frac{\partial}{\partial v_i}\left[\left(\frac{u_i - v_i}{t_u} + F_{Ai} + F_{Bi} + F_{Li} + F_{gi}\right) P\right]\\ &= \frac{\partial}{\partial v_i}\left(\frac{\langle u_i' p\rangle}{t_u} + \langle f_{Ai}' p\rangle + \langle f_{Bi}' p\rangle + \langle f_{Li}' p\rangle\right),\end{aligned} \tag{7.5}$$

where

$$F_{Ai} = A\frac{D_0 U_i}{Dt}, \quad \frac{D_0}{Dt} = \frac{\partial}{\partial t} + U_i\frac{\partial}{\partial X_k},$$

$$F_{Bi} = k_B\int_0^t \frac{d(U_i - V_i)}{dt_1}\frac{dt_1}{\sqrt{t - t_1}}, \quad F_{Li} = L(U_i - V_i)\left(\frac{\partial U_j}{\partial X_i} - \frac{\partial U_i}{\partial X_j}\right).$$

The terms on the right-hand side of Eq. (7.5) stand for the interactions of particles with turbulent eddies, which are described by the interfacial forces appearing on the right-hand side Eq. (7.2). These terms need to be determined. We start with establishing the correlation between velocity fluctuations in the carrier flow and the particles PDF $\langle u_i' p\rangle$. We shall model the velocity field of the carrier flow by a Gaussian process with a given autocorrelation function. Then, using the Furutsu–Donsker–Novikov formula (1.201) for Gaussian random processes, we obtain

$$\langle u_i' p\rangle = \iint u_i'(\boldsymbol{X}, t) u_k'(\boldsymbol{X}_1, t_1)\left\langle \frac{\delta p(\boldsymbol{X}, t)}{\delta u_k(\boldsymbol{X}_1, t_1)}\right\rangle d\boldsymbol{X}_1 dt_1, \tag{7.6}$$

where

$$\frac{\delta p(\boldsymbol{X},t)}{\delta u_k(X_1,t_1)} = -\frac{\partial}{\partial X_j}\left\langle p(\boldsymbol{X},t)\frac{\delta R_{pj}(t)}{\delta u_k(X_1,t_1)}\right\rangle - \frac{\partial}{\partial v_j}\left\langle p(\boldsymbol{X},t)\frac{\delta v_{pj}(t)}{\delta u_k(X_1,t_1)}\right\rangle.$$

To find the functional derivatives here, one needs to solve the equation of motion (7.2) for the particle. The main difficulty is that the expression for $\boldsymbol{f}_A$ contains $D\boldsymbol{u}/Dt$ – a substantial derivative of fluid velocity along the trajectory of the carrier flow. We take it to be equal to the derivative along the particles trajectory $d\boldsymbol{u}/dt$. We also assume that the averaged velocity slip at the interface is zero and the transverse force f_A is negligibly small.

The first assumption is based on the hypothesis that by knowing the duration of the particles interaction with turbulent eddies, we can take into account the effect of the averaged slip at the interface on the fluctuational motion of the disperse phase (the so-called effect of trajectory intersection [13]) in the proper manner. As for the second assumption, insignificance of the lifting force caused by the velocity shear can be proved rigorously only for the case of low Reynolds numbers. Contribution of this effect to the correlation $\langle u_i' p\rangle$ will still be insignificant at finite, but moderate, values of Re_p.

Applying the Laplace transform to Eq. (7.2), we obtain for the particles velocity image:

$$\tilde{v}_{pi}(s) = \frac{1 + At_u s + k_B\tau_u\sqrt{\pi s}}{1 + t_u s + k_B\tau_u\sqrt{\pi s}}\tilde{u}_i(s) = \phi(s)\tilde{u}_i(s). \tag{7.7}$$

The transition to the original is carried out by using the convolution formula

$$v_{pi}(t) = \int_0^t \gamma(t-t_1)u_i(\boldsymbol{R}_p(t_1),t_1)dt_1,\ \boldsymbol{R}_{pi}(t) = \int_0^t v_{pi}(\boldsymbol{R}_p(t)dt, \tag{7.8}$$

where $\gamma(t)$ is the Green function for Eq. (7.2).

Applying functional differentiation to the second equation (7.8), one obtains a system of integral equations in functional derivatives:

$$\frac{\delta v_{pi}(t)}{\delta u_j(\boldsymbol{X}_1,t_1)} = \delta_{ij}\delta(\boldsymbol{X}_1 - \boldsymbol{R}_p(t_1))\gamma(t-t_1)H(t-t_1)$$

$$+ \int_{t_1}^t \gamma(t-t_2)\frac{\partial u_i(\boldsymbol{R}_p(t_2),t_2)}{\partial X_n}\frac{\delta R_{pn}(t_2)}{\delta u_j(X_1,t_1)}dt_2, \tag{7.9}$$

$$\frac{\delta R_{pi}(t)}{\delta u_j(\boldsymbol{X}_1,t_1)} = \delta_{ij}\delta(\boldsymbol{X}_1 - \boldsymbol{R}_p(t_1))\int_{t_1}^t \gamma(t-t_2)dt_2$$

$$+ \int_{t_1}^t\int_{t_2}^t \gamma(t-t_3)dt_3\frac{\partial u_i(\boldsymbol{R}_p(t_2),t_2)}{\partial X_n}\frac{\delta R_{pn}(t_2)}{\delta u_j(X_1,t_1)}dt_2, \tag{7.10}$$

where $H(x)$ is the Heaviside function.

Integral equations (7.9) and (7.10) are solved by the iterative method. As the first approximation, we take the first term on the right-hand side of the equation. This results in a solution that corresponds to a homogeneous, shearless flow. The second approximation takes into account the inhomogeneity of the flow up to the first order spatial derivatives. This yields

$$\frac{\delta R_{pi}(t)}{\delta u_j(\boldsymbol{X}_1,t_1)} = \left(\frac{\delta R_{pi}(t)}{\delta u_j(\boldsymbol{X}_1,t_1)}\right)_1 + \left(\frac{\delta R_{pi}(t)}{\delta u_j(\boldsymbol{X}_1,t_1)}\right)_2,$$

$$\left(\frac{\delta R_{pi}(t)}{\delta u_j(\boldsymbol{X}_1,t_1)}\right)_1 = \delta_{ij}\delta(\boldsymbol{X}_1-\boldsymbol{R}_p(t_1))\int_{t_1}^{t}\gamma(t-t_2)dt_2$$

$$\left(\frac{\delta R_{pi}(t)}{\delta u_j(\boldsymbol{X}_1,t_1)}\right)_2 = \delta(\boldsymbol{X}_1-\boldsymbol{R}_p(t_1)) \times \int_{t_1}^{t}\int_{t_2}^{t}\gamma(t-t_3)dt_3\int_{t_1}^{t_2}\gamma(t_2-t_4)dt_4\frac{\partial u_i(\boldsymbol{R}_p(t_2),t_2)}{\partial X_j}dt_2. \tag{7.11}$$

Then Eq. (7.9) takes the form

$$\begin{aligned}\frac{\delta v_{pi}(t)}{\delta u_j(\boldsymbol{X}_1,t_1)} &= \delta_{ij}\delta(\boldsymbol{X}_1-\boldsymbol{R}_p(t_1))\gamma(t-t_1)H(t-t_1)\\ &+\int_{t_1}^{t}\gamma(t-t_2)\frac{\partial u_i(\boldsymbol{R}_p(t_2),t_2)}{\partial X_n}\left(\frac{\delta R_{pn}(t_2)}{\delta u_j(X_1,t_1)}\right)_1 dt_2\\ &+\int_{t_1}^{t}\gamma(t-t_2)\frac{\partial u_i(\boldsymbol{R}_p(t_2),t_2)}{\partial X_n}\left(\frac{\delta R_{pn}(t_2)}{\delta u_j(X_1,t_1)}\right)_2 dt_2.\end{aligned} \tag{7.12}$$

Approximating the last term in Eq. (7.12) with the accuracy up to the first order spatial derivatives and making use of the second equation (7.8), we obtain

$$\int_{t_1}^{t}\gamma(t-t_2)\frac{\partial u_i(\boldsymbol{R}_p(t_2),t_2)}{\partial X_n}\left(\frac{\delta R_{pn}(t_2)}{\delta u_j(X_1,t_1)}\right)_2 dt_2 = \frac{\partial v_{pi}(\boldsymbol{R}_p(t_2),t_2)}{\partial X_n}\left(\frac{\delta R_{pn}(t)}{\delta u_j(X_1,t_1)}\right)_2. \tag{7.13}$$

We now introduce a two-time autocorrelation function of velocity fluctuations in the carrier flow along the particle trajectory,

$$\Psi_{Lp}(t-t_1) = \frac{\langle u_i'(\boldsymbol{X},t)u_j'(\boldsymbol{R}_p(t_1),t_1)\rangle}{\langle u_i'(\boldsymbol{X},t)u_j'(\boldsymbol{X},t)\rangle}.$$

In view of rapid decrease of the function $\psi_{Lp}(\xi)$ with increase of ξ, we assume that the major contribution to the integrals comes from the region in the vicinity of $\xi = 0$.

Then it follows from Eq. (7.6), Eqs. (7.11)–(7.13)(see [12]) that

$$\langle u_i' p\rangle = \langle u_i' u_k'\rangle \left(f_u \frac{\partial P}{\partial v_k} + t_u g_u \frac{\partial P}{\partial X_k} + t_u l_u \frac{\partial U_n}{\partial X_k}\frac{\partial P}{\partial v_n} \right.$$

$$\left. + t_u^2 h_u \frac{\partial U_n}{\partial X_k}\frac{\partial P}{\partial X_n} + t_u^2 h_u \frac{\partial U_n}{\partial X_k}\frac{\partial V_j}{\partial X_k}\frac{\partial P}{\partial v_j} \right). \tag{7.14}$$

The first two terms on the right-hand side describe the interactions of particles with turbulent eddies in a homogeneous shearless carrier flow, and the last three terms describe the influence of the velocity gradient. The coefficients in Eq. (7.14) are called the involvement factors. They are equal to

$$f_u = \int_0^\infty \psi_{Lp}(\xi)\gamma(\xi)d\xi, \quad g_u = \frac{1}{t_u}\int_0^\infty \psi_{Lp}(\xi)\int_0^\xi \gamma(\xi_1)d\xi_1 d\xi,$$

$$l_u = \frac{1}{t_u}\int_0^\infty \psi_{Lp}(\xi)\int_0^\xi \gamma(\xi-\xi_1)\int_0^{\xi_1} \gamma(\xi_1-\xi_2)d\xi_2 d\xi_1 d\xi, \tag{7.15}$$

$$h_u = \frac{1}{t_u^2}\int_0^\infty \psi_{Lp}(\xi)\int_0^\xi\int_{\xi_1}^\xi \gamma(\xi-\xi_3)d\xi_3 \int_0^{\xi_1} \gamma(\xi_1-\xi_2)d_2 d\xi_1 d\xi.$$

The relations (7.14) and (7.15) are valid for times t much greater than the Lagrangian integral time scale $T_p^{(L)}$ of fluctuations of the carrier flow velocity along the particle trajectory. This time can be considered as the characteristic time of particle interaction with highly energetic turbulent eddies of the carrier flow. The coefficients f_u, g_u, l_u, h_u tell us how strongly the particle is involved into fluctuational motion of the turbulent carrier flow. In order to obtain these coefficients, one needs to know the autocorrelation function $\psi_{Lp}(\xi)$. It is often approximated by the expression

$$\psi_{Lp}(\xi) = \exp(-\xi/T_p^{(L)}).$$

Then, in view of Eq. (7.7), the involvement factors (7.15) are

$$f_u = \phi(s = (T_p^{(L)})^{-1}) = \frac{1 + A\Omega_u + B\Omega_u}{1 + \Omega_u + B\Omega_u}, \quad g_u = \frac{\phi(s = (T_p^{(L)})^{-1})}{\Omega_u}, \tag{7.16}$$

$$l_u = \frac{\phi^2(s = (T_p^{(L)})^{-1})}{\Omega_u}, \quad h_u = \frac{\phi^2(s = T_p^{(L)})}{\Omega_u^2}, \quad \Omega_u = \frac{t_u}{T_p^{(L)}}, \quad B = k_B\sqrt{\pi T_p^{(L)}}.$$

Eq. (7.14) gives us the mixed correlation moments between velocity fluctuations of the continuous and disperse phases:

$$\begin{aligned}\langle u_i' v_j' \rangle &= \frac{1}{\varphi}\left(\int \langle u_i' p\rangle v_j d\mathbf{v} - V_j \int \langle u_i' p\rangle d\mathbf{v}\right) \\ &= f_u \langle u_i' u_j' \rangle + t_u \langle u_i' u_k' \rangle \left(l_u \frac{\partial U_j}{\partial X_k} - g_u \frac{\partial V_j}{\partial X_k} \right), \qquad (7.17) \\ \varphi &= \int p d\mathbf{v}, \quad v_i = \frac{1}{\varphi}\int v_i P d\mathbf{v},\end{aligned}$$

where φ and v_i are, respectively, the averaged volume concentration and the averaged velocity of the disperse phase.

The expression (7.16) connects the mixed correlation moments between velocity fluctuations of the continuous and disperse phases and the Reynolds stresses. In the case of a homogeneous turbulent flow, the expression (7.17) simplifies and, in view of Eq. (7.16), takes the form

$$\langle u_i' v_j \rangle = f_u \langle u_i' u_j' \rangle = \frac{1 + A\Omega_u + B\Omega_u}{1 + \Omega_u + B\Omega_u} \langle u_i' u_j' \rangle. \qquad (7.18)$$

With the Basset force neglected, $B=0$, it follows from Eq. (7.18) that due to the action of non-stationary forces caused by the pressure gradient in the fluid (the buoyant force) and by the effect of the virtual mass, the involvement factor f_u for light particles (at $A>1$) could be greater than 1 and, consequently, the mixed correlation moments could exceed the Reynolds stresses in the fluid. The Basset force increases the instantaneous resistance to the flow and thus exerts a smoothing effect, and in doing so it increases f_u for heavy particles and decreases it for light ones.

Having determined the correlation $\langle u_i' p\rangle$ associated with the resistance force, we now need to find the correlation moments of the particle PDF with the other forces acting on particles. We shall neglect the third- and higher-order derivatives of the PDF. Making use of the continuity equation for incompressible fluid $\partial u_k/\partial X_k = 0$, we can write the correlation moment due to the fluctuations of the buoyancy force and the virtual mass force:

$$\langle f_{Ai}' p \rangle = A\left(\frac{D_0 \langle u_i' p\rangle}{Dt} - \left\langle u_i' \frac{Dp}{Dt} \right\rangle + \langle u_i' p \rangle \frac{\partial U_i}{\partial X_k} + \frac{\partial \langle u_i' u_k' p\rangle}{\partial X_k} \right). \qquad (7.19)$$

$$\begin{aligned}\left\langle u_i' \frac{Dp}{Dt} \right\rangle &= -\left\langle u_i' \frac{\partial}{\partial v_k} \left\{ \left[\frac{u_k - v_{pk}}{t_u} + f_{Ak} + f_{Bk} + f_{Lk} + f_{gk} \right] p \right\} \right\rangle \\ &+ \left\langle u_i'(u_k - v_{pk}) \frac{\partial p}{\partial X_k} \right\rangle = -\left(\frac{U_k - V_k}{t_u} + F_{Ak} + F_{Bk} + F_{Lk} + F_{gk} \right) \frac{\partial \langle u_i' p\rangle}{\partial X_k} \\ &- \frac{\partial}{\partial v_k}\left(\frac{\langle u_i' u_k' p\rangle - \langle u_i' v_k' p\rangle}{t_u} + \langle u_i' f_{Ak}' p\rangle + \langle u_i' f_{Bk}' p\rangle + \langle u_i' f_{Lk}' p\rangle \right) \\ &+ (U_k - V_k)\left\langle u_i' \frac{\partial p}{\partial X_k} \right\rangle + \left\langle u_i' u_k' \frac{\partial p}{\partial X_k} \right\rangle - \left\langle u_i' v_k' \frac{\partial p}{\partial X_k} \right\rangle.\end{aligned} \qquad (7.20)$$

Dropping the terms in Eq. (7.20) that contain second- and higher-order derivatives of p and neglecting contributions from the correlation moments of fluid velocity fluctuations with fluctuations of the Basset force $\langle u'_i f'_{Bk}\rangle$ and from the buoyant force $\langle u'_i f'_{Lk}\rangle$, we get

$$\left\langle u'_i \frac{Dp}{Dt}\right\rangle = -\left(\frac{\langle u'_i u'_k\rangle - \langle u'_i v'_k\rangle}{t_u} + \langle u'_i f'_{Ak}\rangle + \langle u'_i f'_{Bk}\rangle + \langle u'_i f'_{Lk}\rangle\right)\frac{\partial P}{\partial v_k} + \left(\langle u'_i v'_k\rangle - \langle u'_i v'_k\rangle\right)\frac{\partial P}{\partial X_k}. \tag{7.21}$$

$$\langle u'_i f'_{Ak}\rangle = \left(A\left\langle u'_i \frac{\partial u'_k}{\partial t}\right\rangle + U_n\left\langle u'_i \frac{\partial u'_k}{\partial X_n}\right\rangle + \langle u'_i u'_n\rangle \frac{\partial U_k}{\partial X_n} + \left\langle u'_i \frac{\partial u'_k u'_n}{\partial X_n}\right\rangle\right). \tag{7.22}$$

$$\langle u'_i u'_k p\rangle = \langle u'_i u'_k\rangle P. \tag{7.23}$$

When determining $\langle u'p\rangle$ in the first and the third term in Eq. (7.19), we shall confine ourselves to the first term in Eq. (7.14). Then, in view of Eqs. (7.21)–(7.23), the expression (7.19) takes the form

$$\begin{aligned}\langle f'_i p\rangle = A\Bigg[&-\frac{D_0}{Dt}\left(f_u\langle u'_i u'_k\rangle \frac{\partial P}{\partial v_k}\right) + \frac{\langle u'_i u'_k\rangle - \langle u'_i v'_k\rangle}{t_u}\frac{\partial P}{\partial v_k}\\ &+ A\left(\left\langle u'_i \frac{\partial u'_k}{\partial t}\right\rangle + U_n\left\langle u'_i \frac{\partial u'_k}{\partial X_n}\right\rangle + \langle u'_i u'_n\rangle \frac{\partial U_k}{\partial X_n} + \left\langle u'_i \frac{\partial u'_k}{\partial X_n}\right\rangle\right)\frac{\partial P}{\partial v_k}\\ &+ \langle u'_i v'_k\rangle \frac{\partial P}{\partial X_k} - f_u\langle u'_k v'_n\rangle \frac{\partial U_i}{\partial X_k}\frac{\partial P}{\partial v_n} + \frac{\partial\langle u'_i v'_k\rangle}{\partial X_k} P\Bigg].\end{aligned} \tag{7.24}$$

To simplify our analysis of the correlation, we can drop the term $\langle f'_i p\rangle$ in Eq. (7.5), because the Basset force is, as a rule, insignificant and, as opposed to non-stationary forces caused by the pressure gradient in the carrier flow and by the virtual mass acceleration, does not cause any qualitative changes but only slightly attenuates non-stationary effects. Then the correlation between fluctuations of the buoyant force and the particle PDF is

$$\begin{aligned}\langle f'_i p\rangle = L\Bigg[&(U_j - V_j)\left\langle\left(\frac{\partial u'_j}{\partial X_i} - \frac{\partial u'_i}{\partial X_j}\right)p\right\rangle + \left(\frac{\partial U_j}{\partial X_i} - \frac{\partial U_i}{\partial X_j}\right)\langle u'_j p\rangle\\ &+ \left\langle (u'_j - v'_j)\left(\frac{\partial u'_j}{\partial X_i} - \frac{\partial u'_i}{\partial X_j}\right)p\right\rangle\Bigg].\end{aligned} \tag{7.25}$$

At large Reynolds number of the turbulent carrier flow, one can neglect any correlations between fluid vorticity fluctuations and velocities of the continuous and disperse phases. Then, retaining only the first term in the relation (7.14) for $\langle u_i' p\rangle$, we can represent Eq. (7.25) as

$$\langle f_{Li}' p\rangle = -L f_u \langle u_j' u_k'\rangle \left(\frac{\partial U_j}{\partial X_i} - \frac{\partial U_i}{\partial X_j}\right)\frac{\partial P}{\partial v_k}. \tag{7.26}$$

Using the above assumptions and the relations (7.14), (7.24), (7.26), we obtain from Eq. (7.5) a closed kinetic equation for the PDF of particle velocity in a turbulent flow:

$$\begin{aligned}
&\frac{\partial P}{\partial} + v_i\frac{\partial P}{\partial X_i} + \frac{\partial}{\partial v_i}\left\{\left[\frac{U_i - v_i}{t_u} + A\frac{\partial\langle u_i' u_k'\rangle}{\partial X_k} + F_{Ai} + F_{Bi} + F_{Li} + F_{gi}\right]P\right\}\\
&= \frac{\partial}{\partial v_i}\left\{\langle u_i' u_k'\rangle\left(f_u\frac{\partial P}{\partial v_k} + t_u g_u\frac{\partial P}{\partial X_k} + t_u l_u\frac{\partial U_n}{\partial X_k}\frac{\partial P}{\partial v_n} + t_u^2 h_u^2\frac{\partial U_n}{\partial X_k}\frac{\partial P}{\partial X_n}\right.\right.\\
&\left. + t_u^2 h_u\frac{\partial U_n}{\partial X_k}\frac{\partial V_j}{\partial X_k}\frac{\partial P}{\partial v_j}\right) + A\left[\frac{D_0}{Dt}\left(f_u\langle u_i' u_k'\rangle\frac{\partial P}{\partial v_k}\right) + \frac{\langle u_i' v_k'\rangle - \langle u_i' u_k'\rangle}{t_u}\frac{\partial P}{\partial v_k}\right.\\
&- A\left(\left\langle u_i'\frac{\partial u_k'}{\partial t}\right\rangle + U_n\left\langle u_i'\frac{\partial u_k'}{\partial t}\right\rangle + \langle u_i' u_n'\rangle\frac{\partial U_k}{\partial X_n} + \left\langle u_i'\frac{\partial u_k' u_n'}{\partial X_n}\right\rangle\right)\frac{\partial P}{\partial v_k}\\
&\left.\left. - \langle u_i' u_k'\rangle\frac{\partial P}{\partial X_k} + f_u\langle u_k' u_n'\rangle\frac{\partial U_i}{\partial X_k}\frac{\partial P}{\partial v_n}\right] + L f_u\langle u_j' u_k'\rangle\left(\frac{\partial U_j}{\partial X_i} - \frac{\partial U_i}{\partial X_j}\right)\frac{\partial P}{\partial v_k}\right\}.
\end{aligned} \tag{7.27}$$

Terms on the left-hand side describe the convection of the PDF in the phase space of velocities and coordinates, while terms on the right-hand side characterize the diffusional transport due to particle interactions with turbulent eddies of the carrier flow in the same phase space.

The equation (7.27) enables us to obtain a system of continuum equations for the averaged characteristics (moments) of the disperse phase. Integrating over the whole volume of the velocity space, we get the mass conservation equation

$$\frac{\partial\varphi}{\partial t} + \frac{\partial\varphi V_k}{\partial X_k} = 0. \tag{7.28}$$

Multiplication of both sides of Eq. (7.27) by v_i and a further integration over v yields a balance equation for the momentum:

$$\begin{aligned}
\frac{\partial V_i}{\partial t} + V_k\frac{\partial V_i}{\partial X_k} &= -\frac{\partial}{\partial X_k}\left(\langle v_i' v_k'\rangle - A\langle u_i' u_k'\rangle\right)\\
&+ \frac{U_i - V_i}{t_u} + F_{Ai} + F_{Bi} + F_{Li} + F_{gi} - \frac{D_{pik}}{t_u}\frac{\partial\ln\varphi}{\partial X_k},
\end{aligned} \tag{7.29}$$

$$\langle v_i' v_j'\rangle = \frac{1}{\varphi}\int (v_i - V_i)(v_j - V_j) P d\boldsymbol{v},$$

where $\langle v_i' v_j' \rangle$ are turbulent stresses in the disperse phase caused by the participation of particles in the fluctuational motion of the carrier flow. The first term on the right-hand side of Eq. (7.29) describes turbulent migration of particles caused by the emergence of turbulent stresses in the disperse phase and by the action of turbulent stresses in the carrier flow. The relation between these two terms defines the direction of turbulent migration. The last term describes turbulent diffusion of particles, which is characterized by the diffusion coefficient

$$D_{pij} = t_u \left(\langle v_i' v_j' \rangle + g_u \langle u_i' u_j' \rangle + t_u h_u \langle u_i' u_k' \rangle \frac{\partial U_j}{\partial X_k} - A \langle u_i' u_j' \rangle \right). \tag{7.30}$$

It should be noted that the particles turbulent diffusion tensor, as well as the mixed correlation moment of velocity fluctuations in the continuous and disperse phases (7.17), is non-symmetric in shear flows.

The equation for the second moments of velocity fluctuations is obtained from Eq. (7.27) by multiplying both sides of this equation by $v_i v_j$ and integrating over v:

$$\begin{aligned}
&\frac{\partial \langle v_i' v_j' \rangle}{\partial t} + V_k \frac{\partial \langle v_i' v_j' \rangle}{\partial X_k} + \frac{1}{\varphi} \frac{\partial \varphi \langle v_i' v_j' v_k' \rangle}{\partial X_k} = -\langle v_i' v_k' \rangle \frac{\partial V_j}{\partial X_k} - \langle v_j' v_k' \rangle \frac{\partial V_i}{\partial X_k} \\
&+ (1+A)\langle u_i' u_k' \rangle \left(l_u \frac{\partial U_j}{\partial X_k} - g_u \frac{\partial V_i}{\partial X_k} \right) + (1+A)\langle u_j' u_k' \rangle \left(l_u \frac{\partial U_i}{\partial X_k} - g_u \frac{\partial V_i}{\partial X_k} \right) \\
&+ \frac{2}{t_u} \left(f_u \langle u_i' u_j' \rangle - \langle v_i' v_j' \rangle \right) + A \left\{ \frac{2(f_u - 1)}{t_u} \langle u_i' u_j' \rangle + f_u \right. \\
&\times \left[\langle u_i' u_k' \rangle \left(\frac{\partial U_j}{\partial X_k} + \frac{\partial V_j}{\partial X_k} \right) + \langle u_j' u_k' \rangle \left(\frac{\partial U_i}{\partial X_k} - \frac{\partial V_i}{\partial X_k} \right) \right] + \frac{2}{\varphi} \frac{D_0}{Dt} \left(f_u \varphi \langle u_i' u_j' \rangle \right) \\
&\left. - A \left(\frac{D_0 \langle u_i' u_j' \rangle}{Dt} + \langle u_i' u_k' \rangle \frac{\partial U_j}{\partial X_k} + \langle u_j' u_k' \rangle \frac{\partial U_i}{\partial X_k} + \frac{\partial \langle u_i' u_j' u_k' \rangle}{\partial X_k} \right) \right\} \\
&+ L \left[\left(f_u \langle u_i' u_k' \rangle - \langle v_i' v_k' \rangle \right) \left(\frac{\partial U_k}{\partial X_j} - \frac{\partial U_j}{\partial X_k} \right) \right. \\
&\left. + \left(f_u \langle u_j' u_k' \rangle - \langle v_j' v_k' \rangle \right) \left(\frac{\partial U_k}{\partial X_i} - \frac{\partial U_i}{\partial X_k} \right) \right].
\end{aligned} \tag{7.31}$$

Eq. (7.31) describes convective and diffusional transport, emergence of fluctuations within the averaged shear flow, generation of fluctuations due to the involvement of particles into fluctuational motion of the carrier flow, and dissipation of turbulent energy of the disperse phase due to the work of interfacial interaction forces. In a uniform shearless turbulent flow, or for small particles in the framework of the locally homogeneous approximation, there follows from Eq. (7.31) a simple algebraic expression for the stress tensor in the disperse phase:

$$\langle v_i' v_j' \rangle = \left(f_u (1+A) - A \right) \langle u_i' u_j' \rangle = \frac{1 + A^2 \Omega_u}{1 + \Omega_u} \langle u_i' u_j' \rangle. \tag{7.32}$$

The relation (7.32) is identical to a known formula for the fluctuational energy of a particle in a homogeneous, isotropic turbulent flow [14]. Under these conditions, the coefficient of particles turbulent diffusion (7.30) takes the form

$$D_{pij} = t_u(\langle v_i' v_j' \rangle + g_u \langle u_i' u_j' \rangle + A\ f_u \langle u_i' u_j' \rangle) = T_p^{(L)} \langle u_i' u_j' \rangle, \tag{7.33}$$

which corresponds to the solution obtained in [14].

In view of Eq. (7.32), we can describe acceleration resulting from the migration force that is caused by particle interactions with turbulent eddies of the carrier flow by the following expression:

$$F_{Mi} = -\frac{\partial}{\partial X_k}(\langle v_i' v_k' \rangle - A\langle u_i' u_k' \rangle) = -\frac{\partial}{\partial X_k} M\langle u_i' u_k' \rangle, \quad M = \frac{(1-A)(1-A^2\Omega_u)}{1+\Omega_u}.$$

The migration coefficient M is defined by the particle inertia parameters Ω_u and A, which, in their turn, depend on the density ratio ρ^0 and on the virtual mass factor C_A commonly taken to be equal to 0.5. For heavy particles ($A \to 0$), we have $M > 0$ and consequently, these particles move under the action of turbulent migration from highly turbulent regions into regions of low turbulence. But at $A = 1$ and $A = \Omega_u^{-1}$ the sign of M and thereby the direction of the migration force changes. Thus, at $A < 1$, migration of large (inertial) particles and at $A > 1$ – migration of small (inertialess) particles (bubbles) is directed from the regions of low turbulence toward the regions of high fluctuation level. It is interesting to note that turbulent migration will displace large bubbles into low-turbulence regions. At $A = 1$ ($\rho^0 = 1$), both the factor M and the migration force vanish.

One of the most important characteristics of particle behavior in a turbulent flow is the duration of particle interaction with high-energy eddies. For very small inertialess particles, the interaction time $T_p^{(L)}$ coincides with Lagrangian scale of turbulence $T^{(L)}$ measured along the trajectories of liquid particles. For large particles, the time $T_p^{(L)}$ may significantly differ from $T^{(L)}$, and the ratio $T_p^{(L)}/T^{(L)}$ may be greater or smaller than 1 depending on the values of parameters responsible for the particles inertia and for the averaged interfacial velocity slip.

To get an explicit connection between Lagrangian and Eulerian characteristics of turbulence, one should use the Corrsin hypothesis [15] about the possibility of independent representation (and consequently, of averaging) of random velocity fluctuations and fluid particle displacement fields. In accordance with this hypothesis, the Lagrangian time autocorrelation functions of fluid velocity fluctuations calculated along the particle trajectories are connected with Eulerian spacetime autocorrelation functions in a stationary homogeneous isotropic turbulent field by the following relations:

$$\Psi_{Lp}^{(L)}(\boldsymbol{t}) = \int \Psi_L^{(E)}(\boldsymbol{r}, t)\Phi_p(\boldsymbol{r}, t)d\boldsymbol{r}, \Psi_{Np}^{(L)}(t) = \int \Psi_N^{(E)}(\boldsymbol{r}, t)\Phi_p(\boldsymbol{r}, t)d\boldsymbol{r}. \tag{7.34}$$

Here L and N indicate the directions parallel and transverse to the particles relative velocity vector $\mathbf{W} = \mathbf{V} - \mathbf{U}$, and $\Phi_p(\mathbf{r}, t)$ is the probability density of particle displacement over the distance r during the time t that is given by the delta function

$$\Phi_p(\mathbf{r},\boldsymbol{t}) = \delta\left(\mathbf{r} - \mathbf{W}t - \frac{u_0\psi(\boldsymbol{t})}{\sqrt{3}}\mathbf{s}\right), \tag{7.35}$$

where $u_0\psi(t)$ is the effective free path of a particle in its random motion, $\mathbf{s}$ is a unit vector in the direction $\mathbf{r}$, $u_0^2 = \langle u'_k u'_k \rangle/3$ is the intensity of turbulence. Eulerian spacetime correlation function is represented as

$$\Psi_{ij}^{(E)}(\boldsymbol{r},t) = \Psi_{ij}^{(E)}(\boldsymbol{r})\Psi^{(E)}(t), \tag{7.36}$$

and if the turbulence is isotropic, then

$$\begin{aligned} \Psi_{ij}^{(E)}(\boldsymbol{r}) &= [f(r) - g(r)]\frac{r_i r_j}{r^2} + g(r)\delta_{ij}, \\ g(r) &= f(r) + \frac{r}{2}f'(r), r = |r|. \end{aligned} \tag{7.37}$$

Combining Eq. (7.34) with Eqs. (7.35)– (7.37), we see that

$$\begin{aligned} \Psi_{Lp}^{(L)}(t) &= \left[\frac{f(r) - g(r)}{r^2}\left(Wt + \frac{u_0\psi(t)}{\sqrt{3}}\right)^2 + g(r)\right]\Psi^{(E)}(t), \\ \Psi_{Lp}^{(L)}(t) &= \left[\frac{f(r) - g(r)}{r^2}\frac{u_0^2\psi^2(t)}{\sqrt{3}} + g(r)\right]\Psi^{(E)}(t), \\ r &= \sqrt{\left(Wt + \frac{u_0\psi(t)}{\sqrt{3}}\right)^2 + \frac{2}{3}u_0^2\psi^2(t)}, \ W = |\boldsymbol{W}|. \end{aligned} \tag{7.38}$$

The function $\psi(t)$ characterizes the particles effective free path associated with its involvement into fluctuational motion of the carrier flow; it can be determined by solving the particles equation of motion (7.2). Retaining only the two most essential terms on the right-hand side of this equation, one obtains the approximation

$$\psi(r) = t + (1 - A)t_u[\exp(-t/t_u)]. \tag{7.39}$$

For Eulerian correlation functions, one commonly uses the following exponential dependences:

$$f(r) = \exp(-r/L), \Psi^{(E)}(t) = \exp(-t/T^{(E)}), \tag{7.40}$$

where L and $T^{(E)}$ are the Eulerian spatial integral and time integral scales. The formulas (7.40) give a good approximation of correlation functions at large Reynolds numbers, even though they are not correct when r and t tend to zero.

Eqs. (7.38)–(7.40) can be used to determine Lagrangian time integral scales of fluid velocity fluctuations along the particle trajectories:

$$T_{Lp}^{(L)} = \int_0^\infty \Psi_{Lp}^{(L)}(t)dt, \quad T_{Np}^{(L)} = \int_0^\infty \Psi_{Np}^{(L)}(t)dt. \tag{7.41}$$

Formulas (7.41), together with (7.38) and (7.39), tell us how the duration of particle interaction with turbulent eddies is influenced by the particles inertia, by the density ratio of the carrier and disperse phases, and by the effect of trajectory intersection due to the averaged interfacial velocity slip. As a consequence, there follows the relation between the Lagrangian and Eulerian integral time scales of turbulence:

$$\frac{T^{(L)}}{T^{(E)}} = \frac{3+2m}{3(1+m)}, \quad m = \frac{u_0 T^{(E)}}{L}. \tag{7.42}$$

It follows from Eq. (7.42) that the Eulerian time macroscale defined in a reference frame that is moving with the average velocity of the carrier flow is greater than the corresponding Lagrangian scale. Numerical solution allows to trace the influence of the Stokes number $\mathrm{St} = t_u/T^{(E)}$ and the parameter A on the duration of particle interaction with turbulent eddies (Fig. 7.1). For heavy particles ($A < 1$), $T_p^{(L)}$ increases monotonously with St; at $A = 1$, we have $T_p^{(L)} = T^{(L)}$; and for light particles (bubbles) ($A > 1$), the value of $T_p^{(L)}$ decreases as the Stokes number St gets larger. The effect of trajectory intersection is characterized by the parameter $\zeta = W/u_0$. For all values of A, an increase of the average interfacial slip (that is, an increase of ζ), causes $T_p^{(L)}$ to fall (see Fig. 7.2). A decrease of the density ratio of the carrier and disperse phases (i.e., an increase of A) would be qualitatively similar to increased averaged slip.

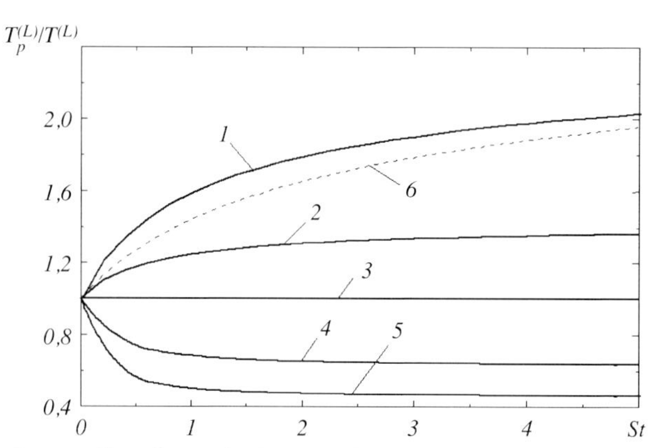

Fig. 7.1 Effect of St on the duration of particles interactions with turbulent eddies: 1–5 – A = 0.05, 1, 2, 3; 6 – [16].

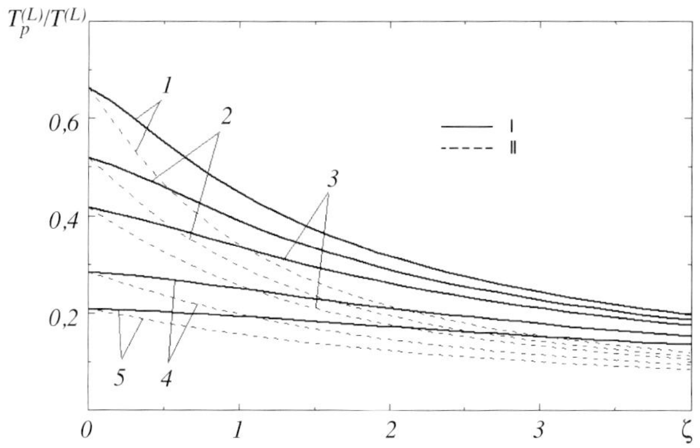

Fig. 7.2 Effect of the averaged slip on the duration of particles interaction with turbulent eddies: 1–5 – A=0.05, 1, 2, 3; I, II – directions parallel and transverse to the relative velocity vector.

In the limit of extremely inertial particles, that is, at $\mathrm{St} \to \infty$, we find from Eqs. (7.37)–(7.42):

$$\frac{T_{Lp}^{(L)}}{T^{(E)}} = \frac{1 + 0.5\, mN + m(\zeta + A/\sqrt{3})^2 (2N)^{-1}}{(1 + mN)^2},$$

$$\frac{T_{Np}^{(L)}}{T^{(E)}} = \frac{1 + 0.5mN + mA^2 (6N)^{-1}}{(1 + mN)^2},$$

$$N = \sqrt{\frac{2A^2}{3} + \left(\zeta + \frac{A}{\sqrt{3}}\right)^2}. \tag{7.43}$$

In the other limiting case, specifically, that of strong influence of the effect of trajectory intersection (in other words, for large values of the drift parameter ζ), we have for all values of A

$$T_{Lp}^{(L)} = \frac{L}{W}, \quad T_{Np}^{(L)} = \frac{L}{2W}. \tag{7.44}$$

Based on the obtained equation of motion for the disperse phase, we can now examine the distribution of bubbles over the cross section of a long vertical tube of diameter D. The flow is assumed to be stationary and hydrodynamically developed. The disperse phase is assumed to have a low volume concentration so that we can neglect the feedback influence of bubbles on turbulent characteristics of the carrier flow as well as the effect of particle collisions. In the present case we can take $\rho^0 \to \infty$.

Parameters of a hydrodynamically developed flow vary only with the radial coordinate r and are independent of the longitudinal coordinate X, while the averaged radial components of velocity of the continuous and disperse phases are equal to zero, $U_r = V_r = 0$. Then the distribution of bubbles over the tubes cross section can be found by solving Eq. (7.29) projected onto the r-direction. We represent fluctuational energy and the coefficient of turbulent diffusion in Eq. (7.29) in the framework of the locally homogeneous approximation (7.32) and (7.33). The intensities of turbulent fluctuations of the radial and azimuthal velocity components of the disperse phase are assumed to be equal. Then it follows from Eq. (7.29) that

$$\frac{T_p^{(L)}}{t_u}\langle u_r'2\rangle\frac{d\ln\varphi}{dr}=-\frac{dM\langle u'2\rangle}{dr}+\frac{C_L}{C_A}(U_x-V_x)\frac{dU_x}{dr}. \qquad (7.45)$$

It is readily seen that the concentration profile of bubbles in the cross section of the tube is shaped by turbulent migration and by the buoyant force caused by velocity shear. To simplify the analysis, we assume that the average phase slip (drift velocity) in Eq. (7.45) is defined by the buoyant force only, while the effect of turbulent transport can be neglected in the first approximation. Then

$$U_x-V_x=\mp g t_v, \qquad (7.46)$$

where the $\mp$ signs relate to the upward and downward flows respectively.

In the bubble size range for which the resistance factor C_D is approximately constant, we have instead of Eq. (7.46)

$$U_x-V_x=\mp\sqrt{\frac{4d_p g}{3C_D}}. \qquad (7.47)$$

According to Eq. (7.44), the duration $T_p^{(L)}$ of bubbles interaction with turbulent eddies under the condition $|U_x-V_x|\gg u_*$ (where u_* is the dynamic velocity) can be written as

$$T_p^{(L)}=\frac{L}{2|U_x-V_x|}, \quad L=0.1\,D, \qquad (7.48)$$

where D is the diameter of the tube.

The parameter

$$\Omega_u=\frac{t_u}{T_p^{(L)}}=\frac{80C_A}{3C_D}\tilde{d}, \quad \tilde{d}=\frac{d_p}{D} \qquad (7.49)$$

is responsible for the bubbles inertia and depends only on the ratio of diameters of the bubble and the tube.

The gradient of the averaged fluid velocity in a hydrodynamically developed flow is given by

$$\frac{U_x}{dr}=\frac{u_*^2 r}{R(\nu_e+\nu_t)}, \qquad (7.50)$$

where ν_t is the coefficient of turbulent viscosity and $R=D/2$ is the radius of the tube.

In view of Eq. (7.47) and Eq. (7.50), we can write Eq. (7.45) in the dimensionless form as

$$\langle\tilde{u}'^2_r\rangle\frac{d\ln\varphi}{d\tilde{r}} = -\Omega_u\frac{dM\langle\tilde{u}'^2\rangle}{d\tilde{r}} \mp \frac{\tilde{C}\sqrt{d^0_p\,Ga\,\Omega_u\tilde{r}}}{1+\tilde{\nu}_t}, \quad \varphi(0)=\varphi_0 \tag{7.51}$$

where $\tilde{r}=r/R$, $\langle\tilde{u}'^2_r\rangle=\langle u'_r 2\rangle/u^2_*$, $\tilde{\nu}_t=\nu_t/\nu$, $\tilde{C}=C_L/(C_A\sqrt{3C_D})$, and $Ga=gD^3/\nu^2$ is the Galileo number.

The solution of Eq. (7.51) is

$$\tilde{\varphi}=\frac{\varphi}{\varphi_0}=\tilde{\varphi}_M\tilde{\varphi}_L,$$

$$\tilde{\varphi}_M=\left[\frac{(M\langle\tilde{u}'^2_r\rangle)_{r=0}}{M\langle\tilde{u}'^2_r\rangle}\right]^{\Omega_u M}\exp\left[\int_0^{\tilde{r}}\ln M\langle\tilde{u}'^2_r\rangle\frac{d\Omega_u M}{d\tilde{r}}d\tilde{r}\right], \tag{7.52}$$

$$\tilde{\varphi}_L=\exp\left[\mp\tilde{C}Ga^{1/2}\int_0^{\tilde{r}}\frac{(\tilde{d}_p)^{1/2}\Omega_u\tilde{r}}{\langle\tilde{u}'^2_r\rangle(1+\tilde{\nu}_t)}d\tilde{r}\right].$$

The expression for bubble concentration is represented as a product of two factors, one of which describes turbulent migration, while the other is responsible for the lifting effect (buoyancy) due to the velocity shear. The extent to which the concentration profile is affected by turbulent migration depends strongly on the bubble size, which determines the sign of the migration factor M and thereby the direction of the migration force; on the other hand, the direction of bubbles motion (upward or downward) is inessential. If the size of bubbles does not vary over the tube cross section, the expression for $\tilde{\varphi}_M$ in Eq. (7.52) simplifies to

$$\tilde{\varphi}_M=\left[\frac{(M\langle\tilde{u}'^2_r\rangle)_{r=0}}{M\langle\tilde{u}'^2_r\rangle}\right]^{\Omega_u M} \tag{7.53}$$

In contrast, the extent to which the concentration profile is affected by the lifting force is qualitatively independent from the bubble size and is primarily determined by the direction of motion.

In calculations, one can use the approximate formula for turbulent viscosity (the van Driest–Reichard formula)

$$\tilde{\nu}_t=\frac{1}{6}\left\{\sqrt{1+4\left[1-\exp\left(\frac{-y_+}{A}\right)\right]^2\kappa^2 y_+^2}-1\right\}(1+\tilde{r}^2), \tag{7.54}$$

where $y_+=yu_*/\nu_e=(1-\tilde{r})R_+$, $R_+=Ru_*/\nu_e$, $\kappa=0.4$, $A=26$.

The intensity of turbulent fluctuations of the radial component of fluid velocity and the Lagrangian scale of turbulence are determined by the relations

$$\langle \tilde{u}'^2_r \rangle = \frac{D_t}{T_L} = \frac{\nu_t}{\mathrm{Sc}_t T_L}, \quad T_L = \frac{\nu}{u_*^2}\sqrt{100 + \left(\frac{l u_*}{\nu_e}\right)^2}, \\ l = R(0.14 - 0.08\tilde{r}^2 - 0.06\tilde{r}^4), \tag{7.55}$$

where l is the Prandtl–Nikuradze mixing length, and Sc_t is the turbulent Schmidt number for the diffusion of an inertless (passive) impurity in a fluid; the latter is taken to be equal to 0.9.

Shown in Figures 7.3 and 7.4 are the distributions of bubbles over the tube cross section calculated on the basis of Eq. 7.52 while taking into account Eq. (7.54) and Eq. (7.55), and the experimental data for the two bubble flow regimes (downward and upward) taken, respectively, from [17] and [18] for different reduced velocities j_f and j_g of the fluid and the gas. The lifting force factor C_L depends on Re_p as well as on the parameter of velocity shear. However, as distinct from the case of low Reynolds numbers, in a flow bypassing a particle at moderate and high Reynolds numbers, these dependencies become sufficiently weak [19]. With increase of Re_p, the factor C_L approaches 0.5 for a non-viscous fluid. Thus, for the case under consideration, if the flow bypassing a bubble has a relatively high Reynolds number of $\mathrm{Re}_p \sim 100$, then $C_L = const$ will be an acceptable approximation. In calculations, the experimental value $C_L = 0.05$ [17] has been used.

Hence, the distribution of particles over the tube cross section is non-uniform, with bubble concentration reaching its maximum on the tubes symmetry axis in a downward flow, whereas in an upward flow, the peak concentration is observed near the walls of the tube.

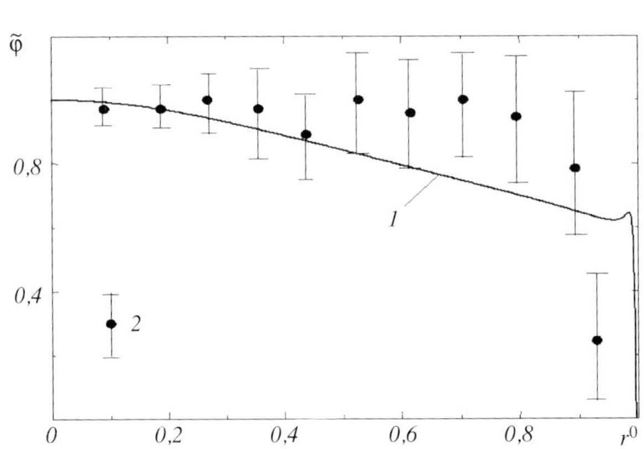

Fig. 7.3 Calculated (1) and measured (2) in [17] bubble distribution in a downward flow ($j_f = 0.71$ m/s, $j_g = 0.10$ m/s).

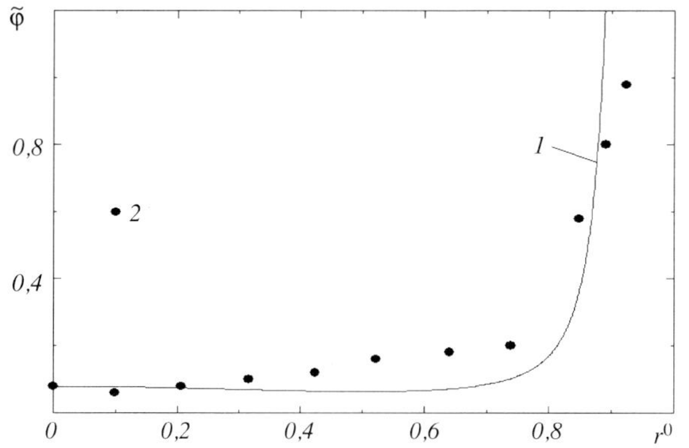

Fig. 7.4 Calculated (1) and measured (2) in [18] bubble distribution in an upward flow ($j_f = 1.391$ m/s, $j_g = 0.180$ m/s).

7.2 Motion of Particles with Mutual Collisions

In concentrated disperse media, particle collisions often play a considerable and sometimes a crucial role. The influence of particle collisions on turbulent transport of momentum of the disperse phase at low volume concentrations has been taken into account in [3,4,10], to which we should add that [4] is devoted to the case of inertial particles whose interaction with turbulent fluctuations can be ignored. The best way to describe disperse phase dynamics while taking into account particle collisions and particle interactions with turbulent fluctuations of the carrier flow is to use the kinetic equation for the PDF of particle velocity. This kinetic equation is derived in the same way as in the previous section. To simplify the analysis, we shall limit ourselves to consideration of particle collisions.

The motion of a heavy particle subject to an external force and to collisions with the other particles in a turbulent flow is described by the Langevin equation

$$\frac{d\boldsymbol{v}_p}{dt} = \frac{\boldsymbol{u} - \boldsymbol{v}_p}{t_v} + \boldsymbol{F} + \boldsymbol{W}, \qquad (7.56)$$

$$\frac{d\boldsymbol{R}_p}{dt} = \boldsymbol{v}_p,$$

where $\mathbf{v}_p$ is particles velocity, $\boldsymbol{u}$ is the velocity of the carrier fluid, and $\boldsymbol{F}$ is the external force.

The first term on the right-hand side of Eq. (7.56) defines the force of interfacial hydrodynamic resistance. The dynamic relaxation time t_v for a particle depends on the Reynolds number of the flow bypassing this particle, and thereby takes into account the influence of both the averaged and the fluctuational component of interfacial slip on the flow regime. The last term describes particle interactions caused by random collisions.

We shall consider only pair interactions, which is consistent with the case of low volume concentration φ of particles. The velocities of two particles after the collision, $\boldsymbol{v}'_p$, $\boldsymbol{v}'_{p1}$, are related to the same velocities before the collision, $\boldsymbol{v}_p$, $\boldsymbol{v}_{p1}$ by

$$\boldsymbol{v}'_p = \boldsymbol{v}_p - \frac{1}{2}(1+e)(\mathbf{w}\cdot\mathbf{k})\mathbf{k}, \quad \boldsymbol{v}'_{p1} = \boldsymbol{v}_{p1} + \frac{1}{2}(1+e)(\mathbf{w}\cdot\mathbf{k})\mathbf{k}, \tag{7.57}$$

where e is the restitution coefficient of the relative velocity (momentum) in the collision of two particles; dependence of the material of the colliding particles is baked into this coefficient ($e=1$ for elastic impact, $e=0$ for perfectly inelastic impact, and $0<e<1$ otherwise). Also present in this equation are the relative velocity of the colliding particles $\mathbf{w}=\boldsymbol{v}_p-\boldsymbol{v}_{p1}$ and the unit vector $\mathbf{k}$ directed from the center of the first particle to the center of the second one.

In order to affect a transition from dynamic trajectory modeling of the individual particles on the basis of stochastic equations (7.56) and (7.57) to continual statistical modeling of the disperse phase as a whole, we introduce the PDF $P(\mathbf{X}, \boldsymbol{v}, t)$ in the phase space of particle coordinates and velocities. An equation for a single-particle PDF is derived from Eq. (7.56) in the same manner as in Section 7.1:

$$\frac{\partial P}{\partial t} + v_k \frac{\partial P}{\partial X_k} + \frac{\partial}{\partial v_k}\left[\left(\frac{U_k - v_k}{t_v} + F_k\right)P\right] = -\frac{1}{t_v}\frac{\partial \langle u'_k p\rangle}{\partial v_k} + J_c, \tag{7.58}$$

where U_k and u'_k are the averaged and fluctuational components of the carrier phase velocity.

The terms on the right-hand side of Eq. (7.58) describe, respectively, particle interactions with turbulent eddies of the carrier flow and particle interactions due to collisions. Assuming the random velocity field of the continuous phase to be Gaussian, one obtains an explicit expression for the correlator $\langle u'_k p\rangle$ in an inhomogeneous turbulent flow, accurate to the first-order spatial derivatives:

$$\langle u'_i p\rangle = \langle u'_i u'_k\rangle \left(f_u \frac{\partial P}{\partial v_k} + t_v g_u \frac{\partial P}{\partial X_k} + t_v l_u \frac{\partial U_n}{\partial X_k}\frac{\partial P}{\partial v_n} + t_v^2 h_u \frac{\partial U_n}{\partial X_k}\frac{\partial P}{\partial X_n} + t_v^2 m_u \frac{\partial U_n}{\partial X_k}\frac{\partial V_j}{\partial X_k}\frac{\partial P}{\partial v_j}\right), \tag{7.59}$$

where

$$f_u = f_{u0},\ g_u = \Psi_{u0} - f_u,\ l_u = g_u - f_{u1},\ h_u = f_{u1} + \Psi_{u1} - 2g_u,$$

$$m_u = f_{u2} + \Psi_{u1} + 2f_{u1} - 3g_u,$$

$$f_{un} = \frac{1}{n!\,t_v^{n+1}}\int_0^\infty \Psi_u(\xi)\xi^n \exp\left(-\frac{\xi}{t_v}\right)d\xi,\ \Psi_{un} = \frac{1}{n!\,t_v^{n+1}}\int_0^\infty \Psi_u(\xi)\xi^n d\xi.$$

The coefficients f_u, g_u, l_u, h_u, m_u characterize the response of particles to turbulent velocity fluctuations of the carrier flow and are defined by the autocorrelation function of velocity fluctuations of the continuous phase along the particle trajectory $\psi_u(\xi)$. A common convention is to take $\Psi_u(\xi) = \exp(-\xi/T_p^{(L)})$. Then the involvement coefficients become

$$f_u = \frac{1}{1+\Omega_u}, \quad g_u = \frac{1}{\Omega_u(1+\Omega_u)}, \quad l_u = \frac{1}{\Omega_u(1+\Omega_u)^2},$$

$$h_u = \frac{1}{\Omega_u^2(1+\Omega_u)^2}, \quad m_u = \frac{1}{\Omega_u^2(1+\Omega_u)^3}, \quad f_{u1} = \frac{1}{(1+\Omega_u)^2}, \quad \Omega_u = \frac{t_v}{T_p^{(L)}}. \tag{7.60}$$

The Lagrangian time scale $T_p^{(L)}$ for an inertialess impurity coincides with the Lagrangian integral time scale of turbulence $T^{(L)}$, whereas for inertial particles, especially in the presence of the averaged velocity slip, $T_p^{(L)}$ may differ substantially from $T^{(L)}$.

The last three terms on the right-hand side of Eq. (7.59) play an important role in shear flows because they are closely related to derivatives of the averaged velocity. In the absence of these terms the expression for $\langle u'p\rangle$ reduces to a corresponding relation obtained in [1,2] for an unbounded homogeneous flow.

The condition $\varphi \ll 1$ makes it possible to neglect the direct contribution of particle collisions to the stresses and the fluctuational energy flux of the disperse phase. Also, the particles are assumed to be sufficiently small to neglect any change of the averaged characteristics of the flow over distance of the order of one particle size. The is the case when $d_p\dot{\gamma}e_p^{1/2} < <1$, where d_p is the particle diameter, $\dot{\gamma}$ – the characteristic shear rate, $e_p = \langle v_k'v_k'\rangle/2$ – the particles fluctuational energy density. Under these conditions, the collision operator in Eq. (7.58) has the form (see [6,20])

$$J_c = \frac{d_p^2}{4}\iint P_2(\boldsymbol{v}, \boldsymbol{v}_1)(\boldsymbol{w}\cdot\boldsymbol{k})d\boldsymbol{k}\, d\boldsymbol{v}_1, \tag{7.61}$$

where $P_2(\boldsymbol{v}, \boldsymbol{v}_1)$ is the two-particle PDF of velocity.

Integration of the kinetic equation (7.58) over the velocity subspace of the phase space gives us a set of continual equations for moments of the PDF: the continuity equation, the equation of conservation of momentum, and the balance equations for turbulent stresses in the disperse phase:

$$\frac{\partial\varphi}{\partial t} + \frac{\partial\varphi V_k}{\partial X_k} = 0. \tag{7.62}$$

$$\frac{\partial V_i}{\partial t} + V_k\frac{\partial V_i}{\partial X_k} = -\frac{\partial\langle v_i'v_k'\rangle}{\partial X_k} + \frac{U_i - v_i}{t_v} + F_i - \frac{D_{pik}}{t_v}\frac{\partial \ln\varphi}{\partial X_k}, \tag{7.63}$$

$$D_{pij} = t_v\left(\langle v_i'v_j'\rangle + g_u\langle u_i'u_j'\rangle + t_v h_u\langle u_i'u_k'\rangle\frac{\partial U_j}{\partial X_k}\right). \tag{7.64}$$

$$\frac{\partial\langle v_i'v_j'\rangle}{\partial t}+V_k\frac{\partial\langle v_i'v_j'\rangle}{\partial X_k}+\frac{1}{\varphi}\frac{\partial\varphi\langle v_i'v_j'v_k'\rangle}{\partial X_k}=-\langle v_i'v_k'\rangle\frac{\partial V_j}{\partial X_k}-\langle v_j'v_k'\rangle\frac{\partial V_i}{\partial X_k}$$
$$-\langle u_i'u_k'\rangle\left[g_u\frac{\partial V_j}{\partial X_k}-l_u\frac{\partial U_j}{\partial X_k}+t_v\frac{\partial U_n}{\partial X_k}\left(h_u\frac{\partial V_j}{\partial X_n}-m_u\frac{\partial U_j}{\partial X_n}\right)\right]$$
$$-\langle v_j'v_k'\rangle\left[g_u\frac{\partial V_i}{\partial X_k}-l_u\frac{\partial U_i}{\partial X_k}+t_v\frac{\partial U_n}{\partial X_k}\left(h_u\frac{\partial V_i}{\partial X_n}-m_u\frac{\partial U_i}{\partial X_n}\right)\right]$$
$$+\frac{2}{t_v}(f_u\langle u_i'u_j'\rangle)-\nu\langle v_i'v_j'\rangle)+J_{ij},\tag{7.65}$$

$$\varphi=\int P d\boldsymbol{v},\quad V_i=\frac{1}{\varphi}\int v_i P d\boldsymbol{v},\quad \langle v_i'v_j'\rangle=\frac{1}{\varphi}\int (v_i-V_i)(v_j-V_j)P d\boldsymbol{v}.$$

Here D_{pij} is the tensor of turbulent diffusion of particles. Eq. (7.65) includes terms that describe the time evolution of the system: convection, diffusion, generation of fluctuations by the averaged shear flow, generation of fluctuations due to the involvement of particles into fluctuational motion of the carrier phase, and dissipation of turbulent energy of the disperse phase that goes into work of the hydrodynamic resistance force. The last term J_{ij} describes particle collisions.

From the expression (7.59) for the correlator $\langle u'v'\rangle$ one can determine the correlation moment of velocity fluctuations of the continuous and disperse phases, which is necessary in order to calculate the feedback action of particles on the turbulent parameters of the carrier flow. Thus, it follows from Eq. (7.59) that

$$\langle u_i'v_j'\rangle=\frac{1}{\varphi}\int\langle u_i'\rangle(v_{pi}-V_i)P d\boldsymbol{v}=\frac{1}{\varphi}\left(\int\langle u_i'p\rangle v_{pj}d\boldsymbol{v}-V_j\int\langle u_i'p\rangle d\boldsymbol{v}\right)$$
$$=f_u\langle u_i'u_j'\rangle+t_v\langle u_i'u_k'\rangle\left[l_u\frac{\partial U_j}{\partial X_k}-g_u\frac{\partial V_j}{\partial X_k}+t_v\frac{\partial U_n}{\partial X_k}\left(m_u\frac{\partial U_j}{\partial X_n}-h_u\frac{\partial V_j}{\partial X_n}\right)\right].\tag{7.66}$$

It remains for us to find the collisional term J_{ij}. To this end, we need to determine the two-particle PDF of velocity at the instant when the collision takes place. By analogy with the molecular chaos hypothesis in the kinetic theory of gases that leads us to the Boltzmann equation, we adopt the assumption that all particle motions are statistically independent. According to this assumption, the two-particle PDF is represented as a product of two single-particle PDFs, and the resulting expressions describing particle collisions in a turbulent flow turn out to be similar to the corresponding relations in the kinetic theory of gases [3,10,21]. But this approach works only for relatively big particles whose dynamic relaxation time is much longer than the integral scale of turbulence, meaning that their relative motion is similar to the chaotic motion of molecules and is uncorrelated. So we resort to a different approach first suggested in [22] that yields a simple explicit expression for the collisional term when used in combination with the Grad expansion [5].

Let us represent the two-particle PDF as a generalized Grad expansion:

$$P_2(\boldsymbol{v}, \boldsymbol{v}_1) = P_{20}(\boldsymbol{v}, \boldsymbol{v}_1) + P_{21}(\boldsymbol{v}, \boldsymbol{v}_1), \tag{7.67}$$

where the first term is the correlated normal distribution [22.23]

$$\begin{aligned} P_{20}(\boldsymbol{v}, \boldsymbol{v}_1) = & \frac{N^2}{(1-R^4)^{3/2}}\left(\frac{3}{4\pi e_p}\right)^3 \\ & \times exp\left[-\frac{3}{4(1-R^4)\pi e_p}(v'_k v'_k - 2R^2 v'_k v'_{1k} + v'_{1k} v'_{1k})\right], \end{aligned} \tag{7.68}$$

$$R = e_i/(e_p e_k)^{1/2}, \quad e_k = \langle u'_k u'_k\rangle/2, \quad e_i = \langle u'_k v'_k\rangle/2, \tag{7.69}$$

N is the number of particles per unit volume (number concentration), R is the correlation coefficient of colliding particles, e_k is the fluctuational energy density of the continuous phase, and e_i is the fluctuational energy density of interfacial interactions.

It should be noted that the expression (7.69) for the correlation coefficient of colliding particles is a local parameter, because it takes into account only that correlation of colliding particles motions that is caused by particle interactions with the velocity field of the carrier flow, and fails to account for the spatial correlation of velocities of particles moving toward the point of collision along different trajectories. Nevertheless, the expression for R correctly describes the behavior of the correlation coefficient in the limiting cases of small and large particles, that is, at $\Omega_u \to 0$ and at $\Omega_u \to \infty$, and is consistent with the results obtained from direct numerical calculations [22].

It follows from Eq. (7.68) that

$$\frac{1}{N}\int P_{20}(\boldsymbol{v}, \boldsymbol{v}_1)d\boldsymbol{v}_1 = P_0(\boldsymbol{v}), \quad \frac{1}{N^2}\iint v'_i v'_j P_{20}(\boldsymbol{v}, \boldsymbol{v}_1)d\boldsymbol{v}d\boldsymbol{v}_1 = \frac{2}{3}e_p\delta_{ij},$$

$$\frac{1}{N^2}\iint v'_i v'_{1j} P_{20}(\boldsymbol{v}, \boldsymbol{v}_1)d\boldsymbol{v}d\boldsymbol{v}_1 = \frac{2}{3}R^2 e_p\delta_{ij},$$

where $P_0(v)$ is the Maxwellian velocity distribution for a single particle.

The second term in Eq. (7.67) represents an expansion over Hermit–Sonin polynomials (which are widely used in the kinetic theory of gases), with only one expansion term being retained [24]:

$$\begin{aligned} P_{21}(\boldsymbol{v}, \boldsymbol{v}_1) = P_{20}(\boldsymbol{v}, \boldsymbol{v}_1)\frac{9}{8e_p^2(1-R^2)}\left(\langle v'_i v'_j\rangle - \frac{2}{3}e_p\delta_{ij}\right)\left[\left(v'_i v'_j - \frac{v'_k v'_k}{3}\delta_{ij}\right)\right. \\ \left. + \left(v'_{1i} v'_{1j} - \frac{v'_{1k} v'_{1k}}{3}\delta_{ij}\right) - 2R^2\left(\frac{v'_i v'_{1j} + v'_j v'_{1i}}{3} - \frac{v'_k v'_{1k}}{3}\delta_{ij}\right)\right]. \end{aligned} \tag{7.70}$$

The distribution $P_{21}(\boldsymbol{v}, \boldsymbol{v}_1)$ satisfies the conditions

$$\frac{1}{N^2}\iint v_i' v_j' P_{21}(\boldsymbol{v}, \boldsymbol{v}_1) d\boldsymbol{v} d\boldsymbol{v}_1 = \langle v_i' v_j' \rangle - \frac{2}{3} R^2 e_p \delta_{ij},$$

$$\frac{1}{N^2}\iint v_i' v_{1j}' P_{21}(\boldsymbol{v}, \boldsymbol{v}_1) d\boldsymbol{v} d\boldsymbol{v}_1 = R^2 \left(\langle v_i' v_j' \rangle - \frac{2}{3} R^2 e_p \delta_{ij} \right).$$

It then follows from Eqs. (7.67)–(7.70) that

$$\begin{aligned} J_{ij} &= \frac{16(1-e^2)\varphi k_p}{3d_p} \left(\frac{2e_p}{3\pi} \right)^{1/2} (1-R^2)^{3/2} \delta_{ij} \\ &= \frac{16(1+e)(3-e)\varphi e_p}{5d_p} \left(\frac{3e_p}{3\pi} \right)^{1/2} (1-R^2)^{3/2} \left(\langle v_i' v_j' \rangle - \frac{2}{3} e_p \delta_{ij} \right). \end{aligned} \tag{7.71}$$

The first term in Eq. (7.71) describes dissipation of turbulent fluctuations of the disperse phase via inelastic collisions, and the second term represents the redistribution of turbulent energy between different components via collisions. This redistribution reflects the tendency of the system to reach an isotropic state. At $R=0$ the expression (7.71) reduces to a relation for uncorrelated chaotic motion of particles [20].

It is readily seen from Eq. (7.71) that the presence of correlation between particle motions results in an increase of fluctuational energy dissipation and thereby of the rate at which the system is approaching its isotropic state. Thus the role of particle collisions diminishes as their motions become more correlated.

Let us now apply the obtained equations to the problem of time evolution of a homogeneous shear layer. Consider a flow along the X-axis with a constant velocity gradient along the Y-axis ($dU_x/dY=const$). Then it follows from Eq. (7.62) and Eq. (7.63) that the volume concentration of particles in a homogeneous shear layer in the absence of external forces should be constant ($\varphi = const$), and the velocity of particles should coincide with that of the carrier flow ($V_x = U_x$). It is known [25] that stationary solutions of equations for the Reynolds stresses in the continuous phase cannot be obtained for a homogeneous shear layer. Therefore the analysis of turbulent stresses in the disperse phase should be carried out on the basis of non-stationary equations for the second moments. Then from Eq. (7.65) combined with Eq. (7.71) there follows a system of equations for stress tensor components:

$$\begin{aligned} \frac{d\langle v_x' 2\rangle}{dt} &= -2(\langle v_x' v_y' \rangle + f_{u1}\langle u_x' u_y' \rangle) \frac{dU_x}{dY} \\ &\quad + \frac{2}{t_v}(f_u \langle u_x' 2\rangle - \langle v_x' 2\rangle) - \frac{2}{3} Q_c - \frac{2}{t_c}\left(\langle v_x' 2\rangle - \frac{2}{3} e_p \right), \\ \frac{d\langle v_y' 2\rangle}{dt} &= \frac{2}{t_v}(f_u \langle u_y' 2\rangle - \langle v_y' 2\rangle) - \frac{2}{3} Q_c - \frac{2}{t_c}\left(\langle v_y' 2\rangle - \frac{2}{3} e_p \right), \end{aligned} \tag{7.72}$$

$$\frac{d\langle v_z'2\rangle}{dt}=\frac{2}{t_v}(f_u\langle u_z'2\rangle-\langle v_z'2\rangle)-\frac{2}{3}Q_c-\frac{2}{t_c}\left(\langle v_z'2\rangle-\frac{2}{3}e_p\right),\frac{d\langle v_x'v_y'\rangle}{dt}$$
$$=-(\langle v_y'2\rangle+f_{u1}\langle u_y'2\rangle)\frac{dU_x}{dY}$$
$$+\frac{2}{t_u}(f_u\langle u_x'u_y'\rangle-\langle v_x'v_y'\rangle)-\frac{2}{t_c}(\langle v_x'v_y'\rangle,$$

$$Q_c=\frac{8(1-e^2)\varphi k_p}{d_p}\left(\frac{2e_p}{3\pi}\right)^{1/2}(1-R^2)^{3/2},$$

$$t_c=\frac{5d_p}{8(1+e)(3-e)\varphi e_p}\left(\frac{2\pi}{2e_p}\right)^{1/2}(1-R^2)^{3/2},$$

where Q_c is the intensity of fluctuational energy dissipation due to particle collisions, and t_c is the effective time between particle collisions.

The expressions for mixed correlation moments of velocity fluctuations of the continuous and disperse phases (7.66) take the form

$$\begin{aligned}&\langle u_x'v_x'\rangle)=f_u\langle u_x'2\rangle-t_v f_{u1}\langle u_x'u_y'\rangle\frac{dU_x}{dY},\langle u_y'v_y'\rangle)=f_u\langle u_y'2\rangle,\\&\langle u_z'v_x'\rangle)=f_u\langle u_z'2\rangle,\langle u_x'v_y'\rangle)=f_u\langle u_x'u_y'\rangle,\\&\langle u_y'v_x'\rangle)=f_u\langle u_x'u_y'\rangle-t_v f_{u1}\langle u_y'2\rangle\frac{dU_x}{dY}.\end{aligned}\qquad(7.73)$$

The results of numerical solution of these equations are presented in [24]. The Reynolds stresses in the disperse phase are determined by the method of large eddy simulation (LES) [26–28]. The initial conditions correspond to an isotropic state. The average velocity shear is 50 s^{-1}, the particle diameter is $d_p=60$ μm, the density ratio between particles and the carrier flow (gas) is $\rho_p/\rho_e=2000$. Thanks to the absence of averaged velocity slip, the duration of eddy–particle interactions (particle interactions with eddies with a high energy content) is taken to be equal to the integral Lagrangian scale $T_p^{(L)}=e_k/(2.075\varepsilon)$, where e_k is the turbulent energy of the continuous phase and ε is the rate of turbulent energy dissipation. The dynamic relaxation time and the Reynolds number are taken in the form

$$t_v=\frac{\rho_p d_p^2}{18\mu_e}(1+0.15Re_p^{0.687}),\quad Re_p=\sqrt{2(e_k+e_p-2e_i)}d_p/v_e.\qquad(7.74)$$

The numerical results are depicted in Fig. 7.5–Fig. 7.8. Fig. 7.5–Fig. 7.7 show the solutions of Eq. (7.72) and Eq. (7.73) (curves 4, 5) juxtaposed to the normal and tangential components calculated in [26]. Fig. 7.8 demonstrates the effect of particle collisions on turbulent stresses in the disperse phase. The stresses are derived from Eq. (7.72) for the case of elastic particles ($e=1$) with $d_p=656$ μm, $\rho_p/\rho_e=85$, and the average velocity shear of 50 s^{-1}. The solution of Eq. (7.72) is illustrated by the curves

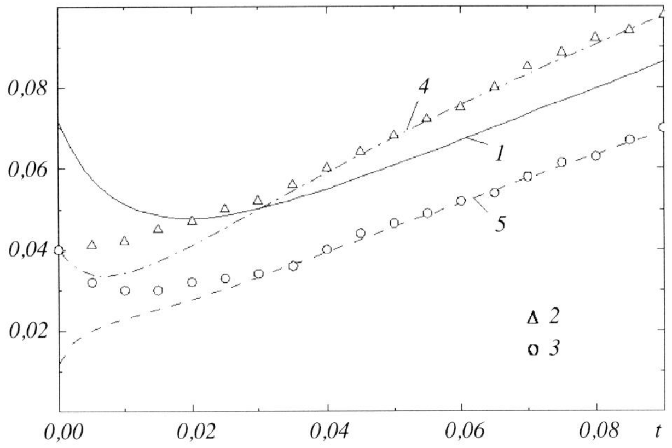

Fig. 7.5 Variation of longitudinal velocity fluctuations (m^2/s^2) with time (s) 1 – $\langle u'_x 2\rangle$; 2,4 – $\langle v'_x 2\rangle$; 3,5 – $\langle u'_x v'_x\rangle$.

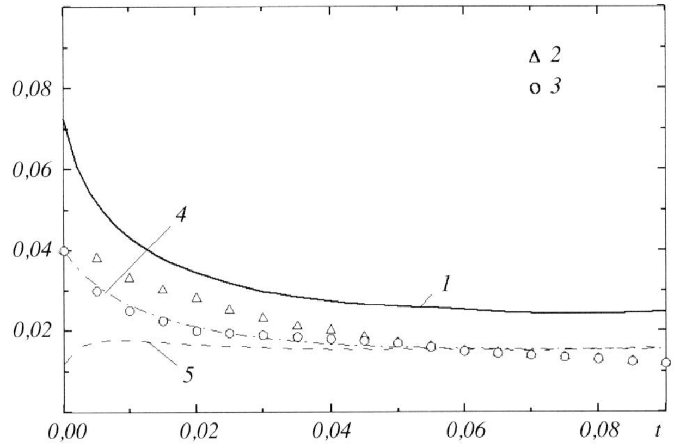

Fig. 7.6 Variation of transverse velocity fluctuations (m^2/s^2) with time (s) 1 – $\langle u'_y 2\rangle$; 2,4 – $\langle v'_y 2\rangle$; 3,5 – $\langle u'_y v'_y\rangle$.

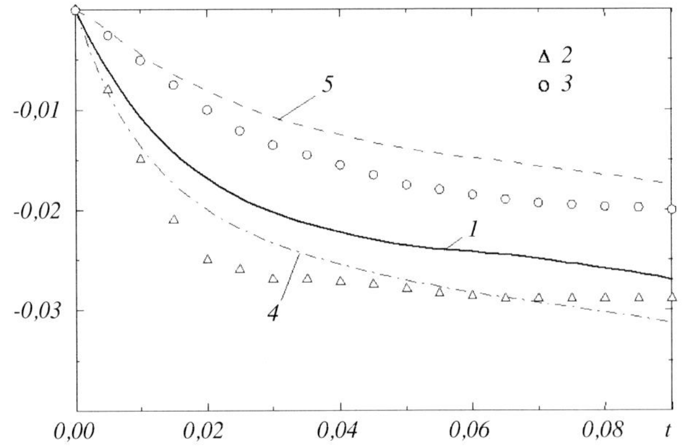

Fig. 7.7 Variation of tangential velocity fluctuations (m^2/s^2) with time (s) 1 – $\langle u'_x u'_y\rangle$; 2,4 – $\langle v'_x v'_y\rangle$; 3,5 – $\langle u'_x v'_y\rangle$.

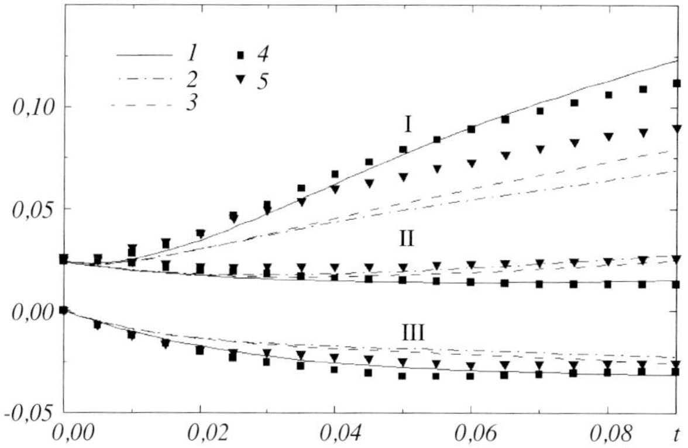

Fig. 7.8 The effect of collisions on velocity fluctuations (m^2/s^2) 1 – $\langle v_x' 2\rangle$; 2,4 – $\langle v_y' 2\rangle$; 3,5 – $\langle v_x' v_y'\rangle$.

1 (no collisions), 2 and 3 (collisions are taking place, $\varphi = 0.0125$). The direct stochastic simulation [27] is illustrated by the curves 4 (no collisions) and 5 (collisions are taking place, $\varphi = 0.0125$).

7.3
Frequency of Collisions of Particles

In many technological and meteorological processes, it is essential that we should know the rate of particle coagulation. When we try to determine this rate, the main challenge is to find collision frequency (or the average time between particle collisions) in a turbulent flow. There exists a large body of theoretical works devoted to particle collisions and coagulation in turbulent flows. Relatively simple (and best known) solutions of this problem have been obtained in the approximation of homogeneous, isotropic turbulence for the limiting cases of very small (inertialess) [29] and very large (inertial) particles [30]. The solution [29] is valid for particles whose dynamic relaxation time t_ν is less than the Kolmogorov microscale of turbulence ($t_\nu < t_{\lambda_0}$); such particles are fully involved into the turbulent motion of the carrier flow. In this case the effort to determine collision frequency ω_{12} can be restricted to particle interactions with small-scale turbulent fluctuations that are responsible for turbulent energy dissipation. The solution [30] holds for the opposite limiting case and refers to particles whose dynamic relaxation time is much greater than Lagrangian time macroscale of turbulence, that is, $t_\nu \gg T^{(L)}$. Motions of such particles are statistically independent. It means that their relative motion is uncorrelated and is similar to chaotic motion of molecules in the kinetic theory of rarefied gases. In this case it is sufficient to consider particle interactions with turbulent eddies with high energy content (large-scale fluctuations), ignoring the contribution of particle interactions with small-scale turbulence to the collision frequency.

The difficulties arise when we try to determine the collision frequency for the case of finite ratios of particle relaxation time to the microscale and macroscale of turbulence (in other words, at $t_{\lambda_0} < t_v < T^{(L)}$), when one needs to take into account particle interactions with both small-scale and with large-scale fluctuations (eddies) while being aware of the correlativity of particle motions. A solution that takes these factors into account has been obtained in [31], but its description of particle interactions with small-scale eddies is not sufficiently consistent and therefore does not provide a smooth transition to the solution obtained in [29] for inertialess particles.

Simultaneous interaction of particles with small-scale and large-scale (i.e., energy-carrying) eddies has been considered in [32]. However, the correlation coefficient of velocities of two particles was not determined correctly, and its suggested value goes to 1 when the particles are identical. As a consequence, particle interactions with energy-carrying eddies do not contribute to the collision frequency of identical particles, and a transition to the limit of very large particles (solution [30]) cannot be made.

Correct expressions for the correlation coefficient and the time interval between particle collisions have been obtained in [22] for the case of identical inertial particles ($t_v \gg T^{(L)}$) under the assumption that the carrier flow velocity and the disperse phase velocity are both normally distributed. Determination of collision frequency with due consideration of particle interactions with both large-scale and small-scale eddies is carried out in [23] by solving the diffusion equation that follows from the kinetic equation for the two-particle PDF of velocity. However, the solution thus obtained predicts a correlation coefficient that approaches 1 (rather than 0) as the particle inertia gets larger, and thus does not allow for a continuous transition to the solution [30].

Interactions of particles with both large-scale and as small-scale eddies has been considered in [33] to obtain the time interval between collisions. This paper assumes normal distributions of velocities and their derivatives. The solution obtained in [33], as well as the one obtained in [32], is based on the so-called cylindrical formulation, which is commonly used in the kinetic theory of gases to calculate the collision frequency of molecules ω_{12} and relate it to the average relative velocity of a particle pair. As shown in [34], a more suitable way to find the collision frequency of low-inertia particles that results from their involvement into small-scale turbulent motion of the carrier flow is the so-called spherical formulation of the problem, which expresses ω_{12} in terms of the average radial component of the particles relative velocity.

It should be noted that both formulations lead to identical results for those characteristics of inertial particle collisions that are caused by the involvement of particles into large-scale turbulent motion. If the particles are small, the outcomes of the two formulations are no longer identical. This is explained by the fact that the longitudinal and transverse correlation functions as well as spatial scales for the two-point velocity field are going to be different to these two formulations. This difference manifests itself as we try to determine the longitudinal and transverse components of the turbulent energy dissipation tensor [14]. The respective expressions for the collision frequency of low-inertia particles and highly inertial particles have been derived in [35] based on separate treatment of particle interactions with small-scale eddies and energy-carrying eddies; while being qualitatively correct, these expressions are not very accurate. As of today, we do not have any single analytic expression for the collision frequency that would be valid in the whole spectrum of values of particle inertia, and that would provide a correct continuous transition to the limiting case solutions [29] and [30].

Numerical solutions of the problem under consideration for relatively small particles with finite (nonzero) relaxation time based on the DNS method [28] are presented in [35,36] (for homogeneous isotropic turbulence) and in [37] (for a flat channel flow). The collision velocity of small particles in a turbulent flow with a constant transverse velocity gradient is calculated in [38]. If there is a velocity shear, the flow cannot be called isotropic. However, because of the local isotropy of its small-scale structure, it presents some interest for theoretical studies of the combined effect of turbulence and velocity shear on the collision frequency of inertialess particles. To simulate collisions of large particles in a turbulent flow $t_v \gg T^{(L)}$; interactions with small-scale fluctuations do not play any considerable role), the method of large eddy simulation (LES) may be employed as being less cumbersome and costly as compared to the DNS method. This method was used in [22] and [39]; the latter publication considered a binary mixture, where relative drift of particles of different densities in the gravitational field plays an important role.

In this section, we present an analytic model that can be used to determine the collision frequency of particles in the entire range of particle inertias. It is assumed that particle density is much higher than the density of the carrier phase (e.g., gas). In addition, the model is generalized for the case when the contribution of the averaged velocity component caused by the relative motion of particles having different inertias (e.g., in the gravitational field or in the velocity field of a shear flow) becomes of special importance. We shall consider only those collisions that are induced by turbulent velocity fluctuations or by the averaged relative motion of particles. Brownian motion and hydrodynamic, molecular, and electrostatic interactions between the particles will not be taken into account.

Let us look at the collisions of spherical particles of radii R_1 and R_2 caused by turbulent fluctuations of the carrier flow velocity. In the framework of the spherical formulation, the collision frequency of particles of types 1 and 2 is defined by the following relation [33]:

$$\omega_{12} = 2\pi R_c^2 \langle |w_r(R_c)| \rangle N_2 = K N_2, \tag{7.75}$$

where $Rc = R_1 + R_2$ is the so-called coagulation radius; R_α is the radius of the $\boldsymbol{\alpha}$-particle; N_α is the number of α -particles per unit volume; $w_r = \boldsymbol{w} \cdot \boldsymbol{r}$ is the radial component of relative velocity $\boldsymbol{w} = \boldsymbol{v}_1 - \boldsymbol{v}_2$ of two particles of types 1 and 2; $\boldsymbol{v}_\alpha$ is the velocity of an α -particle; r is a unit vector pointing from the center of the first particle to the center of the second one; $\mathcal{K}$ is the probability of collision of two particles (collision kernel); $\alpha = 1, 2$.

By convention, probability density of the fluctuational component of radial relative velocity is defined by the Gaussian distribution

$$P(w_r) = \frac{1}{\sqrt{2\pi \langle w_r'2 \rangle}} \exp\left(-\frac{w_r'2}{2\langle w_r'2 \rangle}\right), \tag{7.76}$$

where $w_r = W_r + w_r'$, and W_r, w_r' are the averaged and fluctuational components of radial velocity.

According to Eq. (7.76), in the absence of relative drift of particles ($W_r = 0$), the average absolute value of the radial relative velocity is related to its root-mean-square by the equation

$$\langle |w_r| \rangle = \int_0^\infty |w_r| P(w_r) dw_r = \left(-\frac{2}{\pi}\langle w_r'2 \rangle\right)^{1/2}. \tag{7.77}$$

It follows from here that the turbulent component of the collision kernel in Eq. (7.75) is equal to

$$K_t = \left(8\pi \langle w_r'2 \rangle\right)^{1/2} R_c^2. \tag{7.78}$$

If particle diameters are smaller than the inner scale of turbulence, (the Kolmogorow spatial microscale), the mean square of fluctuations of radial relative velocity is represented as [32.33]

$$\begin{aligned} \langle w_r'2 \rangle &= \langle v_{1r}' \rangle + \langle v_{2r}' \rangle - 2\langle v_{1r}' v_{2r}' \rangle \\ &+ R_1^2 \left\langle \left(\frac{\partial v_{1r}'}{\partial r}\right) \right\rangle + R_2^2 \left\langle \left(\frac{\partial v_{2r}'}{\partial r}\right) \right\rangle + 2R_1 R_1 \left\langle \frac{\partial v_{1r}'}{\partial r} \frac{\partial v_{2r}'}{\partial r} \right\rangle, \end{aligned} \tag{7.79}$$

where r is the distance between the particles centers.

In a homogeneous, isotropic turbulence the quantity $\langle v_{1r}' \rangle$ is directly related to the turbulent energy of the carrier flow $e_k = \langle u_k' u_k' \rangle / 2$ (see [40]):

$$\langle v_{1r}' \rangle = \frac{2}{3} e_p = \frac{2}{3} f_u e_k, \; f_u = \frac{1}{t_v} \int_0^\infty \Psi_u(\xi) \exp\left(-\frac{\xi}{t_v}\right) d\xi, \tag{7.80}$$

where $e_p = \langle v_k' v_k' \rangle / 2$ is the kinetic fluctuational energy of a particle, f_u is the particles involvement factor, and $\psi(\xi)$ is the Lagrangian autocorrelation function of velocity fluctuations in the carrier flow along the particle trajectory.

The expression for the involvement factor f_u in Eq. (7.80) is valid for heavy particles whose density is much higher than the density of the carrier medium (gas), so the only essential force among the interfacial forces is that of aerodynamic resistance. In order to calculate the correlation moment of two particles velocities $\langle v'_{1r} v'_{2r} \rangle$, the joint PDF of the particle velocity and the gas velocity is taken in the form of a Gaussian distribution [22]. In accordance with this assumption, the joint PDF at the point of collision is described by the correlated normal distribution

$$P(\boldsymbol{v}_1, \boldsymbol{v}_2) = \frac{27 N_1 N_2}{64\pi^3 (1 - f_{u1} f_2)^{1/2}} \times \left[-\frac{3}{4(1 - f_{u1} f_2)} \left(\frac{v_{1k} v_{1k}}{k_{p1}} - \frac{f_{u1}^{1/2} f_{u2}^{1/2} v_{1k} v_{2k}}{e_{p1}^{1/2} e_{p2}^{1/2}} \right) \right]. \tag{7.81}$$

Then the correlation moment of two particles velocities is

$$\langle v'_{1r} v'_{2r} \rangle = \frac{2}{3} R_{12} e_{p1}^{1/2} e_{p2}^{1/2}, \quad R_{12} = f_{u1}^{1/2} f_{u2}^{1/2}. \tag{7.82}$$

The correlativity of particle velocities at the point of collision through their interaction with the velocity field of gas at this point is taken into account by the coefficient R_{12}.

The contribution of terms in Eq. (7.79) that contain velocity derivatives is important only for very small particles whose relaxation time t_v is of the same order as the microscale t_{λ_0}. In the interests of simplicity, we determine these terms under the assumption that the particles are fully involved into small-scale motion of the carrier flow. Then

$$\left\langle \left(\frac{\partial v'_r}{\partial r} \right)^2 \right\rangle = \left\langle \frac{\partial v_{1r}}{\partial r} \frac{\partial v_{2r}}{\partial r} \right\rangle = \left\langle \left(\frac{\partial u'_r}{\partial r} \right)^2 \right\rangle.$$

For isotropic turbulence, we have [41]

$$\left\langle \left(\frac{\partial u'_r}{\partial r} \right)^2 \right\rangle = \frac{\bar{\varepsilon}}{15 \nu_e}, \tag{7.83}$$

where $\bar{\varepsilon}$ is the specific dissipation of turbulent energy. In view of Eq. (7.79), Eq. (7.80), and Eqs. (7.82)–(7.84,) there follows from Eq. (7.78) an expression for the turbulent component of the collision kernel:

$$K_t = \left(\frac{8\pi}{3} \right)^{1/2} R_c^2 \left[2(e_{p1} + e_{p2} - 2R_{12} e_{p1}^{1/2} e_{p2}^{1/2}) + \frac{\bar{\varepsilon}}{15 \nu_e} R_c^2 \right]^{1/2}. \tag{7.84}$$

The first two terms in square brackets account for the contribution from the effect of particle involvement into large-scale motion, whereas the third and fourth terms describe the contribution from small-scale motion. As far as stochastic motion of large particles with $t_v \gg T^{(L)}$ is uncorrelated ($R_{12} = 0$) and their involvement into small-scale turbulent motion does not produce a noticeable contribution to the kernel, Eq. (7.84) leads us to the following expression for the turbulent collision kernel that is applicable to large inertial particles [30]:

$$K_t^0 = 4\left(\frac{\pi}{3}\right)^{1/2} R_c^2 (e_{p1} + e_{p2})^{1/2}. \tag{7.85}$$

For very small inertialess particles, which can be completely involved into fluctuational motion of the carrier flow, we have $f_u = R_{12} = 0$, and the contribution of the first two terms is small. In this case the collision kernel takes the form [29]

$$K_t = \left(\frac{8\pi\bar{\varepsilon}}{15\nu_e}\right)^{1/2} R_c^3. \tag{7.86}$$

To determine the involvement coefficient f_u, one has to know the autocorrelation function $\psi_u(\xi)$. It is usually represented by a single-scale exponential function

$$\Psi_u(\xi) = \exp\left(-\frac{\xi}{t_\nu}\right) \tag{7.87}$$

which, generally speaking, could be also used at large Reynolds numbers for sufficiently inertial particles with $t_p > t_t$, where t_t is the Taylor time microscale. To determine f_u in the region of small values of t_p/t_t at moderate Reynolds numbers, we can use the two-scale parabolic exponential function [42]

$$\Psi_u(\xi) = \begin{cases} 1 - \dfrac{\xi^2}{t_t^2}, & \xi > \xi_0, \\ \dfrac{2\xi_0 T_{L0}^2}{t_t^2} \exp\left(-\dfrac{\xi - \xi_0}{T_{L0}}\right), & \xi > \xi_0, \end{cases} \tag{7.88}$$

which at $\xi_0 = \sqrt{T_{L0}^2 + t_t^2} - T_{L0}$ satisfies the conditions

$$\Psi_u(\xi_0 - 0) = \Psi_u(\xi_0 + 0), \quad \Psi_u'(\xi_0 - 0) = \Psi_u'(\xi_0 + 0).$$

In view of Eq. (7.88), the Lagrangian integral scale is

$$T^{(L)} = \int_0^\infty \Psi_u(\xi) d\xi = \frac{t_0^3}{3t_t^2} + \frac{2t_0 T_{L0}^2}{3t_t^2}. \tag{7.89}$$

The Taylor time microscale is equal to

$$t_t = \left(\frac{2Re_t \nu_e}{\sqrt{15}\alpha_0 \bar{\varepsilon}}\right)^{1/2}, \tag{7.90}$$

where $Re_t = (20e_k^2/3\bar{\varepsilon}\nu_e)^{1/2}$ is the Reynolds number determined by the Taylor spatial microscale, and α_0 is connected with the amplitude of acceleration fluctuations in isotropic turbulence trough the relation $\langle\alpha_i\alpha_j\rangle = \alpha_0(\bar{\varepsilon})^{1/2}\nu_e^{3/2}\delta_{ij}$. In the range of values $20 < \mathrm{Re}_t < 100$, one can use the formula $\alpha_0 = 0.13\, Re_t^{0.64}$ [43], which approximates the results of DNS [44].

The parameter T_{L0} is taken as

$$T_{L0} = 0.19\frac{e_k}{\bar{\varepsilon}}\left(1 + \frac{17}{Re_t}\right), \tag{7.91}$$

so that at moderate Reynolds numbers Re_t, the DNS results presented in [44] would be consistent with the formula (7.89) for the Lagrangian integral scale, and at $\mathrm{Re}_t \to \infty$ – with the asymptotic expression $T_\infty^{(L)} = 4e_k/3C_0\bar{\varepsilon}$ (where $C_0 = 7$ in accordance with [43]).

In view of Eq. (7.88), the involvement coefficient in Eq. (7.80) is equal to

$$f_u = 1 + \frac{2t_\nu^2}{t_t^2} = \left[\exp\left(-\frac{\sqrt{T_{L0}^2 + t_t^2} - T_{L0}}{t_\nu}\right)\left(1 + \frac{\sqrt{T_{L0}^2 + t_t^2} - T_{L0}}{t_\nu + T_{L0}}\right) - 1\right]. \tag{7.92}$$

The asymptotic relations for small and large values of t_ν follow directly from Eq. (7.192):

$$\lim_{t_\nu \to 0} f_u = 1 - \frac{2t_\nu^2}{t_t^2}, \quad \lim_{t_\nu \to \infty} f_u = \frac{2T_{L0}^3}{3t_t^2 t_\nu}\left[\left(1 + \frac{t_t^2}{T_{L0}^2}\right)^{3/2} - 1\right]. \tag{7.93}$$

At high Reynolds numbers $\mathrm{Re}_t \to \infty$ and $t_t/T_{L0} \to 0$, Eq. (7.92) provides an expression for the involvement coefficient f_u associated with the autocorrelation function (7.87):

$$f_u = 1 + \frac{t_\nu}{T^{(L)}}. \tag{7.94}$$

For identical particles, the collision kernel (7.84), in view of Eq. (7.80) and Eq. (7.82), takes the form

$$K_t = \left(\frac{8\pi}{3}\right)^{1/2} R_c^2\left[4f_u(1 - f_u)e_k + \frac{\bar{\varepsilon}}{5\nu_e}R_c^2\right]^{1/2}. \tag{7.95}$$

For low-inertia particles at $t_\nu \ll t_t$, the expression (7.84) reduces to

$$K_t = \left(\frac{8\pi}{15}\frac{\bar{\varepsilon}}{\nu_e}\right)^{1/2} R_c^3\left[1 + 30\alpha_0\left(\frac{t_\nu}{t_k}\right)^2\left(\frac{\lambda_0}{R_c}\right)^2\right]^{1/2}, \tag{7.96}$$

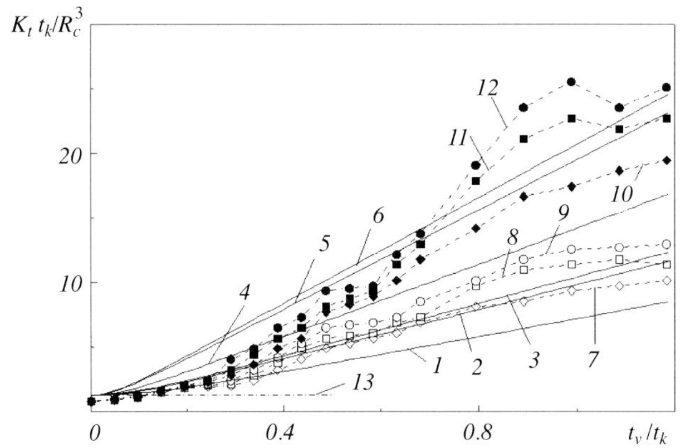

Fig. 7.9 Collision kernel of low-inertia particles: 1–6 – formula (7.117); 7–12 – results of [35]; 1,7 – $\mathrm{Re}_t = 45$, $\lambda_0/\boldsymbol{R}_c = 1$; 2,8 – $\mathrm{Re}_t = 59$, $\lambda_0/\boldsymbol{R}_c = 1$; 3,9 – $\mathrm{Re}_t = 75$, $\lambda_0/\boldsymbol{R}_c = 1$; 4,10 – $\mathrm{Re}_t = 45$, $\lambda_0/\boldsymbol{R}_c = 2$; 5,11 – $\mathrm{Re}_t = 59$, $\lambda_0/\boldsymbol{R}_c = 2$; 6,12 – $\mathrm{Re}_t = 75$, $\lambda_0/\boldsymbol{R}_c = 2$; 13 – formula (7.86)

where $t_k = (\nu_e/\bar{\varepsilon})^{1/2}$ is the Kolmogorov time microscale and $\lambda_0 = (\nu_e^3/\bar{\varepsilon}^{1/4}$ – the Kolmogorov spatial microscale.

Fig. 7.9–Fig. 7.12 compare the expression (7.95) with the DNS results for an isotropic turbulent field [35,36]. It should be noted that among the outcomes of comparison of Eq. (7.95) with numerical data from [35] there are variants that fall outside of the models applicability range $\boldsymbol{R}_c/\lambda_0 < 1$. Therefore a comparison with variants corresponding to $\boldsymbol{R}_c/\lambda_0 > 1$ is not quite correct. Fig. 7.9 presents the above comparison for the case of low-inertia particles $t_\nu \ll t_t$, when the collision kernel has the form (7.96). It can be seen that the dependence shown in Fig. 7.9 gives a qualitatively accurate description of the DNS results [35], showing an increase of K_t with t_ν that is close to linear at $30\alpha_0(t_\nu/t_k)^2(\lambda_0/R_c)^2 \gg 1$.

According to the data [35], the dependence (7.96) predicts growth of the collision kernel with the decrease of the ratio between the particle diameter and the spatial microscale of turbulence (i.e., of R_c/λ_0). Fig. 7.10 shows the dependence of K_t on the ratio of the particles relaxation time t_ν to the temporal Kolmogorov scale t_k, whereas Fig. 7.11 shows the dependence of K_t on the ratio between t_ν and the Eulerian integral scale $T^{(E)} = 2e_k/3\bar{\varepsilon}$. Fig. 7.12 demonstrates the influence of particle inertia on the kinetic energy e_p related to the energy of turbulence e_k, and on the collision kernel K_t. The latter is normalized by the collision kernel K_t^0 obtained from Eq. (7.85) by neglecting motion correlativity, that is, by using the kinetic theory approach. One can see that analytical relations (7.95) and (7.80) are in good agreement with numerical results [36]. The fact that K_t/K_t^0 tends to unity reflects a decrease of particle motion correlativity with the growth of particle inertia.

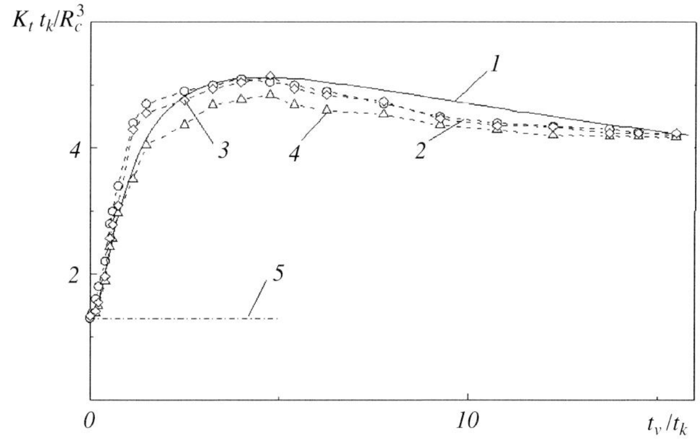

Fig. 7.10 Effect of particle inertia on the collision kernel at $Re_t = 24$ and $\lambda_0/R_c = 1.78$: 1 – formula (7.96); 2,3,4 – results of [35]; 5 – formula (7.96)

Fig. 7.13 compares the dependence (1.95) with the results of DNS [37] in the near-axis zone of a flat channel, where the flow characteristics are close to those observed in isotropic turbulence. The relaxation time and the particle diameter are made dimensionless by introducing $\tau_+ = t_v u_*^2/\nu_e$ and $R_+ = R_c u_*/\nu_e$.

Let us now determine the collision frequency and collision kernel of particles due to the combined effect of turbulence and of the averaged component of particles relative velocity induced by the velocity shear of the carrier flow or by the force of gravity. To calculate the average radial component of relative velocity $\langle |w_r| \rangle$ in Eq. (7.75), it is necessary to average this equation over the random distribution $|w_r|$ and over the solid angle that characterizes spatial orientation of

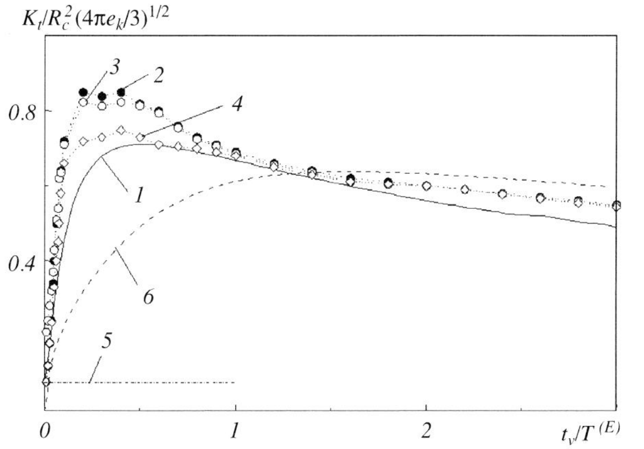

Fig. 7.11 Effect of particle inertia on the collision kernel at $Re_t = 45$ and $\lambda_0/\boldsymbol{R}_c = 1$: 1 – formula (7.95); 2,3,4 – results of [35]; 5 – formula (7.86); 6 – analytical dependence [35].

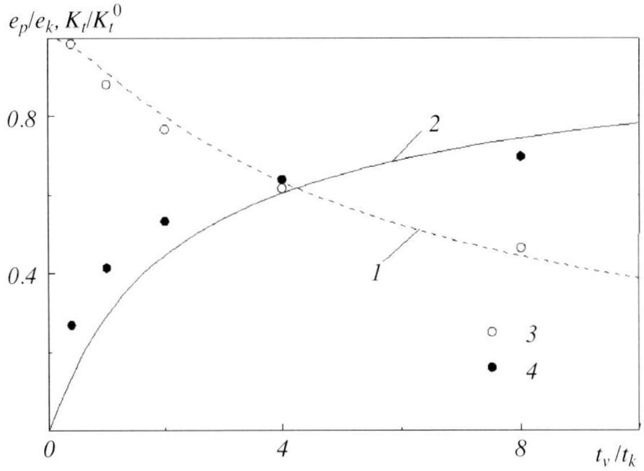

Fig. 7.12 Effect of particle inertia on the kinetic energy of particles and on the collision kernel at $\mathrm{Re}_t = 54.2$ and $\lambda_0/\boldsymbol{R}_c = 0.36$: 1,3 – e_p/e_k; 2,4 – $\boldsymbol{K}_t/\boldsymbol{K}_t^0$; 1,2 – formulas (7.80) and (7.95); 3,4 – results of [36].

the velocity vector w relative to the vector $\boldsymbol{r}$ connecting the centers of colliding particles:

$$\langle |w_r| \rangle = \frac{1}{4\pi} \int_0^{2\pi} \int_0^{\pi} \int_{-\infty}^{\infty} |w_r| P(w_r) \sin\Phi d\psi d\Phi dw_r, \tag{7.97}$$

where φ is the polar angle between the vector $\boldsymbol{r}$ and the Z-axis pointing upward, and Φ is the azimuthal angle in the (X, Y)-plane.

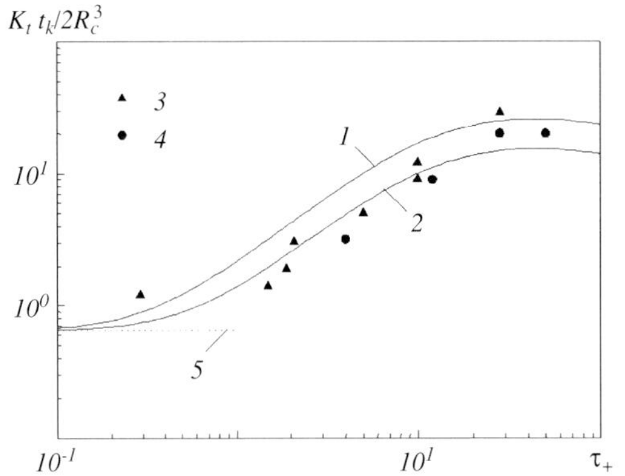

Fig. 7.13 Effect of particle inertia on the collision kernel in the near-axis zone of the channel: 1,2 – formula (7.95); 3,4 – results of [37]; 1,3 – $\boldsymbol{R}_+ = 0.498$; 2,4 – $\boldsymbol{R}_+ = 0.840$; 5 – formula (7.86).

Taking w_r to be normally distributed (see Eq. 7.76), we obtain

$$\langle |w_r| \rangle = \frac{1}{4\pi} \int_0^{2\pi} \int_0^{\pi} \left[\left(\frac{2\langle w_r' 2 \rangle}{\pi} \right)^{1/2} \exp\left(-\frac{W_r^2}{2\langle w_r' 2 \rangle} \right) \right.$$

$$\left. + W_r \mathrm{erf}\left(\frac{W_r^2}{2\langle w_r' 2 \rangle} \right) \right] \sin\Phi d\psi d\Phi. \tag{7.98}$$

In the absence of the averaged relative velocity ($w_r = 0$), Eq. (7.98) reduces to Eq. (7.77).

Suppose the velocity field of the carrier flow is a uniform shear field $\boldsymbol{U} = (\dot{\gamma} Z, 0, 0)$ and the particles are fully involved into the averaged motion of the carrier flow, in other words, $\boldsymbol{V} = \boldsymbol{U}$. Then the averaged radial component w_r resulting from the velocity shear and/or gravity will be

$$W_r = \dot{\gamma} R_c \cos\psi \sin\Phi \cos\Phi + W_g \cos\Phi, \tag{7.99}$$

where $W_g = |t_{v1} - t_{v2}| g$ is the difference between the sedimentaion velocities of the two particles.

In the absence of gravity ($W_g = 0$), the equations (7.75) and (7.98), together with Eq. (7.99), lead to the following expression for the collision kernel in a shear flow (shear coagulation kernel):

$$K_{ts} = (8\pi \langle w' 2 \rangle)^{1/2} R_c^2 \left\{ \sum_{n=0}^{\infty} (-1)^n \left[\frac{\Gamma(2n+1)\Gamma(n+1/2)}{2^{3n+1}\Gamma^2(n+1)\Gamma(2n+3/2)} \left(\frac{\dot{\gamma}^2 R_c^2}{\langle w_r' 2 \rangle} \right)^n \right. \right.$$

$$\left. \left. + \frac{\Gamma(2n+3)\Gamma(n+3/2)}{2^{3n+3}(2n+1)\Gamma(n+1)\Gamma(n+2)\Gamma(2n+7/2)} \left(\frac{\dot{\gamma}^2 R_c^2}{\langle w_r' 2 \rangle} \right)^{n+1} \right] \right\}. \tag{7.100}$$

When $(\dot{\gamma} \boldsymbol{R}_c)^2 / \langle \boldsymbol{w}_r' 2 \rangle \to \infty$, the series (7.100) converges to the Smoluchowski solution [45]

$$K_s = \frac{4}{3} \dot{\gamma} R_c^3. \tag{7.101}$$

The expression (7.100) is cumbersome and inconvenient to use. It can be approximated by the simple formula

$$K_{ts} = (K_t^2 + K_s^2)^{1/2}, \tag{7.102}$$

where K_t is the component of the collision kernel defined by Eq. (7.85), Eq. (7.86), and Eq. (7.95) or Eq. (7.96).

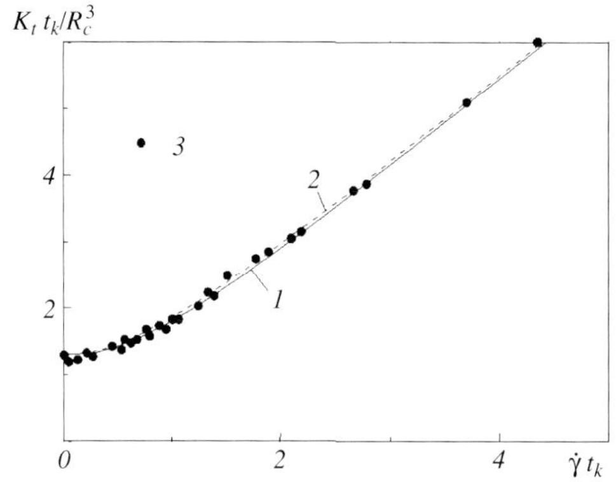

Fig. 7.14 Collision kernel in a turbulent flow with uniform shear: 1, 2 – formulas (7.100) and (7.102); 3 – numerical calculations [38].

Fig. 7.14 shows the dependence (7.100) next to the results of numerical calculation for low-inertia particles when the turbulent component of the collision kernel is defined by the formula (7.86). One can see that Eq. (7.100) is in good agreement with numerical results.

In the absence of shear ($\dot\gamma = 0$), we get the following relation for the collision kernel that describes the combined effect of turbulence and gravity:

$$K_{tg} = \left(8\pi\langle w'2\rangle\right)^{1/2} R_c^2 \left[\frac{1}{2}\exp(-\Sigma^2) + \frac{1}{2}\sqrt{\pi}\left(\Sigma + \frac{1}{2\Sigma}\right)\operatorname{erf}(\Sigma)\right], \tag{7.103}$$

where $\Sigma = W_r/\left(2\langle w_r'2\rangle\right)^{1/2}$ is a parameter that characterizes the relative importance of gravity and turbulence in terms of their effect on the collision kernel. When the influence of gravity is weak, in other words, at small values of Σ, the expression (7.103) transforms into [29]

$$K_{tg} = K_t(1 + \Sigma^2/3). \tag{7.104}$$

At $\Sigma \to \infty$ the relation (7.103) reduces to the collision kernel for the particles subject to gravity only:

$$K_g = \pi R_c^2 W_g \tag{7.105}$$

The obtained formulas for the collision kernel allow us to estimate the time interval between collisions of a particle of type 1 with particles of type 2, and the time interval between collisions of a particle of type 2 with particles of type 1:

$$t_{12} = \omega_{12}^{-1} = (KN_2)^{-1}, \quad t_{21} = \omega_{21}^{-1} = (KN_1)^{-1}. \tag{7.106}$$

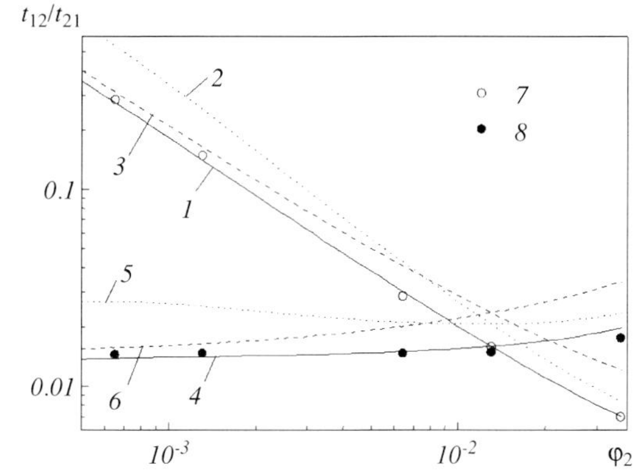

Fig. 7.15 Time between particle collisions in a binary mixture: 1,2,3,7 – $\boldsymbol{t}_{12}$; 4,5,6,8 – $\boldsymbol{t}_{21}$; 1,4 – Eq. (7.106), Eq. (7.103); 2,5 – Eq. (7.106), Eq. (7.85); 3,6 – Eq. (7.106), Eq. (7.105); 7,8 – results of (39).

In Fig. 7.15, the time interval between collisions of particles of different types calculated on the basis of Eq. (7.103) is compared with the results of direct numerical integration of stochastic equations of motion of particles in a turbulent field using the LES method [39]. The considered mixture contains two types of particles of equal radii $r_\alpha = 0.325$ mm but having different densities $\rho_{p1} = 117.5\,\text{kg/m}^3$ and $\rho_{p1} = 235\,\text{kg/m}^3$. Volume concentration $\varphi_\alpha = 4\pi r_\alpha^3 N_\alpha/3$ is fixed at $\varphi_1 = 1.3{\cdot}10^{-2}$ for particles of type 1 and is variable for particles of type 2.

Since the considered particles are sufficiently large, the correlativity of their motion and their interaction with small-scale turbulence do not play any noticeable role, and therefore we can set in Eq. (7.103)$\langle w_r'2\rangle = 2(e_{p1} + e_{p2})/3$. Also shown in Fig. 7.15 is the time interval between collisions found from the expressions (7.85) by neglecting the relative drift, and from Eq. (7.105) by taking into account the gravity force while ignoring the contribution from turbulence. It is readily seen that each one of these two dependencies overestimates t_{12} and t_{21} as compared to the values calculated in [39]. In Eq. (7.85), this overestimation occurs at low volume concentrations, and in Eq. (7.105) – at relatively large values, say, $\varphi_1 = 1.3{\cdot}10^{-2}$, as shown in Fig. 7.15. It should be noted that as we increase φ^2, the collision frequency ω_{12} also increases, and the difference between the average velocities of particles of types 1 and 2 (that is, the relative drift velocity $\boldsymbol{W}_g$) gets smaller as a result. Thus Eq. (7.105) is more suitable for the estimation of t_{12} and t_{21} at small φ_2, whereas Eq. (7.85) works better at great values of φ_2. The equation (7.103), which takes into account the effects of turbulence and gravity, is well consistent with the results of [39].

7.4
Preferential Concentration of Particles in Isotropic Turbulence

In many experimental and theoretical studies the phenomenon of increased particle concentration in certain regions of gradient turbulent flows (e.g., in the near-wall region of a channel), has been observed. The phenomenon of non-uniform particle distribution in inhomogeneous turbulent flows is explained by the turbulent migration (turbophoresis) of particles from the regions of highly intense turbulent velocity fluctuations to the regions of low turbulence [46]. DNS results [34,36,47–50] show that the tendency of particles to prefer certain regions, which leads to their accumulation and further clustering and coagulation, is also discernible in homogeneous turbulence, where there the carrier flow has zero gradient of velocity fluctuations, and particle transport via turbophoresis is not possible. Calculations show that particles are accumulated in the regions of high vorticity because of the action of centrifugal forces. Increased concentration of particles can lead to a noticeable increase of their settling velocity [48] as well as their coagulation frequency [51] in a homogeneous turbulence field.

In view of this phenomenon, there arises a pressing need to find the appropriate methods for modeling of binary dispersion medium and for modeling the process of accumulation of inertial particles in an isotropic turbulent field with given statistical parameters.

Our attention will be limited to small particles whose sizes are smaller than the Kolmogorov fluctuation scale. For such particles, small-scale turbulence, which can be considered as local, isotropic, and homogeneous, exerts a dominating influence. Then the turbulent field can also be assumed isotropic, homogeneous, stationary, incompressible, and having the zero average velocity. Thus the chosen model can be used in calculations of binary dispersion or coagulation of particles whose relaxation time is of the same order as the microscale of turbulence in real turbulent flows.

In order to describe particle interactions with turbulence, we need to determine second-order single-point and two-point correlation moments of velocity fluctuations in the carrier flow.

Lagrangian single-point correlation function is defined as

$$B^0_{Lij}(\tau) = \langle u_i(x,t)u_j(\mathbf{R}(t+\tau,t+\tau))\rangle = u_0^2\Psi_L(\tau)\delta_{ij}, \quad \mathbf{R}(t) = \mathbf{x}, \tag{7.107}$$

where $\mathbf{R}$ is a vector describing the trajectory of a volume element of the continuous medium, u_0^2 is the intensity of velocity fluctuations of the continuous phase, $\psi_L(\tau)$ is the dimensionless autocorrelation function characterized by the Lagrangian integral time scale

$$T^{(L)} = \int_0^\infty \Psi_L(\tau)d\tau. \tag{7.108}$$

Euleruian single-time, two-point correlation and structure functions are introduced through the expressions

$$\begin{aligned} B_{ij}(\mathbf{r}) &= \langle u_i(\mathbf{x},t)u_j(\mathbf{x}+\mathbf{r},t)\rangle, \quad B_{ij}(0) = u_0^2\delta_{ij}, \\ b_{ij}(\mathbf{r}) &= \langle u_i(\mathbf{x}+\mathbf{r},t)-u_i(\mathbf{x},t)\rangle\langle u_j(\mathbf{x}+\mathbf{r},t)-u_j(\mathbf{x},t)\rangle = 2[B_{ij}(0)-B_{ij}(\mathbf{r})]. \end{aligned} \tag{7.109}$$

To describe the behavior of a particle pair, it is necessary to introduce Lagrangian two-point correlation and structure functions,

$$\begin{aligned} B_{Lij}(\boldsymbol{r},\tau) &= \langle u_i(\boldsymbol{R}_1(t)u_j(\boldsymbol{R}_2(t+\tau,t+\tau)\rangle, \quad \boldsymbol{R}_1(t)=x, \quad \boldsymbol{R}_2(t)=\boldsymbol{x}+\boldsymbol{r}, \\ b_{Lij}(\boldsymbol{r},\tau) &= \langle (u_i(\boldsymbol{R}_2(t),t)-u_j(\boldsymbol{R}_1(t),t))(u_j(\boldsymbol{R}_2(t+\tau,t+\tau)) \\ &\quad -u_j(\boldsymbol{R}_1(t+\tau,t+\tau)))\rangle = 2[B_{ij}(0)-B_{ij}(\boldsymbol{r})]. \end{aligned} \tag{7.110}$$

Lagrangian two-point correlation function is connected with Lagrangian single-point correlation function and Eulerian two-point correlation function through the relations

$$B_{Lij}(0,\tau) = B^0_{Lij}(\tau), \quad B_{Lij}(\boldsymbol{r},0) = B_{ij}(\boldsymbol{r}). \tag{7.111}$$

Therefore it can be approximated by

$$B_{Lij}(\boldsymbol{r},\tau) \approx B^0_{Lij}(\tau) + [B_{Lij}(\boldsymbol{r})-B_{Lij}(0)]\Psi_{Lr}(\tau|r), \tag{7.112}$$

where $\psi_{Lr}(\tau|r)$ is the Lagrangian autocorrelation function characterizing relative motion of two particles initially separated by the distance $r=|\mathbf{r}|$. Together with the condition $\psi_{Lr}(0)=1$, the approximation (7.112) satisfies the relations (7.111). For simplicitys sake, this autocorrelation function is given by the expression

$$\Psi_{Lr}(\tau|r) = \exp\left\{-\frac{\tau}{T_r^{(L)}}\right\}, \quad T_r^{(L)} = \int_0^\infty \Psi_{Lr}(\tau)d\tau, \tag{7.113}$$

frequently used in the theory of turbulence. Here $T_r^{(L)}$ is the two-point time integral scale.

Because the approximation (7.113) is not correct at $r \to 0$, it should be considered only in the region of relatively large values of τ.

The approximation (7.113) makes it possible to represent the Lagrangian two-point structure function of velocity fluctuations as the product

$$b_{Lij}(\boldsymbol{r},\tau) = b_{Lij}(\boldsymbol{r})\Psi_{Lr}(\tau|r). \tag{7.114}$$

Since the coefficient of relative diffusion of two particles has the form of an integral of Lagrangian two-point correlations [52],

$$D^r_{ij} = 2\int_0^\tau [B^0_{Lij}(\tau_1)-B_{ij}(\boldsymbol{r},\tau_1)]d\tau_1 = \int_0^\tau b_{Lij}(\boldsymbol{r},\tau_1)d\tau_1, \tag{7.115}$$

substitution of Eq. (7.114) into Eq. (7.115) leads to an expression for the coefficient of relative diffusion of two particles at large values of τ:

$$D_{ij}(\boldsymbol{r}, \tau \to \infty) = b_{ij} T_r^{(L)}. \tag{7.116}$$

To determine $T_r^{(L)}$, consider the behavior of structure functions and of the coefficient of relative diffusion in three spatial intervals: viscous, inertial, and external. It should be noted that in the field of isotropic homogeneous turbulence, any second-rank tensor quantity can be represented as

$$\boldsymbol{M}_{ij}(\mathbf{r}, \tau) = \boldsymbol{M}_{NN}(\boldsymbol{r}, \tau)\delta_{ij} + [\boldsymbol{M}_{LL}(\boldsymbol{r}, \tau) - \boldsymbol{M}_{NN}(\boldsymbol{r}, \tau)]\frac{\boldsymbol{r}_i \boldsymbol{r}_j}{\boldsymbol{r}^2} \tag{7.117}$$

(recall Eq. 7). Here M_{LL} and M_{NN} are the longitudinal and transverse (with respect to the vector $\boldsymbol{r}$) components of tensor M.

In the viscous interval ($r < \lambda_0$), the first terms of the Taylor series expansion of Eulerian structure functions are as follows [55]:

$$b_{LL} = \frac{\bar{\varepsilon} r^2}{15\nu_e}, \quad b_{NN} = \frac{2\bar{\varepsilon} r^2}{15\nu_e}, \tag{7.118}$$

where $\bar{\varepsilon}$ is the specific energy dissipation and ν_e is the coefficient of kinematic viscosity.

At small values of r the difference between velocity fluctuations at two points can be represented as a linear function of the vector connecting these points, namely,

$$\Delta u_i(\boldsymbol{r}, t) = u_i(\boldsymbol{x} + \boldsymbol{r}, \tau) - u_i(\boldsymbol{x}, \tau) = \gamma_{ij}(t) r_j, \tag{7.119}$$

where $\gamma_{ij} = \nabla_i u_j = \partial u_i / \partial x_j$ is the gradient of velocity fluctuations. In a linear isotropic field, the correlation functions of strain and rotation tensors E and Λ are [53,54]

$$\begin{aligned}
\langle E_{ik}(\boldsymbol{x}, t) E_{jn}(\boldsymbol{x} + \boldsymbol{r}, \tau)\rangle &= \frac{\varepsilon}{20\nu_e}\left(\delta_{ij}\delta_{kn} + \delta_{in}\delta_{jk} - \frac{2}{3}\delta_{ik}\delta_{jn}\right)\exp\left\{-\frac{\tau}{\tau_E}\right\}, \\
\langle_{ik}(\boldsymbol{x}, t)\Lambda_{jn}(\boldsymbol{x} + \boldsymbol{r}, \tau)\rangle &= \frac{\varepsilon}{12\nu_e}[\delta_{ij}\delta_{kn} - \delta_{in}\delta_{jk}]\exp\left\{-\frac{\tau}{\tau}\right\}, \\
E_{ij} = \frac{\gamma_{ij} + \gamma_{ji}}{2}, &\quad \Lambda_{ij} = \frac{\gamma_{ij} - \gamma_{ji}}{2}.
\end{aligned} \tag{7.120}$$

The two correlation functions decrease exponentially with their respective characteristic times τ_E and τ_Λ, which are proportional to the inner temporal microscale $t_{\lambda_0} = (\nu_e/\bar{\varepsilon})^{1/2}$. An expression for the Lagrangian two-point correlation function follows from Eq. (7.119) and Eq. (7.120) under the assumption that the distribution of the separation vector between the considered points $\mathbf{r}_i$ is statistically independent from the distribution of the velocity fluctuation gradient tensor γ_{ij}:

$$\boldsymbol{b}_{LLL} = \frac{\bar{\varepsilon}\boldsymbol{r}^2}{15\nu_e}\exp\left\{-\frac{\tau}{\tau_E}\right\},$$
$$\boldsymbol{b}_{LNN} = \frac{2\bar{\varepsilon}\boldsymbol{r}^2}{15\nu_e}\left[\exp\left\{-\frac{\tau}{\tau_E}\right\}+\frac{1}{3}\exp\left\{-\frac{\tau}{\tau_\Lambda}\right\}\right]. \qquad (7.121)$$

Substituting Eq. (7.121) into Eq. (7.115), we obtain the expressions for the longitudinal and transverse components of the relative diffusion coefficient:

$$D_{LL}^r = \frac{\bar{\varepsilon}\tau_E r^2}{15\nu_e}, \quad D_{NN}^r = \frac{\bar{\varepsilon}}{4\nu_e}\left\{\frac{\tau_E}{5}+\frac{\tau_\Lambda}{3}\right\}r^2. \qquad (7.122)$$

At $\tau_E = \tau_\Lambda$, the relation (7.122) is in agreement with the expression for the relative diffusion coefficient obtained in [52] for the viscous interval. On the other hand, from Eq. (7.116) and Eq. (7.118) there follows

$$D_{LL}^r = \frac{\bar{\varepsilon}T_r^{(L)}r^2}{15\nu_e}, \quad D_{NN}^r = \frac{2\bar{\varepsilon}T_r^{(L)}r^2}{15\nu_e}. \qquad (7.123)$$

A comparison of Eq. (7.122) with Eq. (7.123) shows that both expressions coincide at $T_r^{(L)} = \tau_\Lambda = \tau_E$. So, in the viscous interval, the two-point time scale $T_r^{(L)}$ is defined as

$$T_r^{(L)} = \tau_E = A_1 t_{\lambda_0}. \qquad (7.124)$$

The constant $A_1 = \sqrt{5}$ is obtained theoretically in [52] and is close to the value 2.3 found in [53,54] by the DNS method.

Consider now the behavior of turbulence characteristics of the carrier flow in the inertial interval ($\lambda_0 < r < L$), where L is the spatial macroscale. The Kolmogorov similarity hypothesis [55] leads to the following self-similar representation for the second-order structure function:

$$b_{LL} = C(\bar{\varepsilon}r)^{2/3}, \quad b_{NN} = \frac{4}{3}C(\bar{\varepsilon}r)^{2/3}, \qquad (7.125)$$

where $C = 2$ according to [55,56].

As evident from dimensionality considerations applied to the inertial interval, the only temporal scale that can be constructed from the available physical quantities is $(\bar{\varepsilon})^{-1/3}r^{2/3}$. Therefore the temporal scale $T_r^{(L)}$ should be equal to

$$T_r^{(L)} = A_2(\bar{\varepsilon})^{-1/3}r^{2/3}, \quad A_2 = const. \qquad (7.126)$$

On the other hand, the two-point temporal scale may be taken in the form

$$T_r^{(L)} = l b_{LL}^{-1/2}, \tag{7.127}$$

where l is some length scale analogous to the mixing length for near-wall turbulence. In the viscous and inertial intervals, this scale will be taken proportional to the distance between the two points

$$l = \alpha r, \quad \alpha = const, \tag{7.128}$$

which, again, is justified by dimensionality considerations. Then, comparing Eq. (7.124) with Eq. (7.127) in the viscous interval and making use of Eq. (7.118), Eq. (7.125), Eq. (7.126), and Eq. (7.128), we write

$$\alpha = \frac{A_1}{15^{1/2}}, \quad A_2 = \frac{A_1}{(15C)^{1/2}}. \tag{7.129}$$

A substitution of Eq. (7.125) and Eq. (7.126) into Eq. (7.116) leads to the following expressions for the coefficient of relative diffusion in the inertial interval:

$$D_{LL}^r = CA_2(\bar{\varepsilon})^{1/3} r^{4/3} = 0.816(\bar{\varepsilon})^{1/3} r^{4/3}, \quad D_{NN}^r = \frac{4}{3} D_{LL}^r = 1.09(\bar{\varepsilon})^{1/3} r^{4/3},$$
$$C = 2, \quad A_2 = 1/\sqrt{6},$$

which are in good agreement with theoretical dependences obtained in [52]:

$$D_{LL}^r = 2C^{1/2} f_0(\bar{\varepsilon})^{1/3} r^{4/3} = 0.854(\bar{\varepsilon})^{1/3} r^{4/3}, \quad D_{NN}^r = \frac{5}{3} D_{LL}^r = 1.42(\bar{\varepsilon})^{1/3} r^{4/3},$$
$$C = 1.77, \quad f_0 = 0.321.$$

In the external interval ($r > L$), the distance between the two test points where fluctuations are observed is sufficiently large to consider the fluctuations as independent. Hence the correlation functions vanish in the external interval, and the structure functions are equal to

$$b_{LL} = b_{NN} = 2u_0^2. \tag{7.130}$$

In addition, at large values of r the two-point temporal scale converts into an ordinary Lagrangian temporal scale,

$$T_r^{(L)} = T^{(L)}, \tag{7.131}$$

and the coefficient of relative diffusion transforms into

$$D_{ij}^r = 2u_0^2 T^{(L)} \delta_{ij}. \tag{7.132}$$

According to DNS data [44], Lagrangian integral time scale in the Reynolds number range of $\mathrm{Re}_t = 38 - 93$ is approximated by the expression

$$T_0^{(L)} = \frac{T^{(L)}}{t_{\lambda_0}} = 0.06 Re_\lambda + 3, \quad Re_\lambda = \left[\frac{15 u_0^2}{\bar{\varepsilon} \nu_e}\right]^{1/2}. \tag{7.133}$$

Let us now turn to the derivation of the kinetic equation for the PDF of relative velocity of a particle pair. Consider the motion of two identical heavy particles in an isotropic turbulent field in the absence of gravity. Equations describing the motion of each particle are

$$\frac{d\boldsymbol{R}_{p\alpha}}{dt} = \boldsymbol{v}_{p\alpha}, \quad \frac{d\boldsymbol{v}_{p\alpha}}{dt} = \frac{\boldsymbol{u}(\boldsymbol{R}_{p\alpha}, t) - \boldsymbol{v}_{p\alpha}}{t_v}, \tag{7.134}$$

where $\boldsymbol{R}_{p\alpha}$ and $\boldsymbol{v}_{p\alpha}$ are the position and velocity of the particle, $\boldsymbol{u}(\boldsymbol{R}_{p\alpha}, t)$ is the velocity of the continuous phase at the point $\boldsymbol{x} = \boldsymbol{R}_{p\alpha}(t)$, t_v is the dynamic relaxation time for the particle, and α denotes the particles index ($\alpha = 1, 2$). Equations (7.134) hold for particles whose density is way above the density of the continuous phase while their size is smaller than the Kolmogorov microscale. In this case, the only substantial interfacial force is that of viscous resistance.

The equations for the particle pair follow from Eq. (7.134):

$$\frac{d\boldsymbol{r}_p}{dt} = \boldsymbol{w}_p, \quad \frac{d\boldsymbol{w}_p}{dt} = \frac{\Delta \boldsymbol{u}(\boldsymbol{r}_p, t) - \boldsymbol{w}_p}{t_v}, \tag{7.135}$$

where $\mathbf{r}_p = \mathbf{R}_{p2} - \mathbf{R}_{p1}$, $\mathbf{w}_p = \mathbf{v}_{p2} - \mathbf{v}_{p1}$. Since the turbulent velocity field is considered as a random process, Eq. (7.135) is an equation of the Langevin type. In order to make a transition from the stochastic equation (7.135) to a statistical description of relative velocity distribution, let us introduce the PDF of a particle pair,

$$P(\boldsymbol{r}, \boldsymbol{w}, t) = \langle p \rangle = \langle \delta(\boldsymbol{r} - \boldsymbol{r}_p(t)) \delta(\boldsymbol{w} - \boldsymbol{w}_p(t)) \rangle. \tag{7.136}$$

The operation of averaging of Eq. (7.136) is carried out over the ensemble of random realizations of the continuous phases velocity field. The function $P(\mathbf{r}, \mathbf{w}, t)$ is defined as the probability of finding the relative velocity vector equal to w for two particles separated by the distance r at the moment t. Differentiating both parts of Eq. (7.136) with respect to t, one obtains the transport equation for the PDF of the particle pair:

$$\frac{dP}{dt} = w_k \frac{\partial P}{\partial r_k} - \frac{1}{t_v} \frac{\partial w_k P}{\partial w_k} = -\frac{1}{t_v} \frac{\partial \langle \Delta u_k P \rangle}{\partial w_k}. \tag{7.137}$$

The left-hand side of Eq. (7.137) describes the time evolution of the distribution and the convection in the $(\mathbf{r}, \mathbf{w})$ phase space, whereas the right-hand side describes interaction of particles with turbulent eddies of the continuous phase. To determine the correlation $\langle \Delta u\, p \rangle$ describing the interaction of particles with the turbulence, we model the field of relative velocities of the carrier flow by a Gaussian random process with known correlation moments. Then, in view of the Furutzu–Donsker–Novikov formula [57], we obtain:

$$\langle \Delta u_i\, p \rangle = \iint \langle \Delta u_i(\boldsymbol{r}, t)\Delta u_k(\boldsymbol{r}_1, t_1)\rangle \left\langle \frac{\delta p(\boldsymbol{r}, t)}{\delta \Delta u_k(\boldsymbol{r}_1, t_1)} \right\rangle d\boldsymbol{r} d\boldsymbol{r}_1,$$
$$\left\langle \frac{\delta p(\boldsymbol{r}, t)}{\delta \Delta u_k(\boldsymbol{r}_1, t_1)} \right\rangle = \frac{\partial}{\partial r_j}\left\langle p(\boldsymbol{r}, t)\frac{\delta r_{pj}(t)}{\delta \Delta u_k(\boldsymbol{r}_1, t_1)} \right\rangle - \frac{\partial}{\partial w_j}\left\langle p(\boldsymbol{r}, t)\frac{\delta w_{pj}(t)}{\delta \Delta u_k(\boldsymbol{r}_1, t_1)} \right\rangle. \tag{7.138}$$

In order to obtain the functional derivatives entering Eq. (7.138), one should use the solution of Eq. (7.135),

$$r_{pi}(t) = r_{pi}(0) + \int_0^t w_{pi}(t_1)dt_1,$$
$$w_{pi}(t) = w_{pi}(0)\exp\left\{-\frac{t}{t_v}\right\} + \frac{1}{t_v}\int_0^t \Delta u_i(\boldsymbol{r}_p(t_1), t_1)\exp\left\{-\frac{t-t_1}{t_v}\right\}dt_1. \tag{7.139}$$

Applying the operator of functional differentiation to Eq. (7.139) and remembering that the initial conditions $r_{pi}(0)$ and $w_{pi}(0)$ are independent of Δu_i, we write

$$\frac{\delta r_{pi}(t)}{\delta \Delta u_j(\boldsymbol{r}_1, t_1)} = \delta_{ij}\left[1-\exp\left\{-\frac{t-t_1}{t_v}\right\}\right]\delta(\boldsymbol{r}_1-\boldsymbol{r}_p(t_1))H(t-t_1),$$
$$\frac{\delta w_{pi}(t)}{\delta \Delta u_j(\boldsymbol{r}_1, t_1)} = \frac{\delta_{ij}}{t_v}\exp\left\{-\frac{t-t_1}{t_v}\right\}\delta(\boldsymbol{r}_1-\boldsymbol{r}_p(t_1))H(t-t_1), \tag{7.140}$$

where $H(x)$ is the Heaviside function.

In view of Eq. (7.140), the expression (7.138) transforms to

$$\langle \Delta u_i\, p \rangle = -\int_0^t \langle \Delta u_i(\boldsymbol{r}, t)\Delta u_k(\boldsymbol{r}_p(t_1), t_1)\rangle \left[1-\exp\left\{-\frac{t-t_1}{t_v}\right\}\right]dt_1 \frac{\partial P}{\partial r_k}$$
$$-\frac{1}{t_v}\int_0^t \langle \Delta u_i(\boldsymbol{r}, t)\Delta u_k(\boldsymbol{r}_p(t_1), t_1)\rangle \exp\left\{-\frac{t-t_1}{t_v}\right\}dt_1 \frac{\partial P}{\partial w_k}$$
$$= -t_v G_{ik}(\boldsymbol{r}, t)\frac{\partial P}{\partial r_k} - F_{ik}(\boldsymbol{r}, t)\frac{\partial P}{\partial w_k}, \tag{7.41}$$

where $G_{\mathrm{ik}}(\boldsymbol{r},t)$ and $F_{\mathrm{ik}}(\boldsymbol{r},t)$ stand for the integrals of the two-point structure function of continuous phases velocity fluctuations; integration is carried out along the trajectory that describes the relative motion of the two particles. Using the approximation (7.114) for $b_{Lij}(\boldsymbol{r},\tau)$, we can write these functions in the explicit form:

$$G_{\mathrm{ik}}(\boldsymbol{r},t)=\frac{b_{\mathrm{ik}}(\boldsymbol{r})}{t_v}\int_0^t \Psi_{Lr}(\tau|r)\left[1-\exp\left\{-\frac{\tau}{t_v}\right\}\right]d\tau,$$
$$F_{\mathrm{ik}}(\boldsymbol{r},t)=\frac{b_{\mathrm{ik}}(\boldsymbol{r})}{t_v}\int_0^t \Psi_{Lr}(\tau|r)\exp\left\{-\frac{\tau}{t_v}\right\}d\tau. \qquad (7.142)$$

After the substitution of Eq. (7.142) into Eq. (7.141), the correlator $\langle\Delta u_i p\rangle$ takes the form

$$\langle\Delta u_i\, p\rangle=-b_{\mathrm{ik}}\left[t_v g_r\frac{\partial P}{\partial r_{\mathrm{k}}}+f_r\frac{\partial P}{\partial w_{\mathrm{k}}}\right], \qquad (7.143)$$

$$g_r=\frac{1}{t_v}\int_0^\infty \Psi_{Lr}(\tau|r)\left[1-\exp\left\{-\frac{\tau}{t_v}\right\}\right]d\tau,$$
$$f_r=\frac{1}{t_v}\int_0^\infty \Psi_{Lr}(\tau|r)\exp\left\{-\frac{\tau}{t_v}\right\}d\tau. \qquad (7.144)$$

The expression (7.143) is valid for times much greater than the Lagrangian integral time macroscale $T^{(L)}$. The coefficients g_r and f_r characterize the involvement of a particle pair separated by the distance r into fluctuational motion of the carrier flow. If the autocorrelation function is given by the exponential dependence (7.113), these coefficients take the form

$$g_r=\frac{T_r^{(L)}}{t_v}-f_r=\frac{\left(T_r^{(L)}\right)^2}{t_v\left(t_v+T_r^{(L)}\right)},\qquad f_r=\frac{T_r^{(L)}}{t_v+T_r^{(L)}}. \qquad (7.145)$$

Substitution of Eq. (7.143) into Eq. (7.137) leads to a closed kinetic equation for the PDF of relative velocity of the two particles in an isotropic homogeneous turbulence

$$\frac{dP}{dt}+w_{\mathrm{k}}\frac{\partial P}{\partial r_{\mathrm{k}}}-\frac{1}{t_v}\frac{\partial w_{\mathrm{k}}P}{\partial w_{\mathrm{k}}}=b_{\mathrm{ik}}\left(g_r\frac{\partial^2 P}{\partial r_i\partial w_{\mathrm{k}}}+\frac{f_r}{t_v}\frac{\partial^2 P}{\partial w_i\partial w_{\mathrm{k}}}\right). \qquad (7.146)$$

Terms on the right-hand side of Eq. (7.146) describe the diffusional transport in the (**r**, **w**) phase space that is caused by the interaction of particles with turbulent eddies of the carrier flow. Modeling turbulent velocity fluctuations by a Gaussian random process, we can describe this interaction by a second-order operator of the Fokker–Planck type. It is true that, according to the DNS results [51,58], the PDF of relative velocity v for a particle pair differs substantially from the normal distribution, especially at high values of v. However, the tail of a distribution does not significantly affect its average characteristics. Therefore a Gaussian random process predicts the two-point moments of relative velocities with a satisfactory accuracy [34].

It should be remarked that, on the face of it, Eq. (7.146) resembles the kinetic equation for a single-point PDF of velocity in a homogeneous turbulent flow [1,2]. But this resemblance between the single-point and two-point kinetic equations is only apparent, because a single-point kinetic model operates with a single-point PDF in the phase space and thus is unable to take into account spatial correlativity of motions of the two particles. In contrast, the two-point approach allows to examine the correlated motion of particles resulting from their interaction with the turbulent eddies and is thus capable of describing the phenomenon of coagulation (clustering).

Integration of the kinetic equation (7.146) over the velocity space yields a system of equations for two-point moments of the particle pair velocity PDF. The equations for the density of particle pairs, for the average relative density, and for the second-order structure function of particle velocity fluctuations are given below:

$$\frac{dN}{dt}+\frac{\partial N W_k}{\partial r_k}=0, \tag{7.147}$$

$$\frac{dW_i}{dt}+W_k\frac{\partial N W_i}{\partial r_k}=-\frac{\partial b_{pik}}{\partial r_k}-\frac{W_i}{t_v}-\frac{D^k_{pik}}{t_v}\frac{\partial \ln N}{\partial r_k}, \tag{7.148}$$

$$\begin{aligned}\frac{\partial b_{pij}}{\partial t}+W_k\frac{\partial b_{pij}}{\partial r_k}+\frac{1}{N}\frac{\partial N\langle w'_i w'_j w'_k\rangle}{\partial r_k}&=-(b_{pik}+g_r b_{ik})\frac{\partial W_i}{\partial r_k}\\ -(b_{pjk}+g_r b_{jk})\frac{\partial W_i}{\partial r_k}&+\frac{2}{t_v}(f_r b_{ij}-b_{pij}),\end{aligned} \tag{7.149}$$

$$N=\int P d\boldsymbol{w},\quad w_i=\frac{1}{N}\int w_i P d\boldsymbol{w},\quad b_{pij}=\langle w'_i w'_j\rangle=\int (w_i-w_i)(w_j-w_j)P d\boldsymbol{w},$$

$$D^r_{pij}=t_v(b_{pij}+g_r b_{ij}).$$

Equations (7.147)–(7.149) do not form a closed system, because the third equation contains a third-order moment. An attempt to close the system by adding equations for higher-order moments would be pointless, because each new equation would introduce moments of still higher orders. The result would be an infinite chain of equations. To close any finite subsystem of moment equations, one needs closure relations. For example, a system of equations for third-order moments can be closed by using a well-known quasi-normal hypothesis stating that the fourth-order cumulants are equal to zero, which allows to represent the fourth-order moment as a sum of products of second-order correlations. In doing so, we get the following equation for the third-order structure function:

$$\frac{\partial\langle w'_i w'_j w'_k\rangle}{\partial t}+W_n\frac{\partial\langle w'_i w'_j w'_k\rangle}{\partial r_n}+\langle w'_i w'_j w'_n\rangle\frac{\partial W_k}{\partial r_n}+\langle w'_i w'_j w'_n\rangle\frac{\partial W_j}{\partial r_n}+\langle w'_i w'_j w'_n\rangle\frac{\partial W_i}{\partial r_n}$$
$$+\frac{D^r_{pin}}{t_v}\frac{\partial\langle w'_j w'_k\rangle}{\partial r_n}+\frac{D^r_{pjn}}{t_v}\frac{\partial\langle w'_i w'_k\rangle}{\partial r_n}+\frac{D^r_{pkn}}{t_v}\frac{\partial\langle w'_i w'_j\rangle}{\partial r_n}+\frac{3}{t_v}\langle w'_i w'_j w'_k\rangle=0. \tag{7.150}$$

Equations (7.147)–(7.150) form a closed system and describe the two-point statistics of relative velocity of a particle pairs in terms of the third moments. In order to limit ourselves to a second-order approximation, we shall ignore the terms in Eq. (7.150) that are responsible for time evolution, convection, generation due to the gradients of averaged velocity. As a result, we obtain an algebraic equation for the third-order structure function:

$$\langle w'_i w'_j w'_k\rangle=\frac{1}{3}\left(D^r_{pin}\frac{\partial\langle w'_j w'_k\rangle}{\partial r_n}+D^r_{pjn}\frac{\partial\langle w'_i w'_k\rangle}{\partial r_n}+D^r_{pkn}\frac{\partial\langle w'_i w'_j\rangle}{\partial r_n}\right). \tag{7.151}$$

Triple correlations describe the diffusional transport of velocity fluctuations. The form of Eq. (7.151) is consistent with the relations obtained in [59] for triple correlations in a single-phase turbulent flow and in [60,61] for the single-point third-order moments of particle velocity fluctuations in a two-phase turbulent flow. The system of equations (7.147)–(7.149) and (7.151) allows us to model the two-point statistics of the relative velocity of a particle pair by equations for second-order moments.

The averaged relative velocity vector W_i in an isotropic turbulence can be written in terms of its radial component W_r:

$$W_i=r_i W_r/r,$$

and thanks to the spherical symmetry, which implies that the relative velocities as well as the particle pair PDF may depend only on the absolute value of the vector $\boldsymbol{r}$, rather than on its spatial orientation, the system of equations (7.147)–(7.149) and (7.151) reduces to

$$\frac{\partial N}{\partial t}+\frac{1}{r^2}\frac{\partial}{\partial r}(r^2 N W_r)=0,$$

$$\frac{\partial W_r}{\partial t}+W_r\frac{\partial W_r}{\partial r}+\frac{2(b_{pLL}-b_{pNN})}{r}=-\frac{\partial b_{pLL}}{\partial r}-\frac{W_r}{t_v}-(b_{pLL}+g_r b_{LL})\frac{\partial \ln N}{\partial r},$$

$$\frac{\partial b_{pLL}}{\partial t}+W_r\frac{\partial b_{pLL}}{\partial r}=\frac{t_v}{r^2N}\frac{\partial}{\partial r}\left(r^2N(b_{pLL}+g_rb_{LL})\frac{\partial b_{pLL}}{\partial r}\right)$$

$$-\frac{4t_v}{3r}\left((b_{pLL}+g_rb_{LL})\frac{\partial b_{pNN}}{\partial r}+\frac{2}{r}(b_{pNN}+g_rb_{NN})(b_{pLL}-b_{NN})\right)$$

$$-2(b_{pLL}+g_rb_{LL})\frac{\partial W_r}{\partial r}+\frac{2}{t_v}(f_rb_{LL}-b_{pLL}). \quad (7.152)$$

$$\frac{\partial b_{pNN}}{\partial t}+W_r\frac{\partial b_{pNN}}{\partial r}=\frac{t_v}{3r^4N}\left[\frac{\partial}{\partial r}\left(r^4N(b_{pLL}+g_rb_{LL})\frac{\partial b_{pNN}}{\partial r}\right)\right.$$

$$\left.+2\frac{\partial}{\partial r}(r^3N(b_{pNN}+g_rb_{NN})(b_{pLL}-b_{pNN}))\right]-2(b_{pNN}+g_rb_{NN})\frac{W_r}{r}$$

$$+\frac{2}{t_v}(f_rb_{NN}-b_{pNN}).$$

As an application of the above-proposed model and its ramifications, consider a stationary disperse medium with a fixed total number of particles. The stationarity requirement implies a balance of particle fluxes toward and away from the origin, in other words, the averaged radial relative velocity W_r must vanish. Introduce the dimensionless variables: the ratios r/λ_0, v/u_{λ_0}, $N(r)/N(\infty)$, and so on. Here λ_0 is the Kolmogorov microscale and $N(r)$ is the particles radial distribution function. Dimensionless variables will be denoted by a tilde placed over the corresponding symbol. As a result of this transition to dimensionless variables, there appears a dimensionless parameter $St_0 = t_v/t_{\lambda_0}$ called the Stokes number and characterizing particles inertia. The problem reduces to the following system of ordinary nonlinear differential equations for the radial distribution function of particle pairs N and the longitudinal b_{pLL} and transverse b_{pNN} structure functions of particle velocity fluctuations:

$$\frac{2(\tilde b_{pLL}-\tilde b_{pNN})}{\tilde r}+\frac{d\tilde b_{pLL}}{d\tilde r}+(\tilde b_{pLL}-g_r\tilde b_{LL})\frac{\partial \ln\tilde N}{\partial\tilde r}=0,$$

$$St_0^2\left\{\frac{1}{\tilde r^2\tilde N}\frac{d}{d\tilde r}\left(\tilde r^2N(\tilde b_{pLL}+g_r\tilde b_{LL})\frac{d\tilde b_{pLL}}{d\tilde r}\right)-\frac{4}{3\tilde r}\left((\tilde b_{pLL}+g_r\tilde b_{LL})\frac{\partial\tilde b_{pNN}}{\partial\tilde r}\right.\right.$$

$$\left.\left.+\frac{2}{\tilde r}(\tilde b_{pNN}+g_r\tilde b_{NN})(\tilde b_{pLL}-\tilde b_{pNN})\right)\right\}+2(f_r\tilde b_{LL}-\tilde b_{pLL})=0,$$

$$\frac{St_0^2}{3\tilde r^4N}\left\{\frac{d}{d\tilde r}\left(\tilde r^4\tilde N(\tilde b_{pLL}+g_r\tilde b_{LL})\frac{d\tilde b_{pNN}}{d\tilde r}\right)\right.$$

$$\left.+2\frac{d}{d\tilde r}(\tilde r^3\tilde N(\tilde b_{pNN}+g_r\tilde b_{NN})(\tilde b_{pLL}-\tilde b_{pNN}))\right\}+2(f_r\tilde b_{NN}-\tilde b_{pNN})=0.$$

(7.153)

The boundary conditions for the equations (7.153) are as follows:

$$\frac{d\tilde{b}_{pLL}}{d\tilde{r}} = \frac{d\tilde{b}_{pNN}}{d\tilde{r}} = 0 \quad \text{at} \quad \tilde{r} = 0, \tag{7.154}$$

$$\tilde{b}_{pLL} = f_r \tilde{b}_{LL}, \quad \tilde{b}_{pNN} = f_r \tilde{b}_{NN} \quad , \quad \tilde{N} = 1 \quad \text{at} \quad \tilde{r} \to \infty. \tag{7.155}$$

The conditions (7.154) describe the balance of particle fluxes directed toward and away from the origin of the coordinate system and are valid for particle sizes much smaller than λ_0. The conditions (7.155) reflect the lack of correlation between particle motions when the particles are spaced far apart, that is to say, when the particles are randomly distributed in space and their relative velocities no longer depend on r.

Structure functions of velocity fluctuations in the continuous phase $\tilde{b}_{LL}$, $\tilde{b}_{NN}$ in Eq. (7.153) are given by the approximations that combine the relations (7.18), (7.125), and (7.130):

$$\begin{aligned} \frac{1}{\tilde{b}_{LL}^{k}} &= \left(\frac{15}{\tilde{r}^2}\right)^k + \frac{1}{(C\tilde{r}^{2/3})^k} + \left(\frac{\sqrt{15}}{2Re_\lambda}\right)^k, \\ \frac{1}{\tilde{b}_{NN}^{k}} &= \left(\frac{15}{\tilde{r}^2}\right)^k + \left(\frac{3}{4C\tilde{r}^{2/3}}\right)^k + \left(\frac{\sqrt{15}}{2Re_\lambda}\right)^k. \end{aligned} \tag{7.156}$$

The two-point time scale is approximated in a similar way by the relation that combines the expressions (7.27) and (7.31):

$$\frac{1}{(\tilde{T}_r^{(L)})^k} = \left(\frac{\tilde{b}_{LL}^{1/2}}{\alpha \tilde{r}}\right)^k + \frac{1}{(\tilde{T}^{(L)})^k}. \tag{7.157}$$

The constants in Eq. (7.156) and Eq. (7.157) are equal to $C = 2$ and $\alpha = 1/\sqrt{3}$. In order for the results to be independent of k, the latter should be much greater than 1. Numerical estimations show that the results cease to depend on k at $k > 10$. When doing the calculations, the value $k = 20$ was used.

Before we discuss the numerical results, consider some asymptotic solutions. In the case of inertialess particles ($St_0 = 0$), it follows from Eq. (7.153) that

$$\tilde{b}_{pLL} = \tilde{b}_{LL}, \quad \tilde{b}_{pNN} = \tilde{b}_{NN}, \quad \tilde{N} = 1. \tag{7.158}$$

According to (7.158), inertialess particles are fully involved into fluctuational motion of the carrier flow, which is why the particle distribution in space is uniform.

A solution of the system (7.153)–(7.157) can be obtained for low-inertia particles ($0 < St_0 < 1$) by doing a series expansion over the small parameter St_0. The first terms of this expansion are

$$\tilde{b}_{pLL} = \left(\frac{1}{15} + \frac{St_0^2}{75}\right)\tilde{r}^2, \quad \tilde{b}_{pNN} = \left(\frac{2}{15} - \frac{2St_0}{15\sqrt{5}} + \frac{28St_0^2}{675}\right)\tilde{r}^2,$$
$$\tilde{N} \sim (\tilde{r})^{-4St_0^2/5}. \tag{7.159}$$

Consider the inertial space interval ($1 \ll \tilde{r} \ll \tilde{L}$, where $\tilde{L} = L/\lambda_0$). Suppose furthermore that the particle relaxation time belongs to the inertial interval ($1 \ll St_0 \ll \tilde{T}^{(L)}$), and take the limiting case $Re_\lambda \to \infty$, meaning that both $\tilde{T}^{(L)}$ and $\tilde{L}$ go to infinity. Then the problem (7.153)–(7.155) will reduce to

$$\frac{2(\sigma_{pLL}-\sigma_{pNN})}{\rho} + \frac{d\sigma_{pLL}}{d\rho} + (\sigma_{pLL}-g_r\sigma_{LL})\frac{\partial \ln\tilde{N}}{\partial\rho} = 0,$$
$$\frac{1}{\rho^2\tilde{N}}\frac{d}{d\rho}\left(\rho^2 N(\sigma_{pLL}+g_r\sigma_{LL})\frac{d\sigma_{pLL}}{d\rho}\right) - \frac{4}{3\rho}(\sigma_{pLL}+g_r\sigma_{LL})\frac{\partial\sigma_{pNN}}{\partial\rho}$$
$$+\frac{2}{\rho}(\sigma_{pNN}+g_r\sigma_{NN})(\sigma_{pLL}-\sigma_{pNN}) + 2(f_r\sigma_{LL}-\sigma_{pLL}) = 0, \tag{7.160}$$

$$\frac{1}{3\rho^4\tilde{N}}\left[\frac{d}{d\rho}\left(\rho^4\tilde{N}(\sigma_{pLL}+g_r\sigma_{LL})\frac{d\sigma_{pNN}}{d\rho}\right)\right.$$
$$\left. +2\frac{d}{d\rho}(\rho^3\tilde{N}(\sigma_{pNN}+g_r\sigma_{NN})(\sigma_{pLL}-\sigma_{pNN}))\right] + 2(f_r\sigma_{NN}-\sigma_{pNN}) = 0.$$

$$\frac{d\sigma_{pLL}}{d\rho} = \frac{d\sigma_{pNN}}{d\rho} = 0 \quad \text{at} \quad \rho = 0, \tag{7.161}$$

$$\sigma_{pLL} = f_r\sigma_{LL}, \quad \sigma_{pNN} = f_r\sigma_{NN}, \quad \tilde{N} = 1 \quad \text{at} \quad \rho \to \infty. \tag{7.162}$$

$$\sigma_{LL} = C\rho^{2/3}, \quad \sigma_{NN} = \frac{4}{3}C\rho^{2/3}, \quad f_r = \frac{A_2\rho^{2/3}}{1+A_2\rho^{2/3}}, \quad g_r = \frac{A_2^2\rho^{4/3}}{1+A_2\rho^{2/3}},$$

$$\rho = \frac{\tilde{r}}{St_0^{3/2}}, \quad \sigma_{pLL} = \frac{\tilde{b}_{pLL}}{St_0}, \quad \sigma_{pNN} = \frac{\tilde{b}_{pNN}}{St_0}, \quad \sigma_{LL} = \frac{\tilde{b}_{LL}}{St_0}, \quad \sigma_{NN} = \frac{\tilde{b}_{NN}}{St_0}.$$

In the case of highly inertial particles, when their dynamic relaxation time is much greater than the Lagrangian integral scale of turbulence ($St_0 > \tilde{T}^{(L)}$), the particles longitudinal and transverse structure functions become equal and uniform as a result of intensive diffusional transport, whereas the radial distribution function is almost identically equal to unity:

$$\tilde{b}_{pLL} = \tilde{b}_{pNN} = \frac{2\tilde{T}^{(L)}Re_\lambda}{\sqrt{15}St_0}, \quad \tilde{N} = 1. \tag{7.163}$$

The particles collision frequency depends on $\langle|w_r|\rangle$. As shown in [51], the form of the PDF of relative velocity depends on particle inertia, and it is only at large values of the Stokes number that the PDF becomes Gaussian. But even for inertialess particles the quantity $\langle|w_r|\rangle/\langle|w_r'2|\rangle^{1/2}$ is equal to 0.77 (according to the results of DNS [27]), which is sufficiently close to the value $\sqrt{2/\pi}=0.798$ predicted by the normal distribution. Therefore we can state that the relative velocity of a particle pair is distributed according to the normal law. Then

$$\langle|w_r|\rangle=\left(\frac{2}{\pi}\langle|w_r'2|\rangle\right)^{1/2}=\left(\frac{2}{\pi}b_{pLL}\right)^{1/2}. \tag{7.164}$$

In general, equations (7.153) with the boundary conditions (7.154) and 7.155() are solved numerically and the results are compared with those obtained in [49–51] by the DNS method.

In Fig. 7.16, the particles structure functions, structure functions of the continuous phase, and the radial distribution function are plotted against the parameter ρ obtained by solving Eq. (7.160) with the boundary conditions (7.161) and (7.162). One can see that at $\rho\gg 1$, when diffusional transport of velocity fluctuations does not play any significant role, the longitudinal and transverse structure functions of the particles are smaller than the respective structure functions of the continuous phase, and appoximately are valid relations $\sigma_{pLL}=f_r\sigma_{LL}$, $\sigma_{pNN}=f_r\sigma_{NN}$. At small distances the relations structure functions of particles exceed structure functions of the continuous phase. This effect is caused by the diffusional mechanism of velocity fluctuation transport, and it takes place only for sufficiently inertial particles. The

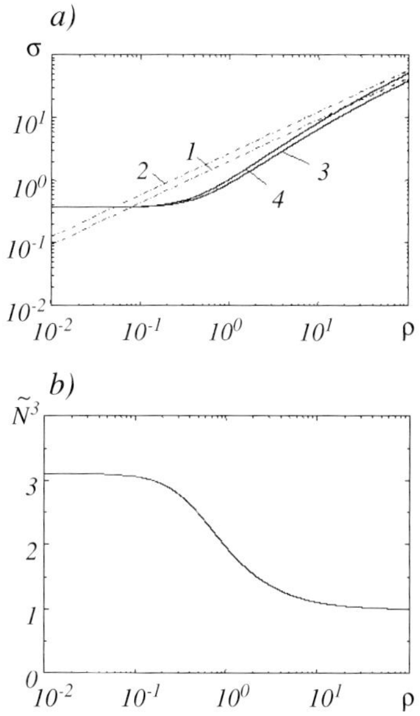

Fig. 7.16 Structure functions (a) and radial distribution function (b) of inertial particles in the inertial spatial interval at $Re_\lambda\to\infty$: 1 – σ_{LL}; 2 – σ_{NN}; 3 – σ_{pLL}; 4 – σ_{pNN}.

values of σ_{pLL} and σ_{pNN} approach each other as ρ gets smaller, and the radial distribution function approaches some limit. At the origin of coordinates, we have

$$\sigma_{pLL}(0) = \sigma_{pNN}(0) = 0.37, \quad \tilde{N}(0) = 3.13. \tag{7.165}$$

Structure functions and the radial distribution function of particles obtained by solving Eqs. (7.153)–(7.155) at a fixed Reynolds number and for different values of the Stokes number are shown in Fig. 7.17.

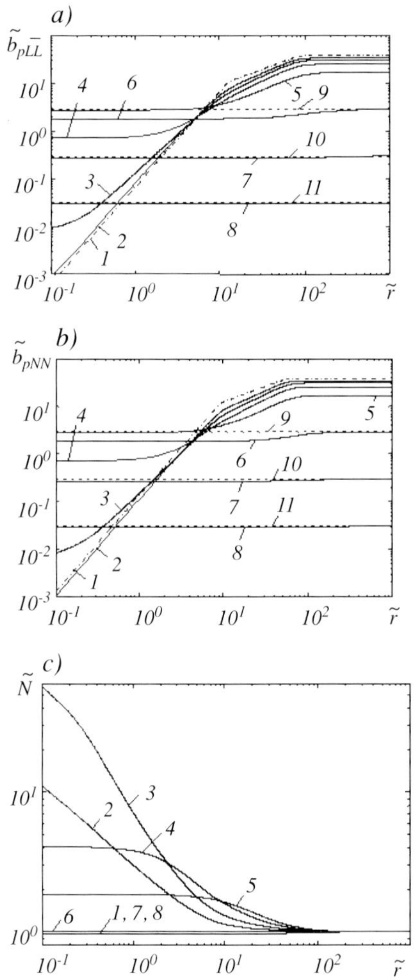

Fig. 7.17 Longitudinal (a) and transverse (b) structure functions and radial distribution function (c) at $Re_\lambda = 75$: 1 – $St_0 = 0$; 2 – $St_0 = 1$; 3 – $St_0 = 2$; 4 – $St_0 = 4$; 5 – $St_0 = 10$; 6,9 – $St_0 = 100$; 7,10 – $St_0 = 1000$; 8,11 – $St_0 = 10000$.

It is easy to see that as St_0 increases, structure functions of the particles deviate more and more from the structure functions of the continuous phase (curve 1) and approach the asymptotic uniform distributions (7.163) for highly inertial particles (curves 9–11). Although $b_{LL} = b_{NN} = 0$ at $\tilde{r} = 0$, the diffusional transport causes structure functions of sufficiently inertial particles at the origin of coordinates to differ from zero. Fig. 7.17, c shows that the radial distribution function of low-inertia particles becomes singular at $\tilde{r} = 0$. With increase of particle inertia, the singularity at the origin disappears, and $\tilde{N} \to 1$.

The dependences of $\langle|\tilde{w}_r|\rangle$ and $\tilde{N}$ on particle inertia, that is, on the Stokes number St_0 (Figs. 7.18–7.21) present the greatest interest. At small values of St_0, the function $\langle|\tilde{w}_r|\rangle$ increases with particle inertia, which corresponds to the solution (7.159). Then it reaches its maximum value at a certain $St_0 = St_0'$. This maximum reflects the growth of $\langle|\tilde{w}_r|\rangle$ with relaxation time, since the motions of particles become less correlated. As St_0 grows further, $St_0 > St_0'$, $\langle|\tilde{w}_r|\rangle$ begins to decrease, which is explained by the lower intensity of particle velocity fluctuations (particles become more inertial and less capable of participating in turbulent motion of the carrier flow). With growth of Re_λ the results get closer to the asymptotic relation that follows from Eqs. (7.160)–(7.162) when the relaxation time belongs to the inertial interval,

$$\langle|\tilde{w}_r|\rangle = \left(\frac{2}{\pi}\sigma_{pLL}(0)\right)^{1/2} St_0^{1/2}.$$

Calculations of the radial distribution function $\tilde{N}$ show that in the limiting cases of low-inertia particles and highly inertial particles the concentration field is statistically homogeneous and $\tilde{N} = 1$. However, according to the DNS data [49–51], the radial distribution function peaks sharply at $St_0 \propto 1$. As shown in Fig. 7.19, at small

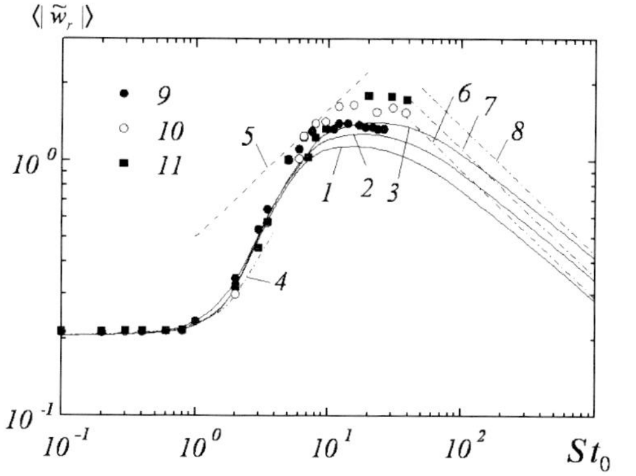

Fig. 7.18 Effect of Re_λ on $\langle|\tilde{w}_r|\rangle$: 1–3 – calculations; 9–11 – [51]; 1,6 – $Re_\lambda = 45$; 2,7 – $Re_\lambda = 58$; 3,8 – $Re_\lambda = 75$.

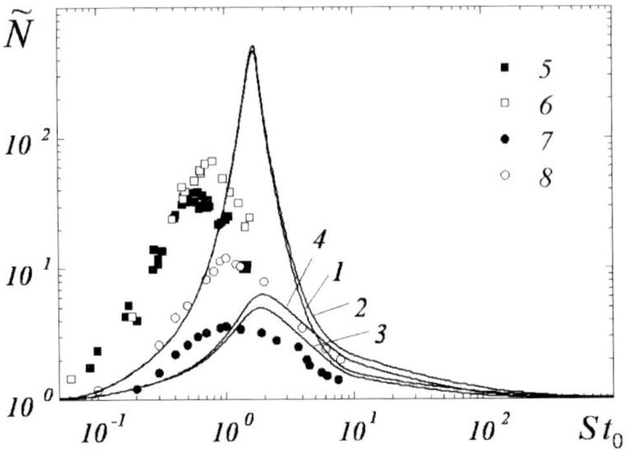

Fig. 7.19 Effect of St_0 on $\tilde{N}$: 1–4 – calculations; 5,6 – [49]; 7,8 – [51]; 1,5 – $\tilde{r} = 0.025$, $Re_\lambda = 37$; 2,6 – $\tilde{r} = 0.025$, $Re_\lambda = 82$; 3,7 – $\tilde{r} = 1$, $Re_\lambda = 24$; 4,8 – $\tilde{r} = 1$, $Re_\lambda = 75$.

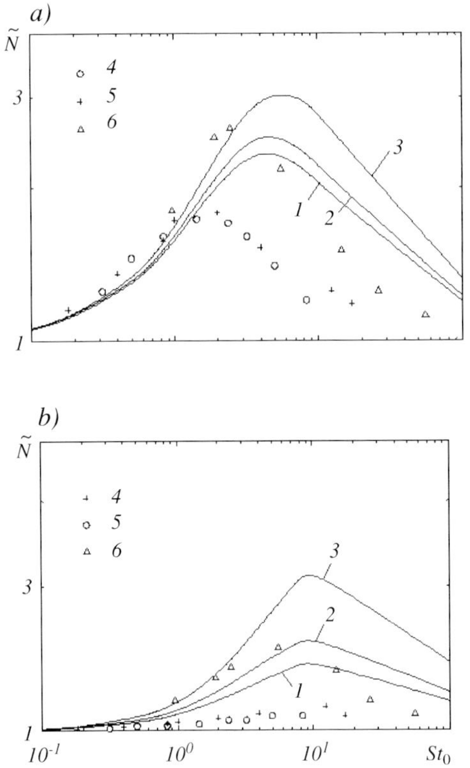

Fig. 7.20 Effect of St_0 and $\tilde{r}$ on $\tilde{N}$: 1–3 – calculations; 4–6 – [50]; (a) $-\tilde{r}=6$; (b) $-\tilde{r}=-24$; 1,4 – $Re_\lambda = 53$; 2,5 – $Re_\lambda = 69$; 3,6 – $Re_\lambda = 134$.

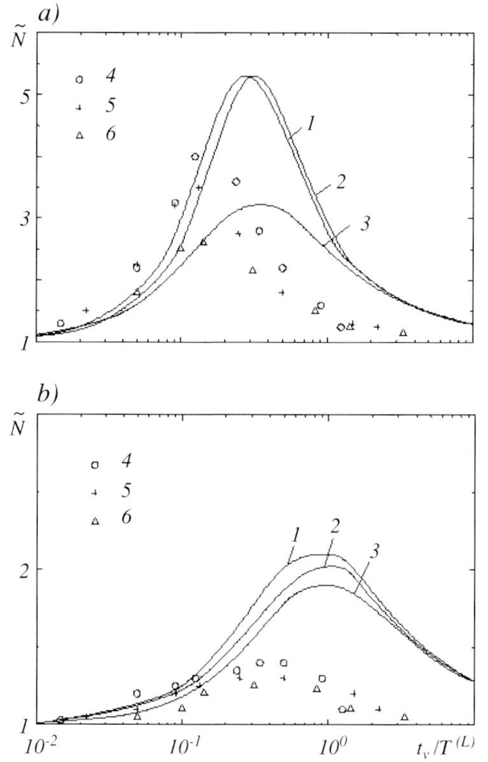

Fig. 7.21 Effect of St_0 and $\tilde{r}$ on $\tilde{N}$: 1–3 – calculations; 4–6 – [50]; (a) $-r/L$=0.05; (b) $-r/L$=–0.3; 1,4 – $Re_\lambda = 53$; 2,5 – $Re_\lambda = 69$; 3,6 – $Re_\lambda = 134$.

distances between the particles, the position of the maximum matches neatly with the Kolmogorov microscale, which indicates the crucial role of small-scale turbulent structures in the formation of particle clusters. With increase of interparticle distances, the peak of $\tilde{\mathbf{N}}$ falls in intensity and shifts to the larger values of particle relaxation time. As shown in [7], the particles can also form spatial clusters at $\mathbf{St}_0 \gg 1$. But since the motion of highly inertial particles is controlled by large-scale turbulent structures, it is better to use integral scales of turbulence rather then Kolmogorovs microscales. This conclusion is readily supported by the comparison of Fig. 7.20 to Fig. 7.21.

The obtained results, as well as the DNS data, prove that the phenomenon of particle clustering manifests itself most strongly for the particles whose relaxation time is close to the Kolmogorov temporal microscale, but is also noticeable, albeit to lesser degree, for highly inertial particles separated by large distances [62,63].

References

1 Derevich, I.V. and Zaichik, L.I. (1988) Particle Deposition from a Turbulent Flow. *Fluid Dynamics*, **23** (5), 722–729 (Ref. 6 in Chapter 5).

2 Reeks, M.W. (1991) On a Kinetic Equation for the Transport of Particles in Turbulent Flows. *Phys. Fluids. A*, **3** (3), 446–456.

3 Simonin, O. (1991) *Second-Moment Prediction of Disperse Phase Turbulence in Particle-Laden Flows*. Proc. 8-th Symp. on Turbulent Shear Flows, Munich. pp. 7.4.1–7.4.6.

4 Louge, M.Y., Mastarakos, E. and Jenkins, J.T. (1991) The Role of Particle Collisions in Pneumatic Transport. *J. Fluid Mech.*, **231**, 345–359.

5 Hirschfelder, J.O., Curtiss, C.P. and Bird, R.B. (1954) *Molecular Theory of Gases and Liquids*, Wiley, New York.

6 Lun, C.K.K., Savage, S.B., Jeffrey, D.J. and Chepurnity, N. (1984) Kinetic Theories for Granular Flow: Inelastic Particles in Couette Flow and Slightly Inelastic Particles in a General Flow Field. *J. Fluid Mech.*, **140**, 223–256.

7 Ding, J. and Gidaspow, D. (1990) A Bubbling Fluidization Model Using Kinetic Theory of Granular Flow. *AIChE J.*, **36** (4), 523–538.

8 Zaichik, L.I. (1992) A Kinetic Model of Particle Transport in Turbulent Flows with Allowance for Collisions. *Eng.-Phys. J.*, **63** (1), 44–50 (in Russian) (Ref. 8 in Chapter 5)

9 Alipchenkov, V.M. and Zaichik, L.I. (2000) Modelling of the Motion of Particles of Arbitrary Density in a Turbulent Flow on the Basis of a Kinetic Equation for the Probability Density Function. *Fluid Dynamics*, **35** (6), 883–900. (Ref. 10 in Chapter 5).

10 Zaichik, L.I. and Pershukov, V.A. (1995) Modelling of Particle Motion in a Turbulent Flow with Allowance for Collisions. *Fluid Dynamics*, **30** (1), 49–63 (Ref. 9 in Chapter 5).

11 Alipchenkov, V.M. and Zaichik, L.I. (2001) Particle Collision Rate in Turbulent Flow. *Fluid Dynamics*, **36** (4), 608–618 (Ref. 11 in Chapter 5).

12 Alipchenkov, V.M. and Zaichik, L.I. (2003) Particle Clustering in Isotropic Turbulent Flow. *Fluid Dynamics*, **38** (3), 417–432.

13 Maxey, M.R. and Riley, J.J. (1983) Equation of Motion for a Small Rigid Sphere in a Nonuniform Flow. *Phys. Fluids. A*, **26** (4), 883–888.

14 Hinze, J.O. (1975) *Turbulence*, 2nd ed., McGraw-Hill, New York.

15 Corrsin, S. (1959) Progress Report on Some Turbulent Diffusion Research. *Advances in Geophysics*, Academic Press, N.Y., **6**, 161–164.

16 Wang, L.-P. and Stock, D.E. (1993) Dispersion of Heavy Particles in Turbulent Motion. *J. Atmosph. Sci.*, **50** (13), 1897–1913.

17 Wang, S. K., Lee, S.J., Jones, O.C. and Lahey, R.T. Jr. (1987) 3-D Turbulence Structure and Phase Distribution Measurements in Bubbly Two-Phase Flows. *Intern. J. Multiphase Flow*, **13** (3), 327–343.

18 Liu, T.J. and Bankoff, S.G. (1993) Structure of Air-Water Bubbly Flow in a Vertical Pipe. I. Liquid Mean Velocity and Turbulence Measurements. II. Void Fraction, Bubble Velocity and Bubble Size Distribution. *Intern. J. Heat Mass Transfer*, **36** (4), 1049–1072.

19 Auton, T.H. (1987) The Lift Force on a Spherical Body in a Rotation Flow. *J. Fluid Mech.*, **183**, 199–218.

20 Jenkins, J.T. and Richman, M.W. (1985) Grad's 13-Moment System for a Dense Gas of Inelastic Spheres. *Arch. Rat. Mech. Anal.*, **87** (4), 355–377.

21 Sommerfeld, M. and Zivkovic, G. (1992) Recent Advances in the Numerical Simulation of Pneumatic

Conveying through Pipe Systems, *Comput. Methods in Appl. Sci*, Elsevier, Amsterdam, pp. 201–212.

22 Laviéville, J., Deutsch, E. and Simonin, O. (1995) Large Eddy Simulation of Interactions Between Colliding Particles and a Homogeneous Isotropic Turbulence Field. In *Proc. 6th Int. Symp. on Gas–Particle Flows. ASME FED* **228**, 347–357.

23 Derevich, I.V. (1996) Particle Collisions in Turbulent Flow. *Fluid Dynamics.* **31** (2), 249–260.

24 Alipchenkov, V.M. and Zaichik, L.I. (1998) Modelling the Dynamics of Colliding Particles in a Turbulent Shear Flow. *Fluid Dynamics,* **33** (4), 552–558 (Ref. 22 in Chapter 5).

25 Durbin, P.A. (1993) On Modelling Three-Dimensional Turbulent Wall Layers. *Phys. Fluids. A,* **5** (5), 1231–1238.

26 Simonin, O., Deutsch, E. and Boivin, M Comparison of Large Eddy Simulation and Second-Moment Closure of Particle Fluctuating Motion in Two-Phase Turbulent Shear Flows. Proc. 9th Symp. on Turbulent Shear Flows, 5.2.1–15.2.6.

27 Laviéville, J. and Simonin, O. (1996) Large Eddy Simulation of Colliding Particles Suspended in a Simple Turbulent Shear Flow. In 8-th Workshop on Two-Phase Flow Predictions, Merseburg.

28 Pope, S.P. (2000) *Turbulent Flows,* Cambridge Univ. Press, Cambridge.

29 Saffman, P.G. and Turner, J.S. (1956) On the Collision of Drops in Turbulent Clouds. *J. Fluid Mech.,* **1**, 16–30.

30 Abrahamson, J. (1975) Collision Rates of Small Particles in a Vigorously Turbulent Fluid. *Chem. Eng. Sci.,* **30** (30), 1371–1379.

31 Williams, J.J. and Crane, R.J. (1983) Particle Collision Rate in Turbulent Flow. *Int. J. Multiphase Flow,* **9** (4), 421–435.

32 Yuu, S. (1984) Collision Rate of Small Particles in a Homogeneous and Isotropic Turbulence. *AIChE J.,* **30** (5), 802–807.

33 Zaichik, L.I. (1998) Estimation of Time Between Particle Collisions in Turbulent Flow. *High Temperature,* **36** (3), 433–437.

34 Wang, L.-P., Wexler, A.S. and Zhou, Y. (1998) Statistical Mechanical Description of Turbulent Coagulation. *Phys. Fluids,* **10** (10), 2647–2651.

35 Zhou, Y., Wexler, A.S. and Wang, L.-P. (1998) On the Collision Rate of Small Particles in Isotropic Turbulence. II. Finite Inertia Case. *Phys. Fluids,* **10** (5), 1206–1216.

36 Sundaram, S. and Collins, L.R. (1997) Collision Statistics in Isotropic Particle-Laden Turbulent Suspension. Part 1. Direct Numerical Simulation. *J. Fluid Mech.,* **335**, 75–109.

37 Chen, M. and Kontomaris, K. (1998) Direct Numerical Simulation of Droplet Collision in a Turbulent Channel Flow. Part II: Collision Rates. *Int. J. Multiphase Flow,* **24** (7), 1105–1138.

38 Mei, R. and Hu, K.C. (1999) On the Collision Rate of Small Particles in Turbulent Flows. *J. Fluid Mech.,* **391**, 67–89.

39 Gourdel, C., Simonin, O. and Brunier, E. (1998) Modelling Simulation of Gas-Solid Turbulent Flows with a Binary Mixture of Particles. *Proc. 3-rd Int. Conf. on Multiphase Flow* 1–8, Lyon.

40 Zaichik, L.I. and Pershukov, V.A. (1996) Problems of Modelling Gas–Particle Turbulent Flows with Combustion and Phase Transitions Review. *Fluid Dynamics,* **31** (5), 635–646 (Ref. 5 in Chapter 5).

41 Landau, L.D. and Lifshitz, E.M. (1987) *Fluid Mechanics,* Pergamon Press, Oxford.

42 Derevich, I.V., Yeroshenko, V.M. and Zaichik, L.I. (1995) The Influence of Particles on Turbulent Flow in

Channels. *Fluid Dynamics*, (1), 40–48 (in Russian).

43 Sawford, B.L. (1991) Reynolds Number Effects in Lagrangian Stochastic Models of Turbulent Dispersion. *Phys. Fluids A*, **3** (6), 1577–1586.

44 Yeung, P.K. and Pope, S.B. (1989) Lagrangian Statistic from Direct Numerical Simulations of Isotropic Turbulence. *J. Fluid Mech.*, **207**, 531–586.

45 Smoluchowski, M.V. (1917) Versuch einer mathemetlschen Theorle der Koagulationskinetik kolloider *Ltfsungen. Z. Phys. Chem.*, **B. 2**, S. 129–168.

46 Reeks, M.W. (1983) The Transport of Discrete Particles in Inhomogeneous Turbulence. *J. Aerosol Sci.*, **14** (6), 729–739.

47 Squires, K.D. and Eaton, J.K. (1991) Preferential Concentration of Particles by Turbulence. *Phys. Fluids A*, **3** (5), 1169–1178.

48 Wang, L.-P. and Maxey, R.M. (1993) Settling Velocity and Concentration Distribution of Heavy Particles in Homogeneous Turbulence. *J. Fluid Mech.*, **256**, 27–68.

49 Read, W.C. and Collins, L.R. (2000) Effect of Preferential Concentration on Turbulent Collision Rate. *Phys. Fluids*, **12** (10), 2530–2540.

50 Fevrier, P., Simonin, O. and Legendre, D. (2001) Particle Dispersion and Preferential Concentration Dependence on Turbulent Reynolds Number from Direct and Large-Eddy Simulation of Isotropic Homogeneous Turbulence. Proc. 4-th Intern. Conf. on Multiphase Flow, 1–8, New Orleans.

51 Wang, L.-P., Wexler, A.S. and Zhou, Y. (2000) Statistical Mechanical Description and Modeling of Turbulent Collision of Inertial Particles. *J. Fluid Mech.*, **415**, 117–153.

52 Lundgren, T.S. (1981) Turbulent Pair Dispersion and Scalar Diffusion. *J. Fluid Mech.*, **111**, 27–57.

53 Girimaji, S.S. and Pope, S.B. (1990) A Diffusion Model for Velocity Gradients in Turbulence. *Phys. Fluids A*, **2** (2), 242–256.

54 Drunk, B.K., Koch, D.L. and Lion, L.W. (1998) Turbulent Coagulation of Colloidal Particles. *J. Fluid Mech.*, **364**, 81–113.

55 Monin, A.S. and Yaglom, A.M. (1975) *Statistical Fluid Mechanics: Mechanics of Turbulence*, MIT Press,Cambridge, MA.

56 Sreenivasan, K.R. (1995) On the Universality of the Kolmogorov Constant. *Phys. Fluids*, **7** (11), 2778–2784.

57 Klyatskin, V.I. (1980) *Stochastic Equations and Waves in Random Inhomogeneous Media*, Nauka, Moscow (in Russian) (Ref. 3 in Chapter 1).

58 Kuznetsov, V.R. and Sabel'nikov, V.A. (1990) *Turbulence and Combustion*, Hemisphere, New York.

59 Hanjalić, K. and Launder, B.E. (1972) A Reynolds Stress Model of Turbulence and its Application to Thin Shear Flows. *J. Fluid Mech.*, **52**, 609–638.

60 Zaichik, L.I. (1999) A Statistical Model of Particle Transport and Heat Transfer in Turbulent Shear Flows. *Phys. Fluids*, **11** (6), 1521–1534.

61 Wang, Q., Squires, K.D. and Simonin, O. (1998) Large Eddy Simulation of Turbulent Gas–Solid Flows in a Vertical Channel and Evolution of Second-Order Models. *Intn. J. Heat Fluid Flow*, **19** (5), 505–511.

62 Alipchenkov, V.M. and Zaichik, L.I. (2003) Particle Clustering in Isotropic Turbulent Flow. *Fluid Dynamics*, **38** (3), 417–432 (Ref. 12 in Chapter 7).

63 Zaichik, L.I., Simonin, O. and Alipchenkov, V.M. (2003) Two Statistical Models for Predicting Collision Rates of Inertial Particles in Homogeneous Isotropic Turbulence. *Phys. Fluids*, **15** (10), 2995–3005.

Author Index

Statistical Microhydrodynamics. Emmanuil G. Sinaiski and Leonid Zaichik

ISBN: 978-3-527-40656-2

Subject Index

Statistical Microhydrodynamics. Emmanuil G. Sinaiski and Leonid Zaichik

ISBN: 978-3-527-40656-2